Coordinated Multi-Point in Mobile Communications
From Theory to Practice

A self-contained guide to coordinated multi-point (CoMP), this comprehensive book covers everything from theoretical basics to practical implementation. Addressing a wide range of topics, it highlights the potential gains of CoMP, the fundamental degrees of freedom involved, and the key challenges of using CoMP in practice. The editors and contributors bring unique real-world experience from running the world's first and largest test beds for LTE-Advanced, and recent field trial results are presented. With detailed insight into the realistic potential of CoMP as a key technology for LTE-Advanced and beyond, this is a must-read resource for professionals and students who want the big picture on CoMP or require in-depth knowledge of how to build cellular communication systems for the future.

Patrick Marsch was the technical project coordinator of the research project EASY-C, where the world's largest research test beds for LTE-Advanced were established and the first live demonstrations of CoMP were performed. He received his Dr.-Ing. degree from Technische Universität Dresden, where he later headed the system level group at the Vodafone Chair, focusing on optimizing spectral efficiency and energy efficiency in heterogeneous cellular deployments. He currently heads a radio research team within Nokia Siemens Networks in Wrocław, Poland.

Gerhard P. Fettweis is the Vodafone Chair Professor at Technische Universität Dresden, with 20 companies from around the world currently sponsoring his research on wireless transmission and chip design. An IEEE Fellow, he runs the world's largest cellular research test beds, coordinated the EASY-C project, and has received numerous awards. He began his career at IBM Research and has since developed nine start-up companies (so far).

Interference is the limiting factor in cellular communications and smart coordination of transmission can lead to significant improvements in quality of service. This book provides a strong outline and lays out some of the fundamental assumptions and theoretical models to treat the subject and supports the theory with results from system-level test benches and field measurements. I recommend this book to everyone interested in the topic.

Siavash M. Alamouti, Vodafone Group R&D Director

Coordinated Multi-Point in Mobile Communications

From Theory to Practice

Edited by

PATRICK MARSCH
Nokia Siemens Networks, Wrocław, Poland

GERHARD P. FETTWEIS
Technische Universität Dresden, Germany

CAMBRIDGE UNIVERSITY PRESS
Cambridge, New York, Melbourne, Madrid, Cape Town, Singapore,
São Paulo, Delhi, Tokyo, Mexico City

Cambridge University Press
The Edinburgh Building, Cambridge CB2 8RU, UK

Published in the United States of America by Cambridge University Press, New York

www.cambridge.org
Information on this title: www.cambridge.org/9781107004115

© Cambridge University Press 2011

This publication is in copyright. Subject to statutory exception
and to the provisions of relevant collective licensing agreements,
no reproduction of any part may take place without the written
permission of Cambridge University Press.

First published 2011

Printed in the United Kingdom at the University Press, Cambridge

A catalogue record for this publication is available from the British Library

ISBN 978-1-107-00411-5 Hardback

Additional resources for this publication at www.cambridge.org/9781107004115

Cambridge University Press has no responsibility for the persistence or
accuracy of URLs for external or third-party internet websites referred to in
this publication, and does not guarantee that any content on such websites is,
or will remain, accurate or appropriate.

Contents

List of Contributors	*page* xiii
Acknowledgements	xvii
List of Abbreviations	xviii
Nomenclature and Notation	xxiv

Part I Motivation and Basics   1

1 Introduction   3
     1.1 Motivation   3
     1.2 Aim of this Book   5
     1.3 Classes of CoMP Considered   5
     1.4 Outline of this Book   6

2 An Operator's Point of View   7
     2.1 The Mobile Internet - A Success Story so far   7
     2.2 Requirements on Future Networks and Upcoming Challenges   8
     2.3 The Role of CoMP   9
     2.4 The Role of Field Trials   10

3 Information-Theoretic Basics   11
     3.1 Observed Cellular Scenarios   11
     3.2 Usage of OFDMA for Broadband Wireless Communications   11
     3.3 Multi-Point Frequency-Flat Baseband Model Considered   13
     3.4 Uplink Transmission   14
         3.4.1 Basic Uplink Capacity Bounds   15
         3.4.2 Full Cooperation in the Uplink   17
         3.4.3 No Cooperation in the Uplink   17
         3.4.4 Numerical Example   19
     3.5 Downlink Transmission   19
         3.5.1 Basic Downlink Capacity Bounds   20
         3.5.2 Full Cooperation in the Downlink   22
         3.5.3 No Cooperation in the Downlink   22
         3.5.4 Numerical Example   23

	3.6	Summary	24
4		**Gains and Trade-Offs of Multi-Cell Joint Signal Processing**	**25**
	4.1	Modeling Imperfect Channel State Information (CSI)	25
		4.1.1 Imperfect CSI in the Uplink	25
		4.1.2 Imperfect CSI in the Downlink	27
	4.2	Gain of Joint Signal Processing under Imperfect CSI	29
	4.3	Trade-Offs in Uplink Multi-Cell Joint Signal Processing	32
		4.3.1 Different Information Exchange and Cooperation Schemes	32
		4.3.2 Numerical Results	35
		4.3.3 Parallels between Theory and Practical Cooperation Schemes	37
	4.4	Degrees of Freedom in Downlink Joint Signal Processing	37
	4.5	Summary	38

Part II Practical CoMP Schemes   39

5		**CoMP Schemes Based on Interf.-Aware Transceivers or Interf. Coord.**	**41**
	5.1	DL Multi-User Beamforming with IRC	41
		5.1.1 Introduction	41
		5.1.2 Downlink System Model	43
		5.1.3 Linear Receivers	44
		5.1.4 Imperfect Channel Estimation	45
		5.1.5 Resource Allocation and Fair User Selection	46
		5.1.6 Single-Cell Performance	48
		5.1.7 Multi-Cell Performance under Perfect CSI	50
		5.1.8 Multi-Cell Performance under Imperfect CSI	52
		5.1.9 Summary	54
	5.2	Uplink Joint Scheduling and Cooperative Interference Prediction	54
		5.2.1 Interference-Aware Joint Scheduling	56
		5.2.2 Cooperative Interference Prediction	61
		5.2.3 Practical Considerations	64
		5.2.4 Applicability of Both Schemes to the Downlink	66
		5.2.5 Summary	67
	5.3	Downlink Coordinated Beamforming	68
		5.3.1 Introduction	68
		5.3.2 Single Receive Antenna at the Terminal	70
		5.3.3 Multiple Receive Antennas at the Terminal	74
		5.3.4 Summary	80
6		**CoMP Schemes Based on Multi-Cell Joint Signal Processing**	**81**
	6.1	Uplink Centralized Joint Detection	81
		6.1.1 Introduction	81
		6.1.2 Joint Detection Algorithms	82

		6.1.3	Local BS Processing with Limited Backhaul Constraint	87
		6.1.4	Local or Partial Decoding with Limited Backhaul Constraint	90
		6.1.5	Provisions for Uplink Joint Processing in WiMax and LTE	92
		6.1.6	Summary	93
	6.2	Uplink Decentralized Joint Detection		94
		6.2.1	Practical Decentralized Interference Cancelation Scheme	95
		6.2.2	Performance Assessment	104
		6.2.3	Summary	108
	6.3	DL Distributed CoMP Approaching Centralized Joint Transmission		108
		6.3.1	System Model	110
		6.3.2	Theoretical Limits for Static Clustering and DPC	111
		6.3.3	Practical (Linear) Precoding	113
		6.3.4	Scheme for Distributed, Centralized Joint Transmission	115
		6.3.5	Summary	121
	6.4	Downlink Decentralized Multi-User Transmission		121
		6.4.1	Decentralized Beamforming with Limited CSIT	122
		6.4.2	Multi-cell Beamforming with Limited Data Sharing	130
		6.4.3	Summary	136

Part III Challenges Connected to CoMP   137

7	**Clustering**			139
	7.1	Static Clustering Concepts		141
		7.1.1	Non-Overlapping Clusters	142
		7.1.2	Overlapping Clusters	145
		7.1.3	Resulting Geometries	146
	7.2	Self-Organizing Clustering Concepts		148
		7.2.1	Self-Organizing Network Concepts in 3GPP LTE	148
		7.2.2	Adaptive Clustering Algorithms	149
		7.2.3	Simulation Results	152
		7.2.4	Signaling and Control Procedures	157
	7.3	Summary		159

8	**Synchronization**			161
	8.1	Synchronization Concepts		161
		8.1.1	Synchronization Terminology	161
		8.1.2	Network Synchronization	163
		8.1.3	Satellite-Based Synchronization	165
		8.1.4	Endogenous Distributed Wireless Carrier Synchronization	166
		8.1.5	Summary	169
	8.2	Imperfect Sync in Time: Perf. Degradation and Compensation		170
		8.2.1	MIMO OFDM Transmission with Asynchronous Interference	173
		8.2.2	Interf.-Aware Multi-User Joint Detection and Transmission	176

	8.2.3	System Level SINR Analysis	178
	8.2.4	Summary	181
8.3	Imperfect Sync in Frequency: Perf. Degradation and Compensation		181
	8.3.1	Downlink Analysis	182
	8.3.2	Uplink Analysis	189
	8.3.3	Summary	192

9 Channel Knowledge — 193

9.1	Channel Estimation for CoMP		193
	9.1.1	Channel Estimation - Single Link	194
	9.1.2	Channel Estimation for CoMP	202
	9.1.3	Multi-Cell Channel Estimation	204
	9.1.4	Uplink Channel Estimation	206
	9.1.5	Summary	208
9.2	Channel State Information Feedback to the Transmitter		208
	9.2.1	Transmission Model	210
	9.2.2	Sum-Rate Performance Measure	211
	9.2.3	Channel Vector Quantization (CVQ)	211
	9.2.4	Minimum Euclidean Distance Based CVQ	213
	9.2.5	Maximum SINR Based CVQ	214
	9.2.6	Pseudo-Maximum SINR based CVQ	215
	9.2.7	Application to Zero-Forcing (ZF) Precoding	216
	9.2.8	Resource Allocation	216
	9.2.9	Simulation Results	216
	9.2.10	Summary	218

10 Efficient and Robust Algorithm Implementation — 219

10.1 Robust and Flexible Base Station Precoding Implementation		219
10.1.1 System Model		220
10.1.2 Transmit Filter Eigendecomposition		221
10.1.3 Transmit Filter Computations		222
10.1.4 The Order-Recursive Filter in Details		224
10.1.5 Example: SINR as Function of the Condition Number		226
10.1.6 Summary		227
10.2 Low-Complexity Terminal-Side Receiver Implementation		227
10.2.1 Introduction to Interference Rejection Combining (IRC)		228
10.2.2 IRC with Known Channel and Interference Covariance		231
10.2.3 Implementation Losses from Imperfect Channel Estimation		233
10.2.4 Losses from Spatial Interf.-and-Noise Covariance Estimation		237
10.2.5 Losses from Channel and Interference Estimation Errors		241
10.2.6 Summary		241

11	**Scheduling, Signaling and Adaptive Usage of CoMP**	**243**
	11.1 Centralized Scheduling for CoMP	243
	11.1.1 Introduction	243
	11.1.2 System Model	244
	11.1.3 Centralized Scheduling Problems	246
	11.1.4 Analyses and Results	251
	11.1.5 Summary	254
	11.2 Decentralized Radio Link Control and Inter-BS Signaling	254
	11.2.1 Resource Allocation	255
	11.2.2 Link Adaptation	256
	11.2.3 Radio Link Measurements	257
	11.2.4 Uplink Power Control	258
	11.2.5 Uplink Timing Advance	259
	11.2.6 HARQ-related Timing Constraints for UL CoMP	259
	11.2.7 Handover	262
	11.2.8 Inter-BS Signaling	262
	11.2.9 Summary	266
	11.3 Ad-hoc CoMP	266
	11.3.1 Introduction	267
	11.3.2 Ad-Hoc CoMP With More Accurate CSI	269
	11.3.3 Ad-Hoc CoMP with CSI Impairments	273
	11.3.4 Ad-Hoc CoMP and HARQ	275
	11.3.5 Summary	276
12	**Backhaul**	**277**
	12.1 Fund. Limits of Interf. Mitigation with Limited Backhaul Coop.	277
	12.1.1 Introduction	278
	12.1.2 Uplink Scenario: Receiver Cooperation	281
	12.1.3 Downlink Scenario: Transmitter Cooperation	286
	12.1.4 UL-DL Reciprocity and Generalized Degrees of Freedom	287
	12.1.5 Summary	291
	12.2 Backhaul Requirements of Practical CoMP Schemes	291
	12.2.1 Types of Backhaul Data and Scaling Laws	291
	12.2.2 Specific Backhaul Requirements of Exemplary CoMP Schemes	294
	12.2.3 Backhaul Latency Requirements	299
	12.2.4 Backhaul Topology Considerations	300
	12.2.5 Summary	300
	12.3 CoMP Backhaul Infrastructure Concepts	301
	12.3.1 Ethernet	301
	12.3.2 Passive Optical Network	303
	12.3.3 Digital Subscriber Line	305
	12.3.4 Microwave	306
	12.3.5 The X2 Interface	307
	12.3.6 Backhaul Topology Concepts	307

12.3.7 Summary — 310

Part IV Performance Assessment — 311

13 Field Trial Results — 313
13.1 Real-time Impl. and Trials of Adv. Receivers and UL CoMP — 313
    13.1.1 Real-time Implementation and Lab Tests — 314
    13.1.2 Uplink Successive Interference Cancelation (SIC) Receiver — 314
    13.1.3 Uplink Macro Diversity Trials with Distributed RRHs — 317
    13.1.4 Summary — 319
13.2 Assessing the Gain of Uplink CoMP in a Large-Scale Field Trial — 319
    13.2.1 Measurement Setup — 320
    13.2.2 Signal Processing Architecture and Evaluation Concept — 321
    13.2.3 Noise Estimation — 322
    13.2.4 Channel Equalization — 322
    13.2.5 Field Trial Results — 325
    13.2.6 Summary — 330
13.3 Real-time Implementation and Field Trials for Downlink CoMP — 331
    13.3.1 Introduction — 332
    13.3.2 Enabling Features — 334
    13.3.3 Real-time Implementation — 346
    13.3.4 Field Trials — 347
    13.3.5 Summary — 352
13.4 Predicting Pract. Achievable DL CoMP Gains over Larger Areas — 353
    13.4.1 Setup and Closed-Loop System Design — 353
    13.4.2 Measurement and Evaluation Methodology — 356
    13.4.3 Measurement Campaign — 358
    13.4.4 Summary — 363
13.5 Lessons Learnt Through Field Trials — 364

14 Performance Prediction of CoMP in Large Cellular Systems — 367
14.1 Simulation and Link-2-System Mapping Methodology — 367
    14.1.1 General Simulation Assumptions and Modeling — 368
    14.1.2 Channel Models and Antenna Models — 370
    14.1.3 Transceiver Techniques — 373
    14.1.4 Link-to-System Interface — 373
    14.1.5 Key Performance Indicators — 375
    14.1.6 Summary — 376
14.2 Obtaining Chn. Model Params. via Chn. Sounding or Ray-Tracing — 376
    14.2.1 Large-Scale-Parameters — 377
    14.2.2 Measurement-based Parameter Estimation — 380
    14.2.3 Ray-Tracing based Parameter Simulation — 380
    14.2.4 Comparison between Measurements and Ray-Tracing — 382

	14.2.5 Summary	387
14.3	Uplink Simulation Results	387
	14.3.1 Compared Schemes	387
	14.3.2 Simulation Assumptions and Parameters	389
	14.3.3 Backhaul Traffic	391
	14.3.4 Simulation Results	392
	14.3.5 Summary	395
14.4	Downlink Simulation Results	396
	14.4.1 Compared Schemes	396
	14.4.2 Simulation Assumptions and Parameters	397
	14.4.3 Detailed Analysis of Coordinated Scheduling/Beamforming	398
	14.4.4 Backhaul Traffic	406
	14.4.5 Simulation Results	406
	14.4.6 Summary	408

Part V Outlook and Conclusions 409

15   Outlook 411

15.1 Using CoMP for Terminal Localization 411
    15.1.1 Localization based on the Signal Propagation Delay 412
    15.1.2 Further Localization Methods 416
    15.1.3 Localization in B3G Standards 418
    15.1.4 Summary 422
15.2 Relay-Assisted Mobile Communication using CoMP 423
    15.2.1 Introduction 423
    15.2.2 Reference Scenario 424
    15.2.3 System and Protocol Description 425
    15.2.4 Trade-Offs in Relay Networks 427
    15.2.5 Numerical Evaluation of CoMP and Relaying 428
    15.2.6 Cost/Benefit Trade-Off 428
    15.2.7 Energy/Benefit Trade-Off 429
    15.2.8 Computation/Transmission Power Trade-Off 430
    15.2.9 Summary 432
15.3 Next Generation Cellular Network Planning and Optimization 432
    15.3.1 Introduction 432
    15.3.2 Classical Cellular Network Planning and Optimization 433
    15.3.3 Physical Characterization of Capacity Gains through CoMP 435
    15.3.4 Summary 443
15.4 Energy-Efficiency Aspects of CoMP 444
    15.4.1 System Model 445
    15.4.2 Effective Transmission Rates 447
    15.4.3 Backhauling 448
    15.4.4 Energy Consumption of Cellular Base Stations 449

	15.4.5 System Evaluation	451
	15.4.6 Summary	453
16	**Summary and Conclusions**	**455**
	16.1 Summary of this Book	455
	16.1.1 Most Promising CoMP Schemes and Potential Gains	455
	16.1.2 Key Challenges Identified	457
	16.2 Conclusions	458
	16.2.1 About this Book	459
	16.2.2 CoMP's Place in the LTE-Advanced Roadmap and Beyond	460
	References	461
	Index	479

List of Contributors

Amin, M. Awais	Qualcomm CDMA Technologies GmbH, Nuremberg, Germany
Bachl, Rainer	ST-Ericsson AT GmbH, Nuremberg, Germany
Bhagavatula, Ramya	University of Texas at Austin, TX, USA
Boccardi, Federico	Alcatel-Lucent Bell Labs, Stuttgart, Germany
Brown III, D. Richard	Worcester Polytechnic Institute, MA, USA
Brück, Stefan	Qualcomm CDMA Technologies GmbH, Nuremberg, Germany
Calin, Doru	Alcatel-Lucent Bell Labs, Murray Hill, NJ, USA
Chae, Chan-Byoung	Yonsei University, Korea
Dammann, Armin	Institute of Communications and Navigation, German Aerospace Center (DLR), Germany
Dekorsy, Armin	Institute for Telecomms. and High-Frequency Techniques, University of Bremen, Germany
Dietl, Guido	DOCOMO Euro-Labs, Munich, Germany
Doll, Mark	Alcatel-Lucent Bell Labs, Stuttgart, Germany
dos Santos, Ricardo B.	Federal University of Ceará, Brazil
Dötsch, Uwe	Alcatel-Lucent Bell Labs, Stuttgart, Germany
Droste, Heinz	Deutsche Telekom Laboratories, Darmstadt, Germany
Fahldieck, Torsten	Alcatel-Lucent Bell Labs, Stuttgart, Germany
Falconetti, Laetitia	Ericsson Research, Aachen, Germany
Fehske, Albrecht	Vodafone Chair, Technische Universität Dresden, Germany
Fettweis, Gerhard	Vodafone Chair, Technische Universität Dresden, Germany
Fischer, Erik	Vodafone Chair, Technische Universität Dresden, Germany
Forck, Andreas	Fraunhofer Institute for Telecommunications, Heinrich Hertz Institute, Berlin, Germany

List of Contributors

Frank, Philipp	Deutsche Telekom Laboratories, Berlin, Germany
Fritzsche, Richard	Vodafone Chair, Technische Universität Dresden, Germany
Garavaglia, Andrea	Qualcomm CDMA Technologies GmbH, Nuremberg, Germany
Gesbert, David	EURECOM - Mobile Communications Department, Sophia-Antipolis, France
Giese, Jochen	Qualcomm CDMA Technologies GmbH, Nuremberg, Germany
Grieger, Michael	Vodafone Chair, Technische Universität Dresden, Germany
Haustein, Thomas	Fraunhofer Institute for Telecommunications, Heinrich Hertz Institute, Berlin, Germany
Heath Jr., Robert W.	University of Texas at Austin, TX, USA
Holfeld, Jörg	Vodafone Chair, Technische Universität Dresden, Germany
Hoymann, Christian	Ericsson Research, Aachen, Germany
Irmer, Ralf	Vodafone Group R&D, Newbury, UK
Jäckel, Stephan	Fraunhofer Institute for Telecommunications, Heinrich Hertz Institute, Berlin, Germany
Jandura, Carsten	Actix GmbH, Dresden, Germany
Jungnickel, Volker	Fraunhofer Institute for Telecommunications, Heinrich Hertz Institute, Berlin, Germany
Kadel, Gerhard	Deutsche Telekom Laboratories, Darmstadt, Germany
Klein, Andrew G.	Worcester Polytechnic Institute, MA, USA
Klein, Anja	Technische Universität Darmstadt, Germany
Koppenborg, Johannes	Alcatel-Lucent Bell Labs, Stuttgart, Germany
Kotzsch, Vincent	Vodafone Chair, Technische Universität Dresden, Germany
Maciel, Tarcisio F.	Federal University of Ceará, Brazil
Marsch, Patrick	Nokia Siemens Networks, Wrocław, Poland
Mayer, Hans-Peter	Alcatel-Lucent Bell Labs, Stuttgart, Germany
Mensing, Christian	Institute of Communications and Navigation, German Aerospace Center (DLR), Germany
Molisch, Andreas F.	Department of Electrical Engineering, University of Southern California, Los Angeles, CA, USA
Müller-Weinfurtner, Stefan	ST-Ericsson AT GmbH, Nuremberg, Germany

Müller, Andreas	Institute of Telecommunications, University of Stuttgart, Germany
Olbrich, Michael	Fraunhofer Institute for Telecommunications, Heinrich Hertz Institute, Berlin, Germany
Palleit, Nico	Institute of Communications Engineering, University of Rostock, Germany
Rost, Peter	NEC Laboratories Europe, Heidelberg, Germany
Sand, Stephan	Institute of Communications and Navigation, German Aerospace Center (DLR), Germany
Schellmann, Malte	Huawei Technologies Düsseldorf GmbH, European Research Center, Munich, Germany
Schneider, Christian	Ilmenau University of Technology, Germany
Schulist, Matthias	Qualcomm CDMA Technologies GmbH, Nuremberg, Germany
Thiele, Lars	Fraunhofer Institute for Telecommunications, Heinrich Hertz Institute, Berlin, Germany
Tian, Yafei	School of Electronics and Information Engineering, Beihang University, China
Tse, David	Wireless Foundations, University of California at Berkeley, CA, USA
Utschick, Wolfgang	Associate Institute for Signal Processing, Technische Universität München, Germany
Voigt, Jens	Actix GmbH, Dresden, Germany
Wachsmann, Udo	ST-Ericsson AT GmbH, Nuremberg, Germany
Wahls, Sander	Fraunhofer Institute for Telecommunications, Heinrich Hertz Institute, Berlin, Germany
Wang, I-Hsiang	Wireless Foundations, University of California at Berkeley, CA, USA
Weber, Andreas	Alcatel-Lucent Bell Labs, Stuttgart, Germany
Weber, Ralf	Qualcomm CDMA Technologies GmbH, Nuremberg, Germany
Weber, Tobias	Institute of Communications Engineering, University of Rostock, Germany
Wei, Xinning	Institute of Communications Engineering, University of Rostock, Germany
Wild, Thorsten	Alcatel-Lucent Bell Labs, Stuttgart, Germany
Wirth, Thomas	Fraunhofer Institute for Telecommunications, Heinrich Hertz Institute, Berlin, Germany
Yang, Chenyang	School of Electronics and Information Engineering, Beihang University, China

Zakhour, Randa — Electrical and Electronic Engineering Department, University of Melbourne, Australia

Zirwas, Wolfgang — Nokia Siemens Networks GmbH & Co. KG, Munich, Germany

Acknowledgements

This book is based on the knowledge and effort of a large number of authors, some of whom have been working in the field of CoMP for over a decade. The editors would like to thank all contributors for their great cooperation in the last months, their constructive discussions on contents, notation and nomenclature, and their patience in fine-tuning contents up to the last minute of editing.

The request for searching for new limits of cellular beyond 3G came from Vodafone Group R&D, initiating our research in the area of CoMP. As the sponsor of the Vodafone Chair at Technische Universität Dresden, Vodafone Group R&D has been instrumental in sharpening our view for CoMP schemes with practical impact. In particular Mike Walker, Trevor Gill and Luke Ibbetson among many others have been of great help in serving as a sounding board for our ideas. As a result, we have focused our research on theoretical limits as well as practical implementation challenges. The result of this view on CoMP technology has provided the basis for what has finally led to this book.

Nothing would be possible without interaction with friends, colleagues, fellow researchers and cooperation partners. The mindset and openness of our scientific community is a platform for inspiration and motor for sharpening our minds. In particular, the team at the Vodafone Chair has been of invaluable help in creating scientific results and providing the framework for inspirations, discussions, and many new insights. Thanks to the whole team for this major help!

While most parts of this book were mutually reviewed by the authors themselves, the editors would like to thank the following external reviewers for their valuable feedback: Fabian Diehm, Alexandre Gouraud, Ines Kluge, Marco Krondorf, Eckhard Ohlmer, Simone Redana, Fred Richter, Hendrik Schöneich, Mikael Sternad, Vinay Suryaprakash, Tommy Svensson, Stefan Valentin, Raphael Visoz, Guillaume Vivier and Steffen Watzek. Also, the appearance of the book would not be as it is without the significant work of Katharina Philipp, who adapted the majority of figures in this book to the same look and feel.

Last but surely not least, the editors would like to thank Phil Meyler and Sarah Finlay from Cambridge University Press for making this book possible, and for the great and patient support during its creation.

Patrick Marsch and **Gerhard Fettweis** (Editors), *January 2011*

List of Abbreviations

ACK	acknowledgement
ADC	analog to digital conversion
AGC	automatic gain control
aGW	advanced gateway
ANR	automatic neighbor relation
AoA	angle of arrival
AWGN	additive white Gaussian noise
bpcu	bits per channel use
BC	broadcast channel
BER	bit error rate
BF	beamforming
BLER	block error rate
BPSK	binary phase shift keying
BS	base station
CAZAC	constant amplitude zero autocorrelation codes
CB	coordinated beamforming
CCU	CoMP central unit
CD	Cholesky decomposition
CDF	cumulative distribution function
CDI	channel direction indicator
CDM	code division multiplex
CDMA	code division multiple access
CFO	carrier frequency offset
CGI	cell global identifier
CIF	compressed interference forwarding
CIR	channel impulse response
CoMP	coordinated multi-point
CP	cyclic prefix
CPRI	common public radio interface
CQI	channel quality indicator
CRC	cyclic redundancy check
CRLB	Cramér-Rao lower bound
CRS	common reference signal
CS	coordinated scheduling

CS/CB	coordinated scheduling / coordinated beamforming
CSG	closed subscriber group
CSI	channel state information
CSIR	channel state information at the receiver
CSI RS	CSI reference signal
CSIT	channel state information at the transmitter
CSU	central scheduling unit
CT	conventional transmission
CTF	channel transfer function
CU	central unit
CVQ	channel vector quantization
DAS	distributed antenna system
DBA	dynamic bandwidth assignment
DF	decode-and-forward
DFT	discrete Fourier transform
DIS	distributed interference subtraction
DL	downlink
DM	device manager
DPC	dirty paper coding
DRS	demodulation reference signal
DSL	digital subscriber line
DSLAM	DSL access multiplexer
DSP	digital signal processor
EASY-C	Enables of Ambient Services and Systems - Part C
eNB	enhanced Node B
EOC	eigenmode-aware optimum combiner
ERC	eigenmode-aware receive combining
EPON	Ethernet PON
E-UTRAN	enhanced UMTS terrestrial radio access network
EVD	eigenvalue decomposition
EvDO	evolution data optimized
FDD	frequency division duplex
FDM	frequency division multiplex
FEC	forward error correction
FFT	fast Fourier transform
FIR	finite impulse response
FPGA	field programmable gate array
FTP	file transfer protocol
g.d.o.f.	generalized degrees of freedom
GF	geometry factor
GSM	global system for mobile communications
GPON	Gigabit-capable PON
GPRS	general packet radio service

GTC	GPON transmission convergence
GPS	global positioning system
GTP-U	GTP user plane
HARQ	hybrid automatic repeat request
H-BLAST	Horizontal Bell Laboratories Layered Space-Time Architecture
HK	Han-Kobayashi
HPBW	half-power beamwidth
HSPA	high-speed packet access
IAP	interference-aware precoding
IC	interference channel
ICI	inter-carrier interference
ICIN	inter-cell interference nulling
IDFT	inverse discrete Fourier transform
IF	intermediate frequency
i.i.d.	independently and identically distributed
IEEE	Institute of Electrical and Electronics Engineers
IFFT	inverse fast Fourier transform
INR	interference-to-noise ratio
IP	Internet protocol
IRC	interference rejection combining
ISD	inter-site distance
ISI	inter-symbol interference
JD	joint detection
JT	joint transmission
LAN	local area network
LDC	linear deterministic channel
LLR	log-likelihood ratio
LMMSE	linear minimum mean square error
LO	local oscillator
LOS	line-of-sight
LSP	large-scale parameters
LSU	LTE signal processing unit
LTE	Long Term Evolution
LTE-A	Long Term Evolution – Advanced
MAC	multiple access channel
MAN	metropolitan area network
MCS	modulation and coding scheme
MET	maximum Eigenvalue transmission
MIESM	mutual information equivalent SINR mapping
MIMO	multiple-input multiple-output
MISO	multiple-input single-output
MF	matched filter
ML	maximum likelihood
MLE	maximum likelihood estimator

MME	mobility management entity
MMSE	minimum mean square error
MPC	multi-path component
MRC	maximum ratio combining
MRM	measurement report message
MRT	maximum ratio transmission
MS	multiple stream
MSE	mean square error
MUI	multi-user interference
MU-MIMO	multi-user MIMO
NGMN	next generation mobile networks
NLOS	non line-of-sight
NMEA	National Marine Electronics Association
NR	neighbor relation
NRT	neighbor relation table
NTP	network time protocol
OAM	operation and maintenance
OC	optimum combining
OCXO	oven-controlled crystal oscillator
ODN	optical distribution network
OFDM	orthogonal frequency division multiplex
OFDMA	orthogonal frequency division multiple access
OLT	optical line termination
ONU	optical network unit
PA	power amplifier
PAPR	peak-to-average power ratio
PCI	physical cell identifier
PDF	probability distribution function
PDH	plesiochronous digital hierarchy
PDCCH	physical downlink control channel
PDSCH	physical downlink shared channel
PDP	power delay profile
PIC	parallel interference cancelation
PLL	phase-locked loop
PMI	precoding matrix indicator
ppb	parts per billion
ppm	parts per million
PPS	pulses per second
PON	passive optical network
POTS	plain old telephone service
PRB	physical resource block
PRS	positioning reference signal
PTP	precision time protocol
PUCCH	physical uplink control channel

PUSCH	physical uplink shared channel
QAM	quadrature amplitude modulation
QoE	quality of experience
QoS	quality of service
QPSK	quadrature phase shift keying
RAN	radio access network
RAP	radio access point
RB	resource block
RE	resource element
RF	radio frequency
RI	rank indicator
RHS	right-hand side
RMS	root mean square
RN	relay node
RNTI	radio network temporary identifier
RoF	radio over fibre
RRH	remote radio head
RRM	radio resource management
RS	reference signal
RSS	received signal strength
RSRP	reference signal received power
RTOA	round-trip time of arrival
RTT	round-trip time
SC	sub-carrier
SC-FDMA	single carrier frequency domain multiple access
SCM	spatial channel model
SCME	spatial channel model extended
SCTP	stream control transmission protocol
SDH	synchronous digital hierarchy
SDMA	spatial division multiple access
SDIV	spatial diversity
S-GW	serving gateway
SIC	successive interference cancelation
SINR	signal-to-interference-and-noise ratio
SIR	signal-to-interference ratio
SISO	single-input single-output
SMUX	spatial multiplexing
SON	self-organizing network
SONET	synchronous optical network
SS	single stream
SSB	single side band
SSP	smale-scale parameters
SNR	signal-to-noise ratio
SU-MIMO	single-user MIMO

SVD	singular value decomposition
SynchE	synchronous Ethernet
TB	transport block
TCI	target cell identifier
TDD	time division duplex
TDM	time division multiplex
TDMA	time division multiple access
TDOA	time delay of arrival
THP	Tomlinson-Harashima precoding
TOA	time of arrival
TTI	transmit time interval
UDP	user datagram protocol
UCA	uniform circular array
UE	user equipment
UL	uplink
ULA	uniform linear array
UMTS	universal mobile telecommunications standard
UTRAN	universal terrestrial RAN
VDSL	very-high-speed digital subscriber line
VID	VLAN identifier
VLAN	virtual local area network
VoIP	voice over IP
WAN	wide area network
WCDMA	wideband code division multiple access
WCI	worst companion indicator
WiMAX	Worldwide Interoperability for Microwave Access
WF	Wiener filter
WINNER	Wireless World Initiative New Radio
WSSUS	wide sense stationary uncorrelated scattering
XGPON	10-Gigabit-capable PON
ZF	zero-forcing

Nomenclature and Notation

Nomenclature

In this book, we generally consider the setup and involved nomenclature depicted in Fig. 3.1 on page 12. Please note that we assume a *site* to consist of three *sectors*, which are equivalent to *cells*. Each sector or cell is assumed to be served by one dedicated base station (BS), even though in practice multiple such BSs may be integrated into one physical device.

CoMP Scheme Classification

Throughout the book, CoMP schemes are classified on one hand according to the extent of cooperation between BSs. We here distinguish between

- interference-aware transmission and detection (possibly with estimation of interference, but without explicit BS cooperation)
- interference coordination (e.g. joint multi-cell scheduling, coordinated beamforming etc.)
- multi-cell joint signal processing (e.g. joint detection or joint transmission)

We further distinguish between *decentralized* and *centralized* CoMP schemes, depending on where the subject of cooperation takes place. This classification is applied to various schemes observed in this book in Table 1.

Notation

Unless stated otherwise, the following holds throughout most parts of the book:

- Calligraphic letters (e.g., $\mathcal{M}$) represent *sets*
- Capital, italic letters, (e.g. $P_{\max}$) denote *constants*
- Bold-face, capital letters (e.g. $\mathbf{H}$) represent *matrices*
- Bold-face, lowercase letters (e.g. $\mathbf{h}$) represent *vectors*
- Lowercase, italic letters represent *scalars*.
- Variables with a hat on top (e.g. $\hat{\mathbf{H}}$) denote *estimates*
- Variables with a bar on top (e.g. $\bar{\mathbf{H}}$) denote an *effective* expression
- Variables with a tilde on top (e.g. $\tilde{\mathbf{H}}$ denote expressions in *time domain*, whereas other expressions are usually in *frequency domain*, see Section 8.2).

Table 1. Classification of CoMP schemes.

	Decentralized	Centralized
Interference-aware transmission/ detection	→ DL multi-user beamforming with IRC (Sections 5.1, 13.3) → IRC (Section 10.2)	
Interference coordination	→ UL cooperative interf. prediction (Sections 5.2.2, 14.3) → DL coordinated sched. / beamforming (CS/CB) (Sections 5.3, 14.4.3)	→ UL joint scheduling (Sections 5.2.1, 14.3) → DL centralized joint scheduling (Section 11.1)
Multi-cell joint signal processing	→ UL decentralized joint detection (Sections 6.2, 13.1, 14.3) → UL distr. interference subtraction (Sections 4.3.1, 13.2) → DL distributed joint transmission (Sections 6.3, 13.3, 13.4)	→ UL centralized joint detection (Sections 6.1, 13.2) → DL centralized joint transmission (Sections 6.3, 13.3)

The following variables are frequently used throughout the book:

- **H**, **h**, or h denote (matrices or vectors of) channel coefficients
- **x** or x are signals to be transmitted, *before* precoding
- **s** or s are signals to be transmitted, *after* precoding
- **y** or y are received signals
- **W** or **w** are transmit/receive filters used at the BS side
- **G** or **g** are transmit/receive filters used at the UE side
- c typically denotes a cluster index
- k and j typically denote user indices
- m typically denotes a base station index
- t, τ and i denote time indices, where t and τ are time-continuous, and i is a discrete sample index
- f and q denote frequency indices, where f is frequency-continuous, and q is a discrete sub-carrier index
- o denotes an OFDM symbol index

As in most publications, $(\cdot)^H$ denotes Hermitian matrix transpose, $\text{tr}\{\cdot\}$ denotes the trace of a matrix, $|\cdot|$ denotes set size when applied to a set, or determinant when applied to a matrix. $E\{\cdot\}$ denotes expectation value. **I** is an identity matrix.

Part I

Motivation and Basics

1 Introduction

Patrick Marsch and Gerhard Fettweis

1.1 Motivation

Mobile communication has gained significant importance in today's society. As of 2010, the number of mobile phone subscribers has surpassed 5 billion [ABI10], and the global annual mobile revenue is soon expected to top $1 trillion [Inf10]. While these numbers appear promising for mobile operators at first sight, the major game-changer that has come up recently is the fact that the market is more and more driven by the demand for mobile data traffic [Cis10]. This is simply because Moore's law in semiconductors leads to continuously more powerful mobile devices with larger storage capacity, which in the era of Web 2.0 require regular synchronization with the Internet. Consequently, Moore's law can also be found in the increase of data rates in wireless communications, as illustrated in Fig. 1.1. The main challenge, however, is that mobile users tend to expect the fast and cheap Internet access that they are used from their fixed lines (e.g. ADSL), but anytime and anywhere while being on the move. This puts mobile operators under the pressure to respond to the increasing traffic demand and provide a more homogeneous quality of experience (QoE) over the area (often referred to as improved *fairness*), while continuously decreasing *cost per bit* - and addressing the more and more crucial issue of *energy efficiency* [FMBF10].

But how can mobile data rates and fairness be increased in general? We have to be aware that current cellular systems are mainly limited by inter-cell interference [GK00] - especially in urban areas where the rate demand is largest and hence base station deployment is dense. Here, each point-to-point communication link is characterized by a certain ratio of desired receive signal power over interference and noise power, where Shannon [Sha48] states a clear upper bound on the *capacity* of the link. This then translates to a maximum *spectral efficiency*, i.e. the maximum data rate achievable for a given bandwidth. In fact, the standard Long Term Evolution (LTE) Release 8 [McC07] uses modulation and coding schemes and link adaptation in conjunction with hybrid automatic repeat request (HARQ) that allow to approach Shannon capacity to within less than a dB at reasonable complexity [LS06]. Hence, the increasing rate demand can surely not be met by improving point-to-point links, but requires other innovations. But which further options do we have?

Introduction

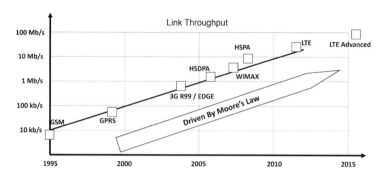

Figure 1.1 Exponential growth of data rates in mobile communications.

- **Use more spectrum**. An option which is currently already being pursued, as visible through recent auctions on spectrum becoming available via the *digital dividend*. Especially *spectrum aggregation* is of interest, i.e. the capability of radio access networks to use non-continuous blocks of spectrum. While capacity grows *linearly* with bandwidth, efficiently usable spectrum is generally a limited resource, and hence cannot be the sole source of rate growth.
- **Use more antennas** - an option that is already used since high-speed packet access (HSPA) and is a main feature of LTE. So-called multiple-input multiple-output (MIMO) techniques allow to obtain additional degrees of freedom, which can be used to spatially separate desired from interfering signals and/or for spatial multiplexing, besides yielding an additional source of diversity. In theory, capacity grows linearly in the minimum number of transmit and receive antennas [Tel99]. However, the number of antennas at base station side is usually limited due to regulatory or site rental issues, and that at the terminal side due to form factor and cost reasons. Further, practical multiplexing gains saturate at some point due to unavoidable antenna correlation.
- **Increase the degree of sectorization**. An alternative is to use *directed* base station antennas in order to obtain a larger quantity of smaller cells with less mutual interference. This is already done since global system for mobile communications (GSM), where 3-fold sectorization is typically applied, but one could principally imagine increasing sectorization [RRMF10], up to the case where each user is served with a dedicated beam.
- **Using more base stations or introducing relays and micro/femto cells** is clearly the strongest driver towards increased data rates, which also allows to improve energy and cost efficiency [MFF10]. Both relays and femto cells further allow to strongly improve indoor coverage, which is a major downside of conventional cellular systems.
- **Introduce coordination or cooperation between cells**. While most previously stated options require the deployment of new equipment, it is known from theory that interference can be overcome and even exploited if coordination or cooperation between cells is introduced. Such schemes are particularly interesting, as they require a fairly small change of infrastructure, and may

lead to a more homogeneous quality of service (QoS) distribution over the area [MKF06]. For this reason, multi-cell coordination or cooperation has been identified as a key technology of LTE-Advanced [PDF+08].

This book focuses on the latter aspect, using the term **coordinated multipoint (CoMP)**. First CoMP approaches were proposed in [BMWT00, SZ01], where the idea was to let multiple base stations jointly transmit to multiple terminals, effectively exploiting interference to obtain large gains in spectral efficiency and fairness [And05, KFVY06]. In the uplink, multiple base stations can cooperatively detect multiple terminals [WBOW00], promising similar gains [MKF06]. While the previous examples are cases of *multi-cell joint signal processing*, CoMP may also refer to schemes with a lesser extent of cooperation between base stations, for example joint scheduling or interference aware transmission and detection. In principle, it is also beneficial to let terminals cooperate [SSS+07b], but it (so far) appears difficult to explain to a mobile user why his or her handset battery is being depleted in order to enhance other users' data rates. Cooperation between terminals is hence not covered by this book.

1.2 Aim of this Book

This book provides a comprehensive overview on various CoMP techniques. It introduces information-theoretic concepts needed to understand and assess the principle degrees of freedom and gains expectable from CoMP, but also covers practical CoMP algorithms and addresses a multitude of challenges connected to their usage. A strong emphasis on implementation aspects and field trial results from the world's largest cellular research test beds gives the reader a detailed insight into the realistic potential of CoMP within the roadmap of LTE-Advanced and beyond, and the associated price and effort that have to be taken into consideration. The book provides the thorough detail required by scholars and professionals from industry or academia who aim at implementing or using CoMP themselves, but also serves as a reference book for the occasional reader.

1.3 Classes of CoMP Considered

As stated before, the term CoMP may refer to a multitude of schemes. All have in common that intra- or inter-cell interference is somehow taken into account or even exploited to enhance data rates and/or fairness. In this work, we classify CoMP schemes on one hand according to the extent of cooperation (or *information exchange*) taking place between cells:

- **Non-cooperative**, but **interference aware transceiver schemes**, where base stations or terminals adjust their transmit or receive strategy according to some knowledge on interference. This does not require explicit information

exchange between cells, but the estimation of interference must be enabled through appropriate reference signal design. This class of schemes includes *single-cell multi-user* signal processing, as used in LTE Release 8 [McC07].
- **Interference coordination** schemes, where limited data is exchanged between cells for the purpose of multi-cell cooperative scheduling, multi-cell interference-aware link adaptation, or multi-cell interference-aware precoding.
- **Joint signal processing** schemes, where user data or (partially) processed transmit or receive signals are exchanged among base stations. One here considers *non-coherent* and *coherent* schemes, where the latter aim at aligning the phases of signals transmitted from or received at different antennas. As we will see, this requires precise synchronization between all involved entities.

We will later also distinguish between *decentralized* and *centralized* CoMP schemes, referring to where the subject of cooperation takes place.

1.4 Outline of this Book

The book is structured into the following 5 parts:

- **Part I - Motivation and Basics** motivates the topic from a technical and economical point of view in Chapters 1 and 2, respectively, and provides information-theoretic basics and a first insight into potential gains and trade-offs of multi-cell joint signal processing in Chapters 3 and 4.
- **Part II - Practical CoMP Schemes** introduces various specific CoMP algorithms, where Chapter 5 focusses on interference-aware transceiver schemes and interference coordination, and Chapter 6 on multi-cell joint signal processing schemes, as classified before in Section 1.3.
- **Part III - Challenges Connected to CoMP** addresses various issues regarding the usage of CoMP in practice. Chapter 7 deals with finding clusters of cells in which CoMP is performed, whereas Chapters 8 and 9 cover the crucial aspects of synchronization and channel estimation and feedback, respectively. Chapter 10 highlights practical implementation aspects such as numerical stability and scalability. Chapter 11 investigates an adaptive, situation-dependent usage of CoMP, while Chapter 12 discusses the additional backhaul infrastructure required for CoMP itself and any required signaling.
- **Part IV - Performance Assessment** discusses CoMP field trial results in Chapter 13. As field trials are usually limited to the observation of exemplary multi-point links under exemplary interference conditions, the prediction of CoMP performance in large-scale systems requires system-level simulations, for which both methodology and results are covered in Chapter 14.
- **Part V - Outlook and Conclusions** finally discusses the usage of CoMP for other purposes than rate and fairness improvements, and elaborates the usage of CoMP in conjunction with relaying or heterogeneous cellular deployment in Chapter 15. The book is then concluded in Chapter 16.

2 An Operator's Point of View

Ralf Irmer

2.1 The Mobile Internet - A Success Story so far

When 3G was launched initially with WCDMA technology (Release 99), it was rather a disappointment with not many services being successful. Some years later, the mobile Internet took off when a number of factors came together:

- HSPA as a technological evolution of 3G with low latency and higher data rate
- Attractive flat-rate price plans by mobile operators
- Availability of mobile broadband hardware in terms of dongles and built-in 3G modules in notebooks
- Smart phones with attractive user interfaces, e.g., iPhone, Android
- Complete country-coverage with HSPA and HSPA+ by mobile operators.

This take-up of the mobile Internet generated substantial additional revenues for mobile operators, at a time when voice and text message revenues started to decline in saturated markets such as Europe. For example, Vodafone had a data revenue growth of 19% in financial year 2009/2010, with more than €4 Billion generated by non-SMS data. Today, only 11% of phones are smartphones, but by 2013 it is expected that more than a third of all active phones within the Vodafone network will be smartphones.

This data revenue growth comes along with a cost for mobile operators - namely data traffic growth. Fig. 2.1 shows the actual and projected traffic growth for Vodafone's European networks in Petabytes/year [Vod10]. It can be seen that data traffic has substantially surpassed voice traffic.

Mobile operators have some levers to cope with the growth in traffic in the short term, including:

- Technology upgrade, i.e. more efficient versions of HSPA or launch of LTE
- Cost reduction, i.e. network sharing, more efficient network operation, and exploitation of economy of scale
- Spectrum re-farming and acquisition of new spectrum
- Traffic management, i.e. enforcement of fair usage policies and launch of differentiated data bundles

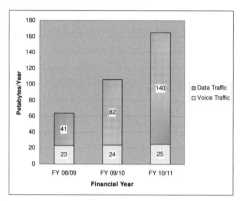

Figure 2.1 Data traffic in Vodafone's European networks in petabytes.

- Network management, i.e. building of new sites, provisioning of additional carriers or offload of traffic to femto cells or WiFi.

In the long term, however, the research community and the industry is required to come up with more fundamental approaches on how to serve mobile data at the right location in the most cost and energy efficient way.

2.2 Requirements on Future Networks and Upcoming Challenges

In 2006, a group of operators published the white paper on *Next Generation Networks beyond HSPA & EvDO* [ABG+06], which lists the high-level requirements on future networks, and an accompanying document listing the detailed technical assessment criteria [IAL+07]. Some of the important requirements are:

- Improved average and cell-edge spectral efficiency
- Low latency
- Simplicity, reliability and total cost of ownership
- Flat architecture

Most of the requirements are already addressed with LTE, which is being commercialized in 2010 in its first release. However, there is a need to develop LTE beyond the first release, in order to address customer and operator requirements. The challenges faced by mobile communications in the second decade of the 21st century are the following:

Exploding data volume - This is driven by attractive services, flat-rate pricing and user-friendly devices. The most prominent example is the iPhone - which resulted in a 10x traffic increase. IPTV, 3D Internet, real-time web, and cloud services will result in step changes in data consumption. IBM is predicting the generation of 16 TB/person/year by 2020. The challenge is that networks need to be structured to cope with data volume explosion without a cost or energy explosion or constant need for equipment upgrades, as illustrated in Fig. 2.2.

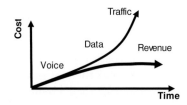

Figure 2.2 The growing gap between traffic and revenue.

Increased data rates - Driven by new services and the evolution from DSL (2 Mbps) to variants of fibre technology (100 Mbps to 1 Gbps), the user expectation of "acceptable" Internet speed will rise substantially in line with the expectation set by fibre networks and thus posing a challenge to wireless technologies.

Ubiquitous indoor coverage - Many data services are important for indoor users and people are usually within buildings. Indoor coverage is therefore important and can be either provided by copper/fibre with local radio distributions (femto cell or WiFi) or from cellular networks.

Ubiquitous outdoor coverage - For voice calls, the user expectation has moved from making calls along major roadways in the 1990s to being reachable all the time in any building. Mobile Internet based on 3G or WiFi today can only be characterized as best-effort, without "continual connectivity" whilst on-the-move and with patchy coverage in many places. In 2020, business customers and consumers will rely more on data connectivity - they will need connectivity anywhere, anytime. Coverage with a minimum guaranteed data rate and hence reliability will be a key differentiator between operators as the world moves from a "nice-to-be-connected" model to one that is "essential-to-be-connected".

There are technical innovations on the horizon to address these challenges:

- Gradual improvements of existing technologies, e.g. better MIMO modes etc.
- Active antennas, which may enable multi-element antennas
- New deployment concepts like femto cells or MetroZone networks. They require innovative backhaul solutions such as in-band and out-band backhaul or mm-wave microwave, and self-organizing principles in order to be manageable
- Miniaturized, flexible, energy-efficient base stations
- Base station cooperation concepts.

2.3 The Role of CoMP

Base station cooperation concepts (CoMP) are especially attractive since they improve the cell-edge data rate and average data rate, and are suitable to increase spectral efficiency (and hence capacity) for much more dense network deployments in urban areas and capacity hotspots. As we will see later, this increase in access capacity with CoMP concepts comes at the cost of more backhaul capacity, i.e. more communication bandwidth between base stations. However, for HSPA+ and LTE, base station sites need high-capacity backhaul (fibre or microwave) anyway, and as the cost of backhaul increases less than linearly with the backhaul capacity, this issue might not be as severe as often stated.

What are the alternatives to CoMP? Different frequency reuse, more spectrum, more sites, more antennas all are very expensive options for an operator. Thus investing into more intelligent baseband (i.e. CoMP algorithms) and backhaul with higher data rate and lower latency requirements seems to be more and more attractive when compared to the other options.

The complicated issue about CoMP concepts is that they are only partially understood from the academic perspective today, and that implementation in a standard at reasonable complexity is difficult. However, let's draw an analogy to MIMO technologies. They are commercially used today in WiFi and cellular communications, but ten years ago there was only limited understanding of MIMO, and the technology was seen by many as too complex to be commercialized.

2.4 The Role of Field Trials

Traditionally, academic innovation is evaluated using analytical models or doing statistical simulations. However, concepts such as CoMP are so complex that it is impossible to come up with models which capture all effects realistically, and to have well-calibrated simulation scenarios. Therefore, it is essential to have field trials of new technologies in an early development stage, in order to

- identify technical challenges early on
- refine simulations and analytical models
- be forced to have and end-to-end view and not pick "interesting" but non-relevant topics
- provide a proof of concept.

For CoMP technology development and evaluation, various authors of this book have set up a cluster of research test beds in Dresden and Berlin, within the EASY-C project led by Vodafone, Deutsche Telekom, Heinrich Hertz-Institute Berlin and Technische Universität Dresden. The significance of these is that enough sites are used to represent typical interference scenarios. More information on these test beds is stated in [I+09], and field trial results will be presented in detail in Chapter 13.

3 Information-Theoretic Basics

Patrick Marsch and David Tse

In this chapter, the reader is made familiar with a set of theoretical concepts to analytically capture the variety of CoMP schemes considered in this book. The reader will obtain a first understanding of the general capacity gains expectable from multi-cell joint signal processing, and the many degrees of freedom involved. The chapter introduces notation that will be reused in most parts of the book.

3.1 Observed Cellular Scenarios

Throughout the book, we generally consider (subsets of) a large cellular system as depicted in Fig. 3.1. Here, a large number of mobile terminals, or *user equipments (UEs)*, is distributed over a set of cells, where we assume that each cell is served by exactly one BS. As this is the case for most currently deployed cellular systems, we further assume that multiple BSs are grouped into so-called *sites*. Note that, differing from some other publications, we consider a *sector* to be equivalent to a *cell*. The term *cluster* is used to indicate a set of cells between which some form of CoMP may take place. Note that we assume that each UE in the system aims at transmitting or receiving dedicated information, i.e. *multi-cast* concepts are *not* covered in the book. As the number of UEs is typically significantly larger than the number of cells, UEs have to be *scheduled* to resources, i.e. to certain transmission windows. In this book, we assume that orthogonal frequency division multiple access (OFDMA) is employed as a media access technique, which allows each UE to be assigned to resources that are (under certain ideal assumptions to be discussed later) orthogonal in time and frequency. As this orthogonality allows us to simplify most of the analytical models and derivations used throughout the book, the basics of OFDMA are stated in the sequel.

3.2 Usage of OFDMA for Broadband Wireless Communications

Three fundamental challenges in mobile communications are the fact that transmission takes place over *a) a shared medium*, which is often subject to *b) rich scattering*, and to which we desire *c) simple and flexible access* of many commu-

Information-Theoretic Basics

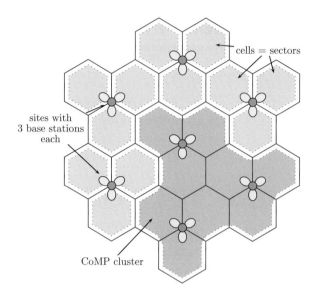

Figure 3.1 Cellular system and key terminology considered.

nicating entities. The first aspect implies that any transmission must be bandlimited in order not to disturb other transmissions on adjacent bands, which requires the design of particular transmit and receive filters. The second aspect implies that any receiver may observe a superposition of multiple differently delayed and attenuated copies of originally transmitted signals, which in the context of broadband transmission may lead to inter-symbol interference (ISI) that has to be dealt with. The third aspect means that we need a low-cost and efficient signal processing solution that can divide a mobile communications system into a large number of flexible bit pipes according to many users' or the applications' needs.

The mobile communications standard LTE Release 8 [McC07] from 3GPP uses an OFDMA approach to address all aspects stated above, where the *baseband* signal processing chain for a downlink example is depicted in Fig. 3.2. Here, the key concept is that the symbols to be transmitted from one BS towards multiple UEs 1..U are modulated in frequency domain, mapped to different *sub-carriers*, and then an inverse discrete Fourier transform (IDFT) is used to generate a time domain signal. A cyclic prefix is inserted before each orthogonal frequency division multiplex (OFDM) symbol (i.e. before each block of samples processed in one IDFT), in order to assure that even a channel with a large delay spread does not cause ISI, and that the transmission leads to a *circularly symmetric convolution* of the transmitted samples with the channel. Each receiving UE can then discard the cyclic prefix, perform a discrete Fourier transform (DFT), and obtain (scaled and noisy) transmitted symbols in frequency domain again. While

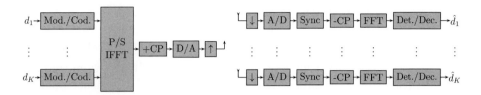

Figure 3.2 OFDMA signal processing chain (downlink example).

being computationally inexpensive and sacrificing only a reasonable extent of capacity for the cyclic prefix, OFDMA has the advantage that it *a)* can easily be designed to fulfill different spectral masks through an appropriate choice of guard bands (i.e. leaving peripheral sub-carriers empty), *b)* avoids ISI and inter-carrier interference (ICI) and hence enables simple OFDM symbol-wise and sub-carrier-wise equalization at the receiver side, and *c)* is highly suitable for multiple access. A detailed explanation of OFDMA can be found in [LP01].

A major *disadvantage* of OFDM is that performing modulation in frequency domain and applying an IDFT leads to signals with a peak-to-average power ratio (PAPR) increasing linearly in the DFT size [WG99]. Especially in the uplink, this aspect is critical, as it implies that a larger power amplifier (PA) back-off is needed, leading to a faster depletion of handset battery. For this reason, 3GPP has decided to employ single carrier frequency domain multiple access (SC-FDMA) in the uplink, where modulation is performed in time domain, after which a small DFT is applied (according to the number of sub-carriers to be occupied by the UE) and the signals are mapped to the sub-carriers to be used by the terminal before the actual large IDFT. The price for a reduced PAPR is the need for ISI cancelation, which may lead to more complex signal processing if used in conjunction with (possibly multi-cell) MIMO equalization. In the remainder of this book, we will assume for simplicity that OFDM is used in both uplink and downlink, knowing that the performance of OFDM and SC-FDMA (with ISI cancelation) is fairly comparable.

3.3 Multi-Point Frequency-Flat Baseband Model Considered

As mentioned before, OFDMA also has the advantage that it enables a simple mathematical notation and analysis, as it is often sufficient to observe the baseband transmission on a single frequency-flat sub-carrier, which can be seen as a transmission over an additive white Gaussian noise (AWGN) channel. Most chapters in this book will make use of this simplification. Only in cases where the correlation of channel realizations in time and frequency is of importance, for example for channel estimation and feedback schemes in Chapter 9, a wideband model will be used. The assumption of an AWGN channel also implicitly requires the OFDM systems of all communicating entities to be perfectly synchronized in time and frequency - an assumption that will later be challenged in Chapter 8.

When observing the transmission on a single frequency-flat OFDMA sub-carrier, we typically consider a subset of a cellular system consisting of M BSs and K UEs, assuming that a scheduling entity has assigned the UEs to the same observed resource in time and frequency. We introduce sets $\mathcal{M} = \{1..M\}$ and $\mathcal{K} = \{1..K\}$. Throughout this book, we consider various antenna setups, where N_{bs} and N_{ue} denote the number of antennas per BS and per UE, respectively, and where $N_{\text{BS}} = MN_{\text{bs}}$ and $N_{\text{UE}} = KN_{\text{ue}}$ denote the overall number of antennas at BS and UE side, respectively. In the sequel, we will now go into details of the uplink and downlink transmission models used throughout the book.

3.4 Uplink Transmission

In the uplink, the precoding, transmission and equalization of each symbol on an OFDMA sub-carrier is illustrated in Fig. 3.3 and stated as

$$\tilde{\mathbf{x}} = \mathbf{W}^H \mathbf{y} = \mathbf{W}^H (\mathbf{H}\mathbf{s} + \mathbf{n}) = \mathbf{W}^H \left(\mathbf{H} \underbrace{\begin{bmatrix} \mathbf{G}_1 & & \mathbf{0} \\ & \ddots & \\ \mathbf{0} & & \mathbf{G}_K \end{bmatrix}}_{\mathbf{G}} \mathbf{x} + \mathbf{n} \right), \quad (3.1)$$

where $\mathbf{x} = [\mathbf{x}_1^T..\mathbf{x}_K^T]^T \in \mathbb{C}^{[N_{\text{UE}} \times 1]}$ are the symbols to be transmitted by the UEs, which we generally assume uncorrelated with $E\{\mathbf{xx}^H\} = \mathbf{I}$. These may then be subject to linear UE-side precoding via matrices $\forall k : \mathbf{G}_k \in \mathbb{C}^{[N_{\text{ue}} \times N_{\text{ue}}]}$, yielding the finally transmitted signals $\mathbf{s} = [\mathbf{s}_1^T..\mathbf{s}_K^T]^T \in \mathbb{C}^{[N_{\text{UE}} \times 1]}$. As we do not consider UE cooperation, the overall transmit covariance $E\{\mathbf{ss}^H\} = \mathbf{\Phi}_{\text{ss}}$ has a block-diagonal structure, i.e. the signals originating from different UEs are uncorrelated. $\mathbf{\Phi}_{\text{ss}}$ is usually subject to a *per-antenna* or *per-UE* power constraint to be stated later. $\mathbf{H} \in \mathbb{C}^{[N_{\text{BS}} \times N_{\text{UE}}]}$ is the instantaneous fast fading realization of the channel on this sub-carrier. We also denote as $\mathbf{H}^m$, $\mathbf{H}_k$, $\mathbf{H}_k^m$ the parts of the channel matrix $\mathbf{H}$ connected to BS m, UE k, or the link from BS m to UE k, respectively, where we use a lower-case $\mathbf{h}$ if the expression becomes a vector (e.g. for $N_{\text{ue}} = 1$). $\mathbf{y} = [\mathbf{y}_1^T..\mathbf{y}_M^T]^T \in \mathbb{C}^{[N_{\text{BS}} \times 1]}$ are the signals received by the BSs, containing zero-mean Gaussian noise $\mathbf{n} \in \mathbb{C}^{[N_{\text{BS}} \times 1]}$ with $E\{\mathbf{nn}^H\} = \sigma^2\mathbf{I}$, where then equalization via a matrix $\mathbf{W} \in \mathbb{C}^{[N_{\text{BS}} \times N_{\text{UE}}]}$ is performed to yield estimates $\tilde{\mathbf{x}} \in \mathbb{C}^{[N_{\text{UE}} \times 1]}$ on the originally generated symbols $\mathbf{x}$. The structure of $\mathbf{W}$ depends on the particular CoMP strategy employed, as we will see later. We also write $\mathbf{W}^m$, $\mathbf{W}_k$ or $\mathbf{W}_k^m$ for the part of $\mathbf{W}$ connected to a particular BS m, UE k, or a specific link, respectively, and use a lower-case $\mathbf{w}$ if this yields a vector.

3.4 Uplink Transmission

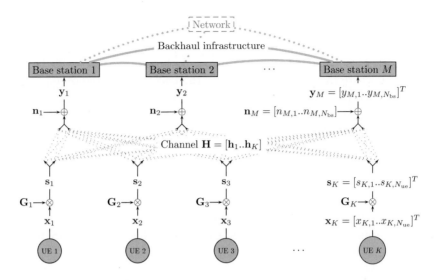

Figure 3.3 Uplink transmission setup.

3.4.1 Basic Uplink Capacity Bounds

For the derivation of information-theoretic bounds in Chapters 3 and 4, we typically use the assumption that the signals **x** are zero-mean Gaussian and belong to long codewords where each symbol sees the same channel realization **H**. Note that the Gaussianity of **x** is not necessarily optimal in terms of capacity [Med00], but strongly simplifies achievable rate derivation. If we now consider as the most simple setup the case where only one UE with $N_{\text{ue}} = 1$ antenna transmits at unit power and is decoded by one BS, then the probability of decoding error decreases exponentially in the codeword length if and only if the transmission rate R in bits per channel use (bpcu) fulfills [CT06]

$$R \leq I(X;Y) = \log_2 \left| \mathbf{I} + \frac{1}{\sigma^2} \mathbf{h}\mathbf{h}^H \right|, \qquad (3.2)$$

where the notation $I(X;Y)$ denotes the *mutual information* between transmitter and receiver side. The rate bound in (3.2) can be proven from both sides. On one hand, one can construct an example coding technique (often based on the idea of *typical sequences*) with a rate equivalent to the right-hand side (RHS) of (3.2), such that any arbitrarily low probability of error can be achieved by simply choosing a sufficiently long codeword. On the other hand, one can prove that regardless of codeword length a non-zero probability of error remains if R exceeds the RHS of (3.2) [CT06], typically making use of *Fano's inequality*. Hence, in this point-to-point case with Gaussian **x**, the *capacity* of the transmission has been precisely established. In the case of $N_{\text{ue}} > 1$, i.e. multiple antennas per UE,

(3.2) changes to

$$R \leq I(X;Y) = \max_{\mathbf{\Phi}_{\text{ss}}} \log_2 \left| \mathbf{I} + \frac{1}{\sigma^2} \mathbf{H} \mathbf{\Phi}_{\text{ss}} \mathbf{H}^H \right|, \qquad (3.3)$$

where the max operation over the transmit covariance $\mathbf{\Phi}_{\text{ss}}$ implies that the UE choses the optimal precoding matrix $\mathbf{G}$ for the current channel realization $\mathbf{H}$, hence requiring transmitter-side channel knowledge. Given perfect such knowledge, and assuming all transmit antennas of the UE to be subject to a *sum power* constraint P_{sum}, a capacity-achieving UE strategy is to perform a singular value decomposition (SVD) of the channel, yielding $\mathbf{H} = \mathbf{U}\mathbf{\Sigma}\mathbf{V}^H$, and choose as precoding matrix $\mathbf{G}$ the RHS eigenvectors $\mathbf{V}$, where the columns are scaled in power such that $\text{tr}\{\mathbf{V}\mathbf{V}^H\} = \text{tr}\{\mathbf{\Phi}_{\text{ss}}\} \leq P_{\text{sum}}$. Assuming that $\mathbf{W} = \mathbf{U}$ is used as BS side receive filter, the transmission from (3.1) can be re-stated as a transmission over $\min(N_{\text{bs}}, N_{\text{ue}})$ independent single-input single-output (SISO) links, often referred to as the *eigenmodes* of the channel. The capacity on each eigenmode can then be proved as in the case of $N_{\text{ue}} = 1$ before. Finding the power scaling for $\mathbf{V}$ that maximizes the sum capacity over all eigenmodes is a convex optimization problem [BV04] that can be solved easily via a *water-filling* algorithm [CT06], but not in closed form. Note that the gap between capacity and rates achievable without channel knowledge at the UE-side (e.g. without precoding and power control, for example $\mathbf{V} = \mathbf{I}$) may be marginal, but then significantly more complex signal processing is required at the receiver side [HTB03].

If now multiple UEs $1..K$ are decoded by the same single BS, the setup resembles a multiple access channel (MAC) [Ahl71], where it becomes interesting to observe the *capacity region*, hence all tuples of rates $R_1..R_K$ at which the UEs can transmit, such that *all* can be decoded at a probability of error decreasing exponentially in the codeword length. The capacity region of the MAC [Ahl71, Lia72] is simply based on the fact that the sum-rate of any subset of UEs is bounded by the joint mutual information between these UEs and the BS, given that all other UEs are turned off. Formally, we can state this as

$$\forall \mathcal{S} \subseteq \{1..K\} : \sum_{k \in \mathcal{S}} R_k \leq \max_{\mathbf{\Phi}_{\text{ss}}} \log_2 \left| \mathbf{I} + \frac{1}{\sigma^2} \mathbf{H} \mathbf{\Phi}_{\text{ss}}(\mathcal{S}) \mathbf{H}^H \right|, \qquad (3.4)$$

where $\mathbf{\Phi}_{\text{ss}}(\mathcal{S})$ is the transmit covariance connected to the subset of UEs in set $\mathcal{S}$. Note that $\mathbf{\Phi}_{\text{ss}}$ is now block-diagonal, as it is connected to multiple UEs. For $K = 2$ UEs, this leads to the well-known pentagon-shaped capacity region illustrated in Fig. 3.4 [TV05]. Interpreting (3.4) from a more practical perspective, it becomes clear that any point on the capacity region can be achieved by applying successive interference cancelation (SIC), hence by successively decoding the transmissions of certain UEs, subtracting the corresponding receive signals, and then decoding other UEs. Each of those cornerpoints of the capacity region where all UEs' rates are non-zero corresponds to one particular SIC order. The optimal $\mathbf{\Phi}_{\text{ss}}$ can be determined via a UE-wise successive SVD and water-filling algorithm [YRBC01].

In the remainder of this chapter, we shift the focus to scenarios with multiple BSs serving adjacent cells, and observe potential capacity gains through multi-cell joint signal processing.

3.4.2 Full Cooperation in the Uplink

Full cooperation, i.e. joint processing of all received signals by all BSs, can under certain idealistic assumptions be modeled as one *virtual* BS with N_{BS} antennas. The capacity region is also given by (3.4), but now based on a channel **H** of higher dimensionality. As opposed to the single-BS case, we have now obtained *array gain* of at most $10\log_{10}(M)$ dB, as all BS antennas receive correlated signals which can be coherently overlapped, while noise terms are uncorrelated, and *spatial multiplexing gain* [ZDZ04], as the larger channel dimensionality improves the eigenvalue distribution and hence orthogonality between UEs. Considering multiple fading realizations, cooperation may also yield spatial diversity gain [TZ03].

3.4.3 No Cooperation in the Uplink

If no cooperation is possible between BSs, the scenario is similar to an *interference channel (IC)* [Ahl74], which is defined as two or more non-cooperative transmitter-receiver pairs that communicate on the same time/frequency resource, subject to mutual interference. Unfortunately, the capacity region of the IC is only known in certain cases, for example for Gaussian signaling, single-antenna transceivers and *very strong interference*, where the interference links are significantly stronger than the transmitter-receiver links, and the capacity region is the same as if there were no interference at all [Car75]. Capacity has also been established for *very weak interference*, where it is optimal if all receivers treat interference as noise [SKC09, MK09]. For all regimes in between, the tightest known inner capacity bound is based on the Han-Kobayashi (HK) transmission scheme, where the transmitters use superimposed transmissions, and the receivers decode both a subset of the interfering transmissions and their desired transmission [HK81], achieving an optimal level of interference cancelation. In all interference cases, capacity is known within one bit [ETW08]. Recent work has been on the IC with multiple antennas at the receiver or transmitter side, introducing the concept of *interference alignment* [WT08, MAMK08, CJ08]. Here, desired signal and interference as seen by each receiver fall into orthogonal signal dimensions, such that each transmitter-receiver pair can use half the system resources for interference-free communication. For the observation of asymptotically large signal-to-noise ratios (SNRs), one often uses the concept of *generalized degrees of freedom* [ETW08], which will later be discussed in Section 12.1.

Note that the communication scenarios taking place on certain resources within a cellular system deviate from the classical IC in the way that it does not matter *where* a UE is decoded, hence transmitter-receiver pairs may be chosen

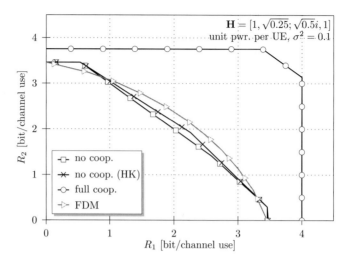

Figure 3.4 (Inner bounds on) capacity regions for no or full cooperation in the uplink.

such that interference is minimized. Simply assigning each UE to its best serving BS clearly renders scenarios of strong or very strong interference [ETW08], or those where HK schemes would be highly beneficial, rather unlikely. Further, the usage of such schemes becomes questionable in the context of practical coding schemes with a gap to capacity [FU98] and under imperfect channel state information (CSI), as we will see in Chapter 4. Instead, the following options appear much more powerful to increase capacity in the non-cooperative case, in particular under imperfect CSI [Mar10]:

- a flexible assignment of BSs to UEs, possibly on a short-term basis, or
- the option of non-cooperative multi-UE detection by one BS.

If we neglect HK schemes (i.e. superimposed transmission and partial decoding of interference), an inner bound on the capacity region for a given BS-UE assignment can be calculated for each BS and set of assigned UEs separately, seeing the remaining UE transmissions not decoded by the BS as spatially colored noise. More precisely, if we denote as $\mathcal{S}_{[m]}$ the set of UEs decoded by BS m, the capacity region of these UEs is inner-bounded as $\forall\, \mathcal{S} \subseteq \mathcal{S}_{[m]}$:

$$\sum_{k \in \mathcal{S}} R_k \leq \max_{\mathbf{\Phi}_{\mathrm{ss}}} \log_2 \left| \mathbf{I} + \left(\mathbf{H}^m \mathbf{\Phi}_{\mathrm{ss}} \left(\mathcal{K} \setminus \mathcal{S}_{[m]} \right) \mathbf{H}_m^H \right)^{-1} \mathbf{H}^m \mathbf{\Phi}_{\mathrm{ss}} \left(\mathcal{S} \right) \left(\mathbf{H}^m \right)^H \right|. \quad (3.5)$$

Note that this inner bound implies that each BS m also has perfect channel knowledge towards interfering UEs, and takes this into account when calculating receive filter $\mathbf{W}^m$, referred to as interference rejection combining (IRC).

3.4.4 Numerical Example

Fig. 3.4 shows inner bounds on capacity regions for no BS cooperation, and the capacity region for full BS cooperation, for an example with $M = K = 2$, $N_{\text{bs}} = N_{\text{ue}} = 1$, $\mathbf{H} = [1, \sqrt{0.25}, \sqrt{0.5}i, 1]$, $\sigma^2 = 0.1$ and unit transmit power limit per UE. In the non-cooperative case, the bound is based on all possible assignments of UEs to BSs, including the option of one BS decoding both UEs. This bound is only marginally extended through HK schemes. For this channel, non-cooperative performance can be improved through frequency division multiplex (FDM), as also shown in the figure, where both UEs are placed on orthogonal resources and hence mutual interference is avoided. Each UE then invests its transmit power into a smaller portion of bandwidth, yielding an improved SNR. As FDM is of little value in connection with BS cooperation [Mar10], however, we will not further observe it in this work.

3.5 Downlink Transmission

In the downlink, the precoding, transmission and equalization of each OFDM symbol on a single sub-carrier can be stated as

$$\tilde{\mathbf{x}} = \mathbf{G}^H \mathbf{y} = \begin{bmatrix} \mathbf{G}_1^H & & 0 \\ & \ddots & \\ 0 & & \mathbf{G}_K^H \end{bmatrix} \left(\mathbf{H}^H \mathbf{W} d(\mathbf{x}) + \mathbf{n} \right) \quad (3.6)$$

where $\mathbf{x} \in \mathbb{C}^{[N_{\text{UE}} \times 1]}$ are the symbols to be transmitted to the UEs, and $d(\cdot)$ can be any arbitrary manipulation of these symbols performed by the BSs. We will see later that a non-linear operation $d(\cdot)$ is in fact required to achieve capacity in the case of multiple UEs. $\mathbf{W} \in \mathbb{C}^{[N_{\text{BS}} \times N_{\text{UE}}]}$ is a precoding matrix applied at the BS side. The transmit covariance is now given as $\mathbf{\Phi}_{\text{ss}} = E\{\mathbf{W}d(\mathbf{x})(d(\mathbf{x}))^H \mathbf{W}^H\}$, which is typically subject to either a *sum*, *per-BS* or *per-antenna* power constraint. The latter is often motivated through the fact that each BS transmit antenna has a separate power amplifier with a limited linear range. In an OFDM context, however, applying a per-antenna power constraint individually on each sub-carrier is rather questionable, as the time-domain signal and its PAPR appear more important. $\mathbf{H} \in \mathbb{C}^{[N_{\text{BS}} \times N_{\text{UE}}]}$ is the channel matrix, as defined for the uplink. $\mathbf{G} \in \mathbb{C}^{[N_{\text{UE}} \times N_{\text{UE}}]}$ is a matrix containing the UE-side receive filters, which is block-diagonal, as we again assume that no cooperation takes place between UEs. $\mathbf{n} \in \mathbb{C}^{[N_{\text{UE}} \times 1]}$ is the thermal noise and background interference present at the receive antennas of the UEs, which we assume zero-mean Gaussian with covariance $E\{\mathbf{nn}^H\} = \sigma^2 \mathbf{I}$. Each UE finally obtains estimates $\tilde{\mathbf{x}} \in \mathbb{C}^{[N_{\text{UE}} \times 1]}$ of the originally transmitted symbols $\mathbf{x}$. The same variable names have been used in both (3.6) and for the uplink in (3.1) to emphasize duality: The receive filter $\mathbf{W}$ in the uplink plays a dual role to the transmit filter in the downlink, and the uplink transmit filters $\mathbf{G}$ a dual role to the downlink receive filters.

3.5.1 Basic Downlink Capacity Bounds

A point-to-point link in the downlink is analog to the uplink, hence (3.3) can be used to state capacity (if the uplink channel $\mathbf{H}$ in (3.3) is replaced by the downlink channel $\mathbf{H}^H$). As stated before, point-to-point capacity under perfect channel knowledge at transmitter and receiver can be achieved with linear precoding and equalization, hence the non-linear operation $d(\cdot)$ is not relevant in this case.

In the case of one BS and *multiple* UEs, the downlink setup resembles a broadcast channel (BC). Note that one must here distinguish between *common information* (i.e. broadcast information to be decoded by all receivers) and *individual information* to be decoded only by single UEs. In this book, we only consider the latter type of information, as it plays the dominant role in cellular systems. As opposed to that of the MAC, the capacity region of the BC is only known for special cases. One of these is the so-called *degraded* BC, where all UE's receive signals can be stated as another UE's signals, but subject to additional noise, and where a superposition coding strategy is capacity achieving [Ber74]. For the general, *non-degraded* BC, which is interesting for us, capacity is only known for the case of Gaussian noise [WSS06]. A major transmission concept in this context is dirty paper coding (DPC) [Cos83], which allows a UE to receive signals free of interference, if this interference is known non-causally to the transmitter. In an example with $K = 2$ UEs, this means that the BSs can transmit to one UE, while simultaneously transmitting to the other UE, but encoded in such a way that the latter transmission is not interfered by the former. This can be applied to an arbitrary number of transmitted streams, and one can see a duality to SIC in the uplink: If the SIC *decoding* order is $1..K$ in the uplink, UE k only sees interference from UEs $k + 1..K$, as the others have already been decoded and their signals removed. Equivalently, if the downlink DPC *encoding* order is $1..K$, UE k only sees interference from UEs $k + 1..K$, as the previously encoded streams can be considered as known interference and be pre-cancelled at the transmitter side. Different from the uplink, however, where SIC can be practically implemented at reasonable effort and requires only receiver-side channel knowledge, DPC is mainly a theoretical construct. Sub-optimal, practical schemes of limited complexity for the downlink are Tomlinson-Harashima precoding (THP) [Tom71, HM72] and sphere-precoding [HRF06], but these generally require highly precise channel knowledge at the transmitter-side. In Chapters 3 and 4, we consider both BC capacity (i.e. DPC performance) and the rates achievable with linear precoding, where residual interference between streams is simply accepted as noise, knowing that a practical transmission scheme may then perform somewhere in between.

The capacity region of the BC can principally be stated as

$$\mathcal{R}_{\text{BC}} = \text{conv} \left\{ \bigcup_{\mathbf{W}, \pi} \langle R_1(\mathbf{W}, \pi), .., R_K(\mathbf{W}, \pi) \rangle \right\}, \quad (3.7)$$

where $\bigcup$ states the *union* of multiple rate regions, and $\mathrm{conv}(\cdot)$ is a *convex hull* operation [BV04], in this case over all choices of precoding matrix $\mathbf{W}$ and encoding order π, where for each fixed parameter choice the UE rates are bounded as

$$R_k(\mathbf{W}, \pi) \leq \log_2 \left| \mathbf{I} + \left(\sigma^2 \mathbf{I} + \sum_{\pi(j) > \pi(k)} \mathbf{H}_k^H \mathbf{W}_j \mathbf{W}_j^H \mathbf{H}_k \right)^{-1} \mathbf{H}_k^H \mathbf{W}_k \mathbf{W}_k^H \mathbf{H}_k \right|. \quad (3.8)$$

Unfortunately, finding the optimal $\mathbf{W}$ for a certain point on the capacity region (or, equivalently, the precoder maximizing a particularly weighted sum of UE rates), is not trivial, as any sum of UE rates as given in (3.8) is typically non-convex in $\mathbf{W}$ [BS02]. It has been observed in [JVG04], however, that an interesting duality can be exploited between uplink and downlink that we want to briefly illustrate in the sequel. Let us state $\mathbf{A}_k = \sigma^2 \mathbf{I} + \sum_{\pi(j) > \pi(k)} \mathbf{H}_k^H \mathbf{W}_j \mathbf{W}_j^H \mathbf{H}_k$ and $\mathbf{B}_k = \sigma^2 \mathbf{I} + \sum_{\pi(j) < \pi(k)} \mathbf{H}_j \mathbf{G}_j \mathbf{G}_j^H \mathbf{H}_j^H$ as the interference terms in downlink and uplink, respectively. We then re-state the rate bound in the BC from (3.8) as

$$R_k(\mathbf{W}, \pi) \leq \log_2 \left| \mathbf{I} + \mathbf{A}_k^{-1} \mathbf{H}_k^H \mathbf{W}_k \mathbf{W}_k^H \mathbf{H}_k \right| \quad (3.9)$$

$$= \log_2 \left| \mathbf{I} + \mathbf{A}_k^{-\frac{1}{2}} \mathbf{H}_k^H \mathbf{B}_k^{-\frac{1}{2}} \mathbf{B}_k^{\frac{1}{2}} \mathbf{W}_k \mathbf{W}_k^H \mathbf{B}_k^{\frac{1}{2}} \mathbf{B}_k^{-\frac{1}{2}} \mathbf{H}_k \mathbf{A}_k^{-\frac{1}{2}} \right| \quad (3.10)$$

$$= \log_2 \left| \mathbf{I} + \mathbf{B}_k^{-\frac{1}{2}} \mathbf{H}_k \underbrace{\mathbf{A}_k^{-\frac{1}{2}} \mathbf{B}_k^{\frac{1}{2}} \mathbf{W}_k}_{=\mathbf{G}_k} \mathbf{W}_k^H \mathbf{B}_k^{\frac{1}{2}} \mathbf{A}_k^{-\frac{1}{2}} \mathbf{H}_k^H \mathbf{B}_k^{-\frac{1}{2}} \right| \quad (3.11)$$

$$= \log_2 \left| \mathbf{I} + \mathbf{B}_k^{-1} \mathbf{H}_k \mathbf{G}_k \mathbf{G}_k^H \mathbf{H}_k^H \right|, \quad (3.12)$$

which is equivalent to the uplink rate bound for a MAC, given fixed transmit filters $\mathbf{G}_k$ and an *opposite* decoding order $\bar{\pi}$, as this can be derived from (3.4). The equality in (3.10) is based on the fact that $|\mathbf{I} + \mathbf{A}\mathbf{B}| = |\mathbf{I} + \mathbf{B}\mathbf{A}|$, and that in (3.11) based on the idea of *channel flipping* [JVG04]. The authors in [JVG04] have furthermore shown that the above equalities hold for all UEs if and only if the sum power is the same in both cases, i.e. if $\mathrm{tr}\{\sum_k \mathbf{G}_k \mathbf{G}_k^H\} = \mathrm{tr}\{\sum_k \mathbf{W}_k \mathbf{W}_k^H\}$. Hence, we can conclude that the capacity region of the MIMO BC under a sum power constraint is equivalent to that of the MIMO MAC (obtained through the reciprocal channel $\mathbf{H}$) under the same sum power constraint. As the standard uplink is typically subject to a *per-UE* power constraint, we can obtain the BC capacity region by taking the convex hull around many MAC regions with different per-UE powers summing up to the same overall power. This is illustrated in Fig. 3.5 for the same example channel as before. It was shown in [WSS06] that the obtained BC rate region corresponds to the *Sato upper bound* [Sat78], proving that there can indeed be no scheme that performs better. Hence, capacity has been established for the BC case of Gaussian noise.

Equations (3.9)-(3.12) also suggest that we can calculate the optimal precoding matrix $\mathbf{W}$ if the dual uplink transmit filters $\mathbf{G}$ are known. This is possible by calculating $\forall k$ $\mathbf{B}_k$ directly from $\mathbf{G}_1..\mathbf{G}_K$, and then determining $\mathbf{A}_k$ and $\mathbf{W}_k = \mathbf{B}_k^{-1/2} \mathbf{A}_k^{1/2} \mathbf{G}_k$ iteratively, starting with UE K [JVG04]. As the dual uplink

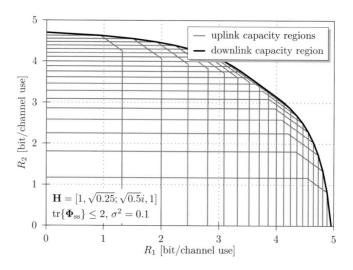

Figure 3.5 Illustration of uplink/downlink duality.

is subject to a sum power constraint, however, the calculation of $\mathbf{G}_k$ requires not only the optimization of each UE's individual transmit covariance as in Section 3.4.1, but also the power distribution among UEs. Under the assumption of non-linear precoding (DPC), this is a convex optimization problem and can be solved via a gradient-based algorithm as stated in, e.g., [VVH03].

Uplink/downlink duality can also be used to calculate capacity regions and the precoding matrix $\mathbf{W}$ for a BC under *per-antenna-(group) power constraints* [YL07]. Under these constraints, the dual uplink is subject to a least favorable noise covariance constrained within a polyhedron. Capacity region calculation then becomes more complex, as the update of dual uplink transmit covariances $\mathbf{G}_k$ and uplink noise covariance have to be performed iteratively, and convergence may become an issue.

3.5.2 Full Cooperation in the Downlink

As in the uplink, full cooperation in the downlink means that the rate expression from (3.8) can be used, but is now applied to a larger compound channel $\mathbf{H}$. As in the uplink, this yields spatial multiplexing, array and diversity gain.

3.5.3 No Cooperation in the Downlink

Without BS cooperation in the downlink, we are again facing an IC, where the best known transmission scheme is based on superposition coding and partial decoding of interference (HK). Analog to the uplink, however, the usage of HK techniques is of little relevance if we allow for a fast-fading dependent assignment of UEs to BSs, and consider the option of local, non-cooperative transmission

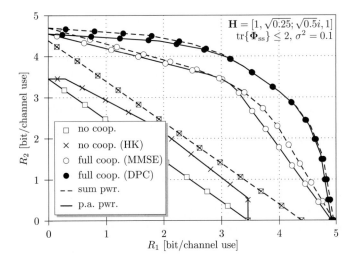

Figure 3.6 (Inner bounds on) capacity regions for no or full cooperation in the downlink.

from one BS to multiple UEs. In general, the non-cooperative case implies that each UE k receives desired signals from only one BS, which means that its associated precoding vector $\mathbf{w}_k$ may have non-zero entries only on the elements connected to the antenna of this BS. It has been shown in [MF09a, Mar10] that duality can now still be applied, and inner bounds on capacity regions can be determined by observing a dual uplink where for each UE only the *receive* antennas connected to one BS may be used for detection.

3.5.4 Numerical Example

Fig. 3.6 shows inner bounds on capacity regions for no BS cooperation and capacity regions for full BS cooperation, for the same example channel as before, i.e. $M = K = 2$, $N_{\text{bs}} = N_{\text{ue}} = 1$, $\mathbf{H} = [1, \sqrt{0.25}, \sqrt{0.5}i, 1]$, $\sigma^2 = 0.1$ and either a unit per-antenna power constraint, or a sum power contraint of 2 (i.e. in both cases $\text{tr}\{\mathbf{\Phi}_{\text{ss}}\} \leq 2$). In the non-cooperative case, an inner bound on the capacity region is in principal based on all possible assignments of UEs to BSs, including the option of one BS transmitting to both UEs, though this is not beneficial for this particular channel. We again also observe HK schemes, but can see that these are only interesting under a per-antenna power constraint. In general, the difference between sum and per-antenna power constraint is only visible at the sides of the capacity regions, while the sum-rate remains largely unaffected, especially under non-linear precoding (DPC).

3.6 Summary

In this chapter, we have formalized the uplink and downlink transmissions considered throughout the remainder of the book, and introduced the basic information-theoretic concepts inherent in the many degrees of freedom of CoMP. We have seen how *capacity regions* can be computed for uplink and downlink under full base station cooperation, and inner bounds on these (or *achievable rate regions*) can be computed for cases of no BS cooperation, as capacity remains unknown here. While all computations are rather straight-forward for the uplink, we have seen that uplink/downlink duality can be used to also make the downlink more mathematically amenable. The results in this chapter already suggest substantial rate gains through multi-cell joint signal processing, but this will be analyzed in more detail in the next chapter.

4 Gains and Trade-Offs of Multi-Cell Joint Signal Processing

Patrick Marsch

In this chapter, we focus on CoMP schemes based on multi-cell joint signal processing. We extend the transmission models from Chapter 3 to incorporate imperfect channel knowledge at the transmitter and receiver side, which strongly limits the interference regimes in which joint signal processing is reasonably beneficial. Regarding the uplink, we then explore different types of information exchange between cooperating base stations, revealing general trade-offs that have to be considered here if backhaul is a limited resource. For the downlink, different principle joint transmission concepts are introduced.

4.1 Modeling Imperfect Channel State Information (CSI)

In this chapter, we observe an exemplary frequency-flat OFDM sub-carrier, assuming that scheduling of users to resources has already taken place.

4.1.1 Imperfect CSI in the Uplink

Let us first observe the uplink, where the main issue is usually the channel state information at the receiver (CSIR), in particular if only $N_{\text{ue}} = 1$ transmit antenna per user equipment (UE) is assumed, as in LTE Release 8 [McC07]. Let us assume that the channel introduced in (3.1) is re-written as

$$\mathbf{H} = \hat{\mathbf{H}} + \mathbf{E}, \qquad (4.1)$$

where $\hat{\mathbf{H}}$ is an *un-biased* minimum mean square error (MMSE) estimate of the channel, and $\mathbf{E}$ is the channel estimation error. We can then also re-write the complete transmission from (3.1) (without the receive filter $\mathbf{W}$) to

$$\mathbf{y} = \left(\hat{\mathbf{H}} + \mathbf{E}\right)\mathbf{s} + \mathbf{n} = \hat{\mathbf{H}}\mathbf{s} + \underbrace{\mathbf{Es}}_{\text{channel estimation related noise term}} + \mathbf{n}, \qquad (4.2)$$

where we can see that the channel estimation error leads to an additional noise term $\mathbf{Es}$. Equation (4.1) implies that if the channel and its estimate are assumed block-static (as defined for the former in Chapter 3), then the estimation error is also block-static. If we now observe the *average* capacity of the transmission over many transmission blocks, the impact of channel estimation noise can be overestimated due to Jensen's inequality, if the term $\mathbf{Es}$ is treated as a random variable with a different realization *in each channel access*. We further overestimate its impact by modeling it as a spatially white Gaussian random variable, as this has the largest entropy for a given variance [Med00], i.e. by observing

$$\mathbf{y} = \hat{\mathbf{H}}\mathbf{s} + \mathbf{v} + \mathbf{n} \text{ with } \mathbf{v} \sim \mathcal{N}_\mathbb{C}(\mathbf{0}, \mathbf{\Phi}_{\text{hh}}) \quad (4.3)$$

with $\mathbf{\Phi}_{\text{hh}} = E\{\mathbf{E}\mathbf{\Phi}_{\text{ss}}\mathbf{E}^H\}$, which is diagonal under the realistic assumption that the estimation errors of each channel coefficient are uncorrelated. It is shown in [YG06a] that the assumption of Gaussianity has little impact on the mutual information unless $N_{\text{BS}} \gg N_{\text{UE}}$. In other words, in the CoMP scenarios considered in this book, there is not much rate improvement possible if receiver-side algorithms treat term $\mathbf{Es}$ as anything else than just spatially white Gaussian noise. With (4.3), we can now formulate a fairly tight inner bound on the multiple access channel (MAC) capacity region by changing (3.4) to

$$\forall \mathcal{S} \subseteq \{1..K\}: \sum_{k \in \mathcal{S}} R_k \leq \max_{\mathbf{\Phi}_{\text{ss}}} \log_2 \left| \mathbf{I} + \left(\sigma^2 \mathbf{I} + \mathbf{\Phi}_{\text{hh}}\right)^{-1} \hat{\mathbf{H}} \mathbf{\Phi}_{\text{ss}}(\mathcal{S}) \hat{\mathbf{H}}^H \right|. \quad (4.4)$$

Note that, different from the MAC capacity region under imperfect channel state information (CSI), it is now not optimal anymore to let all UEs transmit at maximum power. This is because $\mathbf{\Phi}_{\text{hh}}$ itself is a function of the transmit covariance $\mathbf{\Phi}_{\text{ss}}$, hence increasing one UE's power will lead to the fact that the residual channel estimation related noise impairing successive interference cancelation (SIC) performance is also increased. This is the reason why the MAC capacity region under imperfect CSI is not a pentagon anymore, see [MF09b].

The question is now how $\mathbf{\Phi}_{\text{hh}}$ and $\hat{\mathbf{H}}$ can be modeled for a channel realization $\mathbf{H}$ and a particular channel estimation scheme. One option is to observe the variance of the *absolute* (i.e. link-independent) channel estimation error variance, which can be obtained via the Cramér-Rao lower bound [Kay93] as

$$\sigma_E^2 = \frac{\sigma_{\text{p}}^2}{N_{\text{p}} \cdot p_{\text{pilots}}}. \quad (4.5)$$

Here, σ_{p}^2 denotes the variance of the noise the channel estimation is subject to, N_{p} is the number of pilots used to obtain CSI, and p_{pilots} is the pilot power. Note that σ_{p}^2 may deviate from σ^2 if, e.g., pilot sequences of multiple cells are designed to be orthogonal, while data transmission in these cells is subject to mutual interference not addressed by CoMP (hence leading to an increased background noise $\sigma^2 > \sigma_{\text{pilots}}^2$). With the definition of $\mathbf{E}$ in (4.1), we can now state

$$E\left\{e_{i,j}\left(e_{i,j}\right)^H\right\} = \frac{E\left\{|h_{i,j}|^2\right\} \cdot \sigma_E^2}{E\left\{|h_{i,j}|^2\right\} + \sigma_E^2}, \quad (4.6)$$

from which the calculation of $\mathbf{\Phi}_{\text{hh}}$ is straightforward. Note that some authors derive the impact of channel estimation noise using a different model where $\hat{\mathbf{H}} = \mathbf{H} + \mathbf{E}$ [PSS04, MF09b], i.e. where the estimated channel and estimation error are assumed correlated, but obtain the same final result as in (4.4). Now we still have the problem that (4.4) states an inner bound on the capacity region for a given channel estimate $\hat{\mathbf{H}}$, assuming that the actual channel $\mathbf{H}$ is 'fluctuating' around this. In most cases, however, we want to observe the opposite case, i.e. the capacity of the transmission over an *actual* channel $\mathbf{H}$ under imperfect CSI. It is discussed in [Mar10] that this can be *approximated* by replacing $\hat{\mathbf{H}}$ in (4.4) by the *expectation value* of the channel estimate, which under the assumption of an unbiased MMSE detector is simply an element-wise scaled version $\bar{\mathbf{H}}$ of the actual channel $\mathbf{H}$ with [MF09b]

$$\forall\, i,j\ :\ \bar{h}_{i,j} = \frac{h_{i,j}}{\sqrt{1+\sigma_{\mathrm{E}}^2/E\{|h_{i,j}|^2\}}}. \qquad (4.7)$$

The approximation is very accurate as long as the average channel gain $E\{|h_{i,j}|^2\}$ is at least a few dB larger than the absolute channel estimation error variance σ_{E}^2, below which channel estimation is questionable, anyway.

While the Cramér-Rao lower bound from (4.5) refers to channel estimation of a frequency-flat block-static channel, the transmission of each symbol in an orthogonal frequency division multiple access (OFDMA) system is of course subject to a slighty different channel, as this varies over time and frequency. Here, pilots are typically scattered across these two dimensions, and channel estimation (including inter- and extrapolation of the channel for all data symbols) is ideally performed such that it takes the correlation of the channel realization in time and frequency into account, e.g. via 2D MMSE channel estimation [HKR97a]. Hence, each detected symbol is subject to a different accuracy of CSI, depending on its location w.r.t. the pilot positions and the concrete interpolation scheme applied. One way to keep an analytical study simple is to calculate capacity bounds based on a frequency-flat block-static channel as before, but use a value of N_{p} representative for the *average* performance for a concrete pilot and channel estimation scheme, coherence bandwidth and time (i.e. a particular delay spread and UE speed) [MRF10, Mar10].

4.1.2 Imperfect CSI in the Downlink

In the downlink, especially under base station (BS) cooperation, also channel state information at the transmitter (CSIT) becomes a crucial issue [CS03]. As in frequency division duplex (FDD) systems CSIT is typically realized via CSI feedback, which requires quantization and is subject to feedback delay, CSIT is usually (significantly) less accurate than CSIR. It is here possible to re-write (4.1) as

$$\mathbf{H} = \underbrace{\hat{\mathbf{H}}^{\mathrm{BS}} + \mathbf{E}^{\mathrm{BS}}}_{=\hat{\mathbf{H}}} + \mathbf{E}, \qquad (4.8)$$

where the channel knowledge $\hat{\mathbf{H}}$ at the receiver side (in the downlink hence the UE side) is expressed as the channel knowledge $\hat{\mathbf{H}}^{\mathrm{BS}}$ at the transmitter side plus an uncorrelated noise component $\mathbf{E}^{\mathrm{BS}}$. The latter hence denotes the *additional* uncertainty the transmitter side has on the channel knowledge in comparison to the receiver side. A simple way to model this is to assume that each channel coefficient is quantized with a certain number of quantization bits N_b and then fed back to the transmitter side, so that rate-distortion theory yields [CT06]

$$\forall\, i,j\, :\, \hat{h}^\mathrm{BS}_{i,j} = \sqrt{1-2^{-N_\mathrm{b}}} \cdot \hat{h}_{i,j} \text{ and } E\left\{\left|e^\mathrm{BS}_{i,j}\right|^2\right\} = 2^{-N_\mathrm{b}} \cdot E\left\{\left|\hat{h}_{i,j}\right|^2\right\}. \quad (4.9)$$

The downlink transmission equation from (3.6) can then be modified to

$$\mathbf{y} = \hat{\mathbf{H}}^\mathrm{BS}\mathbf{s} + \mathbf{v} + \mathbf{u} + \mathbf{n}, \quad (4.10)$$

where $\mathbf{v} \sim \mathcal{N}_\mathbb{C}(\mathbf{0}, \mathbf{\Phi}_\mathrm{hh})$ is a noise term connected to imperfect CSI at both UE and BS side, while $\mathbf{u} \sim \mathcal{N}_\mathbb{C}(\mathbf{0}, \mathbf{\Phi}^\mathrm{BS}_\mathrm{hh})$ with $\mathbf{\Phi}^\mathrm{BS}_\mathrm{hh} = E\{\mathbf{E}^\mathrm{BS}\mathbf{\Phi}_\mathrm{ss}(\mathbf{E}^\mathrm{BS})^H\}$ is a noise term connected to the *additional* CSI uncertainty at the transmitter side. Assuming both terms to be spatially white Gaussian again yields an overestimation of their impact. While the impact of $\mathbf{v}$ on capacity is straightforward (as explained for the uplink), this is not the case for $\mathbf{u}$. Intuitively, one would expect that for a given CSIT, there should be a benefit of an improved CSIR. And indeed, the case of *linear precoding* can be modeled such that each UE k is *not* negatively affected by the noise term $E\{\mathbf{E}^\mathrm{BS}\mathbf{\Phi}_\mathrm{ss}(\{k\})(\mathbf{E}^\mathrm{BS})^H\}$ connected to inaccurate CSIT and the desired signal covariance $\mathbf{\Phi}_\mathrm{ss}(\{k\})$ of this UE [Mar10]. This is different if dirty paper coding (DPC) is employed, which requires the transmitter side to know the *exact* overlap of desired signal and interference to be canceled at the UE side. Under imperfect CSIT, DPC can hence only be performed w.r.t. $\hat{\mathbf{H}}^\mathrm{BS}$, and there is no benefit at all if the receiver side has better CSI. This leads to the fact that beyond some point of decreasing CSIT, DPC is not superior to linear precoding any more, as we will see later.

Note that (4.10) becomes highly inaccurate under strongly degraded or no CSIT at all. While capacity tends to zero in our model, it is known that even without CSIT non-zero rates can be achieved in a broadcast channel (BC) if the BSs transmit sequentially to single UEs from one antenna (under a sum power constraint) or using unit precoders (for per-antenna power constraints). This, however, is a regime of operation which is surely not relevant for CoMP.

With (4.10) and the discussion before, an inner bound for the BC capacity region under imperfect CSIT and CSIR can now be computed by considering the convex hull over all precoding matrices $\mathbf{W}$ and encoding orders π, where [MF09a]

$$R_k(\mathbf{W}, \pi) \leq \log_2 \left| \mathbf{I} + \left(\sigma^2 \mathbf{I} + \mathbf{\Phi}^k_{ii} + \mathbf{\Phi}^k_\mathrm{hh} + \mathbf{\Phi}^{\mathrm{BS},k}_\mathrm{hh}\right)^{-1} \left(\hat{\mathbf{H}}^\mathrm{BS}_k\right)^H \mathbf{W}_k \mathbf{W}^H_k \hat{\mathbf{H}}^\mathrm{BS}_k \right|$$

$$\text{with } \mathbf{\Phi}^k_{ii} = \sum_{\pi(j)>\pi(k)} \left(\hat{\mathbf{H}}^\mathrm{BS}_k\right)^H \mathbf{W}_j \mathbf{W}^H_j \hat{\mathbf{H}}^\mathrm{BS}_k. \quad (4.11)$$

Here, $\mathbf{\Phi}^k_{ii}$ is the residual inter-user interference, $\mathbf{\Phi}^k_\mathrm{hh}$ is the part of matrix $\mathbf{\Phi}_\mathrm{hh}$ that is connected to UE k, denoting noise due to imperfect CSI at transmitter and receiver side, and $\mathbf{\Phi}^{\mathrm{BS},k}_\mathrm{hh}$ is noise due to additional CSI imperfectness

4.2 Gain of Joint Signal Processing under Imperfect CSI

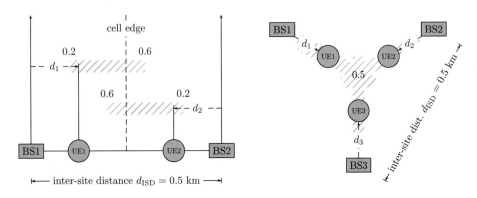

(a) Setup for $M = K = 2$ BSs and UEs. (b) Setup for $M = K = 3$ BSs and UEs.

Figure 4.1 Interference scenarios considered in this chapter.

at the transmitter side. As in the uplink, the capacity region can again be approximated by replacing $\hat{\mathbf{H}}^{\mathrm{BS}}$ in (4.11) by the *average* BS-side channel estimate $\bar{\mathbf{H}} = E\{\hat{\mathbf{H}}^{\mathrm{BS}}\}$ with $\forall\, \bar{h}_{i,j} = h_{i,j} \cdot \sqrt{1 - 2^{-N_\mathrm{b}}}/\sqrt{1 + \sigma_\mathrm{E}^2/E\{|h_{i,j}|^2\}}$. It has been shown in [MF09a, Mar10] that uplink/downlink duality is still applicable to the capacity region in (4.11), where the dual uplink is then subject to a particular extent of CSIR. Unfortunately, calculation of dual uplink precoders $\mathbf{G}_k$ and the power distribution among UEs is now not a convex problem any more, but still numerically more tractable (e.g. through a brute-force search) than trying to solve (4.11) directly. Duality can also be used to observe non-cooperative performance under imperfect CSI and various power constraints [Mar10].

In the rest of this chapter, we use values of $N_\mathrm{p} = 2$ and $N_\mathrm{b} = 6$, which have shown in [MRF10, Mar10] to be representative for the performance in an OFDMA system with the pilot structure of LTE Release 8, a UE speed of 3 km/h, a maximum delay spread of 1 μs, and a CSI feedback delay of 3 ms.

4.2 Gain of Joint Signal Processing under Imperfect CSI

Based on the modified transmission equations from the last section, we now observe the gains to be expected from joint signal processing in uplink and downlink in different interference scenarios, and under perfect and imperfect CSI. We consider setups with $M = K = 3$, $N_\mathrm{ue} = 1$ antenna per UE, and $N_\mathrm{bs} = 2$ antennas per BS. All UEs are moved simultaneously from their cell-center to the common cell-edge, as depicted in Fig. 4.1(b), with an inter-site distance (ISD) of 500 m. Here, the values d_i state the normalized distance of the terminals to their assigned BS, where a value of 0.5 reflects the cell-edge case. Exemplary channel matrices **H** are generated according to a log-linear pathloss model with pathloss coefficient of 3.5, and are constructed to be of *average orthogonality*.

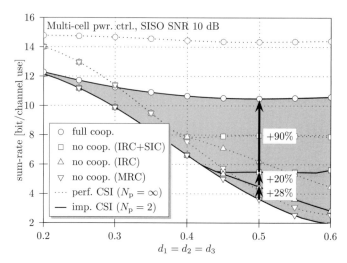

Figure 4.2 Uplink joint signal processing gains for scenarios with $M = K = 3$.

In the *uplink*, shown in Fig. 4.2, we assume multi-cell power control, where the transmit power of each UE is adjusted w.r.t. a certain target *average* receive power at all BSs. Target receive power and noise variance are chosen such that a single-input single-output (SISO) signal-to-noise ratio (SNR) of 10 dB is obtained at the cell-edge. We compare the following schemes:

- non-cooperative detection, based on maximum ratio combining (MRC) (i.e. considering interference as spatially white noise),
- non-cooperative detection, based on interference rejection combining (IRC) (i.e. taking the spatial properties of interference into account),
- non-cooperative detection, allowing a flexible assignment of UEs to BSs and the joint detection of multiple UEs at the same BS, and
- fully-cooperative joint detection of all UEs by all BSs.

The strongest rate gains can be obtained at the cell-edge [KRF07]. Here, using IRC to exploit the spatial color of interference (but without BS cooperation) already yields 28% rate increase. Further 20% are possible if local SIC is used, i.e. interference subtraction also not requiring BS cooperation. The strongest gains, however, with an additional 90%, are visible if the BSs jointly process all UEs, profiting from array and spatial multiplexing gain and yielding MAC performance. The gain of local, non-cooperative interference subtraction disappears quickly as we move away from the cell-edge, as enabling the decoding and subtraction of interference poses constraints on the rates of interferers [HK81]. Han-Kobayashi (HK) techniques (superposition coding and partial interference decoding) would yield marginal rate improvements here, and are strongly sensitive to imperfect CSI. Towards the cell-centers, all stated gains strongly diminish, especially under imperfect CSI, as then the interference links cannot be estimated well enough to be exploited. At the cell-edge, however, the *relative* rate improve-

4.2 Gain of Joint Signal Processing under Imperfect CSI

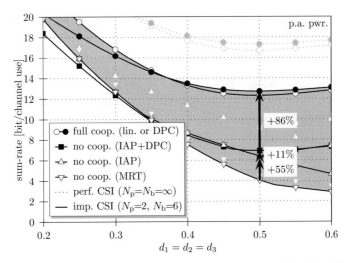

Figure 4.3 Downlink joint signal processing gains for scenarios with $M = K = 3$.

ments due to full cooperation increase for decreasing CSI, as additional diversity alleviates the impact of channel estimation errors [Mar10].

In the *downlink*, shown in Fig. 4.3, we assume that the transmit power is fixed, such that we obtain a SISO SNR at the cell-edge of 10 dB. The compared schemes are analog to those in the uplink, namely

- non-cooperative transmission, based on maximum ratio transmission (MRT),
- non-cooperative transmission, based on interference-aware precoding (IAP),
- non-cooperative transmission, allowing a flexible assignment of UEs to BSs and joint transmission from one BS to multiple UEs, possibly using local DPC,
- fully-cooperative joint transm. from all BSs to all UEs, possibly with DPC.

The difference between linear precoding and DPC is shown through empty and filled markers, respectively. We can see similar effects as in the uplink. The cooperation gain is again largest at the cell-edge, with 55% improvement due to interference-aware precoding, 11% additionally due to the option of local, non-cooperative multi-UE transmission with DPC, and another 86% if full joint transmission is employed. Under imperfect CSI, DPC is only (marginally) superior to linear precoding at the cell-edge, as interference links can otherwise not be estimated accurately enough. This is also the reason why local multi-UE transmission with DPC is less beneficial than its counterpart (local SIC) in the uplink. The small gap between linear and non-linear techniques under full cooperation, also under perfect CSI, is due to the fact that the compound channel already enables a fairly good spatial separation of the UEs without DPC. In general, the *relative* gain of cooperation remains more or less the same, regardless of the extent of CSIT [Mar10]. The reason is that both cooperative and also non-cooperative transmission degrades equally for diminishing CSIT.

4.3 Trade-Offs in Uplink Multi-Cell Joint Signal Processing

4.3.1 Different Information Exchange and Cooperation Schemes

We now explore the gray area in Fig. 4.2, i.e. the regime between no and full BS cooperation in the uplink, by introducing different kinds of BS information exchange and cooperation for a small setup with $M = K = 2$. While these are information-theoretic concepts, we will later see many parallels to practical CoMP algorithms. We initially focus on schemes with one phase of information exchange between BSs, and briefly address iterative cooperation in Section 4.3.3.

Distributed Interference Subtraction (DIS)
Here, one BS decodes one UE's transmission and forwards the decoded data to the other BS for interference cancelation [MF08d], as shown in Fig. 4.4(a). The best-known scheme from information theory is to apply source coding, i.e. to compress the information handed over the backhaul, so that the benefiting BS can reconstruct it only by using its own received signals as side-information [SW73a]. This BS then re-encodes the interfering UE transmission, and subtracts the corresponding interference from its received signals before decoding an assigned UE. If a backhaul capacity of β is available and information exchange takes place from BS 1 to BS 2, then the UE rates can be lower-bounded as

$$R_1 \leq \log_2 \left| \mathbf{I} + \left(\sigma^2 \mathbf{I} + \mathbf{\Phi}_{hh}^1 + \bar{\mathbf{h}}_2^1 p_2 \left(\bar{\mathbf{h}}_2^1\right)^H\right)^{-1} \bar{\mathbf{h}}_1^1 p_1 \left(\bar{\mathbf{h}}_1^1\right)^H \right| \quad (4.12)$$

$$R_1 \leq \beta + \underbrace{\log_2 \left| \mathbf{I} + \left(\sigma^2 \mathbf{I} + \mathbf{\Phi}_{hh}^2 + \bar{\mathbf{h}}_2^2 p_2 \left(\bar{\mathbf{h}}_2^2\right)^H\right)^{-1} \bar{\mathbf{h}}_1^2 p_1 \left(\bar{\mathbf{h}}_1^2\right)^H \right|}_{\text{zero if source coding is not considered}} \quad (4.13)$$

$$R_2 \leq \log_2 \left| \mathbf{I} + \left(\sigma^2 \mathbf{I} + \mathbf{\Phi}_{hh}^2\right)^{-1} \bar{\mathbf{h}}_2^2 p_2 \left(\bar{\mathbf{h}}_2^2\right)^H \right|. \quad (4.14)$$

The inequality in (4.12) is based on the fact that UE 1 is decoded by BS 1 under the full interference from UE 2, and that in (4.13) due to the backhaul constraint. The underbraced term corresponds to the rate at which BS 2 could decode UE 1 without cooperation. Eq. (4.14) is finally based on the fact that BS 2 can decode UE 2 free of interference from UE 1, as this has been subtracted.

In regimes of weak interference and low backhaul, the rate/backhaul trade-off of DIS can be improved if only a *portion* of the decoded data is forwarded to the other BS [MF08a]. In information theory, this can be modeled via superposition coding. Instead of decoded data, one might also consider forwarding quantized interference directly, having the advantage that BSs benefiting from cooperation need not re-encode interference [SSPS09c, GMFC09]. This so-called compressed interference forwarding (CIF) has practical advantages, but is always inferior to DIS with superposition coding [Mar10]. We will later also provide simulation results for CIF, but refer the interested reader to [GMFC09] for details.

4.3 Trade-Offs in Uplink Multi-Cell Joint Signal Processing

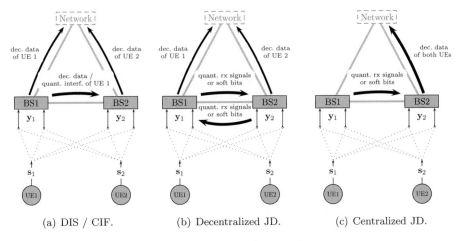

(a) DIS / CIF.   (b) Decentralized JD.   (c) Centralized JD.

Figure 4.4 Different uplink multi-cell joint signal processing concepts.

Decentralized Multi-Cell Joint Detection

In multi-cell joint detection (JD), cooperation is based on an exchange of *quantized receive signals* between BSs. Let us first consider a *decentralized* approach, where both BSs quantize their received signals and forward these simultaneously to the partnering BS, as shown in Fig. 4.4(b). Both BSs can then independently decode their assigned UE, making use of the receive signals from all N_{BS} antennas in the cluster. The UE rates can be bounded as

$$R_1 \leq \log_2 \left| \mathbf{I} + \left(\sigma^2 \mathbf{I} + \mathbf{\Phi}_{\text{hh}} + \bar{\mathbf{h}}_2 p_2 \left(\bar{\mathbf{h}}_2 \right)^H \begin{bmatrix} \mathbf{0} & \mathbf{0} \\ \mathbf{0} & \mathbf{\Phi}_{\text{qq}}^2 \end{bmatrix} \right)^{-1} \bar{\mathbf{h}}_1 p_1 \left(\bar{\mathbf{h}}_1 \right)^H \right| \quad (4.15)$$

$$R_2 \leq \log_2 \left| \mathbf{I} + \left(\sigma^2 \mathbf{I} + \mathbf{\Phi}_{\text{hh}} + \bar{\mathbf{h}}_1 p_1 \left(\bar{\mathbf{h}}_1 \right)^H \begin{bmatrix} \mathbf{\Phi}_{\text{qq}}^1 & \mathbf{0} \\ \mathbf{0} & \mathbf{0} \end{bmatrix} \right)^{-1} \bar{\mathbf{h}}_2 p_2 \left(\bar{\mathbf{h}}_2 \right)^H \right| . (4.16)$$

The terms $\mathbf{\Phi}_{\text{qq}}^1$ and $\mathbf{\Phi}_{\text{qq}}^2$ in (4.15) denote covariances of the quantization noise introduced by the BSs before signal exchange. The best backhaul efficiency would again be obtained via source coding. As quantization and source coding can be modeled separately under Gaussian signaling [WZ76], we state that each BS m creates a discrete representation $\tilde{\mathbf{y}}_m$ of its receive signals, and then passes a source encoded version over the backhaul, so that the other BS $l \neq m$ can reconstruct $\tilde{\mathbf{y}}_m$ by exploiting its own received signals $\mathbf{y}_l$. Assuming a total backhaul capacity of β, the quantization noise terms then need to fulfill [dS08]

$$\sum_{m=1}^{2} \log_2 \left| \mathbf{I} + \left(\mathbf{\Phi}_{\text{qq}}^m \right)^{-1} \mathbf{\Phi}_{\text{yy}}^{m|l \neq m} \right| \leq \beta, \quad (4.17)$$

where $\mathbf{\Phi}_{\text{qq}}^{m|l}$ is the receive signal covariance at BS m, *conditioned* on the signals received by BS l. Optimal quantization noise covariances may be calculated via a Karhunen-Loève transform [GDV02] succeeded by water-filling [dS08], where the backhaul is optimally invested into the spatial dimensions of the received signals. For decentralized JD, this is equivalent to letting each BS locally equalize the interfering UE to obtain a scalar value which is then quantized. As source coding might be regarded infeasible in practice, we also consider the case where this is omitted, or a practical quantizer, where one quantization bit is lost per real signal dimension [LBG80]. In these cases, (4.17) changes to [Mar10]

$$\sum_{m=1}^{2} \log_2 \left| \mathbf{I} + \left(\mathbf{\Phi}_{qq}^m\right)^{-1} \mathbf{\Phi}_{yy}^m \right| \leq \beta \text{ or } \forall m : \mathbf{\Phi}_{qq}^m = \left(2^{\max\left(\frac{\beta_m}{N_{bs}} - 2, 0\right)} - 1 \right)^{-1} \Delta\left(\mathbf{\Phi}_{yy}^m\right) \quad (4.18)$$

with $\beta_1 + \beta_2 \leq \beta$. The rate/backhaul trade-off of decentralized JD can be improved if the backhaul is used *successively*, and not simultaneously. One BS could forward quantized receive signals, after which the other BS would decode its assigned UE and subtract the corresponding signals from its receive signals *before* quantizing and forwarding the remaining signals to the former BS. However, this only yields (marginal) gains in interference regimes where the following scheme is superior, anyway, while increasing *latency* [Mar10].

Centralized Multi-Cell Joint Detection

Let us finally consider the case where one BS quantizes its received signals, and forwards these to the other BS, where *both* UEs are then jointly decoded. Assuming that received signals are forwarded from BS 1 to BS 2, the UE rates can be stated for a given quantization noise covariance as

$$\sum_{k \in \mathcal{S}} R_k \leq \log_2 \left| \mathbf{I} + \left(\sigma^2 \mathbf{I} + \mathbf{\Phi}_{hh} + \begin{bmatrix} \mathbf{\Phi}_{qq}^1 & \mathbf{0} \\ \mathbf{0} & \mathbf{0} \end{bmatrix} \right)^{-1} \sum_{k \in \mathcal{S}} \tilde{\mathbf{h}}_k p_k \left(\bar{\mathbf{h}}_k\right)^H \right| \quad (4.19)$$

One benefit of centralized JD, becoming evident from (4.19), is the option of SIC at the decoding BS. The quantization noise covariance $\mathbf{\Phi}_{qq}^1$ has to fulfill

$$\log_2 \left| \mathbf{I} + \left(\mathbf{\Phi}_{qq}^1\right)^{-1} \mathbf{\Phi}_{yy}^{1|2} \right| \leq \beta,$$

$$\log_2 \left| \mathbf{I} + \left(\mathbf{\Phi}_{qq}^1\right)^{-1} \mathbf{\Phi}_{yy}^1 \right| \leq \beta \text{ or } \mathbf{\Phi}_{qq}^1 = \left(2^{\max\left(\frac{\beta}{N_{bs}} - 2, 0\right)} - 1 \right)^{-1} \Delta\left(\mathbf{\Phi}_{yy}^1\right), \quad (4.20)$$

with or without source coding, or based on practical quantization, respectively. Note that even under perfect CSIR, the rate region is not a polygon anymore, as we have the degree of freedom of assigning different portions of backhaul to the two UEs. This is treated analytically and illustrated in [dS08].

In [SSS07a], centralized JD has been investigated in conjunction with partial *local decoding*. For above cooperation direction, this would mean that BS 1 decodes part of its assigned UE's transmission itself, and forwards the remaining received signals to the other BS for joint decoding of the remaining signals from both UEs. While a benefit regarding the rate/backhaul trade-off was reported in [SSS07a], this is only marginally superior to a simple *time-share* between a decentralized and centralized cooperation strategy [MF08c].

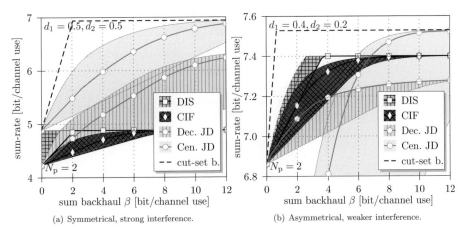

Figure 4.5 Sum-rate vs. backhaul for different uplink cooperation strategies.

4.3.2 Numerical Results

Let us now compare the rate/backhaul trade-offs achievable with the cooperation concepts stated before, again focussing on $M = K = 2$. In Fig. 4.5, the achievable sum-rate of both UEs is plotted as a function of backhaul, under imperfect CSI with $N_\mathrm{p} = 2$. The left case shows a symmetrical interference scenario, where both UEs are at the cell-edge (i.e. $d_1 = d_2 = 0.5$), while the right case resembles an asymmetrical scenario of weaker interference ($d_1 = 0.4$, $d_2 = 0.2$). For each scheme, multiple lines show the range between the best rate/backhaul trade-off achievable in theory (upper left) and under practical considerations (lower right). The dashed line indicates the *cutset-bound* [CT06], i.e. the sum-rate achieved if every bit of backhaul leads to an equivalent sum-rate increase until MAC performance is reached. Only centralized JD asymptotically achieves MAC performance for a large backhaul, due to the full extent of spatial multiplexing, array and interference cancelation gain. At the cell-edge (Fig. 4.5(a)), the scheme outperforms all others, and source coding is highly beneficial due to strong signal correlation. Decentralized JD also shows good asymptotical performance, but lacks the option of SIC. One could argue that each BS could also decode the interference as well, but then it would suffice to perform cooperation only in one direction, i.e. do centralized JD requiring less backhaul. However, such a strategy may still be interesting from a signaling perspective, see Section 11.2. For the cell-edge case, there is no benefit of using DIS or CIF, as both BSs can independently decode the interference and subtract this before decoding their UEs, without requiring backhaul at all. In the asymmetrical case of weaker interference (Fig. 4.5(b)), the story changes. Beside lacking array and spatial multiplexing gain, decentralized schemes can now offer an improved rate/backhaul trade-off in regimes of low backhaul. Especially DIS here appears attractive, as BS 1 can decode its assigned UE at moderate interference, while the extent of interference cancelation enabled by the exchange of decoded bits over the backhaul is large.

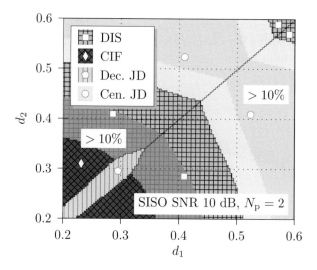

Figure 4.6 Best uplink cooperation concept for a backhaul capacity of 4 bpcu.

This is confirmed in Fig. 4.6, where the best cooperation concept is shown as a function of UE location, for an exemplary backhaul of 4 bpcu. While centralized JD is best for strong, possibly asymmetric interference, DIS is superior for weaker interference and a constrained backhaul. CIF and decentralized JD are only interesting in regimes of very weak interference, where we know from Fig. 4.2 that expected CoMP gains are small, anyway. The results suggest that a practical system should switch between centralized JD and DIS depending on the interference situation. This is emphasized by the darker areas in Fig. 4.6, where one strategy yields more than 10% larger rates than the other.

One may wonder why all schemes presented before actually yield a rate/backhaul trade-off far away from the cut-set bound. This question enables an interesting insight into fundamental properties of the compared schemes:

- While centralized JD asymptotically achieves MAC performance, it fails to meet the slope of the cut-set bound, as a certain extent of backhaul is wasted into the quantization of noise [dS08]. In fact, the cut-set bound is approached if source coding is applied and the SNR approaches infinity [dS08, Mar10].
- DIS, however, usually does not meet the flat part of the cut-set bound, as it lacks spatial multiplexing and array gain, and the first UE decoded does not profit from cooperation at all. It also usually fails to meet the *slope* of the cut-set bound, as the entropy of the data handed over the backhaul is mostly larger than the rate gain due to interference cancelation. An exception is the Z-interference channel [ZY08] with $d_1 = 0.5$, $d_2 = 0$, where (with source coding) every backhaul bit indeed yields exactly one bit of sum-rate increase, until asymptotic DIS performance is reached [GMF09].

Any uplink joint signal processing scheme is hence subject to a trade-off between using backhaul efficiently due to local preprocessing (with limited gain), or wasting backhaul into the quantization of noise (with maximum CoMP gain).

4.3.3 Parallels between Theory and Practical Cooperation Schemes

Practical algorithms applied to non-Gaussian signaling typically combine decentralized and centralized strategies. In [KF08, KRF08], for example, each BS (partially) decodes both the strongest interferer and its own UE, and forwards soft-bits to the other BS. This principally corresponds to centralized JD, but code-awareness and local preprocessing is used to exploit the structure in signals and interference for efficient backhaul usage (but reduced gain). The fact that terminal rates are constrained by the first (partial) decoding process can be alleviated by using *iterative* BS cooperation [BC07, AEH08, WT09a], i.e. starting with coarse decoding and refining this in each iteration. For the case of iterative DIS, however, even under very theoretical considerations, the rate/backhaul trade-off is only marginally improved over one-shot cooperation (though the asymptotic sum-rate is improved) [GMF09, Mar10]. In practice, every backhaul usage will always inherit additional redundancy (and increase latency), hence rendering iterative schemes even more questionable [MJH06].

4.4 Degrees of Freedom in Downlink Joint Signal Processing

As suggested by the title of this section, there are no such fundamental trade-offs to be made in the downlink as in the uplink. This is because transmitter-side signal processing is generally performed on noiseless, deterministic signals (as opposed to noisy observations in the uplink), and the main design choice is whether this processing is to be performed in a centralized, distributed or decentralized way. Fig. 4.7 illustrates these three principle kinds of multi-cell joint transmission (JT), which are briefly explained in the sequel.

In a *centralized case*, shown in Fig. 4.7(a), a central unit (CU) (possibly a BS itself) performs all preprocessing for a cluster of cooperating cells. It then quantizes and forwards the transmit signals to the other. This can be done in either time or frequency domain (where the latter is of course more backhaul efficient), and can be based on a common public radio interface (CPRI). It is also possible to forward *analog* radio frequency (RF) signals over fibre-optic cables, known as radio over fibre (RoF) [DMF10]. In all cases, the transmitting BSs are basically operated as remote radio heads (RRHs). Clearly, source coding is not applicable in the downlink, as there is no side-information to be exploited.

Backhaul requirements can be strongly reduced (while possibly maintaining the same performance) if precoding is performed in a *distributed* way, as shown in Fig. 4.7(b). In this case, the network provides all involved BSs with the same

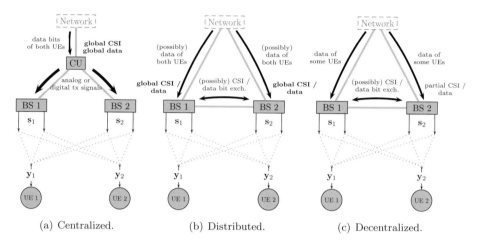

Figure 4.7 General kinds of downlink multi-cell joint signal processing.

data to be transmitted to all jointly served UEs (or the BSs distribute this data among each other), such that all BSs calculate their part of the precoding matrix $\mathbf{W}$ independently. A crucial aspect, however, is the fact that all BSs now require global CSI. This can be assured by either exchanging channel information between BSs, or by designing the CSI feedback from the terminal side such that all involved BSs can individually decode this. A distributed downlink JT scheme is observed in Sections 6.3, 13.3 and 13.4.

Before mentioned CSI requirement can be alleviated if *decentralized* downlink JT is performed, as shown in Fig. 4.7(c). In this case, the involved BSs may have strongly different extents and accuracies of CSI, and different extents of knowledge on UE data bits, but still contribute to the transmission through local precoding. Such a scheme is considered in Section 6.4.

4.5 Summary

In this chapter, we have extended the models from Chapter 3 to observe multi-cell joint signal processing under imperfect channel knowledge. In both uplink and downlink, we have seen that in representative scenarios of up to 3 cooperating base stations, spectral efficiency gains of more than 100% are thinkable at the cell-edge, while these gains strongly decrease towards the cell-center. Further considering a limited backhaul capacity between base stations has revealed a major trade-off in the uplink: Either backhaul is used efficiently, but only a limited extent of capacity gain is achieved, or backhaul is wasted into the quantization of noise, but yielding maximum gain. In the downlink, we have discussed three different joint transmission concepts that differ in the way how user data, channel knowledge and precoding are distributed among the base stations.

Part II

Practical CoMP Schemes

5 CoMP Schemes Based on Interference-Aware Transceivers or Interference Coordination

In this chapter, we introduce CoMP schemes where no or little information is exchanged between cooperating base stations. In Section 5.1, we observe an interference-aware downlink transmission scheme where each base station performs individual intra-cell beamforming, while the terminals are able to mitigate inter-cell interference to a certain extent through a particular interference estimation and rejection concept. The level of base station cooperation is then increased in Sections 5.2 and 5.3, where joint multi-cell scheduling and link adaptation, and multi-cell coordinated beamforming are investigated, respectively.

5.1 Downlink Multi-User Beamforming with Interference Rejection Combining

Lars Thiele, Thomas Wirth, Malte Schellmann, Thomas Haustein and Volker Jungnickel

In this section, we evaluate a *non-cooperative* downlink transmission scheme, i.e. where no explicit cooperation takes place between base stations (BSs), but where interference-aware transmission and reception is performed within cells. The BSs perform intra-cell precoding based on limited feedback from the user equipments (UEs), in conjunction with interference-aware scheduling and interference rejection combining (IRC) at the terminal side. This section is based on "Interference-aware scheduling in the synchronous cellular multi-antenna downlink", by L. Thiele, M. Schellmann, T. Wirth and V. Jungnickel, which appeared in [TSWJ09]. © 2009 IEEE.

5.1.1 Introduction

Transmission with multiple antennas both at the transmitting and receiving ends of a wireless link has become increasingly mature in recent years. From theory, the fundamental capacity gain of the multiple-input multiple-output (MIMO) radio link, being proportional to the minimum of the number of transmit and receive antennas, is well understood for an isolated point-to-point link. Under perfect channel knowledge at transmitter and receiver, a capacity-achieving

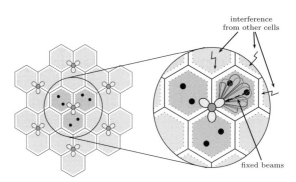

Figure 5.1 Concept of intra-cell beamforming considered in this section.

strategy is to flexibly invest transmit power into the eigenmodes of the channel via water-filling (see Section 3.4.1). In practical systems, however, under imperfect channel knowledge and a limited granularity of power allocations and modulation and coding schemes (MCSs), one typically switches between the two fundamental transmission modes spatial multiplexing (SMUX) and spatial diversity (SDIV) [ZT03] depending on the current channel state, in order to improve the error rate performance for fixed data rate transmission [HP05b] or to increase the spectral efficiency [SAH+04].

To enable ubiquitous broadband wireless access, MIMO transmission must be made robust against multi-cell interference. However, it is not fully evident yet how the potential capacity gains of MIMO can be realized under these conditions. In fact, early results obtained for a small set of linear transceiver settings, i.e. number of antennas, equalization and precoding strategy, indicate only small gains for SMUX over SDIV systems [CDG00]. The achievable spectral efficiency may be enhanced by incorporating multi-user MIMO (MU-MIMO) into system design and thus turning the focus to multi-user links [GKH+07]. However, BSs would require coherent channel state information (CSI) to optimally serve their users in MU-MIMO, which is difficult to obtain in frequency division duplex (FDD) systems, as a high rate feedback link would be required from the terminals to the base stations.

Further, fair resource assignment is mandatory in cellular networks in order to guarantee radio access for all users. The multi-path structure of signal and interference channels may be used beneficially in this interference-aware scheduling process. Supplemental to the time-domain scheduling already used in today's radio systems, groups of frequency resources may be assigned to the users according to their frequency-selective signal-to-interference-and-noise ratio (SINR) conditions. In this case, users may beneficially be assigned to their best resources.

This section targets a practical solution for decentralized interference management. The key to success is a predictable interference scenario at the receiver side, which also helps to improve the link adaptation process. Thus, we consider using fixed beams (i.e., fixed sets of possible precoding vectors) for transmission

5.1 DL Multi-User Beamforming with IRC

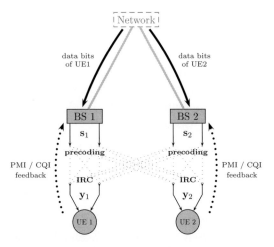

Figure 5.2 Non-cooperative transmission and PMI/CQI feedback concept considered.

as depicted in Fig. 5.1. In particular, terminals are assumed to report their preferred precoding matrix indicators (PMIs) in combination with corresponding post-equalization SINRs via a low-rate feedback channel. For the equalization at the UE, comprehensive channel knowledge on the radio system is required, which may be obtained by multi-cell channel estimation based on pilot symbols, as discussed in Section 9.1. Therefore, downlink transmission has to be synchronized [JWS+08]. With this approach, we demonstrate substantial throughput gains for MIMO systems in multi-cell environments, similar to those known for point-to-point links. We further indicate potential performance gains under the influence of imperfect channel estimation in systems with non-synchronized and synchronized BSs.

5.1.2 Downlink System Model

We extend the multi-antenna downlink model from Section 3.5, observing an orthogonal frequency division multiple access (OFDMA) transmission on a single sub-carrier from M BSs to K UEs that are scheduled to the same resource in time and frequency. The BSs and UEs are equipped with N_{bs} and N_{ue} antennas, respectively, leading to an overall number of $N_{\text{BS}} = MN_{\text{bs}}$ transmit antennas and $N_{\text{UE}} = KN_{\text{ue}}$ receive antennas. This implies that each BS may transmit up to N_{bs} *streams* simultaneously on the same resource, while each UE may receive up to N_{ue} such streams simultaneously. Clearly, there is the degree of freedom that a BS may serve many UEs with fewer streams each, or fewer UEs with more streams each, which we will explore later. As we are observing *non-cooperative* downlink transmission, this means that each stream may only be transmitted from one BS, as illustrated for a setup with $M = K = 2$ in Fig. 5.2. Consequently, the overall precoding matrix $\mathbf{W} \in \mathbb{C}^{N_{\text{BS}} \times N_{\text{UE}}}$ as introduced in Section 3.5 is sparse, as each column connected to one UE and one stream may only have non-zero entries connected to the antennas of one BS.

In the sequel, let us observe one UE k which is served by BS $m = k$. While set $\mathcal{K}$ captures all K UEs, we denote as $\mathcal{K}_m$ the set of all UEs served by BS m simultaneously on the same resource, which is obviously limited to the number of BS transmit antennas, e.g. $|\mathcal{K}_m| \leq N_\text{bs}$. All received signals of our observed UE k can be expressed as

$$\mathbf{y}_k = \underbrace{(\mathbf{H}_k^m)^H \mathbf{W}_k^m \mathbf{x}_k}_{\overline{\mathbf{H}}_k} + \underbrace{\sum_{j \in \{\mathcal{K}_m \setminus k\}} (\mathbf{H}_k^m)^H \mathbf{W}_j^m \mathbf{x}_j}_{\text{Intra-cell intfr. } \zeta_k} + \underbrace{\sum_{j \in \{\mathcal{K} \setminus \mathcal{K}_m\}} (\mathbf{H}_k)^H \mathbf{W}_j \mathbf{x}_j + \mathbf{n}_k}_{\text{Inter-cell intfr. and noise } \mathbf{z}_k}, \tag{5.1}$$

where $\mathbf{H}_k$ is the channel between UE k and all BSs, $\mathbf{W}_k$ is the compound precoding vector used to serve UE k, and $\mathbf{H}_k^m$ and $\mathbf{w}_k^m$ are the sub-portions of these matrices or vectors connected to BS m, as introduced in Chapter 3. We write as $\overline{\mathbf{H}}_k$ the *effective* channel between UE k and its serving BS after precoding, which consists of one column for each of the N_ue streams the UE may potentially receive, i.e. $\overline{\mathbf{H}}_k = [\overline{\mathbf{H}}_{k,1} \ldots \overline{\mathbf{H}}_{k,N_\text{bs}}]$. The corresponding potential data streams stacked in $\mathbf{x}_k$ with $\mathbf{x} \sim \mathcal{N}_\mathbb{C}(\mathbf{0}, \mathbf{I})$ are distorted by the intra-cell and inter-cell interference and noise aggregated in ζ_k and $\mathbf{z}_k$, respectively. Each BS m may select a limited number $Q_m \leq N_\text{bs}$ of active beams to serve one user with multiple beams or multiple users simultaneously. This is done by choosing the corresponding columns of BS m-related precoding matrix $\mathbf{W}^m$ from the columns of a pre-defined beam set $\mathbf{\Omega}_i^m$. In the case of $N_\text{bs} = 2$, beam set size $\omega = 2$ and discrete Fourier transform (DFT)-based precoding, this can be either

$$\mathbf{\Omega}_1^m = \frac{1}{\sqrt{2}} \begin{bmatrix} 1 & 1 \\ i & -i \end{bmatrix} \text{ or } \mathbf{\Omega}_2^m = \frac{1}{\sqrt{2}} \begin{bmatrix} 1 & 1 \\ 1 & -1 \end{bmatrix}. \tag{5.2}$$

Columns in $\mathbf{W}^m$ representing streams that are not used are simply filled with zeros. Note that $\mathbf{W}^m$ has to be scaled depending on the choice of Q_m in order to fulfill a *per base station power constraint*, i.e. $\text{tr}\{\mathbf{W}^m (\mathbf{W}^m)^H\} \leq P_m$. If only one beam is active, i.e. $Q_m = 1$, we name it single stream (SS) mode, while for $Q_m > 1$, we refer to it as multiple stream (MS) mode.

5.1.3 Linear Receivers

Assuming that a linear equalizer $\mathbf{g}_{k,u}$ is employed to extract the useful signal from $\mathbf{y}_k$ connected to stream u, this yields a post-equalization SINR given by

$$\text{SINR}_{k,u} = \frac{\mathbf{g}_{k,u}^H \overline{\mathbf{h}}_{k,u} \overline{\mathbf{h}}_{k,u}^H \mathbf{g}_{k,u}}{\mathbf{g}_{k,u}^H \mathbf{Z}_{k,u} \mathbf{g}_{k,u}}, \tag{5.3}$$

where $\mathbf{Z}_{k,u}$ is the covariance matrix of the streams received by UE k (except stream u) and the interfering signals and noise aggregated in ζ_k and $\mathbf{z}_k$, i.e. $\mathbf{Z}_{k,u} = \sum_{v \neq u} \overline{\mathbf{h}}_{k,v} (\overline{\mathbf{h}}_{k,v})^H + E\{(\zeta_k + \mathbf{z}_k)(\zeta_k + \mathbf{z}_k)^H\}$. For IRC [Win84], the interference-aware minimum mean square error (MMSE) receiver is used, i.e.

$$\mathbf{g}_{k,u}^{\text{MMSE}} = \mathbf{R}_{yy,k}^{-1}\overline{\mathbf{h}}_{k,u}, \tag{5.4}$$

where $\mathbf{R}_{yy,k}$ denotes the covariance matrix of the received signal $\mathbf{y}_k$, i.e.

$$\mathbf{R}_{yy,k} = E\left\{\mathbf{y}_k\left(\mathbf{y}_k\right)^H\right\} = \overline{\mathbf{H}}_k\overline{\mathbf{H}}_k^H + E\left\{\left(\zeta_k + \mathbf{z}_k\right)\left(\zeta_k + \mathbf{z}_k\right)^H\right\}. \tag{5.5}$$

The derivation of the MMSE receiver is discussed in detail in Section 10.2. The MMSE receiver yields a post-equalization SINR

$$\text{SINR}_{k,u}^{\text{MMSE}} = \overline{\mathbf{h}}_{k,u}^H \mathbf{Z}_{k,u}^{-1} \overline{\mathbf{h}}_{k,u}. \tag{5.6}$$

Based on this SINR, the achievable spectral efficiency is evaluated in a downlink OFDMA multi-cellular simulation environment. For reference purposes, we compare these results with the performance achievable by using a maximum ratio combining (MRC) receiver

$$\mathbf{g}_{k,u}^{\text{MRC}} = \overline{\mathbf{h}}_{k,u} \tag{5.7}$$

yielding a post-equalization SINR

$$\text{SINR}_{k,u}^{\text{MRC}} = \frac{\left\|\overline{\mathbf{h}}_{k,u}^H \overline{\mathbf{h}}_{k,u}\right\|^2}{\overline{\mathbf{h}}_{k,u}^H \mathbf{Z}_{k,u} \overline{\mathbf{h}}_{k,u}}. \tag{5.8}$$

5.1.4 Imperfect Channel Estimation

For theoretical investigations, full channel state information at the receiver (CSIR) may be assumed. In order to obtain the achievable data rate in a practical system, we introduce different channel estimation models. In [TSWJ08], IRC was shown to be highly sensitive to estimation errors, since the spatial structure of the interference covariance matrix is utilized for equalization. In the following, we assume quasi-static channel conditions over the observation interval. For evaluation, we assume perfect synchronization between UEs and their serving BSs [MKP07] and a sufficiently large cyclic prefix, which alleviates the effect of inter-symbol interference (ISI). We distinguish between the following cases:

Non-synchronized BSs, i.e. BSs are not synchronized to each other with respect to carrier frequencies and frame start. Therefore, we introduce channel estimation errors according to $\widehat{\overline{\mathbf{h}}}_{k,u} = \overline{\mathbf{h}}_{k,u} + \delta_{k,u}$. Term $\widehat{\overline{\mathbf{h}}}_{k,u}$ denotes the biased estimate of variable $\overline{\mathbf{h}}_{k,u}$, and $\delta_{k,u}$ denotes the zero-mean Gaussian distributed error with variance μ. For SINR estimation, we consider knowledge on frequency-flat and frequency-selective independently and identically distributed (i.i.d.) interference powers σ_{IF}^2 according to (5.9) and (5.10), respectively. Further, we consider the case of full frequency-selective covariance knowledge based on received data signals $\mathbf{y}_k$. In the following, q and n denote the discrete subcarrier and transmit time interval (TTI) index, respectively, and the estimated interference covariance $\widehat{\mathbf{Z}}_{k,u}$ is given as

Frequency-flat i.i.d. interference power σ_{IF}^2

$$\widehat{\mathbf{Z}}_{k,u} = \left(\mathrm{E}_q \left\{ \left(\sum_{\forall j,v} |\overline{\mathbf{h}}_{j,v}(q)|^2 \right) - |\widehat{\overline{\mathbf{h}}}_{k,u}(q)|^2 \right\} + \sigma^2 \right) \cdot \mathbf{I} \qquad (5.9)$$

Frequency-selective i.i.d. interference power σ_{IF}^2

$$\widehat{\mathbf{Z}}_{k,u}(q) = \left(\left(\sum_{\forall j,v} |\overline{\mathbf{h}}_{j,v}(q)|^2 \right) - |\widehat{\overline{\mathbf{h}}}_{k,u}(q)|^2 + \sigma^2 \right) \cdot \mathbf{I} \qquad (5.10)$$

Frequency-selective covariance $\mathbf{Z}_{k,u}$

$$\widehat{\mathbf{Z}}_{k,u}(q) = E_n \left\{ \mathbf{y}_k(q,n) \mathbf{y}_k(q,n)^\mathrm{H} \right\} - \widehat{\overline{\mathbf{h}}}_{k,u}(q) \widehat{\overline{\mathbf{h}}}_{k,u}(q)^\mathrm{H} \qquad (5.11)$$

Synchronized BSs, using multi-cell channel estimation based on virtual pilot sequences $c_{k,u}(n)$ [TSSJ08, HKF10]. These sequences are block-orthogonal and defined over time-domain. For channel estimation, the receiver uses a simple correlator. For simplicity, we drop the sub-carrier index q in the sequel. According to [TSSJ08], we use pilot sequences $c_{k,u}(n)$ which are derived from Hadamard matrices. Hence, the multi-cell channel knowledge degrades with increasing mobility of the UE, and we state

$$\widehat{\overline{\mathbf{h}}}_{k,u} = \frac{1}{N} \sum_{n=0}^{N-1} c_{k,u}^*(n) \mathbf{y}_k(n) \qquad (5.12)$$

$$\widehat{\mathbf{Z}}_{k,u} = \sum_{\forall j,v} \widehat{\overline{\mathbf{h}}}_{j,v} \widehat{\overline{\mathbf{h}}}_{j,v}^\mathrm{H} - \widehat{\overline{\mathbf{h}}}_{k,u} \widehat{\overline{\mathbf{h}}}_{k,u}^H. \qquad (5.13)$$

5.1.5 Resource Allocation and Fair User Selection

Resource allocation and selection of the proper spatial transmission mode (i.e. SS or MS, see Section 5.1.2) is carried out by a score-based scheduling process developed in [STJH07], which is briefly described as follows: In a first step, the UEs evaluate the current channel conditions per physical resource block (PRB) in terms of their achievable SINR conditions. By using (5.6) and (5.8) and a suitable SINR-to-rate mapping function, they can determine for each transmission mode the expected achievable rate per supported beam. This information is conveyed to the BS, where a score-based resource scheduling algorithm is performed: To enable direct comparison of the single per-beam rates from different spatial modes, the stream rates are weighted by a so-called penalty factor, which accounts for the higher power allocated to the SS beam compared to MS mode. In particular, if Q is the number of simultaneously active streams in MS mode, the penalty factor is chosen as Q^{-1}. For each user, the (weighted) per-beam rates from all modes over all PRBs are ranked by their quality, and corresponding scores are assigned. Mode selection and resource assignment is then done

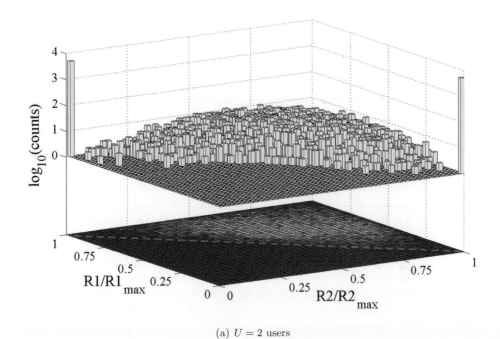

(a) $U = 2$ users

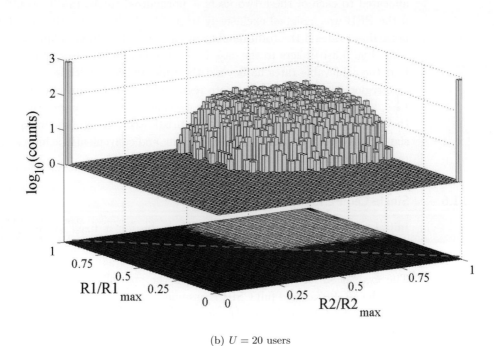

(b) $U = 20$ users

Figure 5.3 Rate allocation across two data streams, if the scheduler may choose from $U = 2$ or $U = 20$ users [TSWJ09]. © 2009 IEEE.

for each PRB individually: First, each beam available per transmission mode is assigned to the user providing the minimum, i.e. best, score for that beam. Then, the mode is selected which corresponds to the minimum overall user score.

The objective of this score-based resource allocation process is to assign each user to his best resources, and the decision on the spatial mode is taken under the premise of achieving a high throughput for each user. Clearly, the process is of heuristic nature, and hence the global scheduling target of assigning each user an equal amount of resources is achieved only on average, or if the number of available resources tends to infinity. However, its convenient property for practical applications is its flexible utilization, as the set of resources can be defined over arbitrary dimensions (time/frequency/space). Thus, fairness w.r.t. an equal amount of resources for all active users can be established on a small time-scale, e.g. even for the scheduling of resources contained within a single orthogonal frequency division multiplex (OFDM) symbol.

An illustration of the performance achievable by the score-based scheduler is given in Fig. 5.3. It depicts the histogram of normalized achievable user rates in the rate region plane for two UEs which may be scheduled in each PRB. In particular, we assume two spatial layers to be available in each PRB (i.e. $N_\mathrm{bs} = 2$), allowing two users to be served simultaneously in MU-MIMO mode. The rate allocated to each of these two users is normalized to the rate it would achieve if the PRB was assigned exclusively to it. Fig. 5.3(a) shows the distribution of normalized rates if the total number of users to select from is limited to $U = 2$, while Fig. 5.3(b) refers to the case $U = 20$. From both figures, it is clearly seen that the achievable rates lie beyond the time division multiple access (TDMA) rate region (dashed line in the rate region plane). For an increasing number of UEs, the histogram is more and more concentrated in the upper right corner of the rate region. This shows that the heuristic score-based scheduling approach significantly outperforms TDMA scheduling and conveniently achieves high user rates by properly utilizing MU-MIMO.

5.1.6 Single-Cell Performance

Initial performance evaluation is carried out for a fixed system setting in an isolated cell (i.e., $\mathbf{z}_k = \mathbf{n}_k$ in (5.1)), where K UEs, each equipped with $N_\mathrm{ue} = 2$ receive antennas, communicate with a dual-antenna BS ($N_\mathrm{bs} = 2$). The evaluation environment is based on the spatial channel model extended (SCME) [3GP07a], and full CSIR is assumed. We investigate the probabilities of mode selection depending on the mean signal-to-noise ratio (SNR) conditions, which are depicted in Fig. 5.4 for 2 or 10 users, respectively. Note that resources where a rate cannot be supported by any user are not assigned by the scheduler. For that reason, the selection probability of SS mode drops down to 75% at $P_s/N_0 = -5\,dB$ in the first case. Three different configurations of the adaptive mode switching system are considered here:

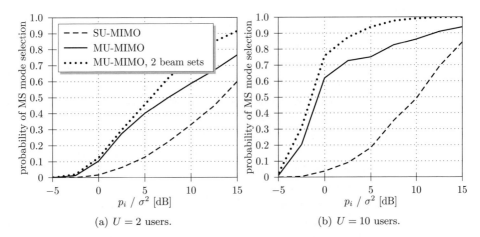

Figure 5.4 Probability of the selection of multiple stream (MS) mode vs. SNR [TSWJ09]. © 2009 IEEE.

1. **SU-MIMO**: MU-MIMO option is switched off, i.e. MS mode reduces to single-user MIMO (SU-MIMO). Now only one user is served per PRB either in diversity or SU-MIMO mode.
2. **MU-MIMO** system as described in Section 5.1.5 with the first beam set Ω_1^m from (5.2) being available. Simultaneously active beams can be assigned independently to different users. The mode per user is selected per PRB, i.e. a user may be served in different modes simultaneously.
3. **MU-MIMO, 2 beam sets**: Adaptive MU-MIMO system with both beam sets from (5.2) being available.

The points where the curves in Fig. 5.4 cross the median highlights the SNR regions where the MS mode becomes the dominantly selected one. From both figures, we observe that going from SU-MIMO to MU-MIMO promotes selection of the MS mode substantially, as the crossing point is shifted by 5 dB in case of 2 users and by more than 10 dB in case of 10 users down towards the low SNR regime. For 10 users, the crossing point falls below an SNR of 0 dB. The support for MU-MIMO mode also results in significant gains in the spectral efficiencies (refer to Fig. 5.7 later). These results strongly emphasize that MU-MIMO is the key for the efficient use of spatial multiplexing transmission even at low SNR, as also discussed in Section 11.1.

Providing an additional beam set shifts the crossing point even further down, which can be attributed to the finer granularity in the quantization of the transmit vector space. For 10 users, the crossing point in Fig. 5.4(b) can be shifted down to about -1.5 dB now. Further, it can be observed that the shape of the probability curves approach that of a step function, highlighting that the system behavior tends towards a hard mode switching at a fixed SNR value.

Table 5.1. Simulation assumptions.

Parameter	Value
channel model	3GPP SCME
scenario	urban-macro with scenario-mix[a]
traffic model	full buffer
carrier frequency f_c	2 GHz, frequency reuse 1
system bandwidth	18 MHz, 100 PRBs
inter-site distance (ISD)	500 m
number of sites	19 having 3 cells each
N_{bs} ; antenna spacing	1,2,4 ; 4λ
transmit power	46 dBm
sectorization	triple, with FWHM of $68°$
BS height	32 m
N_{bs} ; antenna spacing	1,2,4 ; $\lambda/2$
UE height	2 m
CQI granularity	1 PRB
feedback delay	0 ms
channel estimation	as specified in text

[a] Note, a mobile terminal might experience different propagation scenarios, i.e. line-of-sight (LOS) and non line-of-sight (NLOS), to distinct BSs.

5.1.7 Multi-Cell Performance under Perfect CSI

Turning the focus to a multi-cell system, the performance is investigated in a triple-sectorized hexagonal cellular network with $M = 57$ BSs in total, i.e. a center site with three sectors or cells surrounded by two tiers of interfering sites. Simulation parameters are given in Table 5.1. Initial results are based on the assumption of full and perfect CSIR. The SCME with urban macro scenario parameters is used [3GP07a], yielding a user geometry as in [HVKS03]. The UEs are always served by the BS whose signal is received with highest average power over the entire frequency band. For capacity evaluation, only UEs being placed inside the three central cells are evaluated. In this way, BS signals transmitted from the 1st and 2nd tier of sites model the inter-cell interference [TSZJ07]. Performance is evaluated for both the sum-throughput in a sector and the throughput for individual users. Both values are normalized to the signal bandwidth, yielding a sector's overall spectral efficiency and normalized user throughput, respectively. The achievable rates are determined from the SINRs calculated according to expression (5.6) by using a quantized rate mapping function [IST07b], representing achievable rates in a practical system. From these results, cumulative distribution function (CDF) plots are obtained.

Case 1: All BSs provide one fixed unitary beam set. With respect to the single-input single-output (SISO) reference case, Fig. 5.5 (solid lines) indicates a capacity increase of the median sector's spectral efficiency by a factor of $\alpha = 1.95$,

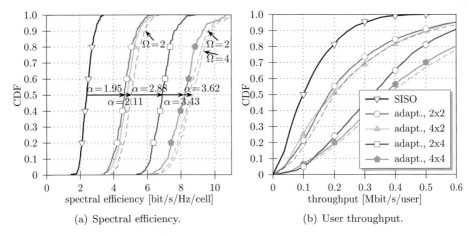

Figure 5.5 Idealistic system performance for the SISO, MIMO 2×2 ($N_{\mathrm{bs}} \times N_{\mathrm{ue}}$), 4×2, 2×4 and 4×4 system for 20 users per cell or sector. Dashed lines indicate the performance achievable with $\omega = \{2, 4\}$ beam sets $\mathbf{\Omega}_i^m$ [TSWJ09]. © 2009 IEEE.

$\alpha = 2.88$ and $\alpha = 3.43$ for the MIMO 2×2 ($N_{\mathrm{bs}} \times N_{\mathrm{ue}}$), 2×4 and 4×4 system. We can observe only small additional capacity gains for systems with $N_{\mathrm{bs}} > N_{\mathrm{ue}}$ compared to a system with $N_{\mathrm{bs}} = N_{\mathrm{ue}}$. This is mainly caused by the constraint of DFT-based precoding, where the total transmit power is distributed evenly over all antennas. In contrast, the system with $N_{\mathrm{bs}} < N_{\mathrm{ue}}$ benefits from advanced capabilities for interference suppression and higher receive diversity. This enables the system to achieve larger scaling factors, e.g. $\alpha = 2.88$ for MIMO 2×4. The 5th percentile of normalized user throughput, which may serve as a measure to represent the throughput of cell-edge users, shows similar scaling.

Case 2: All BSs provide multiple fixed unitary beam sets. Fig. 5.5 (dashed lines) further indicates the potential capacity gains for allowing the users to choose from multiple beam sets. Here, the system may profit from an improved channel quantization, yielding a capacity increase of $\alpha = 2.11$ for MIMO 2×2 with two beam sets. However, it has to be considered that then also the PMI feedback overhead doubles from 1 bit to 2 bit.

Interference prediction: Note that considering independent adaptation of beam sets for all BSs does not influence the received interference covariance matrix $\mathbf{Z}_{k,u}$, since the Wishart product $\mathbf{W}^m (\mathbf{W}^m)^H$ equals the scaled identity matrix if we assume $\mathbf{W}^m$ to be unitary. However, changing the power allocation for different MIMO transmission modes results in a multi-cell system where $\mathbf{Z}_{k,u}$ cannot be predicted at the receiver side. In order to support cell-edge terminals, we suggest to arrange e.g. SS with full base station power in an agreed access scheme known to the users.

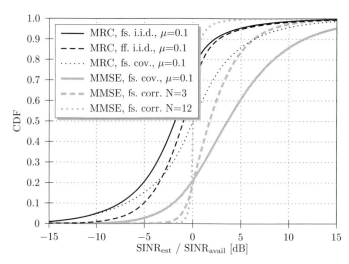

Figure 5.6 SINR estimation errors [TSWJ09]. © 2009 IEEE.

5.1.8 Multi-Cell Performance under Imperfect CSI

In the following, we take channel estimation errors into account, i.e. using (5.9)-(5.13). Fig. 5.6 indicates the estimation error of the SS SINR at the terminal. We compare the ratio of the estimated SINR_{est} to the achievable $\text{SINR}_{\text{avail}}$ under perfect CSIR and estimated equalization weights. Employing either MRC in an asynchronous network or IRC in a synchronized one leads to significantly different estimation errors. For MRC based on (5.10), the estimation suffers in two ways: There is a median shift of -1.9 dB, i.e. SINR_{est} is systematically too low. In addition, the estimation error has a considerable variance. With overestimated SINR conditions, the channel may be overloaded, i.e. the reported channel quality indicator (CQI) and the supported MCS do not match, which results in substantial performance degradation and increased block error rate (BLER). Assuming that strong channel codes as well as hybrid automatic repeat request (HARQ) mechanisms are able to correct errors if 10% of the resources are overloaded, we have to ensure that the 90th percentile of $\text{SINR}_{\text{est}}/\text{SINR}_{\text{avail}}$ is below 0 dB. This can be achieved by introducing a safety factor $S < 1$, shifting all SINR_{est} correspondingly.

For MRC based on (5.10), we can estimate S to be 2.3 dB from Fig. 5.6. Focusing on the median value, there is an overall penalty (offset) of approx. $\text{SINR}_{\text{pen}} = 4.2$ dB at the multiple access channel (MAC) compared to $\text{SINR}_{\text{avail}}$. Averaging the interference power σ_{IF}^2 over the entire frequency band, i.e. using (5.9), reduces the penalty to $\text{SINR}_{\text{pen}} = 3.7$ dB. Covariance estimation, i.e. (5.11), leads to unbiased SINR_{est}, but the S-factor is higher due to the larger variance, resulting in $\text{SINR}_{\text{pen}} = 6.3$ dB. Concentrating on asynchronous downlink transmission, we conclude that an interference estimation scheme assuming a frequency-flat i.i.d. σ_{IF}^2 results in the highest performance.

5.1 DL Multi-User Beamforming with IRC

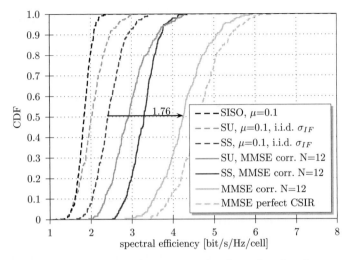

Figure 5.7 MIMO 2×2 system performance under channel estimation errors [TSWJ09]. © 2009 IEEE.

The penalties can be reduced further if the interference is estimated more precisely, e.g. in a synchronous system using an MMSE receiver and the correlation approach as given in (5.12) and (5.13). For a correlation window spanning $N = 3$ pilot symbols, we assume to be able to distinguish between the channels belonging to 3 out of 57 sectors or cells. Hence, interference cannot be separated sufficiently, and thus SINR is systematically overestimated. However, already with a correlation window spanning $N = 12$ pilot symbols, 12 sectors and thus more interferers can be identified, and the SINR is determined more precisely [TSWJ08]. The safety factor is then $S = 0.9$ dB, and the median shift becomes negligible.

Fig. 5.7 shows the achievable sum-rates in the multi-cell system including SINR_{pen}. As a lower bound, we use the performance in the SISO case including the effects of estimation errors for the desired channel $\widehat{\mathbf{h}}_{k,u}$. The upper bound is given by the adaptive transmission system assuming perfect CSIR. Assuming the UE is able to estimate its dedicated channel with $\mu = 0.1$ and $\mathbf{Z}_u$ according to (5.10) and the system is forced to SU-MIMO mode only, results in an inferior performance compared to the SS transmission using MRC. The reason is that the estimation error leads to inter-stream interference in the SU-MIMO case, which is not present with SS transmission.

The next three CDF curves are all based on the estimates (5.12) and (5.13). Although the MMSE receiver can exploit the knowledge of interference, the SS mode using the MMSE receiver outperforms SU-MIMO transmission. Fully adaptive transmission yields a significant system throughput gain, which is mostly related to MU-MIMO scheduling. Note that the gap to the adaptive system with perfect CSIR amounts to 8% only, indicating the robustness of the proposed scheme. Finally, we come to the following conclusion: Synchronized downlink transmission from all BSs in combination with MMSE receivers based on estimates (5.12) and (5.13) outperforms the asynchronous case. However, if the system design would be constrained to non-synchronized BSs, SS transmission in combination with the MRC receiver would be a suitable choice. The difference in the average throughput between both cases is significant and amounts to 76% in our results. Thus, the overall throughput gain achievable with synchronized BSs is still significant even under practical considerations.

5.1.9 Summary

In this section, we have evaluated the gains from using interference-aware, frequency-selective MU-MIMO scheduling in a cellular network with synchronized base stations. Terminals were assumed to be able to estimate their dedicated and a certain number of interfering channel coefficients. Two important observations were made: Efficient MU-MIMO transmission can be achieved by using fixed unitary precoding, i.e. without the requirement of full channel knowledge. Further, proper application of the MU-MIMO mode enables to conveniently serve even users with multiple streams who experience relatively poor SNR conditions. Thus, the MU-MIMO mode establishes a win/win situation for both, low and high rate users. In addition, it was shown that knowledge on the interference channels yields a more precise estimation of the achievable SINR compared to the traditional approach, where interference is assumed white. Thus, CQI feedback and supported modulation and coding scheme can be matched more accurately.

Acknowledgements

The authors are grateful for financial support from the German Ministry of Education and Research (BMBF) in the national collaborative project EASY-C under contract No. 01BU0631.

5.2 Uplink Joint Scheduling and Cooperative Interference Prediction

Philipp Frank, Andreas Müller and Heinz Droste

While the last section introduced non-cooperative, but interference-aware transmission and detection schemes, we now look into CoMP schemes where adjacent base stations (BSs) actually exchange information in order to coordinate resource usage and applied transmission strategies. More precisely, a joint scheduling approach is proposed, where the resource allocation in different cells is performed jointly, as well as a novel approach for cooperation-based interference prediction through which the link adaptation can be significantly improved. Both

5.2 Uplink Joint Scheduling and Cooperative Interference Prediction

schemes are described for the uplink in the following, but in principle they may be employed in the downlink as well.

Joint scheduling generally belongs to the class of so-called interference coordination techniques, which recently have attracted a lot of research attention due to their potential to efficiently mitigate inter-cell interference and hence to realize significant performance gains compared to non-cooperative systems [BPG+09, ADF+09]. The basic idea of interference coordination in general is to let different BSs cooperate with each other in order to control and account for the inter-cell interference originating from the corresponding cooperating cells. This may be done in either a static or dynamic manner. With a static approach, there are usually some pre-configured restrictions regarding the resource allocation, for example, that on some frequency resources no cell-edge users may be scheduled, as it is the case for static fractional frequency reuse [XSX07]. With a dynamic scheme, in contrast, such restrictions are determined on a much shorter time scale and usually by taking the instantaneous channel conditions into account. In case of dynamic fractional frequency reuse, for instance, there would be only restrictions on certain frequency resources when high interference is expected, see for example [FKR+09, MMT08]. Clearly, dynamic interference coordination generally should lead to a better performance than static approaches, but this comes at the cost of a higher complexity and possibly a higher backhaul load [BPG+09].

A general problem of interference coordination with independent, i.e., cell-specific scheduling is that the scheduling of one user in a certain cell may directly impose certain restrictions on other cooperating cells and vice versa. Thus, finding the globally optimal solution becomes hardly feasible in practice. This is also because the imposed restrictions cannot be changed arbitrarily fast due to the inherent BS-BS signaling delay over the backhaul (see Section 12.2). However, this drawback can be overcome with a global scheduling algorithm that is applied across all cooperating BSs, taking into account the channel state information (CSI) of all associated user equipments (UEs) in order to find the optimal or at least close-to-optimal allocation of radio resources. Below, we propose such a centralized cooperative scheduling scheme in Section 5.2.1, with which the resource allocation as well as the link adaptation is performed jointly by a central scheduling unit (CSU) for a set of cooperating BSs [FMDS10]. However, since this requires the signaling of multi-cell CSI from all cooperating BSs to the CSU, it results in a possibly massive backhaul load. In order to reduce this backhaul load as well as the signal processing complexity, we therefore propose in a second step in Section 5.2.2 a novel multi-cell interference prediction scheme. In contrast to the joint scheduling approach, the scheduling process of the interference prediction scheme is still done independently by each BS as in conventional systems, and only the inter-cell interference that is expected to occur during a future data transmission is predicted and then used for improving the link adaptation process [MF10]. This is accomplished by exchanging scheduling

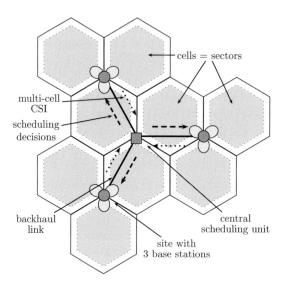

Figure 5.8 Illustration of the joint scheduling concept considered.

information between a set of cooperating BSs combined with multi-cell channel estimation. As will be seen in Section 14.3, this way still considerable performance gains can be achieved, but the generated backhaul load is much smaller than that for joint scheduling.

5.2.1 Interference-Aware Joint Scheduling

The uplink of a cellular network as shown in Fig. 5.8 is considered, where different BS sites are interconnected with a CSU via high-capacity backhaul links, assumed to facilitate a fast information exchange. It should be noted that the depicted CSU in Fig. 5.8 is not necessarily a separate device, but it may also be incorporated in one of the involved BSs. As illustrated in Fig. 5.8, all cooperating BSs periodically send multi-cell CSI of the associated UEs to the corresponding CSU, which thus becomes aware of the interference a certain UE scheduled in one cell would cause to another cell within the same cooperation cluster. This way, strong interference situations—which may occur, for example, if cell-edge UEs of neighboring cells are allocated to the same radio resources—can be avoided by taking the predicted inter-cell interference caused by the various UEs located within the cooperation cluster into account. The avoidance of high interference levels may not only significantly increase the overall system performance in terms of the average cell throughput, but it also contributes to a better fairness since UEs located close to the cell-edge generally benefit most from it.

A flow chart of the considered joint scheduling algorithm is depicted in Fig. 5.9. In a first step, each CSU reserves certain radio resources for the requested retrans-

5.2 Uplink Joint Scheduling and Cooperative Interference Prediction

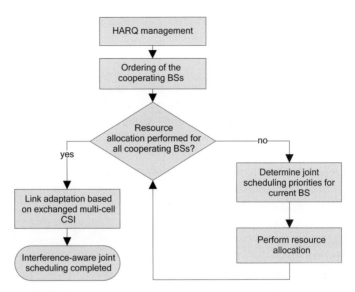

Figure 5.9 Flow chart of the interference-aware joint scheduling algorithm [FMDS10]. © 2010 IEEE.

missions of all associated BSs, and then the actual joint scheduling process is carried out. Since the simultaneous allocation of radio resources to all UEs located within the respective cooperation cluster would cause a tremendous increase in computational complexity, we assume in the following that the joint scheduling procedure is carried out stepwise for each set of UEs assigned to one of the cooperating BSs. This way, the computational effort can be significantly reduced. However, this entails also that the BSs associated to a certain CSU have to be ordered by means of a certain fairness criterion in order to sustain fairness among the various UEs. For that purpose, the long-term cell throughput averaged over the number of assigned UEs is considered as fairness criterion, which can be expressed for the m-th BS by

$$T_{\text{avg},m}(t+1) = \tau\, T_{\text{avg},m}(t) + (1-\tau)\, \frac{T_{\text{inst},m}(t)}{|\mathcal{K}_m|}, \quad (5.14)$$

where $T_{\text{avg},m}(t)$ denotes the long-term throughput for the m-th BS at the time interval t, $T_{\text{inst},m}(t)$ the instantaneous throughput, τ the forgetting factor and $\mathcal{K}_m$ the set of UEs assigned to BS m. The actual BS ordering is then done in such a way that the corresponding average long-term throughputs according to (5.14) are non-decreasing, i.e., the resource allocation always starts with the BS associated with the lowest long-term throughput, then it is done for the one with the second smallest one, etc.

Having determined the ordering of the cooperating BSs, the radio resources are allocated to the various UEs based on the exchanged multi-cell CSI. To this end, not only the current channel conditions between the UEs and their serving BSs are taken into account, but also the expected inter-cell interference caused by assigning these UEs to certain radio resources. Thus, the joint scheduling priority for the b-th radio resource and k-th UE associated to its serving BS m can be expressed by

$$S_{k,b}(t) = G_{k,b,\mathcal{K}_b}(t) + \sum_{j \in \mathcal{K}_b} G_{j,b,\tilde{\mathcal{K}}_b}(t), \quad k \in \mathcal{K}_m, \tag{5.15}$$

where $G_{k,b,\mathcal{K}_b}(t)$ denotes the scheduling priority for the k-th UE allocated to the b-th radio resource on which the UEs in set $\mathcal{K}_b$ are already scheduled. Furthermore, $G_{j,b,\tilde{\mathcal{K}}_b}(t)$ indicates the updated scheduling priority for the already scheduled UE j, taking into account that the k-th UE will be allocated to the b-th radio resource. In this regard, the updated set of interfering UEs allocated to the b-th radio resource for the j-th UE is given by

$$\tilde{\mathcal{K}}_b = (\mathcal{K}_b \setminus j) \cup k. \tag{5.16}$$

In the following, only the calculation of the scheduling priority $G_{j,b,\tilde{\mathcal{K}}_b}(t)$ is explicitly outlined, but the scheduling priority $G_{k,b,\mathcal{K}_b}(t)$ can be determined in a similar way and therefore is not further considered in more detail here. It is assumed that the radio resources are shared between the various UEs by means of the well-known proportional fair approach, but it should be noted that any other scheduling metric may be used in conjunction with our joint scheduling scheme as well. The basic idea of proportional fair scheduling is to realize a reasonable trade-off between the maximal total throughput and cell-edge throughput. Clearly, on the one hand, fair resource allocation among the UEs will lower the overall throughput compared to the maximum possible one, but in return it provides a higher throughput for UEs with relatively poor channel conditions, thus improving the system fairness. In general, the proportional fair metric is given by the ratio between instantaneously supportable and long-term throughput of a certain UE [VTL02], i.e., $G_{j,b,\tilde{\mathcal{K}}_b}(t)$ can be determined by

$$G_{j,b,\tilde{\mathcal{K}}_b}(t) = \frac{R_{j,b,\tilde{\mathcal{K}}_b}(t)}{T_j^\alpha(t)}, \tag{5.17}$$

with $R_{j,b,\tilde{\mathcal{K}}_b}(t)$ as the instantaneous supportable throughput and α as the fairness factor, which determines the trade-off between efficiency in terms of total throughput and fairness. Furthermore, $T_j(t)$ denotes the long-term average throughput given by

$$T_j(t+1) = \begin{cases} \beta T_j(t) & j \notin \mathcal{K}_{\text{total}}(t) \\ \beta T_j(t) + (1-\beta) \bar{R}_j(t) & j \in \mathcal{K}_{\text{total}}(t) \end{cases}, \tag{5.18}$$

where β denotes the forgetting factor and $\mathcal{K}_{\text{total}}(t)$ as well as $\bar{R}_j(t)$ denote the set of all scheduled UEs at time interval t and the aggregated throughput of the scheduled UE j, respectively. The instantaneous supportable throughput $R_{j,b,\tilde{\mathcal{K}}_b}(t)$ may be estimated by means of the Shannon capacity formula

$$R_{j,b,\tilde{\mathcal{K}}_b}(t) = \log_2\left(1 + \gamma_{j,b,\tilde{\mathcal{K}}_b}\right), \tag{5.19}$$

5.2 Uplink Joint Scheduling and Cooperative Interference Prediction

with $\gamma_{j,b,\tilde{\mathcal{K}}_b}$ as the uplink signal-to-interference-and-noise ratio (SINR) of the UE j on the b-th radio resource. Let us assume in the sequel that all BSs are equipped with N_{bs} antenna elements whereas all UEs have only a single antenna element (i.e., $N_{\text{ue}} = 1$). Then, the uplink SINR $\gamma_{j,b,\tilde{\mathcal{K}}_b}$ can be expressed by

$$\gamma_{j,b,\tilde{\mathcal{K}}_b} = \frac{P_{j,b}\, \mathbf{w}_{j,b}^H\, \mathbf{h}_{j,b}\, \mathbf{h}_{j,b}^H\, \mathbf{w}_{j,b}}{\mathbf{w}_{j,b}^H \left(E\left\{ \mathbf{i}_{j,b,\tilde{\mathcal{K}}_b}\, \mathbf{i}_{j,b,\tilde{\mathcal{K}}_b}^H \right\} + \sigma^2 \mathbf{I} \right) \mathbf{w}_{j,b}}, \qquad (5.20)$$

with $P_{j,b}$ as the transmit power of UE j for the b-th radio resource, $\mathbf{h}_{j,b} \in \mathbb{C}^{[N_{\text{bs}} \times 1]}$ as the channel vector from the j-th UE to its serving BS, $\mathbf{w}_{j,b} \in \mathbb{C}^{[N_{\text{bs}} \times 1]}$ as the corresponding weight vector for coherent detection, $\mathbf{i}_{j,b,\tilde{\mathcal{K}}_b} \in \mathbb{C}^{[N_{\text{bs}} \times 1]}$ as inter-cell interference caused by the set of UEs $\tilde{\mathcal{K}}_b$ and σ^2 as thermal noise variance. Based on the exchanged multi-cell CSI, the CSU is able to predict the interference covariance matrix $\boldsymbol{\Phi}_{\text{ii}} = E\{\mathbf{i}_{j,b,\tilde{\mathcal{K}}_b}\, \mathbf{i}_{j,b,\tilde{\mathcal{K}}_b}^H\} \in \mathbb{C}^{[N_{\text{bs}} \times N_{\text{bs}}]}$ in (5.20), which is given by

$$\boldsymbol{\Phi}_{\text{ii}} = \sum_{q \in \tilde{\mathcal{K}}_b} P_{q,b}\, \mathbf{h}_{q,j,b}\, \mathbf{h}_{q,j,b}^H, \qquad (5.21)$$

where $P_{q,b}$ and $\mathbf{h}_{q,j,b} \in \mathbb{C}^{[N_{\text{bs}} \times 1]}$ denote the transmit power of UE q for the b-th radio resource and the channel vector from the q-th UE to the serving BS of UE j on the considered radio resource b, respectively. Clearly, $\boldsymbol{\Phi}_{\text{ii}}$ in (5.21) contains both the inter-cell interference level caused by the already scheduled UEs associated to the cooperating BSs as well as the one that will be generated by assigning the k-th UE to the considered radio resource. As a result, the joint scheduling priorities in (5.15) reflect the weighted sum-throughput taking the current inter-cell interference situation into account. This consequently leads to an interference-aware joint scheduling, aiming at reducing the inter-cell interference within the given cooperation cluster while still taking channel-dependent scheduling as well as user fairness into account.

Having determined the joint scheduling priorities in (5.15) for all UEs associated to a certain BS, the central scheduler generally aims at maximizing the priority for each radio resource. The complexity of the resource allocation process depends on the used access scheme. In case of single carrier frequency domain multiple access (SC-FDMA), for example, which is used in the 3GPP LTE uplink, the allocated radio resources of each UE have to be either adjacent or evenly spaced in frequency in order to achieve a low peak-to-average power ratio (PAPR) [MLG06]. However, this leads to a significantly reduced allocation flexibility and a higher complexity. To overcome this problem, a resource allocation algorithm presented in [CRA+08] may be applied after determining the joint scheduling priorities in (5.15). The basic idea of this algorithm is that adjacent radio resources are assigned to a certain UE until either a different UE has a higher scheduling priority or the maximum transmit power is reached. This way, the allocation constraints due to SC-FDMA can be met, while still exploiting the multi-user diversity and the frequency selectivity of the uplink channel.

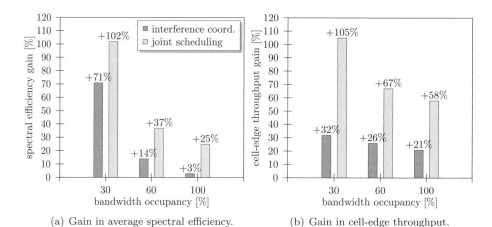

Figure 5.10 Relative uplink performance gains of the presented joint scheduling scheme as well as of a dynamic interference coordination scheme compared to a 3GPP LTE Release 8 system with 500 m inter-site-distance and six cooperating cells per BS.

Finally, after completing the resource allocation of all cooperating BSs, the link adaptation selects for each UE the spectrally most efficient modulation and coding scheme (MCS) that can be supported by its current uplink channel without exceeding a given target block error rate (BLER). To this end, the corresponding SINR is estimated by evaluating the available multi-cell CSI, resulting in a more accurate link adaptation. This is because the knowledge of which UEs are scheduled in the cooperating cells together with the available multi-cell CSI facilitate an accurate prediction of the interference situation that will occur during the actual (future) data transmission. Especially in the uplink, this may lead to significant additional performance gains since the interference situation there is usually rather volatile. This is because from one transmit time interval (TTI) to the other completely different sets of UEs may be scheduled in nearby cells.

An example for the achievable uplink performance of the presented joint scheduling scheme for different bandwidth utilizations is depicted in Fig. 5.10, where the relative gains compared to an LTE Release 8 system in terms of average spectral efficiency as well as cell-edge throughput are shown. The detailed simulation assumptions, parameter settings as well as further results will be introduced later in Section 14.3. In order to achieve a certain bandwidth occupancy, the scheduling is performed until the intended degree of bandwidth utilization is reached. In addition to the joint scheduling results, Fig. 5.10 shows for comparison also the performance of a state-of-the-art dynamic interference coordination scheme based on *high interference indicator* signaling [3GP07c, FMDS10]. First of all, it can be seen that the achievable performance is heavily dependent on the bandwidth occupancy. The gains increase with decreasing bandwidth occupancy, which indicates that the flexibility in assigning radio resources to the various UEs is considerably increased at a low bandwidth occupancy. As a result, severe inter-cell interference situations can be avoided by exploiting the whole bandwidth, i.e. preventing the UEs associated to the cooperating BSs from being allocated to the same radio resources. Furthermore, it is shown in Fig. 5.10 that the joint scheduling scheme outperforms the dynamic interference coordination scheme due to the higher flexibility in jointly allocating radio resources to the various UEs, which consequently leads to an improved avoidance of severe inter-cell interference. The better system performance, however, comes at the cost of an increased backhaul load due to the required exchange of multi-cell CSI [FMDS10].

5.2.2 Cooperative Interference Prediction

Interference-aware joint scheduling as considered in Section 5.2.1 generally causes a relatively high backhaul load since multi-cell CSI of all cooperating BSs has to be signaled to the CSU. Especially in case of lack of fast optical fiber links, this may represent a major barrier for implementing such an approach in practical systems in the short-term. In the following, we therefore present a somewhat more lightweight yet efficient scheme, where the scheduling is performed independently by all BSs as in conventional networks, but then the interference that will occur during the associated data transmission is predicted in a cooperative manner for improving the link adaptation process. To this end, first of all the basic principle of conventional link adaptation in general and a related problem are reviewed, after which the actual interference prediction scheme is explained in detail.

Review and Problem of Conventional Link Adaptation

An indispensable prerequisite for the efficient application of fast link adaptation schemes in cellular networks is the availability of accurate estimates of the current channel quality of the link between the UE for which the adaptation should be done and its associated serving BS. In the uplink, such estimates generally can be readily determined by a BS by evaluating the reference signals that have to be transmitted by the corresponding UEs anyway for facilitating coherent detection and channel-aware scheduling.

Having estimates of the current channel quality, a BS can determine for each scheduled UE the spectrally most efficient MCS for which a given target BLER is not exceeded. Afterwards, the selected MCS is signaled as part of the scheduling grant to the corresponding UE and the actual data transmission then typically starts slightly later so that the UEs have enough time to prepare the respective transport blocks for transmission. Due to the inherent delay between the time when link adaptation is performed and the actual data transmission, link adaptation often becomes inaccurate. This is because the channel conditions during the actual data transmission might differ substantially from the channel conditions during the time when the link adaptation was performed. On one hand, this is because the involved channels naturally change during that time, but at least for

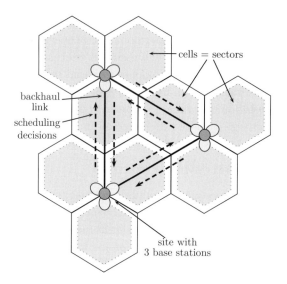

Figure 5.11 Illustration of cooperation between different BSs in case of cooperative interference prediction.

low to moderate user speeds, the impact of this effect should be only marginal. On the other hand and more importantly, the interference situation may have completely changed, since from one TTI to the other completely different sets of users may be scheduled in nearby cells. As a consequence, the selected MCSs are often over- or underestimated, thus leading to a very high BLER or a rather low spectral efficiency, respectively. Without any appropriate countermeasures as proposed in the following, the performance therefore would be often degraded compared to the idealized case with perfect link adaptation based on the channel conditions during the actual data transmission.

Proposed Interference Prediction Scheme
The fundamental idea of the considered interference prediction scheme is to perform the link adaptation not based upon the currently estimated SINR values, but rather based upon *predicted* SINR values likely to occur during the associated future data transmissions. For that purpose, it is necessary that a BS can accurately predict the interference level that it will experience during such future data transmissions already a couple of TTIs in advance. This may be accomplished by means of cooperation between different BSs as illustrated in Fig. 5.11. First of all, every BS performs conventional scheduling and power control, i.e. it determines which UEs should transmit on which radio resources and at which power levels. If the employed scheduling algorithm is channel-aware—which is the case for a proportional-fair scheduler, for example—the corresponding scheduling metrics are calculated as in conventional systems, taking into account only the currently observed channel and interference conditions, respectively.

Afterwards, every BS exchanges the resource allocation tables that have been fixed during the scheduling process with a certain set of cooperating BSs via a fast backhaul network. For the case of a 3GPP LTE system, for example, this could be realized via the X2-interface [HT09]. Note that low-latency backhaul links are a crucial prerequisite for the proposed approach since an additional delay is introduced by exchanging and processing the scheduling information as well as by performing the actual prediction of the interference. Without a fast data exchange the overall latency may increase, resulting in a performance degradation compared to the idealized case without any additional delay [MF10]. Provided that the various BSs have reasonably accurate CSI not only of the channels from the UEs located in their own cell, but also from those associated with any of their cooperating BSs, they can eventually accurately predict the interference level that will be generated by these UEs when the actual data transmission takes place. If, for example, the channel from the k-th interfering UE to the various antenna elements of a particular BS sector m is denoted by $\mathbf{h}_k \in \mathbb{C}^{[N_{\text{bs}} \times 1]}$, the expected contribution of this interferer to the overall interference covariance matrix simply would be given by

$$\boldsymbol{\Phi}_{\text{ii},k} = P_k\, \mathbf{h}_k\, \mathbf{h}_k^H, \tag{5.22}$$

where P_k is the transmit power associated with UE k. The predicted interference is then used as an input to the link adaptation stage, and afterwards the corresponding scheduling grants (including the assigned MCSs) are signaled to all scheduled UEs, which finally transmit their data a couple of TTIs after the reception of these grants. However, note that the scheduling decisions themselves are not updated based on the predicted interference levels, since otherwise the actual future interference situation would change again. Hence, in that case some iterative procedure would be necessary, thus leading to an increased complexity and backhaul load as well as a higher latency. An example for how the accuracy of the link adaptation can be improved with the proposed approach is depicted in Fig. 5.12, where the simulated distribution for certain deviations between the ideal and the used MCSs are shown for the cases with and without interference prediction. In this regard, the BSs may choose between several different MCSs according to [3GP09f]. Furthermore, it is assumed that in case of interference prediction each BS always receives scheduling information with a delay of two TTIs from its six cooperating sectors. It can be seen that with interference prediction the probability that the ideal MCS is selected is almost twice as high as for the case without interference prediction and also the variance of the deviations from the ideal MCS can be considerably reduced. Note that further simulation results can be either found in Section 14.3 or in [MF10]. Furthermore, the simulation assumptions and parameter settings used for generating the results in Fig. 5.12 are the same as the ones that will be used later in Section 14.3.

Clearly, the performance of the approach strongly depends on the number of cooperating BSs. While a BS generally should be able to predict the interference rather accurately with a large number of cooperation partners, it would frequently underestimate the actual interference level if it cooperates with very few other BSs only. This is because with the basic scheme as described above no interference from non-cooperating cells is taken into account. In real-world scenarios, however, the set of cooperating BSs is in most cases very likely restricted to nearby neighbors only—on the one hand in order to keep the backhaul load limited and on the other hand because it is unrealistic that a BS may accurately estimate the channels from all UEs within a large number of cooperating cells, as pointed out in Section 9.1. Therefore, it is essential that the impact of the interference caused by UEs in non-cooperating cells is somehow taken into account as well. An efficient way to do that is to employ an additional outer loop link adaptation scheme, similar to the one presented in [NAV02]. With this scheme, always a UE-specific offset Δ_{offset} is added to the predicted SINR values in dB before performing the actual link adaptation, which is permanently adjusted based on the outcome of previous transmission attempts. In particular, if an (initial) transmission attempt is successful, Δ_{offset} is increased by δ_{up} whereas otherwise it is decreased by δ_{down}. If these two step sizes δ_{up} and δ_{down} are chosen as in [NAV02] such that

$$\delta_{\text{down}}|_{\text{dB}} = \left(\frac{1}{\text{BLER}_{\text{target}}} - 1 \right) \delta_{\text{up}}|_{\text{dB}}, \qquad (5.23)$$

it is eventually possible to adjust the link adaptation in such a way that the obtained average BLER always corresponds to the configured target $\text{BLER}_{\text{target}}$. In the stationary case, i.e., after a sufficiently long warm-up phase, the average interference originating from non-cooperating cells is thus implicitly included in the outer-loop link adaptation offset Δ_{offset}.

5.2.3 Practical Considerations

A crucial prerequisite for both BS cooperation schemes presented in this section is that accurate multi-cell CSI is obtained. For that reason, it is necessary that the reference signals transmitted by different UEs within a certain cooperation cluster can be separated again at the BS side, for example through orthogonal reference signals as in Section 9.1. In any case, all BSs have to be aware of the reference signals assigned to the various UEs. This consequently requires further signaling between cooperating BSs in addition to the necessary information exchange via the backhaul network already outlined before. However, note that this usually does not have to be done during every TTI since the utilized reference signals and hopping patterns are normally assigned in a semi-persistent manner. Therefore, this additional backhaul load is expected to be comparatively small.

In case of cooperative interference prediction as introduced in Section 5.2.2, it is quite obvious that the requirements on the accuracy of the multi-cell channel estimation between a BS and UEs located in other cells are generally much lower

5.2 Uplink Joint Scheduling and Cooperative Interference Prediction

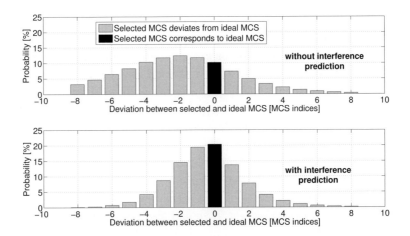

Figure 5.12 Exemplary illustration of the improved link adaptation accuracy with interference prediction for the uplink of a 3GPP LTE Release 8 system with 500 m inter-site-distance and six cooperating cells per BS [MF10]. © 2010 IEEE.

than those for the estimation of the desired link between a certain UE and its serving BS. On one hand, this is because estimation errors made for different interfering channels may compensate each other—particularly if the number of cooperating BSs is relatively high—and on the other hand because it may be already sufficient for achieving a good performance to know whether on a certain radio resource very high or very low interference has to be expected, whereas the exact figures are only of secondary importance. In addition, if the channel from a certain UE in one of the cooperating cells cannot be estimated reliably since it is in a deep fade, this should also not represent a major problem since in such a case this UE would cause only low interference anyway. By contrast, the requirements on the accuracy of the multi-cell channel estimation are more stringent in the case of interference-aware scheduling. This is due to the fact that the resource allocation decisions heavily depend on the predicted inter-cell interference level caused by single UEs, for which reason a high deviation between the predicted and the actual interference levels during a data transmission would lead to rather inaccurate resource allocation decisions.

Another prerequisite for the proposed schemes is that cooperating BSs can quickly exchange the required information, such as the multi-cell CSI or scheduling tables, via a fast backhaul network. However, it is quite clear that even if cooperating BSs are interconnected by means of direct optical fiber links, in general an additional delay is introduced because some time is always required for the processing of the exchanged information. As a consequence, the overall latency increases and the performance may degrade to some extent compared to the idealized case without any additional delay. This is due to an increased mismatch between the channels used as the basis for the scheduling and link adaptation stages and those during the actual data transmission. Besides, the increased delay between scheduling and actual data transmission clearly also affects potential hybrid automatic repeat request (HARQ) retransmissions. In the 3GPP LTE uplink, for example, a synchronous HARQ protocol is used. If one of the previously discussed schemes is to be introduced here, it might therefore be necessary to switch to an asynchronous HARQ or adjust HARQ timing.

Finally, it goes without saying that in case of joint scheduling the generated backhaul load generally should be much higher than for cooperative interference prediction, particularly due to the required exchange of multi-cell CSI in that case. As a concrete example, it will be shown later in Section 14.3 for a particular scenario that the average backhaul load per site can be reduced from 251 Mbit/s in the case of joint scheduling to 26 Mbit/s for cooperative interference prediction. However, it should also be noted that even for joint scheduling, the backhaul capacity requirements are still considerably smaller than those for joint signal processing CoMP schemes, which will be introduced in Chapter 6.

5.2.4 Applicability of Both Schemes to the Downlink

Although the basic principles of the two different BS cooperation schemes presented in the previous sections have been explained focusing on the cellular uplink, they may be readily applied to the downlink as well. In the downlink, one generally has to distinguish between cases where channel reciprocity may be exploited to obtain accurate CSI at the BS side—as it might be the case for time division duplex (TDD) systems, for example—and cases where channel reciprocity cannot be exploited, so that appropriate CSI has to be fed back to the various BSs from the corresponding UEs, as addressed in Section 9.2.

In the first case, the presented joint scheduling scheme basically might be applied in exactly the same manner to the downlink as described before for the uplink. By contrast, the downlink realization of the interference prediction scheme in this case would imply that the cooperating BSs would not only have to exchange their resource allocation tables (including information about the applied precoders, if applicable), but also the multi-cell CSI of all UEs located within the cooperation cluster. This might be realized in a two-step approach in order to reduce the backhaul load, so that a cooperating BS first of all obtains the resource allocation table of the considered BS and then signals the multi-cell CSI, the transmit power levels and the used precoders back, but only for the UEs which are actually scheduled by the considered BS. With this information, each BS is then able to accurately predict the interference and hence improve the link adaptation process.

In case that no explicit CSI is available at the BS side due to lack of channel reciprocity, the UEs would have to periodically send feedback information back to their serving BS. Depending on the applied feedback concept, this feedback information consists of either a recommendation of the transmission parameters to be used or information about the estimated downlink channel. The latter feedback alternative is a crucial prerequisite for performing joint scheduling, since the CSU has to be aware of the downlink channels of the UEs in order to determine the joint scheduling priorities. Clearly, the CSU not only requires the downlink channels between the UEs and their serving BSs, but also those between any of their cooperating BSs, thus the UEs have to acquire those additional interfering channels by performing a multi-cell channel estimation. This information may then be reported either to the serving BSs or directly to the CSU in order to avoid an additional delay and further backhaul signaling. By contrast, the interference prediction scheme may be used in conjunction with both feedback alternatives. In case that CSI feedback is sent to the BSs, the above described two-step approach may be employed for the downlink realization of this scheme. However, if the UEs instead report a recommendation of the transmission parameters only, then the BSs would not only notify the scheduled UEs on which resources data will be transmitted to them, but also signal relevant information about what the main interfering BS will do on these resources. This could include but is not limited to information about the precoders that will be used in the cooperating (=interfering) sectors and the assigned power levels. Having CSI of the interfering BS by performing a multi-cell channel estimation, the UE can then take this information into account and perform a more reliable prediction of the interference situation during the actual data transmission, as done in Section 5.1. Based on the predicted interference, they can determine appropriate transmission parameters again, such as the MCS, for example, and signal these parameters to their serving BS, which then uses these parameters for the actual data transmission.

5.2.5 Summary

Two novel uplink CoMP schemes have been presented where different BSs cooperate with each other via a backhaul network in order to mitigate the effects of inter-cell interference. While the *interference-aware joint scheduling* scheme coordinates the allocation of radio resources to the various UEs by means of periodically exchanged multi-cell CSI between the cooperating BSs and a central scheduling unit, the *cooperative interference prediction* scheme is a more lightweight but yet efficient approach with reduced backhaul load requirements. The latter scheme only requires the exchange of scheduling information between a set of cooperating BSs to predict the inter-cell interference level that will occur during a future data transmission for improving the link adaptation process. Since both proposed schemes are transparent to the UEs and cause only a minor to moderate backhaul load, they represent very attractive options for future LTE-A systems.

5.3　Downlink Coordinated Beamforming

Chan-Byoung Chae, Ramya Bhagavatula, Doru Calin
and Robert W. Heath Jr

In this section, we consider downlink interference coordination schemes where base stations (BSs) exchange channel state information (CSI) in order to adjust their transmission strategies, so that the generated extent of inter-cell interference is reduced. Such schemes, known as *coordinated beamforming* in the 3GPP LTE-A literature, offer a fair balance between ensuring a reasonable load on the backhaul links and attaining the performance gains using cooperation. The shared CSI is used by BSs to design individual precoding matrices (or beamforming vectors for single-stream transmission) to transmit exclusively to users within their own cell [EC05, ZHKA09, LJSL09]. Consequently, this is also known as *distributed beamforming*. CSI exchange over the backhaul has been shown to use a small backhaul bandwidth as compared to data exchange, for moderate Doppler spreads [SH09], which is also shown in Subsection 12.2.2.

5.3.1　Introduction

There are several distributed approaches for coordinated beamforming in the literature. For example, [EC05] proposes an iterative algorithm to minimize transmit power, which does not necessarily maximize sum-rates. The authors of [LJSL09] propose a non-iterative distributed solution to design precoding matrices for multi-cell systems, which will maximize the sum-rates for only a two-cell system at high signal-to-noise ratio (SNR), using a per base station power constraint. Another important partial cooperation-based transmit strategy is inter-cell interference nulling (ICIN) [ZA10, JLD08, LKL09, BH10], in which each BS transmits in the null-space of the interference it is causing to neighboring cells.

The performance of a cooperative transmission strategy is highly dependent on the quality of the CSI fed back by the users. Most of the literature on multi-cell cooperation assumes that full CSI is available at the transmitters [SSPS09c, SZ01, JTS+08a, EC05, LJSL09]. The impact of imperfect CSI was considered in [MF09a, GMF10a, PTW10]. In [MF09a], the authors consider imperfect CSI at the BS due to limited feedback or estimation errors and show that the performance gains from BS cooperation can be obtained even when CSI is imperfect. Noisy CSI estimates were considered in [PTW10], where the objective was to maximize the performance gains that can be obtained using the worst-case CSI perturbations.

Since quantization and feedback is a major source of imperfect CSI, it is important to consider CSI quantization in multi-cell cooperative systems. Limited feedback for multi-cell systems is a topic of ongoing research [BH10]. Unfortunately, results from the well-investigated single-cell limited feedback are not directly

5.3 Downlink Coordinated Beamforming

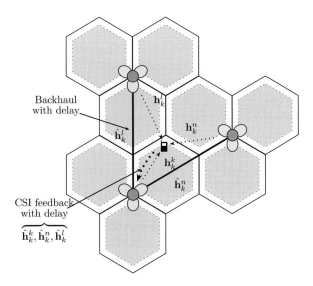

Figure 5.13 CSI feedback and backhauling concept considered for inter-cell interference nulling (ICIN).

applicable to the multi-cell scenario. While the CSI of only one channel is fed back in the single-cell case, cooperative strategies require feedback of CSI from multiple BSs using the same feedback link. Further, in single-cell transmission, quantized CSI reaches the BS after experiencing a delay in the feedback channel [ZHKA09]. In the multi-cell cooperative framework, however, quantized CSI is subject to an additional source of delay in the backhaul link. The impact of delayed CSI on the performance of non-cooperative systems [ZHKA09] has been investigated extensively. The effect of delayed limited feedback on the performance of cooperative systems has received comparatively less attention.

In this section, two different cases are considered: **i)** a single receive antenna and **ii)** multiple receive antennas at the user equipment (UE). For the single receive antenna case, the BSs need to optimize their precoding matrices/beamforming vectors to maximize the sum-rate under given constraints but for a multiple receive antenna case, the precoding/postcoding matrices/vectors should be jointly optimized. In Subsection 5.3.2, we describe ICIN, a low-complexity and non-iterative partial cooperative strategy that uses explicit per-base power constraints and yields reasonable gains in the sum-rate, while resulting in a small burden over the backhaul link. Note that ICIN requires that the total number of antennas per BS be larger than the number of single-antenna terminals considered in one transmission, an aspect we will discuss in detail later. We also describe some limited CSI feedback algorithms for ICIN. In Subsection 5.3.3, we further extend the cooperative strategies to the multiple antenna cases and show performance results.

5.3.2 Single Receive Antenna at the Terminal

We now briefly describe ICIN for the setup in Fig. 5.13 for an M cell system. We assume that on a particular observed resource, the BS in each cell serves a single active user. The received signal power of the desired and interfering signals is a function of the user's location in the cell. A similar approach was adopted in [JTS+08a, SSBN+06]. Each user is assumed to face interference from $M-1$ neighboring BSs. We index the users in each cell by the BS they obtain their desired signal from, i.e. the k-th BS services the k-th user, for $k=1,\ldots,K$. Note that by assuming that there is a single user in each cell on each resource, we fix $K=M$. We assume that all BSs are equipped with N_{bs} antennas, while each user supports a single receive antenna.

As defined in Chapter 3, the channel between BS m and UE k is denoted by $\mathbf{h}_k^m \in \mathbb{C}^{N_{\text{bs}} \times 1}$. The symbol to be transmitted to the k-th user is denoted by x_k, where the transmit power is $E\{x_k x_k^H\} = E_s$, implying a per-base station power constraint. The channels are subject to large-scale fading, which includes distance-dependent path-loss and shadowing effects, and small-scale fading. After averaging over the small-scale fading effects, the received SNR of the desired signal is denoted by ρ_k. The interfering signal SNR from a BS m to UE k is given by $\alpha_k^m \rho_k$, where $\alpha_k^m \in [0,1]$ (i.e. the interfering signal strength can at most be equal to that of the desired signal). A similar parameter is used in [JTS+08a, SSBN+06] to model the SNR of the interfering signal with respect to the received signal. Note that ρ_k and α_k^m are independent of the beamforming vectors. Observing the transmission of a single frequency-flat sub-carrier of an orthogonal frequency division multiple access (OFDMA) system, and assuming that the channels remain constant over the codeword transmission, the signal-to-interference-and-noise ratio (SINR) of the k-th user is given by

$$\text{SINR}_k = \frac{\rho_k \left|\left(\mathbf{h}_k^k\right)^H \mathbf{w}_k\right|^2}{\sum_{j \neq k} \alpha_k^j \rho_k \left|\left(\mathbf{h}_k^j\right)^H \mathbf{w}_j\right|^2 + 1}, \quad (5.24)$$

where $\mathbf{w}_k$ denotes the beamforming vector employed at BS k to transmit to UE k, which is normalized to have unit-norm. We assume that the BSs have perfect knowledge of the involved SNR terms ρ_k and α_k^m.

When full and instantaneous CSI is available at all BSs, the k-th BS has instantaneous knowledge of not only its own desired channel to UE k, $\mathbf{h}_k^k$, but also of the interference caused to neighboring cells, i.e. $\mathbf{h}_j^k$, $j = 1, \ldots, K$, $j \neq k$. The k-th BS then computes the beamforming vector, $\mathbf{w}_k$, as [ZA10, JLD08, LKL09]

$$\mathbf{w}_k = \mathbf{a}_k, \quad \text{where} \quad \mathbf{A}_k = [\mathbf{a}_1 \ldots \mathbf{a}_K] = \left[\tilde{\mathbf{h}}_1^k, \ldots, \tilde{\mathbf{h}}_K^k\right]^\dagger, \quad (5.25)$$

where $(.)^\dagger$ refers to the pseudo-inverse and $\tilde{\mathbf{h}}_k^m = \mathbf{h}_k^m / \|\mathbf{h}_k^m\|$ denotes the normalized channel between UE k and BS m. Since $\mathbf{A}$ is full-rank with high probability if

$N_{\text{bs}} \geq K$, (5.25) ensures perfect interference nulling in most cases, i.e. $\mathbf{h}_k^j \mathbf{w}_j = 0$, for $k, j = 1, \ldots, K$, $j \neq k$. Note that while ICIN is a simple, non-iterative and distributed coordinated beamforming strategy, it suffers from the dimensionality constraints imposed from computing the pseudo-inverse, i.e. $N_{\text{bs}} \geq K$. For the more practical case where the number of users in the system is greater than N_{bs}, scheduling can be employed to enforce $N_{\text{bs}} \geq K$. This implies that there exists a trade-off between increasing the number of users for simultaneous transmission in the cells and perfect interference cancelation. Clustering can also be employed to group cells into clusters of size $\leq N_{\text{bs}}$ each and use intra-cell time division multiple access (TDMA) or a comparable orthogonal transmission strategy, to make sure that $N_{\text{bs}} \geq K$.

Limited Feedback

Practical feedback channels are bandwidth-limited and have delays associated with them. Hence, it is important to investigate the performance of ICIN with delayed limited feedback [BH10]. The channel directions $\tilde{\mathbf{h}}_k^m[t]$ are quantized to the unit-norm vectors given by $\hat{\mathbf{h}}_k^m[t]$ at the k-th user, where we now introduce variable t to capture the time instant, for example a transmit time interval (TTI). We assume that each user can utilize B_{tot} bits for feedback, and that B_k and B_k^j bits are used to quantize $\tilde{\mathbf{h}}_k^k[n]$ and $\tilde{\mathbf{h}}_k^j[t]$, $j \neq k$ respectively, where $B_k + \sum_{j \neq k} B_k^j = B_{\text{tot}}$. The delay associated with quantizing $\tilde{\mathbf{h}}_k^k[t]$ to $\hat{\mathbf{h}}_k^k[t]$ and feeding back the latter to the k-th BS is denoted by D_k. The k-th user also quantizes the interfering channels, $\tilde{\mathbf{h}}_k^j[t]$, $j \neq k$ to $\hat{\mathbf{h}}_k^j[t]$, $j \neq k$ and feeds back the latter to the k-th BS, which then forwards this information to the j-th BS over the backhaul link, incurring an overall delay of D_k^j. The limited feedback model is also shown in Fig. 5.13.

At the time instant t, the k-th BS has knowledge of $\hat{\mathbf{h}}_k^k[t - D_k]$ and $\hat{\mathbf{h}}_j^k[t - D_j^k]$, for all $j \neq k$. The beamforming vector at the t-th time instant, $\mathbf{w}_k[t]$, is designed using the delayed and quantized CSI of the desired channels and the interference caused to other cells [BH10]

$$\mathbf{w}_k[t] = \mathbf{a}_k \text{ where} \tag{5.26}$$

$$\mathbf{A} = \left[\hat{\mathbf{h}}_1^k[t-D_1^k]..\hat{\mathbf{h}}_{k-1}^k[t-D_{k-1}^k], \hat{\mathbf{h}}_k^k[t-D_k], \hat{\mathbf{h}}_{k+1}^k[t-D_{k+1}^k]..\hat{\mathbf{h}}_K^k[t-D_K^k]\right]^\dagger.$$

When $N_{\text{bs}} \geq K$, the beamforming vector lies in the $N_{\text{bs}} - (K-1)$ dimensional null-space of the $K-1$ interfering channels. Hence, when $N_{\text{bs}} = K$, $\mathbf{w}_k[t]$ will lie in a one-dimensional sub-space, independent of $\hat{\mathbf{h}}_k^k$. This implies that if we have $N_{\text{bs}} = K$, it is not necessary to feedback the quantized desired channel back to the BS, i.e. $B_k = 0$. In contrast, when $N_{\text{bs}} > K$, knowledge on $\hat{\mathbf{h}}_k^k$ is desirable to determine the best $\mathbf{w}_k[t]$ in the $N_{\text{bs}} - (K-1)$ dimensional sub-space. By assuming that $\mathbf{h}_k^k[t]$ and $\mathbf{h}_j^k[t]$, $j \neq k$ are constant throughout the codeword transmission, the current and delayed CSI are related by the Gauss-

Markov block fading autoregressive model [TJMW01]

$$\mathbf{h}_k^k[t] = \eta_k \mathbf{h}_k^k[t - D_k] + \sqrt{1 - \eta_k^2}\mathbf{e}_{\mathbf{h}_k^k}[t], \text{ and} \qquad (5.27)$$

$$\mathbf{h}_j^k[t] = \eta_j^k \mathbf{h}_j^k[t - D_j^k] + \sqrt{1 - (\eta_j^k)^2}\mathbf{e}_{\mathbf{h}_j^k}[t], \qquad (5.28)$$

where $\mathbf{e}_{\mathbf{h}_k^k}[t]$ and $\mathbf{e}_{\mathbf{h}_j^k}[t]$ denote the channel knowledge uncertainties, which are uncorrelated with $\mathbf{h}_k^k[t - D_k]$ and $\mathbf{h}_j^k[t - D_j^k]$, respectively. The entries of $\mathbf{e}_{\mathbf{h}_k^k}[t]$ and $\mathbf{e}_{\mathbf{h}_j^k}[t]$ are distributed by $\mathcal{N}_\mathbb{C}(0, 1)$. The correlation coefficients for the desired and interfering channels are denoted by η_k and η_j^k, respectively. Clarke's autocorrelation model is used to determine η_k and η_j^k as [ZHKA09]

$$\eta_k = J_0(2\pi D_k f_d T_s), \text{ and } \eta_j^k = J_0(2\pi D_j^k f_d T_s), \qquad (5.29)$$

where J_0 is the zeroth order Bessel function of the first kind, f_d is the Doppler spread and T_s is the symbol duration. The Doppler spread is given as $f_d = \nu f_c/c$, where ν is the relative velocity of the transmitter-receiver pair, f_c the carrier frequency, and c the speed of light. The mean loss in sum-rate due to delayed limited feedback is bounded in [BH10], as a function of delays, signal strengths, and can be minimized choosing B_k and B_k^j as per Theorems 5.1-5.3.

Theorem 5.1. *Given the total number of bits allocated to quantize all the channels seen by one UE, B_{tot}, the optimum number of bits assigned to the desired channel at the k-th user, B_k, at low SNR is given by*

$$B_k = \frac{B_{tot}}{|\mathcal{K}| + 1} - \frac{(N_{bs} - 1)|\mathcal{K}|}{|\mathcal{K}| + 1} \log_2\left(\rho_k \frac{N_{bs}}{N_{bs} - 1} \prod_{j \in \mathcal{K}} (\alpha_k^j (\eta_k^j)^2)^{\frac{1}{|\mathcal{K}|}}\right),$$

for $N_{bs} > K$, and $B_k = 0$ for $N_{bs} = K$. The optimum number of bits assigned to all the interfering channels at the k-th user is computed as $B_{k,int} = B_{tot} - B_k$ [BH10].

Theorem 5.2. *Given the total number of bits allocated to quantize all the channels seen by one UE, B_{tot}, the optimum number of bits assigned to the desired channel at the k-th user, B_k, at high SNR is given by*

$$B_k = (N_{bs} - 1)\log_2\left((|\mathcal{K}| - 1)\Gamma\left(\frac{N_{bs}}{N_{bs} - 1}\right)\right).$$

The optimum number of bits assigned to all the interfering channels at the k-th user, $B_{k,int} = B_{tot} - B_k$ [BH10].

Theorem 5.3. *The optimum number of bits invested by UE k into the quantization of the channel to the j-th interfering BS, B_k^j is given by*

$$B_k^j = \frac{B_{k,int}}{|\mathcal{K}|} + (N_{bs} - 1)\log_2\left(\frac{\alpha_k^j(\eta_k^j)^2}{\prod_{j \in \mathcal{K}}(\alpha_k^j(\eta_k^j)^2)^{\frac{1}{|\mathcal{K}|}}}\right),$$

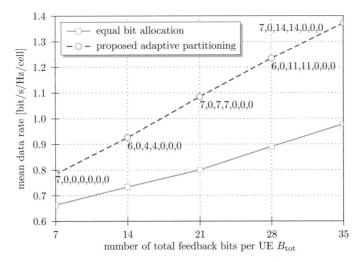

Figure 5.14 Comparison of the mean data-rate at the cell-edge for different values of B_{tot}. Bit assignments are shown corresponding to each B_{tot}.

for $\sum_{j\in\mathcal{K},\ j\neq k} B_k^j = B_{k,int}$ and $B_k^j = 0$ for $j \notin \mathcal{K}$, where $\mathcal{K}$ is the largest set of interferers that satisfies [BH10]

$$\log_2\left(\frac{\prod_{j\in\mathcal{K}}(\alpha_k^j(\eta_k^j)^2)^{\frac{1}{|\mathcal{K}|}}}{(\alpha_k^j(\eta_k^j)^2)}\right) < \frac{B_{tot}}{|\mathcal{K}|(N_{\text{bs}}-1)}.$$

For numerical evaluation, we consider a seven-cell system, i.e. $M = K = 7$. Each BS has eight antennas ($N_{\text{bs}} = 8$) and each user has a single antenna. The system setup is based on the urban micro-cell propagation scenario in the 3GPP spatial channel model (SCM). The inter-site distance (ISD) is assumed to be 800 m. The pathloss between the BSs and the UE is modeled using the COST 231 Walfish-Ikegami non line-of-sight (NLOS) model, adopted for urban micro-cells. Using a carrier frequency of 1.9 GHz, BS and UE heights of 12.5 m and 1.5 m, respectively, a building height of 12 m, building to building distance 50 m and street width 25 m, the path-loss in dB from BS m to a UE k at a height of 1.5 m is given as

$$PL_k^m[dB] = 34.53 + 38\log_{10}(d_k^m), \qquad (5.30)$$

where d_k^m denotes the distance from UE k to BS m. The transmit power is $E_s = 33$ dBm for all BSs, and the noise power is given by -114 dBm. We also model the delay associated with the feedback and backhaul links to be one and two frames, respectively. Note that the α_k^m parameters are obtained as a difference of the path-losses from the desired and interfering BSs. For example, $\alpha_k^m[dB] = PL_k^k[dB] - PL_k^m[dB]$.

It is seen from Fig. 5.14 that while the limited feedback technique in this section outperforms equal-bit allocation for all B_{tot}, the improvement in data rate is about 40 % at $B_{\text{tot}} = 35$. At $B_{\text{tot}} = 7$, the desired channel is given all the 7 bits, while at $B_{\text{tot}} = 35$, the two strong interfering channels are assigned 14 bits. Equal-bit allocation, in contrast, sees an increase in the feedback bits for the strong interfering channels from 1 to 5 bits per channel. Quantizing the strong interferers more finely at the cost of allocating zero bits to the weak channels leads to the significant improvement in data rates using the proposed algorithm.

5.3.3 Multiple Receive Antennas at the Terminal

So far, we have assumed a single receive antenna at each UE, thus only transmit beamforming vectors need to be designed. In this section, we consider multiple receive antennas at the UE, which gives more degrees of freedom to jointly design beamforming vectors. Before explaining multi-cell coordinated beamforming with multiple receive antennas, let us first study a jointly optimized beamforming algorithm for the multiple-input multiple-output (MIMO) broadcast channel (BC) [GKH+07]. It will help us understand multi-cell coordinated beamforming with multiple antennas at the UE side, which will be introduced later.

Consider a multi-user MIMO (MU-MIMO) system with one BS with N_{bs} antennas, and N_{ue} receive antennas for each of the K users. The channel between the BS and UE k is represented by an $N_{\text{bs}} \times N_{\text{ue}}$ matrix $\mathbf{H}_k$. Let x_k be the symbol to be transmitted to the k-th UE, assuming only one stream per UE, and $\mathbf{n}_k$ be the additive white Gaussian noise vector of size $N_{\text{ue}} \times 1$ seen by UE k. Let $\mathbf{w}_k$ and $\mathbf{g}_k$ be the transmit beamforming vector and the receive combining vector for the k-th UE, respectively. Then, the signal at the k-th UE after equalization is given by

$$\hat{x}_k = \sqrt{P}\mathbf{g}_k^H \mathbf{H}_k^H \mathbf{w}_k x_k + \sqrt{P}\mathbf{g}_k^H \mathbf{H}_k^H \sum_{j \neq k} \mathbf{w}_j x_j + \mathbf{g}_k^H \mathbf{n}_k \qquad (5.31)$$

where, P is the transmit power. The transmit beamforming vector is chosen in the null-space of $\mathbf{g}_j^H \mathbf{H}_j$ ($\forall j \neq k$), that is $\mathbf{g}_j^H \mathbf{H}_j^H \mathbf{w}_k = 0$ (the zero inter-user interference constraint). Assume that the receive beamforming vectors are maximum ratio combining (MRC) filters, given by $\mathbf{g}_k = \mathbf{H}_k^H \mathbf{w}_k$. This is a reasonable design (though not necessarily the only one and not the optimal one under residual, spatially colored interference) since it achieves the sum-rate very close to capacity under the zero inter-user interference constraint [CMJH08]. In fact, the achievable sum-rate can be slightly enhanced especially in the low SNR regime by using minimum mean square error (MMSE)-type beamforming/combining, as investigated in Section 10.2. This method requires, however, knowing all users' noise variances at the BS, requiring additional feedback. For simplicity, we focus, in this section, on the zero inter-user interference.

To find the transmit beamforming and the receive combining vectors, a simple iterative algorithm was proposed [CMIH08] based on the assumption of MRC at the UE, i.e. where transmit and receive filters are tied to each other. For an exemplary case of 2 users ($K = 2$), the algorithm may be summarized as such:

the two transmit beamforming vectors ($\mathbf{w}_k$, where $k = 1$ or 2) are initialized to some random vectors or the principal singular vectors of the channel matrices $\mathbf{H}_1$ and $\mathbf{H}_2$. Then, the following operations are repeated with increasing i (iteration index) until a stopping criterion is satisfied:

$$\tilde{\mathbf{H}}^{(i)} = \left[\left(\left(\mathbf{w}_1^{(i)}\right)^H \mathbf{H}_1 \mathbf{H}_1^H \right)^T \left(\left(\mathbf{w}_2^{(i)}\right)^H \mathbf{H}_2 \mathbf{H}_2^H \right)^T \right]^T,$$
$$\mathbf{W}^{(i+1)} = \left(\tilde{\mathbf{H}}^{(i)} \right)^{-1},$$
(5.32)

where $\mathbf{W}^{(i)} = [\mathbf{w}_1^{(i)} \mathbf{w}_2^{(i)}]$, and $\mathbf{w}_k^{(i)}$ is the transmit beamforming column-vector for the k-th UE at the i-th iteration, without normalization. The BS repeats this procedure until the change in $\mathbf{w}_k^{(i)}$ is sufficiently small i.e., $\|\mathbf{w}_k^{(i)} - \mathbf{w}_k^{(i-1)}\| < \epsilon$ where ϵ is an arbitrary small number.

The convergence of the iterative update algorithm is not guaranteed. Nevertheless, it typically converges with a small ϵ in almost all trial cases with more than 20 iterations for two transmit antenna systems [CMIH08]. This algorithm, however, may also affect the system's stability because, at times, the iterative algorithm converges very slowly. To resolve this issue, non-iterative coordinated beamforming algorithms have been proposed [CMJH08]. The authors in [CMJH08] used two methods (power iteration and generalized eigen analysis) to find the optimal transmit/receive beamforming vectors for two transmit antenna systems. Since we now consider a two-user system, we need to solve the optimization problem as follows:

$$\left\{ \mathbf{w}_{1,opt}, \mathbf{w}_{2,opt} \right\} = \arg \max_{\mathbf{w}_1:\|\mathbf{w}_1\|=1, \mathbf{w}_2:\|\mathbf{w}_2\|=1}$$
$$\left\{ \log_2 \left(1 + |\mathbf{w}_1^H \mathbf{R}_1 \mathbf{w}_1| \right) + \log_2 \left(1 + |\mathbf{w}_2^H \mathbf{R}_2 \mathbf{w}_2| \right) \right\}$$
$$\text{s.t. } |\mathbf{w}_1^H \mathbf{R}_1 \mathbf{w}_2| = |\mathbf{w}_2^H \mathbf{R}_2 \mathbf{w}_1| = 0.$$
(5.33)

where $\mathbf{R}_1$ and $\mathbf{R}_2$ are the $N_{\text{bs}} \times N_{\text{bs}}$ normalized matched channel matrices defined by $\mathbf{H}_k \mathbf{H}_k^H / \|\mathbf{H}_k\|_F^2$ and $\mathbf{w}_1, \mathbf{w}_2$ are the transmit beamforming vectors of size $N_{\text{bs}} \times 1$.

Theorem 5.4. *If $N_{\text{bs}} = 2$, $N_{\text{ue}} \geq 2$ and $\mathbf{R}_1$ and $\mathbf{R}_2$ are both invertible, then the following claim holds. If (non-zero) transmit beamforming vectors $\mathbf{w}_1$ and $\mathbf{w}_2$ satisfy the zero inter-user interference conditions, i.e.,*

$$\mathbf{g}_1^H \mathbf{R}_1 \mathbf{w}_2 = 0$$
$$\mathbf{g}_2^H \mathbf{R}_2 \mathbf{w}_1 = 0$$

then $\mathbf{w}_1, \mathbf{w}_2$ are the generalized eigenvectors of $(\mathbf{R}_1, \mathbf{R}_2)$, which means:

$$\mathbf{R}_1 \mathbf{w}_1 = \lambda_1 \mathbf{R}_2 \mathbf{w}_1$$
$$\mathbf{R}_2 \mathbf{w}_2 = \lambda_2 \mathbf{R}_1 \mathbf{w}_2$$

for some scalars λ_1 and λ_2 [CMJH08].

Theorem 5.4 means that for $N_{\text{bs}} = 2$, any zero inter-user interference solution is a generalized eigenvector of $\mathbf{R}_1$ and $\mathbf{R}_2$.

Theorem 5.5. *If $\mathbf{t}_m, \mathbf{t}_n$ are generalized eigenvectors of $(\mathbf{R}_1, \mathbf{R}_2)$ and they correspond to distinct eigenvalues, then any $\mathbf{t}_m, \mathbf{t}_n$ satisfy the zero inter-user interference constraint, where $m, n = 1, 2, \cdots$, the number of generalized eigenvectors, $m \neq n$. In other words, for a two-user system, any set of generalized eigenvectors of $(\mathbf{R}_1, \mathbf{R}_2)$ satisfy the zero inter-user interference condition [CMJH08].*

From Theorem 5.5, it is clear that the generalized eigenvectors of $\mathbf{R}_1$ and $\mathbf{R}_2$ satisfy the zero inter-user interference constraint (5.33). Note that this solution is not sum-capacity optimal for arbitrary antenna configurations. The idea here is to use the transmit beamforming vectors shown in Theorems 5.4 and 5.5 to obtain zero inter-user interference even for more than two antenna systems.

This algorithm can be generalized for a case of three BSs with one transmit antenna each and multiple UEs with multiple antennas [CKH10]. The condition that there is no inter-user interference between UE 1 and UEs 2 and 3 is equivalent to

$$\mathbf{w}_1^H \mathbf{R}_1 \mathbf{w}_2 = 0 = \mathbf{w}_2^H \mathbf{R}_2 \mathbf{w}_1 \tag{5.34}$$
$$\mathbf{w}_1^H \mathbf{R}_1 \mathbf{w}_3 = 0 = \mathbf{w}_3^H \mathbf{R}_3 \mathbf{w}_1. \tag{5.35}$$

In the case when $\overline{\mathbf{R}_1 \mathbf{w}_1} \times \mathbf{R}_2 \mathbf{w}_1 \neq 0$, where $\times$ denotes the cross-product defined in [CKH10], *Lemma* 2, 3) in [CKH10] asserts that (5.34) is equivalent to

$$\mathbf{w}_2 = \lambda(\overline{\mathbf{R}_1 \mathbf{w}_1} \times \mathbf{R}_2 \mathbf{w}_1), \quad \lambda \in \mathbb{C},$$

similarly as in the proof of *Theorem* 5 in [CKH10]. By the same reason, if $\overline{\mathbf{R}_1 \mathbf{w}_1} \times \mathbf{R}_3 \mathbf{w}_1 \neq \mathbf{0}$, (5.35) is equivalent to

$$\mathbf{w}_3 = \mu(\overline{\mathbf{R}_1 \mathbf{w}_1} \times \mathbf{R}_3 \mathbf{w}_1), \quad \mu \in \mathbb{C}.$$

Now the BSs need to nullify inter-user interference between UE 2 and UE 3. That is,

$$\mathbf{w}_2^H \mathbf{R}_2 \mathbf{w}_3 = 0 = \mathbf{w}_3^H \mathbf{R}_3 \mathbf{w}_2$$
$$\Leftrightarrow \mathbf{R}_2 \mathbf{w}_2 \cdot \mathbf{w}_3 = 0 = \mathbf{w}_2 \cdot \mathbf{R}_3 \mathbf{w}_3 \tag{5.36}$$
$$\Leftrightarrow \mathbf{w}_2 \cdot \mathbf{R}_2 \mathbf{w}_3 = 0 = \mathbf{w}_2 \cdot \mathbf{R}_3 \mathbf{w}_3.$$

Again by *Lemma* 2 in [CKH10], (5.36) is equivalent to

$$\mathbf{w}_2 = \nu(\overline{\mathbf{R}_2 \mathbf{w}_3} \times \mathbf{R}_3 \mathbf{w}_3), \quad \nu \in \mathbb{C}, \tag{5.37}$$

as long as $\overline{\mathbf{R}_2 \mathbf{w}_3} \times \mathbf{R}_3 \mathbf{w}_3 \neq \mathbf{0}$.

By expressing $\mathbf{w_2}$ and $\mathbf{w_3}$ in terms of $\mathbf{w_1}$, (5.37) is equivalent to the following:

$$\overline{\mathbf{R}_1\mathbf{w}_1} \times \mathbf{R}_2\mathbf{w}_1 \parallel \overline{\mathbf{R}_2(\overline{\mathbf{R}_1\mathbf{w}_1} \times \mathbf{R}_3\mathbf{w}_1)} \times \mathbf{R}_3(\overline{\mathbf{R}_1\mathbf{w}_1} \times \mathbf{R}_3\mathbf{w}_1). \tag{5.38}$$

Here, $\mathbf{x} \parallel \mathbf{y}$ denotes that the complex vectors $\mathbf{x}$ and $\mathbf{y}$ are parallel.

Define two functions $\phi, \psi : \mathbb{C}^3 \to \mathbb{C}^3$ as

$$\phi(\mathbf{w}) = \mathbf{R}_1\mathbf{w} \times \overline{\mathbf{R}_2\mathbf{w}}$$
$$\psi(\mathbf{w}) = \overline{\mathbf{R}_2(\overline{\mathbf{R}_1\mathbf{w}} \times \mathbf{R}_3\mathbf{w})} \times \mathbf{R}_3(\overline{\mathbf{R}_1\mathbf{w}} \times \mathbf{R}_3\mathbf{w}).$$

where the components ϕ and ψ are polynomials of degrees 2 and 4, respectively, in the components of $\mathbf{w}$. By applying *Lemma* 2, 3 in [CKH10], we see that solving (5.38), with the restriction that $\phi(\mathbf{w}_1) \neq \mathbf{0}$ and $\psi(\mathbf{w}_1) \neq \mathbf{0}$, is equivalent to finding a solution $\mathbf{w}_1$ for the following equation:

$$\phi(\mathbf{w}_1) \times \psi(\mathbf{w}_1) = \mathbf{0}.$$

Note that $\psi(\mathbf{w}_1) \neq \mathbf{0}$ implies $\overline{\mathbf{R}_1\mathbf{w}_1} \times \mathbf{R}_3\mathbf{w}_1 \neq \mathbf{0}$. Therefore, we can find *all* possible $(\mathbf{w}_1, \mathbf{w}_2, \mathbf{w}_3)$ satisfying the no inter-user interference condition, in generic (non-singluar) cases as described below.

Theorem 5.6. *Under the "non-singular" hypothesis* $\phi(\mathbf{w}_1) \neq \mathbf{0}$ *and* $\psi(\mathbf{w}_1) \neq \mathbf{0}$, *no inter-user interference is achieved by* $(\mathbf{w}_1, \mathbf{w}_2, \mathbf{w}_3)$, *if and only if*

$$\phi(\mathbf{w}_1) \times \psi(\mathbf{w}_1) = \mathbf{0}$$
$$\mathbf{w}_2 = \lambda \overline{\phi(\mathbf{w}_1)}$$
$$\mathbf{w}_3 = \mu(\overline{\mathbf{R}_1\mathbf{w}_1} \times \mathbf{R}_3\mathbf{w}_1)$$

for some $\lambda, \mu \in \mathbb{C}$.

Extension to the Two-Cell Case

While considering multiple antennas per UE, we so far constrained ourselves to the case of one BS (hence observing a BC). We now extend this to the two-cell case[1]. Note that the essential difference is that only the antennas connected to one BS may be used for the transmission towards one particular UE. Otherwise, the BSs would also have to exchange the data to be transmitted to the terminals, resembling a joint signal processing CoMP scheme which will be investigated in Sections 6.3 and 6.4. Let us initially focus on a two-cell MIMO system as shown in Fig. 5.15, where two BSs serve two UEs equipped with more than one receive antenna. As usual, the channel between BS m and UE k is denoted as $\mathbf{H}_k^m$. The

---

[1] Optimal M-cell coordinated beamforming algorithm with a zero inter-cell interference constraint is still unknown, thus in the section we mostly focus on a two-cell system.

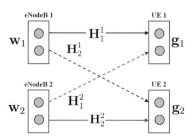

Figure 5.15 Two-user MIMO interference channel.

received signal after equalization at each UE k can be written as

$$\hat{x}_k = \sqrt{\frac{P}{2}} \mathbf{g}_k^H \left(\mathbf{H}_k^k\right)^H \mathbf{w}_k x_k + \sqrt{\frac{P}{2}} \mathbf{g}_k^H \left(\mathbf{H}_k^j\right)^H \mathbf{w}_j x_j + \mathbf{g}_k^H \mathbf{n}_k \qquad (5.39)$$

$$= \underbrace{\sqrt{\frac{P}{2}} \frac{\mathbf{w}_k^H \mathbf{H}_k^k \left(\mathbf{H}_k^k\right)^H \mathbf{w}_k}{\|\left(\mathbf{H}_k^k\right)^H \mathbf{w}_k\|} x_k}_{\text{Desired signal}} + \underbrace{\sqrt{\frac{P}{2}} \frac{\mathbf{w}_j^H \mathbf{H}_k^k \left(\mathbf{H}_k^j\right)^H \mathbf{w}_j}{\|\left(\mathbf{H}_k^k\right)^H \mathbf{w}_k\|} x_j}_{\text{Other-cell interference}} + \frac{\mathbf{w}_j^H \mathbf{H}_k^k \mathbf{n}_k}{\|\left(\mathbf{H}_k^k\right)^H \mathbf{w}_k\|}$$

where $j \neq k$, P is the total transmit power and MRC is also assumed at the UE, i.e. $\mathbf{g}_k = \frac{(\mathbf{H}_k^k)^H \mathbf{w}_k}{\|(\mathbf{H}_k^k)^H \mathbf{w}_k\|}$. Then the design goal is to maximize the desired signal term and to remove the other-cell interference term found in (5.39). Thus we introduce an interference-aware coordinated beamforming with MRC algorithm that satisfies the following condition:

$$\mathbf{g}_1^H \left(\mathbf{H}_1^2\right)^H \mathbf{w}_2 = 0 = \mathbf{g}_2^H \left(\mathbf{H}_2^1\right)^H \mathbf{w}_1$$
$$\Leftrightarrow \mathbf{w}_1^H \left(\mathbf{H}_1^1\right)^H \mathbf{H}_1^2 \mathbf{w}_2 = 0 = \mathbf{w}_2^H \left(\mathbf{H}_2^2\right)^H \mathbf{H}_2^1 \mathbf{w}_1, \qquad (5.40)$$

which implies that the other-cell interference term in (5.39) is perfectly removed; at the same time, the proposed system maximizes the desired effective channel gain $|\mathbf{g}_k^H (\mathbf{H}_k^k)^H \mathbf{w}_k|^2$ by using MRC. Note that it can be guaranteed that there is no inter-user interference thanks to the transmit beamforming vectors. Since we are considering a two-cell environment, this can be interpreted as the two-user MIMO interference channel (IC) illustrated in Fig. 5.15 [CHHT10].

Theorem 5.7. *Under a zero other-cell interference constraint in (5.40), the sufficient and necessary beamforming vectors with MRC for UE ℓ (where k, j are 1 or 2, $j \neq k$) are generalized eigenvectors of $\left(\mathbf{H}_k^k\right)^H \mathbf{H}_k^j$ and $\left(\mathbf{H}_j^k\right)^H \mathbf{H}_j^j$.*

Theorem 5.8. *Given $\mathbf{w}_j$, where each BS has two transmit antennas, the sufficient and necessary beamforming vector (unique up to complex multiplications) for UE k, $\mathbf{w}_k$ can be expressed as*

$$\mathbf{w}_k = \mu \begin{pmatrix} -z_2^* \\ z_1^* \end{pmatrix} \quad or \quad \mathbf{w}_k = \mu \begin{pmatrix} z_2^* \\ -z_1^* \end{pmatrix}, \qquad (5.41)$$

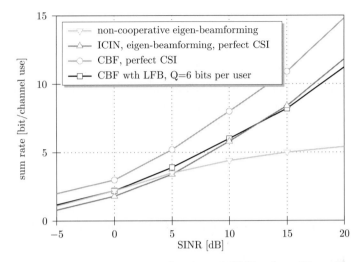

Figure 5.16 Sum-rate comparisons as a function of SINR, where $N_{\text{bs}} = N_{\text{ue}} = 2$. Each BS has the same transmit power $P/2$, where P is the total transmit power.

where

$$\mathbf{z} = \left(\mathbf{H}_k^k\right)^H \mathbf{H}_k^j \mathbf{w}_j = \begin{pmatrix} z_1 \\ z_2 \end{pmatrix},$$

and z_2^* is the complex conjugate of z_2.

To enable practical implementation, we also introduce simple two channel quantization methods. The normalized matched channel matrices can be written as, without the user index,

$$\mathbf{R_H} = \frac{\left(\mathbf{H}_k^k\right)^H \mathbf{H}_k^k}{\|\mathbf{H}_k^k\|_F^2} = \begin{bmatrix} R_{H,11} & R_{H,12} \\ R_{H,21} & R_{H,22} \end{bmatrix} \quad \text{and} \quad \mathbf{R_G} = \frac{\left(\mathbf{H}_k^j\right)^H \mathbf{H}_k^j}{\|\mathbf{H}_k^j\|_F^2} = \begin{bmatrix} R_{G,11} & R_{G,12} \\ R_{G,21} & R_{G,22} \end{bmatrix}.$$

Since $\mathbf{R_H}$ and $\mathbf{R_G}$ are Hermitian matrices, we use the following properties to design scalar quantization.

$$R_{H,11} + R_{H,22} = 1 \quad \text{and} \quad R_{G,11} + R_{G,22} = 1,$$
$$R_{H,12}^H = R_{H,21} \quad \text{and} \quad R_{G,12}^H = R_{G,21}$$

Therefore, only six scalar parts are needed to be quantized for computing the transmit beamforming and the receive combining vectors, i.e., $R_{H,11}$, $R_{G,11}$, $\text{Re}\{R_{H,12}\}$, $\text{Re}\{R_{G,12}\}$, $\text{Im}\{R_{H,12}\}$, and $\text{Im}\{R_{H,12}\}$, where $\text{Re}\{\cdot\}$ and $\text{Im}\{\cdot\}$ denote real and imaginary part, respectively. On the other hand, $\mathbf{R_H}$ and $\mathbf{R_G}$ can be jointly quantized using vector quantization as follows:

$$\mathbf{v_H} = \begin{pmatrix} R_{H,11} \\ \text{Re}\{R_{H,12}\} \\ \text{Im}\{R_{H,12}\} \end{pmatrix} \quad \text{and} \quad \mathbf{v_G} = \begin{pmatrix} R_{G,11} \\ \text{Re}\{R_{G,12}\} \\ \text{Im}\{R_{G,12}\} \end{pmatrix}.$$

Upon receiving the quantized values from the UEs over a control channel, the BSs can estimate $\mathbf{R_H}$ and $\mathbf{R_G}$ and compute the transmit beamforming vectors before transmitting the data.

Fig. 5.16 shows the achievable sum-rate results for i) coordinated beamforming, ii) non-cooperative eigen-beamforming, and iii) interference nulling algorithms introduced in [CHHT10]. For this figure, we model the elements of each UE's channel matrix as independent complex Gaussian random variables with zero mean and unit variance $\mathcal{N}_\mathbb{C}(0, 1)$. Note that the algorithm introduced is not directly related to the channel model. Once the BSs know all channel matrices, the transmit beamforming and receive combining vectors can be computed through Theorems 5.7 and 5.8. As can be seen from Fig. 5.16, the coordinated beamforming algorithm shows reasonably good sum-rate performance compared with other solutions regardless of SNR values. Note that the coordinated beamforming algorithm in the figure uses 6 bits limited feedback per user, i.e., 3 bits each for $\mathbf{R_H}$ and $\mathbf{R_G}$, respectively.

5.3.4 Summary

In this section, we presented some latest results in downlink cooperative beamforming, which is important for interference management in upcoming cellular standards like 3GPP LTE-A. We described the details of several strategies, distinguishing whether each terminal is equipped with one or multiple receive antennas. We also presented simulation results from these strategies to illustrate the potential gains that can be obtained from such kind of CoMP. From an LTE-Advanced point of view, the most likely scenario is to have BSs with 2 or 4 transmit antennas, UEs with 2 receive antennas, thus the solutions introduced in the chapter would be good candidates for the LTE-A systems. A particular implementation of coordinated beamforming will be described and simulated in Section 14.4.3.

6 CoMP Schemes Based on Multi-Cell Joint Signal Processing

In this chapter, we focus on CoMP schemes where user data or received signals connected to multiple users are exchanged between base stations for joint signal processing. Such schemes promise larger spectral efficiency gains than pure interference coordination techniques, but typically come at the price of larger backhaul requirements and (particularly in the downlink) more severe synchronization requirements. After Sections 6.1 and 6.2 introduce centralized and decentralized uplink CoMP schemes, respectively, 6.3 and 6.4 focus on the downlink.

6.1 Uplink Centralized Joint Detection

Chenyang Yang, Yafei Tian and Andreas F. Molisch

In this section, uplink centralized joint detection with infinite or limited backhaul capacity is studied, as already introduced in Section 4.3.1. A cluster of base stations (BSs) sends either raw or preprocessed receive signals from the user equipments (UEs) to a CoMP central unit (CCU), where joint processing is performed to deal with inter-cell interference. The CCU can be a separate entity, or any of the BSs involved in the cooperation. This is practical in systems such as LTE-A employing a flat network architecture. The detection algorithms are applicable to various static or dynamic clustering schemes as described in Chapter 7.

6.1.1 Introduction

When multiple BSs are connected via perfect backhaul links with infinite capacity, uplink centralized joint detection resembles a multiple access channel (MAC) problem, where the CCU is a super-receiver, and the BSs form a distributed antenna system (DAS) [Mol01]. Consequently, various optimal or suboptimal multi-user detectors such as the maximum likelihood (ML), linear minimum mean square error (LMMSE) detector, MMSE detector with successive interference cancelation (SIC) or parallel interference cancelation (PIC), and iterative detectors based on the Turbo principle, can be used for joint detection. By removing the inter-cell interference to a certain extent and obtaining significant array gain and diversity gain as shown in Chapter 4, the system spectral efficiency can be significantly increased [Mol01, DP03].

When joint detection is implemented in practice, the received signals at the cooperative BSs need to be quantized and then forwarded via the backhaul links to the CCU for centralized processing. This entails requirements for large backhaul capacity, typically on the order of Mbps or even Gbps [MF07b, HFG09]. In existing or upcoming systems such as LTE, the backhaul capacity is typically limited. Considering realistic backhaul constraints, we can transfer the locally demapped signals or soft-decoded information to the CCU [FHG09], or forward the locally compressed receive signals exploiting the correlation inherent in the message of the UEs [MF09b].

One aspect of uplink CoMP lies in the fact that the signals received by multiple BSs are correlated, which leads to redundancy. This implies that each BS can exploit this signal correlation and compress its received signals, then transmit the compressed signals to the CCU, which reduces the information needed to be exchanged via the backhaul links. The CCU then decompresses the signals with its own received signal as the side information (if it is a BS itself), and finally estimates the messages of the UEs [dCS09].

Another inherent feature of CoMP systems is their asymmetric channels, i.e., the average channel power from one UE to its local BS and other cooperative BSs are different, and the power from multiple UEs in different cells to one BS also differ. The channel asymmetry cannot be compensated by power control [HYK$^+$10], which is quite different from the near-far effect in single cell multi-user systems. This does not change the structure of the joint detector at the CCU when infinite backhaul capacity is assumed, but it has large impact on the system performance. Moreover, when the backhaul link constraint is taken into account, this feature introduces new degrees of freedom to design uplink CoMP schemes. Depending on the relative locations of multiple UEs (i.e., various user pairings), the CCU may jointly detect only some UEs, and the supporting BSs may quantize, compress, decode or even partially decode their received signals to reduce the information to be transmitted to the CCU [MF08a, SSPS09a].

6.1.2 Joint Detection Algorithms

In the following, we introduce and discuss several joint signal processing algorithms at the CCU, including ML detection, LMMSE, MMSE detection with SIC and PIC, and Turbo detection. The computational complexity of the algorithms will be analyzed. Simulation results are provided to compare the detection performance of these algorithms. To show the benefit of centralized joint detection versus non-cooperative detection in various typical settings, we will also provide simulation results on the system sum-rate. Throughout the remainder of this section, we will focus on a frequency-flat channel, which could be a single sub-carrier of an orthogonal frequency division multiple access (OFDMA) system, as introduced in Chapter 3.

We consider a cluster of M cells, where each cell involves one BS with N_{bs} receive antennas and one UE with N_{ue} transmit antennas. Multiple UEs from different cells operate on the same time/frequency resource, and the received signals at the BSs from $K = M$ UEs in different cells are superimposed. As in Chapter 3, we assume that all entities are perfectly synchronized in time and frequency (the impact of imperfect synchronization is discussed in detail in Sections 8.2 and 8.3), and the received signals are transferred to the CCU for joint detection after BS local processing. The signals from M BSs gathered at the CCU can be expressed as

$$\mathbf{y} = \mathbf{H}\mathbf{s} + \mathbf{n} = \mathbf{H} \underbrace{\begin{bmatrix} \mathbf{G}_1 & & \mathbf{0} \\ & \ddots & \\ \mathbf{0} & & \mathbf{G}_K \end{bmatrix}}_{\mathbf{G}} \mathbf{x} + \mathbf{n}_o + \mathbf{n}_p, \tag{6.1}$$

where $\mathbf{y} = [\mathbf{y}_1^T, \cdots, \mathbf{y}_M^T]^T$, and $\mathbf{y}_m$ are the forwarded symbols from BS m with local processing errors, $\mathbf{s}$ and $\mathbf{x}$ are the transmitted symbols after and before precoding, $\mathbf{H}$ is the composite channel matrix, $\mathbf{G}$ is the (block-diagonal) precoding matrix, $\mathbf{n}_o$ denotes the noise vector at all M BSs including the thermal noise at the receiver front-end and the inter-cluster interference from the cells outside of the cooperative cluster, and $\mathbf{n}_p$ denotes the local processing errors at the BSs due to quantization or compression.

Assuming that the CCU knows the channel matrix $\mathbf{H}$ and precoding matrix $\mathbf{G}$ perfectly, the system capacity region with unlimited backhaul is described in (3.4). We can use (3.4) to calculate the block error rate in fading channels, which can serve as an achievable lower bound for practical systems. Given modulation and coding schemes, the transmission data rate is known. If for one channel realization the rate is larger than the sum capacity computed from (3.4), then the data block can not be decoded correctly, or we can say an *outage* happens.

The ML detection algorithm finds the most probably transmitted symbols, which, under the assumption of Gaussian noise, minimizes the Euclidean distance between the received signals and all possible received symbols,

$$\hat{\mathbf{x}} = \min_{\mathbf{x}} \|\mathbf{y} - \mathbf{H}\mathbf{G}\mathbf{x}\|^2. \tag{6.2}$$

ML detection is optimal if the transmitted symbols are equally probable, but is not feasible for most applications due to its prohibitive complexity. Suppose that each UE transmits N_{ue} streams and M_{s}-QAM is used, then the detection complexity at the CCU is on the order of $O(M_{\text{s}}^{N_{\text{ue}}K})$. For example, when $M_{\text{s}} = 4$, $N_{\text{ue}} = 2$, and $K = 5$, the complexity would be on the order of 2^{20} computational steps per channel access, which is beyond the capability of current systems.

To reduce the complexity of ML detection, sphere detection can be applied. It only searches the candidates when the distance $\|\mathbf{y} - \mathbf{HGx}\|^2$ is less than a given radius, leading to a complexity roughly cubic in $N_{ue}K$ [HTB03]. The LMMSE detection algorithm is a widely used alternative. It minimizes the mean square estimation error between $\mathbf{x}$ and its estimate, which is

$$\hat{\mathbf{x}} = \mathbf{W}^H \mathbf{y} \qquad (6.3)$$

$$\text{where } \mathbf{W} = \left(\mathbf{HGG}^H \mathbf{H}^H + \mathbf{\Phi}_{nn}\right)^{-1} \mathbf{HG}, \qquad (6.4)$$

and $\mathbf{\Phi}_{nn}$ is the covariance matrix of $\mathbf{n}_o$ plus $\mathbf{n}_p$. The LMMSE algorithm has a complexity on the order of $O((N_{ue}K)^2)$. For the example system setting listed earlier, the computational cost is on the order of 10^2 operations. Considering that the interference of a UE to other UEs can be eliminated after its signal has been detected, the MMSE-SIC algorithm is expected to perform better. The MMSE-SIC algorithm is a serially concatenated MMSE detection scheme, which detects and decodes the signals (streams) of the UEs one by one and subtracts their interference from the remaining signals, which is

$$\mathbf{y}_1 = \mathbf{y}, \quad \hat{\mathbf{x}}_1 = \mathbf{W}_1^H \mathbf{y}_1,$$
$$\mathbf{y}_2 = \mathbf{y}_1 - \mathbf{H}_1 \mathbf{G}_1 e\left(d\left(\hat{\mathbf{x}}_1\right)\right), \quad \hat{\mathbf{x}}_2 = \mathbf{W}_2^H \mathbf{y}_2,$$
$$\vdots$$
$$\mathbf{y}_K = \mathbf{y}_{K-1} - \mathbf{H}_{K-1} \mathbf{G}_{K-1} e\left(d\left(\hat{\mathbf{x}}_{K-1}\right)\right), \quad \hat{\mathbf{x}}_K = \mathbf{W}_K^H \mathbf{y}_K, \qquad (6.5)$$

where $\mathbf{H}_k$ is the channel matrix from UE k to all M BSs, $\mathbf{W}_k$ is the receive filter designed for UE k, $d(\cdot)$ denotes decoding and $e(\cdot)$ denotes (re-)encoding. The decoding order determines which vertex of the pentagon-shaped capacity region as illustrated in Fig. 3.4 can be achieved. For example, we can decode and cancel the signals in descending order of signal strength, which will improve fairness among UEs. This algorithm can be seen as an adaptation of Horizontal Bell Laboratories Layered Space-Time Architecture (H-BLAST) [WFGV98].

The MMSE-SIC algorithm has a complexity on the same order of the LMMSE detection algorithm. PIC is another kind of interference cancelation scheme with less detection latency, which cancels the interference of other UEs in parallel. Before $\mathbf{x}_k$ is detected, the estimated signals from all other UEs should be subtracted from $\mathbf{y}$. However, if the initial estimations of $\mathbf{x}_1, \cdots, \mathbf{x}_K$ are not correct, the error propagation will be more severe.

PIC will have better performance if we have *a priori* information of coded bits forwarded from the decoder. With this probability information, soft estimation of the transmitted symbols rather than hard decisions can be formed and canceled. Through iteration between the detector and decoder, detection performance can be improved significantly. This scheme belongs to Turbo detection [WP99], in which various kinds of detectors and decoders can be applied. We will describe an ML detector and MMSE-PIC detector appropriate for Turbo detection in the following [DMP04], while the decoding process itself is described in [WH02].

Extrinsic information exchange is required between the detector and the decoder. For the ML detector, the obtained extrinsic information is

$$L_e(c_{k,i}) = \log \frac{\sum_{\mathbf{x} \in X_{k,i}^+} p(\mathbf{y}|\mathbf{x}) p(\mathbf{x})}{\sum_{\mathbf{x} \in X_{k,i}^-} p(\mathbf{y}|\mathbf{x}) p(\mathbf{x})} - L_a(c_{k,i}), \quad (6.6)$$

where $c_{k,i}$ is the ith coded bit of UE k, $X_{k,i}^+ = \{[\mathbf{x}_1^T, \cdots, \mathbf{x}_K^T]^T : c_{k,i} = 1\}$ and $X_i^- = \{[\mathbf{x}_1^T, \cdots, \mathbf{x}_K^T]^T : c_{k,i} = -1\}$, $p(\mathbf{y}|\mathbf{x})$ is a multivariate Gaussian distribution with the signal model of (6.1), $p(\mathbf{x}) = \prod_{k=1}^{K} \prod_{i=1}^{N_k} p(c_{k,i})$, N_k is the number of bits modulated on N_{ue} sub-streams of UE k, and $L_a(c_{k,i}) = \log(P(c_{k,i} = 1)/P(c_{k,i} = -1))$ denotes *a priori* information from the decoding stage.

For the MMSE-PIC detector, we first need to construct the soft estimation of the transmitted symbols using the *a priori* information transferred from the decoding stage. Since $\mathbf{x}_k$ is a mapping result of $c_{k,i}, i = 1, \cdots, N_k$, the soft estimation of $\mathbf{x}_k$ is the weighted summation of all possible mapping results $\mathbf{x}_k(c_{k,1}, \cdots, c_{k,N_k})$ with their corresponding probabilities,

$$\tilde{\mathbf{x}}_k = \sum_{c_{k,i} \in (1,0), i=1,\cdots,N_k} \mathbf{x}_k(c_{k,1}, \cdots, c_{k,N_k}) \prod_{i=1}^{N_k} p(c_{k,i}). \quad (6.7)$$

Then, for each UE, we can subtract the estimated interference from other UEs,

$$\tilde{\mathbf{y}}_k = \mathbf{y} - \mathbf{H}\mathbf{G}\bar{\mathbf{x}}_k, \quad (6.8)$$

where $\bar{\mathbf{x}}_k = (\tilde{\mathbf{x}}_1, \cdots, \tilde{\mathbf{x}}_{k-1}, \mathbf{0}, \tilde{\mathbf{x}}_{k+1}, \cdots, \tilde{\mathbf{x}}_K)^T$ is the estimated interference symbol vector. An MMSE filter is applied to $\tilde{\mathbf{y}}_k$ to further suppress the residual interference plus noise, which is given by

$$\mathbf{W}_k = E\{\tilde{\mathbf{y}}_k \tilde{\mathbf{y}}_k^H\}^{-1} E\{\tilde{\mathbf{y}}_k \mathbf{x}_k^H\} \quad (6.9)$$
$$= \left(\mathbf{H}_k \mathbf{G}_k \mathbf{G}_k^H \mathbf{H}_k^H + \bar{\mathbf{H}}_k \bar{\mathbf{G}}_k \mathbf{Q} \bar{\mathbf{G}}_k^H \bar{\mathbf{H}}_k^H + \mathbf{\Phi}_{\text{nn}}\right)^{-1} \mathbf{H}_k \mathbf{G}_k, \quad (6.10)$$

where $\bar{\mathbf{H}}_k$ and $\bar{\mathbf{G}}_k$ involve all the vectors of $\mathbf{H}$ and $\mathbf{G}$ except those of $\mathbf{H}_k$ and $\mathbf{G}_k$, $\mathbf{Q}$ is the covariance matrix of the residual interference, i.e.,

$$\mathbf{Q} = \text{diag}\left([\mathbf{q}_1, \cdots, \mathbf{q}_{k-1}, \mathbf{q}_{k+1}, \cdots, \mathbf{q}_K]\right),$$

where $\mathbf{q}_k = \left[1 - |\tilde{x}_{k,1}|^2, \cdots, 1 - |\tilde{x}_{k,N_{\text{ue}}}|^2\right]$, $\tilde{x}_{k,n}$ is the nth element of $\tilde{\mathbf{x}}_k$. This covariance matrix will approach zero when the estimates from the decoding stage are accurate enough. As shown in [WP99], the output of the MMSE filter, i.e., the estimation of the j-th stream of UE k, can be modeled as

$$z_{k,j} = \mathbf{w}_{k,j}^H \tilde{\mathbf{y}}_k = \mu_{k,j} x_{k,j} + \eta_{k,j}, \quad (6.11)$$

where $\mathbf{w}_{k,j}$ is the j-th column of matrix $\mathbf{W}_k$, $\mu_{k,j} = E\{z_{k,j} x_{k,j}^H\} = \mathbf{w}_{k,j}^H \mathbf{H}_k \mathbf{g}_{k,j}$, and $\eta_{k,j}$ is well-approximated by a Gaussian variable with zero mean and variance

$$\nu_{k,j}^2 = E\left\{|z_{k,j} - \mu_{k,j} x_{k,j}|^2\right\} = \mu_{k,j} - |\mu_{k,j}|^2. \quad (6.12)$$

The extrinsic information is given in the same form as in (6.6), except that $\mathbf{y}$ is replaced by $z_{k,j}$, $\mathbf{x}$ is replaced by $x_{k,j}$, and (6.1) is replaced by (6.11). Since the vector operation is replaced by multiple scalar operations for calculating the extrinsic information, the complexity is significantly reduced.

Simulation Settings and Results

In this subsection, we first show the block error rate (BLER) of these detectors, and then show the impact of user positions and pairing on the system sum-rate. We will compare the sum-rate achieved by the centralized joint detection with that by the non-cooperative MMSE-SIC, with which each BS only locally decodes the message of its own user and treats the inter-cell interference as noise.

We consider two cells, each has a 4-antenna BS and a single antenna user. The distance between the two BSs is 500 m, the pathloss in dB follows $35.3 + 37.6\log_{10}d$, and shadowing is not considered. The cell-center user is located on the line connecting the two BSs and is 50 m from its serving BS, and the cell-edge user is 245 m from its serving BS. The small-scale fading is assumed to be i.i.d. Rayleigh fading. The uplink transmit power is 30 dBm to both users, and the noise power at each BS is -99 dBm. We choose BS 1 as the CCU. BS 2 forwards its received signal via an infinite-capacity backhaul link to BS 1, i.e., $\mathbf{n}_p = 0$. The sum-rates achieved by the joint LMMSE and MMSE-SIC are computed using Shannon's capacity formula with their respective signal-to-interference-and-noise ratio (SINR) and are averaged over realizations of small-scale fading. Note that the sum-rate of the MMSE-SIC is the same as the sum capacity shown in (3.4) [SAH+04].

The BLER versus signal-to-noise ratio (SNR) of various joint detection algorithms is shown in Fig. 6.1, where two users are assumed to have identical SNR. Each user employs binary phase shift keying (BPSK) modulation and a rate-1/3 convolutional code with generators $(155, 117, 123)_8$. The data packet length is 256 bits. For ML, LMMSE, and MMSE-SIC detection, the soft decision Viterbi decoder is used. For Turbo detection with MMSE-PIC, the Max-Log-MAP algorithm is used in the soft-input-soft-output decoder [WH02] with 6 iterations. To compare with the BLER lower bound, which is calculated from the achievable sum-rate formula, the data block involves the data packets of two users and therefore the BLER is accounted if any of these two packets are wrong.

From this figure, we can see that the LMMSE detector has a similar performance as the MMSE-SIC detector when the convolutional coding and soft-decision Viterbi decoder are employed. At a BLER of 10^{-2}, the ML detector performs 1 dB better. Even without iteration, (6.6) is used to send the soft decision values to the Viterbi decoder, otherwise, the ML detector with hard decisions cannot outperform the LMMSE detector with soft decisions. The Turbo detection algorithm with MMSE-PIC detector and soft-input-soft-output decoder has 2 dB SNR gain over the non-iterative MMSE-SIC detection algorithm, and the gap to the theoretical lower bound is about 1 dB.

To show the gain of joint detection and observe the impact of user pairing on the performance, Fig. 6.2 gives the sum-rate cumulative distribution

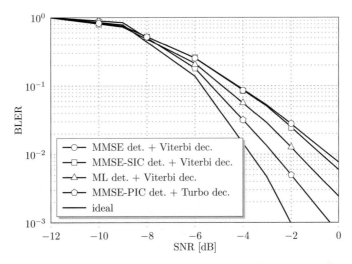

Figure 6.1 Block error rate of various joint detection algorithms.

function (CDF) of joint LMMSE, joint MMSE-SIC and non-cooperative detection in different settings. The legend "center-edge" represents the results where one user is located at the cell-center but another user is at the cell-edge. The legend "edge-edge" and "center-center" respectively represent the cases where both users are at cell-edge or at cell-center. Although in our simulation setting the non-cooperative system has enough degrees of freedom to deal with the interference, it shows that even when the two users are located at the cell-center where each user has high SINR, the joint detection still outperforms non-cooperative detection (though this gain can be expected to become marginal under imperfect channel knowledge, as shown in Section 4.2). The gain of performing joint detection increases when at least one user is at the cell-edge. In these cases, the cell-edge user will benefit from the joint detection significantly thereby joint detection outperforms non-cooperative detection more evidently. On the other hand, since the performance of the cell-center user dominates the sum-rate, the sum-rate will reduce when more users are located at the cell-edge.

6.1.3 Local BS Processing with Limited Backhaul Constraint

In practical systems, the cooperative BSs need some kind of local processing before they forward the received information to the CCU. The BSs can directly quantize the IQ samples of their received signals, but transmitting this information to the CCU requires high capacity backhaul links. To reduce the requirement for the backhaul capacity, the BSs can compress the received signals. Compression can be performed through source coding schemes with compression distortion. As mentioned in Section 4.3.1, we can further exploit the correlation among the received signals from different BSs, where the amount of information transferred is the conditional mutual information between the signals received by one

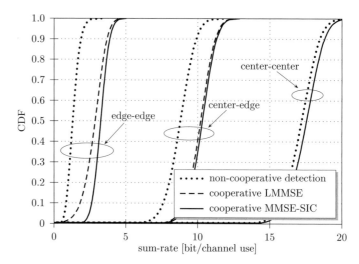

Figure 6.2 Sum-rate CDF under infinite backhaul capacity, which shows the performance gap between joint detection and non-cooperative detection. 9 dB inter-cluster interference is considered.

BS and those already known by the CCU. Both the quantization noise and the compression distortion will deteriorate the joint detection performance.

Local BS Processing Methods

Consider that one of the M BSs serves as the CCU. Without loss of generality, we again consider BS 1 as the CCU. The backhaul link constraint from BS m to the CCU is denoted as C_m. If a per-link capacity constraint is applied, all the links have same capacity, i.e., $C_m = C$. If the sum-capacity constraint is applied, i.e., $\sum_m C_m = C$, the capacity of each link would be allocated by some optimization criteria such as the sum-rate maximization in the cluster.

We first consider the direct quantization scheme. The data symbols received by BS m are quantized with B_m bits each for real and imaginary dimensions separately, where $B_m = C_m/(2N_\text{bs})$. Assume that the signal level is within the range of $[-A, +A]$, then with uniform quantization the quantization noise variance in each dimension is

$$\sigma_{q,m}^2 = \frac{A^2 \cdot 2^{-2B_m}}{3}. \tag{6.13}$$

The covariance matrix of the quantization noise vector is therefore given as $\boldsymbol{\Psi}_m = \text{diag}([2\sigma_{q,m}^2, \cdots, 2\sigma_{q,m}^2])$, representing the processing errors of BS m. We next address the source coding scheme without exploiting the signal correlation between BS m and BS 1. The covariance of the received signal of BS m is

$$\boldsymbol{\Phi}_{\text{yy},m} = \mathbf{H}^m \boldsymbol{\Phi}_{\text{ss}} (\mathbf{H}^m)^H + \sigma_{\text{o},m}^2 \mathbf{I}, \tag{6.14}$$

where $\mathbf{H}^m$ denotes the channel matrix from all K users to BS m, $\boldsymbol{\Phi}_{\text{ss}} = \mathbf{G}\mathbf{G}^H$ is a block diagonal matrix where each block is the covariance matrix of the transmitted signals of one user, and $\sigma_{\text{o},m}^2$ is the variance of the observation noise including the receiver thermal noise and the inter-cluster interference at BS m. According to the backhaul constraint, the distortion matrix $\boldsymbol{\Psi}_m$ should satisfy

$$\log_2 \det \left(\mathbf{I} + (\boldsymbol{\Psi}_m)^{-1} \left(\mathbf{H}^m \boldsymbol{\Phi}_{\text{ss}} (\mathbf{H}^m)^H + \sigma_{\text{o},m}^2 \mathbf{I} \right) \right) \leq C_m. \qquad (6.15)$$

We finally consider the multi-source compression coding exploiting the signal correlation between BS m and BS 1. We only consider the correlation between the signals received in BS m and BS 1, even though BS 1 may have retrieved more information from other BSs. The amount of mutual information that should be transferred depends on the conditional covariance matrix $\boldsymbol{\Phi}_{\text{yy},m|1}$ between the signals received in BS m and in BS 1, which is expressed as [dCS09]

$$\boldsymbol{\Phi}_{\text{yy},m|1} = \mathbf{H}^m \left(\mathbf{I} + \frac{\boldsymbol{\Phi}_{\text{ss}}}{\sigma_{\text{o},m}^2} (\mathbf{H}^1)^H \mathbf{H}^1 \right)^{-1} \boldsymbol{\Phi}_{\text{ss}} (\mathbf{H}^m)^H + \sigma_{\text{o},m}^2 \mathbf{I}. \qquad (6.16)$$

The distortion matrix in this case should satisfy [dCS09]

$$\log_2 \det \left(\mathbf{I} + (\boldsymbol{\Psi}_m)^{-1} \boldsymbol{\Phi}_{\text{yy},m|1} \right) \leq C_m. \qquad (6.17)$$

This is a lossy decentralized multi-source compression with receiver-side information. In practice, we can first use the conditional Karhunen-Loève transform to decompose the vector signal into independent streams, and then compress the resulting scalar streams separately. Given the quantization noise or compression distortion matrix $\boldsymbol{\Psi}_m$ in BS m, the sum-rate achievable with joint detection is

$$R_{\text{sum}} = \log_2 \det \left(\mathbf{I} + (\boldsymbol{\Phi}_{\text{nn}} + \boldsymbol{\Psi})^{-1} \mathbf{H} \boldsymbol{\Phi}_{\text{ss}} \mathbf{H}^H \right), \qquad (6.18)$$

where $\boldsymbol{\Psi} = \text{diag}(\boldsymbol{\Psi}_1, \cdots, \boldsymbol{\Psi}_M)$, $\boldsymbol{\Phi}_{\text{nn}} = \text{diag}\{\sigma_{o1}^2 \mathbf{I}_{N_{\text{bs}}}, \cdots, \sigma_{oM}^2 \mathbf{I}_{N_{\text{bs}}}\}$, and $\mathbf{I}_{N_{\text{bs}}}$ is the identity matrix of size N_{bs}.

Simulation Settings and Results
In Figs. 6.3(a) and 6.3(b), we show the sum-rate of two cell-edge users when MMSE-SIC is used at the CCU (i.e., BS 1), and different local processers are used at the supporting BS (i.e., BS 2), where the backhaul is constrained to 3 and 10 bit/channel use, respectively. Inter-cluster interference is not considered. All simulation settings are the same as in Fig. 6.1, unless otherwise stated.

Comparing with the results shown in Fig. 6.2, we can see that the performance of the joint detection degrades severely due to the processing errors of the supporting BS but still outperforms non-cooperative detection. When increasing the backhaul capacity constraint, the two schemes with compression improve significantly, but the scheme with direct quantization improves much slower. When $C = 10$ bit/channel use, the performance of the two compression schemes is almost the same, which means that it is unnecessary to use multi-source compression coding which is more complex and needs more accurate channel state information (CSI). With such a moderate backhaul capacity, using single source compression rather than using the direct quantization is a good choice.

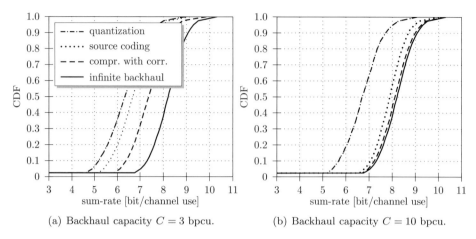

Figure 6.3 The impact of quantization noise and compression distortion on the performance of joint MMSE-SIC.

To observe the impact of the user pairing on the performance, we simulate the sum-rate of the two users in two cases in Fig. 6.4. The legend "center-edge" represents the results where one user in the cell of BS 1, i.e., the CCU, is located at the cell-center but another user is at the cell-edge. The legend "edge-center" represents the opposite case. Again, we do not consider inter-cluster interference.

It is shown that it is critical for the performance which user is located at the cell-center; this is in contrast to the scenario of perfect backhaul considered in Section 6.1.2, Fig. 6.2. Remember that BS 1 is the CCU. Therefore, signals received at this BS do not suffer from compression errors. When the user of BS 1 is located at the cell-center (*center-edge* in the legend), its performance will dominate the sum-rate. Thereby, the impact of the local processing error is negligible. On the other hand, when the user of the supporting BS 2 is located at the cell-center (*edge-center* in the legend), the local processing errors at BS 2 will degrade the performance of the joint detection although the received signal of this user at BS 2 has high SINR.

6.1.4 Local or Partial Decoding with Limited Backhaul Constraint

Alternative to the direct quantization, single-source and multi-source coding, the BSs can also demodulate or decode the signal from every user, and then transfer the original bits or (soft) coded bits of each user to the CCU for joint detection. The bits from different BSs are combined at the CCU to yield final decisions [FHG09].

Local Decoding
For the cell-center users, we can decode their messages locally at their serving BSs. These messages are able to be decoded correctly with high probability.

6.1 Uplink Centralized Joint Detection

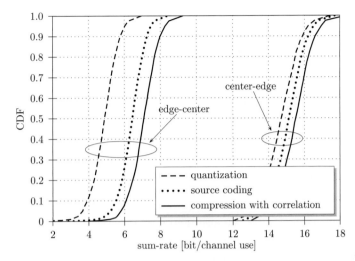

Figure 6.4 The impact of user pairing on the performance of joint MMSE-SIC, backhaul capacity constraint $C = 3$ bit/channel use.

As shown in Fig. 6.2 for the case where both users are at the cell-center, local decoding performs closely to or even the same as joint decoding. These decoded bits can be forwarded to the CCU to facilitate interference cancelation for other cell-edge users. Otherwise, it is shown in Fig. 6.4 that under limited backhaul capacity the cell-center user of the supporting BS will degrade the sum-rate no matter whether we quantize or compress its received signals.

Due to the mutual interference, it is not wise to decode the information of all users at one BS only. On the other hand, if each BS demodulates and transfers all the soft information of multiple users, the transfer rate may even exceed that of the direct quantization scheme. For example, assume that 3 BSs cooperate to serve 3 users, where each user transmits a data stream with 16-QAM modulation. This amounts to 12 information bits for each BS. If 4 bits are used to represent the log-likelihood ratio (LLR) of each information bit, then totally 48 bits need to be transferred to the CCU. However, if we use 8 bits to quantize each received sample, only 16 bits need to be transferred for both the real and imaginary parts of the superposed signals.

Therefore, it is better that the BSs locally decode the information of some UEs and forward to the CCU, but locally process and then transfer the received signals from other UEs to the CCU for joint detection, as also stated in Section 4.3.2. This is in fact a receiver mode switching, which is analogous to the downlink CoMP mode switching, where the BSs serve some UEs without cooperation and jointly transmit to other UEs (see Section 6.4).

Partial Decoding and Rate Splitting
For those users who are between cell-center and cell-edge, we can divide the data streams of each user into two parts. One is decoded by the local BS, and the other

is at the CCU. In particular, for each UE that is in the cells of supporting BSs, we use rate splitting to divide its data streams, which is implemented by superposition coding [MF08a, SSPS09a]. At each supporting BS, the received messages are partially decoded and the residual received signals are compressed and then forwarded to the CCU. At the CCU, MMSE-SIC is used for joint detection. By optimizing the power allocated to the two parts of the data streams, maximum sum-rate can be achieved under a certain constraint of backhaul capacity. Partial decoding at the local BS can reduce the information to be transferred to the CCU, but may also reduce the user's rate since decoding is performed under interference. It is expected that such a multi-level decoding strategy may provide a smooth transition from the two extreme schemes: (i) each BS only compresses its received raw signals, or (ii) each BS only locally decodes the messages of its own users [dCS09], as also pointed out in Section 4.3.2.

However, numerical results show that the performance gain of rate splitting is marginal (e.g. compared to a time-share between different cooperation strategies). The same conclusion is drawn in [MF08a, SSPS09a] with slightly different forms of superposition coding.

This is because CoMP channels are asymmetric, which cannot be compensated by power allocation. Using rate splitting for a user, say UE k, is equivalent to dividing the user into two users with different powers under a sum power constraint, UE $k1$ and UE $k2$. In CoMP systems, from the viewpoint of its local BS, say BS 1, UE $k1$ allocated with more power looks like a cell-center user. Other BSs, however, also receive higher power from UE $k1$. Consider that a true cell-center user in BS 1 will lead to high receive power at its local BS but low receive power at other BSs, which makes local decoding more desirable. Now it is clear that rate splitting is not equivalent to dividing a user into two users with different positions, and hence does not help to improve the performance as expected.

6.1.5 Provisions for Uplink Joint Processing in WiMax and LTE

Advanced WiMax, or IEEE 802.16m, enables CoMP as an option. As defined in Section 16.5.2 of the standard [IEE10a], two types of uplink multi-BS multiple-input multiple-output (MIMO) are defined: (i) single-BS decoding with multi-BS coordination, and (ii) uplink multi-BS joint processing. The former type is similar to the coordinated beamforming scheme introduced in Section 5.3. The latter corresponds to joint detection CoMP as observed in this section, and offers two variants of processing: macro diversity (where multiple BSs receive the signal only from a single UE), or joint detection for multiple UEs.

In order to enable simultaneous reception at multiple BSs, a number of provisions are taken: first of all, 802.16m defines a "zone" (i.e., a time period within each basic "frame duration"), during which multi-BS reception may take place. During this zone, transmission from one or more UEs to the different cooperating

BSs occurs on the same time-frequency resources (i.e., the same OFDM symbols and the same sub-carriers). This is a notable contrast to "standard" transmission in WiMax, where UEs in adjacent cells are transmitting on (mostly) different time-frequency resources, in order to reduce interference. Uplink sounding, i.e., determination of the channel information for the different UEs, is done in such a way that the sounding signals allocated to the different UEs are orthogonal to each other, as in Section 9.1. When macro-diversity is enabled (i.e., only one signal exists on a time-frequency resource, the BSs exchange log-likelihood ratios of the bit decisions over the backhaul network. For cooperative reception (multiple UEs on a time-frequency resource), quantized versions of the received signals are exchanged over the backhaul network.

LTE-A also considers CoMP concepts, where similar provisions are made. For centralized uplink CoMP, there should be backhaul links to connect the eNodeBs to the CCU. Consider the per-link capacity, which is simply $C \cdot B_W$. In an LTE-A system, each eNodeB includes three BSs, the number of the transmit antennas at each BS may be 2, 4 or 8, and the bandwidth B_W is 20 MHz. Here we simply ignore the various overhead, and simply multiply the backhaul constraints in bit/channel use with the system bandwidth to obtain Mbit/s. Therefore, when $C = 3$ bit/channel use or $C = 10$ bit/channel use, the required per-link backhaul is 180 Mbps or 600 Mbps, respectively. When considering direct quantization, this implies $B_k = 0.375$ bits and $B_k = 1.25$ bits per antenna when $N_{\text{bs}} = 4$.

6.1.6 Summary

In this section, uplink centralized joint detection was studied, where multiple BSs forward partially processed receive signals to a CCU or another BS for a joint and centralized decoding of multiple terminals.

Different joint detection concepts were first introduced under the assumption of infinite backhaul capacity between cooperative base stations, and compared in terms of performance and complexity. Then, constrained backhaul links were considered and schemes introduced where local preprocessing and compression are performed before signals are forwarded over the backhaul. It was shown that the impact of compression distortion on the performance depends strongly on user locations and pairing. Furthermore, results confirmed that it is beneficial to let some terminals be decoded locally (and the data bits potentially forwarded to the CCU for interference cancelation), while others are decoded by the CCU itself, which can be seen as a hybrid of a decentralized and centralized cooperation, as discussed in Section 4.3.1.

Finally, the provisions for uplink centralized joint processing in WiMax and LTE-A standardization were discussed.

6.2 Uplink Decentralized Joint Detection

Xinning Wei and Tobias Weber

This section provides a practical decentralized uplink CoMP scheme for inter-cell interference cancelation. As opposed to the schemes introduced in Section 6.1, this scheme does not require a CoMP central unit (CCU), but performs multi-cell cooperative signal processing simultaneously and iteratively at the coordinated base stations (BSs) which exchange information via the backhaul.

The proposed decentralized CoMP scheme is expected to be a practical candidate in the uplink physical layer design for future LTE-A systems applying orthogonal frequency division multiplex (OFDM) and multiple-input multiple-output (MIMO) techniques. Concerning the signal processing implementation architecture, recently more and more attention has been paid to decentralized CoMP schemes [MF07b, KRF08, PHG09, MF09b, WWWS09, dCS09] as opposed to centralized schemes, due to the following advantages:

- From the network operator's point of view, the change to the current architecture of cellular networks is expected to be as little as possible. Based on the existing networks, the decentralized scheme requires only some extra backhaul links between the neighboring BSs.
- The joint signal processing for all the user equipments (UEs) can be distributed into parallel cooperative signal processing among individual UEs at the coordinated BSs. The computational load is shared by the coordinated BSs to avoid a central unit of possibly large computational complexity.
- The decentralized CoMP scheme in this section can make full use of the iterative zero-forcing (ZF) joint detection (JD) algorithm. With a reduced computational cost, the system performance of the proposed decentralized iterative ZF scheme can asymptotically approach that of its centralized counterpart.
- In contrast to the centralized scheme with a fixed central unit responsible for a static cooperative cluster, the decentralized CoMP scheme in this section is more flexible to efficiently make full use of the dynamically selected UE-oriented significant channel state information (CSI). Although the significant channels for a UE are not necessarily limited in a structure-oriented geographical area, in most realistic scenarios they exist in this UE's own cell and adjacent cells. Since only local significant CSI is considered at each BS in the decentralized scheme, only backhaul links connecting adjacent BSs and only the synchronization in the local area are required.

The proposed decentralized CoMP scheme is investigated in a 3-cell reference scenario. This small scenario can be considered as one cooperative cluster in the whole system, for example obtained through static or dynamic clustering techniques explained in Chapter 7. Perfect channel knowledge is assumed in the present section, while channel estimation for obtaining CSI in practical systems will be discussed in Chapter 9.

6.2.1 Practical Decentralized Interference Cancelation Scheme

The decentralized CoMP scheme should make a good compromise between system performance and implementation complexity. On one side, in order to reduce the computational load and backhaul requirements, we consider only part of the full CSI in the cooperative signal processing in each cooperative cluster. On the other side, in order to maintain a good system performance, the considered CSI is that of the UE-oriented significant channels which play a significant role in the system performance of each UE. Based on the UE-oriented significant CSI, a practical JD algorithm suitable for interference cancelation is implemented in a decentralized way. In the following, the decentralized CoMP scheme is discussed from the aspects of significant channel selection, the signal processing algorithm and the implementation architecture.

Significant Channel Selection
Without loss of generality, the significant channel selection is performed in a cooperative cluster of M cells. For example, $M = 3$ is considered in this section. Assuming OFDM is used as motivated in Chapter 3, we investigate a single subcarrier with one active UE in each cell at each time slot. We consider $M = K$ BSs with N_{bs} antennas each and K single-antenna UEs. Altogether, $N_{BS} = MN_{bs}$ antennas with indices $a = 1, \ldots, N_{BS}$ at the BS side and K antennas with indices $k = 1, \ldots, K$ at the UE side are considered. The uplink channel matrix $\mathbf{H}$ in a cooperative cluster is of dimensions $N_{BS} \times K$.

The signal processing algorithm which has an influence on the system performance shall be taken into account in the significant channel selection. In the iterative ZF JD algorithm which will be described in this section in detail, firstly the matched filtering estimate for each UE is computed considering the significant useful channels, and then the interfering signals corresponding to the significant interfering channels are iteratively removed from this estimate. According to the role of the physical channels in the data transmission, two types of significant channels for every UE are distinguished from each other as follows:

- **Significant useful channels** for a certain UE in the uplink are the channels over which we get significant useful contributions when we estimate the data symbols transmitted from this UE.
- **Significant interfering channels** for a certain UE in the uplink are the channels over which we get significant interfering signals from other UEs when we receive the data symbols from this UE.

Correspondingly, we will use a significant useful channel indicator matrix $\widetilde{\mathbf{H}}_U$ and individual significant interfering channel indicator matrices $\widetilde{\mathbf{H}}_I^{(k)}$ to indicate significant useful channels and significant interfering channels, respectively. In the above indicator matrices, "1"s are assigned into the positions corresponding to the selected significant channels, while "0"s are assigned into the positions corresponding to the insignificant channels.

The useful channels for each UE k are the channels between this UE and all the BSs. The selection of significant useful channels for each UE k is performed based on the k-th column vector of the channel matrix $\mathbf{H}$. Obviously, a single matrix $\widetilde{\mathbf{H}}_U$ of dimensions $N_{BS} \times K$ is sufficient to represent all the significant useful channels for all the UEs.

The interfering channels for each UE k are the channels between other UEs and all the BSs. A significant interfering channel for one UE could be considered as an insignificant interfering channel for another UE. Therefore, it is reasonable to separately represent the significant interfering channels for individual UEs k by individual UE-specific significant interfering channel indicator matrices $\widetilde{\mathbf{H}}_I^{(k)}$. Furthermore, for each UE k there are two kinds of channels irrelevant to the interference considered in the proposed decentralized CoMP scheme, and they are indicated as "don't care" elements in each $\widetilde{\mathbf{H}}_I^{(k)}$. Firstly, corresponding to the useful channels for UE k, the elements in the k-th column of $\widetilde{\mathbf{H}}_I^{(k)}$ are certainly "don't care" elements. Secondly, the elements in the rows of $\widetilde{\mathbf{H}}_I^{(k)}$ corresponding to the insignificant useful channels for this UE k are also "don't care" elements. The reason is that in the proposed JD algorithm the received signals at the BS antennas corresponding to the insignificant useful channels for each UE k will not be used in the data estimation of this UE.

Taking the channel group including all channels between all antennas of one BS and one UE as a selection unit, a practical significant channel selection scheme according to the following mathematical criteria is proposed. For each UE k, firstly we select its significant useful channels. Let $\mathcal{A}_m$ denote the set of indices a of the antennas belonging to BS m, $m = 1, \ldots, M$. The channel from UE k to BS antenna a, characterized by the coefficient h_k^a, is selected as a significant useful channel if the channel group gain $\sum_{\forall a \in \mathcal{A}_m} |h_k^a|^2$ covers a significant portion of the sum of all useful channel gains for this UE $\sum_m \sum_{\forall a \in \mathcal{A}_m} |h_k^a|^2$. Then, we select the significant interfering channels for each UE k based on the channel coefficients excluding the "don't care" elements. Let $\mathcal{B}_k$ denote the set of indices of the BSs corresponding to the selected significant useful channel groups for UE k. For each UE k, if a channel with the channel coefficient $h_{k'}^a$, $k' \neq k$, is selected as a significant interfering channel, it has to fulfill the following condition. Namely, the channel group weighting factor magnitude $\sum_{\forall a \in \mathcal{A}_m} |(h_k^a)^H h_{k'}^a|$ corresponding to the scaling of the interference in the matched filtering estimate covers a significant portion of the sum of the channel group weighting factor magnitudes $\sum_{k' \neq k} \sum_{m \in \mathcal{B}_k} \sum_{\forall a \in \mathcal{A}_m} |(h_k^a)^H h_{k'}^a|$ for all the interferences to UE k. In fact, we implicitly select the relevant significant interfering channels based on the selected significant useful channels. As can be seen from the above mathematical criteria, the selection of $h_{k'}^a$ depends on the considered useful channel coefficient h_k^a.

Two UEs have compatible significant interfering channels if all the significant interfering channels selected for one UE are never considered as insignificant interfering channels for the other UE. If all the UEs have compatible significant interfering channels, all the individual UE-specific significant interfering channel

6.2 Uplink Decentralized Joint Detection

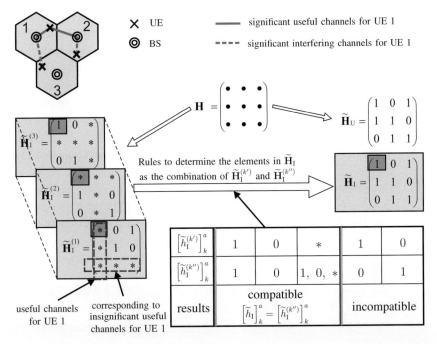

- • : channel coefficients   1 : significant channels   0 : insignificant channels   * : don't care elements

Figure 6.5 Example for significant channel selection and indicator matrix formalism.

indicator matrices $\widetilde{\mathbf{H}}_I^{(k)}$ can be represented by one combined significant interfering channel indicator matrix $\widetilde{\mathbf{H}}_I$. Details of the rules for this combination can be found in Fig. 6.5, in which the proposed significant channel selection scheme and the corresponding significant channel indicator matrix formalism are visualized. For the sake of simplicity, a 3-cell system with one antenna at each BS is used as an exemplary scenario here. Assuming that the relation between the channel magnitudes strongly depends on the corresponding distances between the UEs and the BSs, $\widetilde{\mathbf{H}}_U$ and $\widetilde{\mathbf{H}}_I^{(k)}$, $k = 1, 2, 3$, can be easily obtained. Two kinds of "don't care" elements denoted by "*" in $\widetilde{\mathbf{H}}_I^{(1)}$ are pointed out. Furthermore, the element-wise combination of individual matrices $\widetilde{\mathbf{H}}_I^{(k)}$ to obtain $\widetilde{\mathbf{H}}_I$ is performed. For example, the $(1,1)$-th entries of $\widetilde{\mathbf{H}}_I^{(1)}$, $\widetilde{\mathbf{H}}_I^{(2)}$ and $\widetilde{\mathbf{H}}_I^{(3)}$ are "*", "*", and "1", respectively. Following the combination rules in Fig. 6.5 considering the "don't care" elements, the $(1,1)$-th entry of $\widetilde{\mathbf{H}}_I$ is "1".

In practice, significant channel selection has to be possible at moderate complexity. Therefore, the roughly estimated channel magnitudes instead of the accurate channel coefficients could be used in practice. Further, it is not necessary to perform the significant channel selection at every signal processing time slot. In a certain time interval depending on the characteristics of the time-variant mobile radio channels, the cooperative signal processing could be performed based on the same significant channel selection results.

Iterative ZF JD Algorithm with Significant CSI

We now discuss the signal processing algorithm in the uplink decentralized CoMP scheme. In cellular systems with strong noise, the minimum mean square error (MMSE) algorithm or the matched filter (MF) algorithm could be applied. In the interference-limited cellular systems as considered in this section, the linear multiuser ZF algorithm which can eliminate all the inter-cell interference is a good choice to achieve a good system performance. However, a pseudo-inversion of the channel matrix typically required in the linear ZF algorithm can cause great computation complexity especially in cellular systems with a large number of cells. Meanwhile, the linear ZF algorithm is not suitable for a decentralized implementation. Following the Jacobi method in linear algebra [GVL96] which solves the linear matrix equation requiring matrix inversion in an iterative way, the parallel interference cancelation (PIC) algorithm has been proposed [DSR98, Ver98]. Generally, in a realistic large cellular system, the PIC algorithm has a lower complexity but a similar performance as compared to the conventional linear ZF algorithm. Following the idea of the PIC algorithm, an iterative ZF JD algorithm focusing on interference cancelation with significant CSI instead of full CSI is proposed in this section. The reasons for choosing this algorithm in the practical uplink CoMP scheme are briefly summarized:

- As one practical multiuser detection strategy, the parallel iterative ZF algorithm has a moderate computational complexity.
- This algorithm follows the principle of interference cancelation, and therefore has a fairly good performance in realistic interference-limited cellular systems.
- The parallel signal processing with UE-oriented significant CSI is suitable for a decentralized implementation based on the coordinated BSs.

The significant CSI considered in the signal processing algorithm is taken from the channel coefficients included in the full channel matrix $\mathbf{H}$ according to the significant channel selection results. Corresponding to the significant channel indicator matrices $\widetilde{\mathbf{H}}_U$, $\widetilde{\mathbf{H}}_I^{(k)}$ and $\widetilde{\mathbf{H}}_I$, the significant CSI is described by the significant useful channel matrix

$$\mathbf{H}_U = \mathbf{H} \odot \widetilde{\mathbf{H}}_U, \tag{6.19}$$

the UE-specific significant interfering channel matrices

$$\mathbf{H}_I^{(k)} = \mathbf{H} \odot \widetilde{\mathbf{H}}_I^{(k)}, \tag{6.20}$$

and the combined significant interfering channel matrix

$$\mathbf{H}_I = \mathbf{H} \odot \widetilde{\mathbf{H}}_I, \tag{6.21}$$

where the operator $\odot$ denotes the element-wise multiplication of two matrices. The above significant channel matrices contain channel coefficients of the significant channels, "0"s corresponding to the insignificant channels, and the "don't care" elements.

6.2 Uplink Decentralized Joint Detection

For single-antenna transmitters ($N_{ue} = 1$), the transmitted vector $\mathbf{s}$ as defined in Section 3.4 is equivalent to the data vector $\mathbf{x}$, and hence we obtain from (3.1)

$$\mathbf{y} = \mathbf{H}\mathbf{s} + \mathbf{n} = \mathbf{H}\mathbf{x} + \mathbf{n}. \tag{6.22}$$

Applying the iterative ZF JD algorithm with full CSI, i.e., the PIC algorithm, the estimated data vector $\hat{\mathbf{x}}(i)$ in the i-th iteration can be described by [Ver98]

$$\hat{\mathbf{x}}(i) = \left(\text{diag}\left(\mathbf{H}^H\mathbf{H}\right)\right)^{-1}\left(\mathbf{H}^H\mathbf{y} - \overline{\text{diag}}\left(\mathbf{H}^H\mathbf{H}\right)\hat{\mathbf{x}}(i-1)\right), \tag{6.23}$$

where $\overline{\text{diag}}(\cdot)$ sets all the elements on the diagonal of its argument to zero. In the end of every iteration i, the estimated data vector $\hat{\mathbf{x}}(i)$ could be forwarded to a data estimate refiner applying hard quantization or soft quantization techniques to obtain the refined estimated data vector $\hat{\hat{\mathbf{x}}}(i)$. Now we will apply the significant CSI described by $\mathbf{H}_U$ given in (6.19) and by $\mathbf{H}_I^{(k)}$ given in (6.20) instead of full CSI into the PIC algorithm. According to the functionalities of the channels in data transmission, firstly the significant useful channel coefficients in $\mathbf{H}_U$ will be considered in the matched filtering part. Then, these significant useful channel coefficients and their corresponding significant interfering channel coefficients in $\mathbf{H}_I^{(k)}$ will be considered in the iterative interference cancelation part. In this way, the proposed iterative ZF JD algorithm with significant CSI can be derived as

$$\hat{\mathbf{x}}(i) = \mathbf{\Lambda}^{-1}\left(\mathbf{H}_U^H\mathbf{y} - \overline{\text{diag}}\left(\mathbf{\Phi}_H\right)\hat{\mathbf{x}}(i-1)\right). \tag{6.24}$$

In the above equation, the channel gain scaling matrix $\mathbf{\Lambda}$ is defined as

$$\mathbf{\Lambda} = \text{diag}\left(\mathbf{H}_U^H\mathbf{H}_U\right), \tag{6.25}$$

and the channel correlation matrix $\mathbf{\Phi}_H$ is defined as

$$\mathbf{\Phi}_H = \begin{pmatrix} [\mathbf{H}_U]_1^H \mathbf{H}_I^{(1)} \\ \vdots \\ [\mathbf{H}_U]_K^H \mathbf{H}_I^{(K)} \end{pmatrix}, \tag{6.26}$$

where the matrix operator $[\,]_k$ returns the k-th column vector of its argument. Knowing that a suitable data estimate refinement can further improve the system performance, in this section the iterative algorithm applying the transparent data estimate refinement, i.e., $\hat{\hat{\mathbf{x}}}(i) = \hat{\mathbf{x}}(i)$, is considered. Without loss of generality, it can be treated as a benchmark for this kind of iterative algorithms with different data estimate refinement techniques.

In the case that the linear iterative JD algorithm with significant CSI described by (6.24) converges with $\hat{\mathbf{x}}(i) = \hat{\mathbf{x}}(i-1)$ and the matrix $(\mathbf{\Lambda} + \overline{\text{diag}}(\mathbf{\Phi}_H))$ has full rank, the limiting value of $\hat{\mathbf{x}}(i)$ can be easily calculated from (6.24) as

$$\hat{\mathbf{x}}(\infty) = \left(\mathbf{\Lambda} + \overline{\text{diag}}(\mathbf{\Phi}_H)\right)^{-1}\mathbf{H}_U^H\mathbf{y}. \tag{6.27}$$

Under special conditions, this equation can be simplified step by step in different cases as shown in the following:

- In the case that all the individual UE-specific significant interfering channel matrices $\mathbf{H}_I^{(k)}$ can be combined to one matrix $\mathbf{H}_I$, one obtains

$$\boldsymbol{\Phi}_H = \mathbf{H}_U^H \mathbf{H}_I, \tag{6.28}$$

and the limiting value of this iterative JD algorithm is described by

$$\hat{\mathbf{x}}(\infty) = \left(\boldsymbol{\Lambda} + \overline{\mathrm{diag}\left(\mathbf{H}_U^H \mathbf{H}_I\right)}\right)^{-1} \mathbf{H}_U^H \mathbf{y}. \tag{6.29}$$

This happens when UEs have compatible significant interfering channels.
- In the above case, if the significant interfering channel matrix $\mathbf{H}_I$ covers all the non-zero elements of the significant useful channel matrix $\mathbf{H}_U$, one obtains

$$\boldsymbol{\Lambda} = \mathrm{diag}\left(\mathbf{H}_U^H \mathbf{H}_U\right) = \mathrm{diag}\left(\mathbf{H}_U^H \mathbf{H}_I\right), \tag{6.30}$$

and the limiting value of this iterative JD algorithm can be represented by

$$\hat{\mathbf{x}}(\infty) = \left(\mathbf{H}_U^H \mathbf{H}_I\right)^{-1} \mathbf{H}_U^H \mathbf{y}. \tag{6.31}$$

This case happens when the significant interfering channels for every UE are the significant useful channels of other UEs.
- One special case is the iterative JD algorithm with full CSI, for which

$$\mathbf{H}_U = \mathbf{H}_I = \mathbf{H} \tag{6.32}$$

holds, and the limiting value of this iterative algorithm is

$$\hat{\mathbf{x}}(\infty) = \left(\mathbf{H}^H \mathbf{H}\right)^{-1} \mathbf{H}^H \mathbf{y}. \tag{6.33}$$

This case simply corresponds to the full ZF solution.

Decentralized Signal Processing Scheme

Let us now focus on how the above stated computations can be performed in a decentralized way. It is worth noting that for each UE k, the significant *useful* channel coefficients $[h_U]_k^a$ from $\mathbf{H}_U$ are considered in the part of matched filtering. In the part of iterative interference cancelation, at the BSs corresponding to the significant useful channels, the significant *interfering* channel coefficients $[h_I^{(k)}]_{k'}^a$ from $\mathbf{H}_I^{(k)}$ are considered for every UE k.

Step 1: Initialization for the iterative signal processing of UEs $k = 1, \ldots, K$, at their corresponding BSs $m = k$ in parallel:
  (a) Assign the channel gain scaling factor

$$\lambda_k = \sum_a ([h_U]_k^a)^H [h_U]_k^a = \sum_a |[h_U]_k^a|^2 \tag{6.34}$$

for every UE k at BS $m = k$. λ_k can be previously estimated in practice.
  (b) Assign the initial value of the estimated data symbol as $\hat{x}_k(0) = 0$ for every UE k at BS $m = k$.

Step 2: Matched filtering for every UE k, $k=1,\ldots,K$, in parallel:

6.2 Uplink Decentralized Joint Detection

(a) Compute the matched filtering estimate components

$$\tilde{r}_k^a = ([h_U]_k^a)^H y_a \qquad (6.35)$$

for every UE k at all the BSs corresponding to its significant useful channels.

(b) Collect $\tilde{r}_k^a$ from the coordinated BSs through the backhaul links, and sum them up at BS $m = k$ to obtain the matched filtering estimate for each UE k as

$$r_k = \sum_a \tilde{r}_k^a = \sum_a ([h_U]_k^a)^H y_a. \qquad (6.36)$$

Step 3: Iterative interference cancelation for every UE k, $k = 1, \ldots, K$, in parallel: The following steps are performed iteratively in iterations i, $i = 1, \ldots, L$. The required number L of iterations can be previously assigned.

(a) Compute reconstructed interfering signals

$$\left[f^{(k)}\right]_{k'}^a = ([h_U]_k^a)^H \left[h_I^{(k)}\right]_{k'}^a \hat{x}_{k'}(i-1) \qquad (6.37)$$

from different UEs $k' \neq k$ to UE k at the BSs with antenna indices a corresponding to the significant interfering channels for UE k.

(b) Collect the reconstructed interfering signals $\left[f^{(k)}\right]_{k'}^a$ from coordinated BSs over the backhaul, and subtract them from r_k at BS $m = k$ to obtain the estimated data symbol for every UE k as

$$\hat{x}_k(i) = \frac{1}{\lambda_k} \left(r_k - \sum_{k' \neq k} \sum_a \left[f^{(k)}\right]_{k'}^a \right)$$

$$= \frac{1}{\lambda_k} \left(r_k - \sum_{k' \neq k} \sum_a ([h_U]_k^a)^H \left[h_I^{(k)}\right]_{k'}^a \hat{x}_{k'}(i-1) \right). \qquad (6.38)$$

(c) Forward the estimated data symbols $\hat{x}_k(i)$ which are required at other BSs in the next iteration to compute the reconstructed interfering signals in Step 3(a).

For the sake of simplicity, the 3-cell cellular system with a single antenna at each BS in Fig. 6.5 is taken as the exemplary scenario to visualize the implementation of the decentralized JD with significant CSI in Fig. 6.6. Significant CSI according to the significant channel selection results in Fig. 6.5 is applied in the decentralized signal processing in Fig. 6.6. For example, two significant useful channels for UE 1 corresponding to its neighboring BS 1 and BS 2 are considered to obtain the matched filtering data estimate r_1. Only one significant interfering channel at each involved BS is considered for interference reconstruction and cancelation for each UE.

Additionally, the backhaul communication steps between coordinated BSs in the above exemplary 3-cell scenario applying the proposed decentralized JD are demonstrated in Fig. 6.7. For the computation of data estimates $\hat{x}_k(i)$ of UEs k,

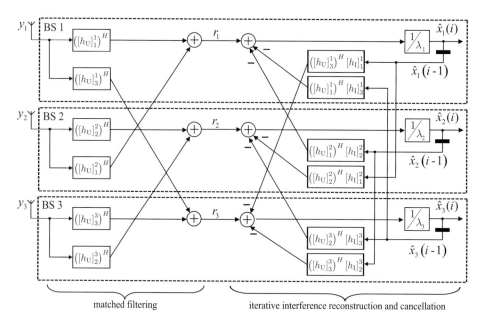

Figure 6.6 Decentralized signal processing with significant CSI in a 3-cell system.

$k=1,\ldots,K$, at their corresponding BSs $m=k$, three kinds of information have to be exchanged between the coordinated BSs as follows:

- "$u_{m,l}$" denotes the matched filtering data estimate which has to be forwarded from BS l to BS m.
- "$z_{m,l}$" denotes the preliminary estimated data symbol in the previous iteration which has to be forwarded from BS l to BS m.
- "$v_{m,l}$" denotes the weighted interfering signal which has to be forwarded from BS l to BS m.

In the CoMP scheme applying JD with full CSI, for each UE k all $u_{m,l}$, $v_{m,l}$ and $z_{m,l}$ from BSs $l=1,\ldots,k-1,k+1,\ldots,K$ have be to forwarded to BS $m=k$. In the proposed CoMP scheme applying JD with significant CSI, the backhaul communication load can be significantly reduced. This advantage is shown in Fig. 6.7, where a few intermediate results are exchanged between BSs.

In this section, the backhaul load is evaluated in terms of the number N_{BL} of intermediate signal processing results which have to be exchanged between the BSs to estimate one data symbol of each UE. It is found that even with the same number of significant channels, the backhaul load varies with different choices of significant channels. Applying the above proposed JD algorithm considering N_{U} significant useful channel groups and N_{I} significant interfering channel groups for each UE in a K-cell cellular system, the lower bound and the upper bound

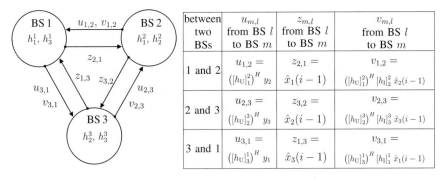

Figure 6.7 Backhaul communication steps for JD with partial CSI in a 3-cell system.

of N_{BL} are derived as

$$N_{\text{BL}}^{\text{lower}} = \underbrace{(N_{\text{U}} - 1)}_{\text{matched filtering}} + \underbrace{N_{\text{I}} \cdot L}_{\text{interference cancellation}} \quad \text{and} \quad (6.39)$$

$$N_{\text{BL}}^{\text{upper}} = \underbrace{(N_{\text{U}} - 1)}_{\text{matched filtering}} + \underbrace{\begin{cases} 2 \cdot N_{\text{I}} \cdot L & \text{for } 1 \leq N_{\text{I}} \leq (K-1) \\ (N_{\text{I}} + K - 1) \cdot L & \text{for } (K-1) < N_{\text{I}} \leq (K-1)^2 \\ (K-1) \cdot K \cdot L & \text{for } (K-1)^2 < N_{\text{I}} \leq (K-1) \cdot K \end{cases}}_{\text{interference cancellation}}.$$

Additional backhaul communication between cooperative BSs is considered as one major drawback of the proposed CoMP scheme. From a signal processing point of view, the backhaul traffic could be greatly reduced by applying significant CSI instead of full CSI in the JD scheme shown above. Furthermore, various approaches could be applied to efficiently exchange the information between BSs in practical systems. For example, the likelihood information on the transmitted bits could be calculated in the above JD scheme and exchanged between BSs to reduce the backhaul traffic [KF08, AEH08]. In [MF08a, MF09b, dCS09], the quantized received signals are first compressed via Wyner-Ziv source coding, exploiting the correlation between received signals [WZ76], and then exchanged between BSs. Similarly, the matched filtering data estimate $u_{m,l}$ and the weighted interfering signal $v_{m,l}$, which have to be exchanged between the BSs, could also be quantized or compressed. In [MF08d], a CoMP scheme based on the exchange of distributively decoded messages is proposed, and Slepian-Wolf source coding [SW73a] is applied for the compression of the decoded bits. In [SSPS09c, GMFC09], a CoMP scheme based on the exchange of quantized transmit sequences derived from the decoded messages is proposed, and Wyner-Ziv source coding is used for the compression of the quantized sequences. Similarly, the preliminary estimated data symbol $z_{m,l}$, the decoded bits or the transmit sequences in above JD scheme could be appropriately compressed via source coding and then exchanged between BSs. Backhaul issues will be discussed in detail in Chapter 12.

6.2.2 Performance Assessment

Analytical Calculations

The system performance of the uplink decentralized CoMP scheme can be investigated based on the limiting value of the iterative ZF JD algorithm with significant CSI. Considering the transparent data estimate refinement, the limiting value of the estimated data vector described by (6.27) can be rewritten as

$$\hat{\mathbf{x}} = \underbrace{\text{diag}\left(\left(\mathbf{\Lambda} + \overline{\text{diag}\left(\mathbf{\Phi}_\text{H}\right)}\right)^{-1} \mathbf{H}_\text{U}^H \mathbf{H}\right) \mathbf{x}}_{\text{useful contribution}} + \underbrace{\overline{\text{diag}\left(\left(\mathbf{\Lambda} + \overline{\text{diag}\left(\mathbf{\Phi}_\text{H}\right)}\right)^{-1} \mathbf{H}_\text{U}^H \mathbf{H}\right)} \mathbf{x}}_{\text{interference}}$$

$$+ \underbrace{\left(\mathbf{\Lambda} + \overline{\text{diag}\left(\mathbf{\Phi}_\text{H}\right)}\right)^{-1} \mathbf{H}_\text{U}^H \mathbf{n}}_{\text{noise}}. \qquad (6.40)$$

Based on the above limiting value, the signal-to-interference-and-noise ratio (SINR) can be calculated as an important performance metric. Furthermore, the system performance in every channel snapshot can be analytically assessed in terms of the bit error rate (BER) and the capacity based on the SINR. It is assumed that the applied modulation scheme can ensure the same average transmitted power P per data symbol, and the zero-mean Gaussian noise $\mathbf{n} \in \mathbb{C}^{[N_\text{BS} \times 1]}$ has the covariance $E\left\{\mathbf{n}\mathbf{n}^H\right\} = \sigma^2 \mathbf{I}$. The data vector $\mathbf{x}$ and the noise vector $\mathbf{n}$ are statistically independent. For one channel snapshot, the SINR of the data estimate $\hat{x}_k$ for UE k can be written as

$$\gamma^{(k)} = \frac{S^{(k)}}{N^{(k)} + I^{(k)}}, \quad \text{where} \qquad (6.41)$$

$$S^{(k)} = P\left[\text{diag}\left(\left(\mathbf{\Lambda} + \overline{\text{diag}\left(\mathbf{\Phi}_\text{H}\right)}\right)^{-1} \mathbf{H}_\text{U}^H \mathbf{H}\right) \text{diag}\left(\mathbf{H}^H \mathbf{H}_\text{U} \left(\mathbf{\Lambda} + \overline{\text{diag}\left(\mathbf{\Phi}_\text{H}^H\right)}\right)^{-1}\right)\right]_{k,k},$$

$$I^{(k)} = P\left[\overline{\text{diag}\left(\left(\mathbf{\Lambda} + \overline{\text{diag}\left(\mathbf{\Phi}_\text{H}\right)}\right)^{-1} \mathbf{H}_\text{U}^H \mathbf{H}\right)}\, \overline{\text{diag}\left(\mathbf{H}^H \mathbf{H}_\text{U} \left(\mathbf{\Lambda} + \overline{\text{diag}\left(\mathbf{\Phi}_\text{H}^H\right)}\right)^{-1}\right)}\right]_{k,k},$$

and $N^{(k)} = \sigma^2 \left[\left(\mathbf{\Lambda} + \overline{\text{diag}\left(\mathbf{\Phi}_\text{H}\right)}\right)^{-1} \mathbf{H}_\text{U}^H \mathbf{H}_\text{U} \left(\mathbf{\Lambda} + \overline{\text{diag}\left(\mathbf{\Phi}_\text{H}^H\right)}\right)^{-1}\right]_{k,k} \qquad (6.42)$

can be calculated from (6.40). Since only partial CSI instead of full CSI is applied in the cooperative signal processing, the data estimates may contain slightly rotated and scaled useful contributions. However, such a rotation or scaling can be easily estimated and compensated at the receiver.

Numerical Simulation Results

In the following, the system performance of the proposed CoMP scheme is assessed with respect to numerical simulation results in Figs. 6.8 and 6.9. A small cellular system including 3 cells with a frequency reuse factor of 1 as shown in Fig. 6.5 is taken as the reference scenario. Some key pre-assumptions for the simulations are listed as follows:

6.2 Uplink Decentralized Joint Detection

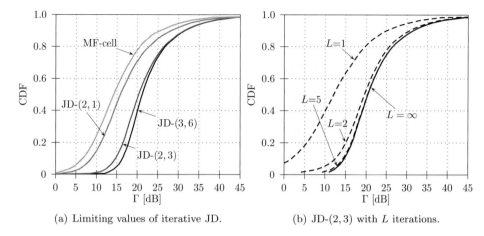

(a) Limiting values of iterative JD.   (b) JD-(2,3) with L iterations.

Figure 6.8 CDF of the output SINR in the uplink with $\Upsilon_N = 20\,\text{dB}$.

- Rayleigh fading and a pathloss with attenuation exponent $\alpha = 3$ with respect to the distance are considered.
- Applying OFDM, in every cell one BS with 3 antennas and one active UE with a single antenna in every time slot and sub-carrier are considered. The UE is randomly and uniformly distributed in its cell.
- The data vector $\mathbf{x} \in \mathbb{C}^{[K \times 1]}$ including independently and identically distributed (i.i.d.) zero-mean Gaussian elements has the covariance $E\{\mathbf{xx}^H\} = P \cdot \mathbf{I}$ with $P = 1$. The noise vector $\mathbf{n} \in \mathbb{C}^{[N_{\text{BS}} \times 1]}$ including i.i.d. zero-mean Gaussian elements has the covariance $E\{\mathbf{nn}^H\} = \sigma^2 \cdot \mathbf{I}$.
- The proposed JD scheme considering N_U significant useful and N_I significant interfering channel groups for every UE is denoted by "JD-(N_U, N_I)".
- The parameter Υ_N is used to indicate the ratio of the average receive power to the noise power in the reference scenario. For a UE at any fixed position, its average matched filtering receive power considering all BS antennas in the 3-cell system can be calculated as a constant parameter. With the unit transmit power per data symbol $P = 1$, the average receive power for this UE is $P_B = E\{\sum_a h_k^a (h_k^a)^H\}$. The noise condition in the reference scenario can be characterized by the parameter $\Upsilon_N(\text{dB}) = 10\log_{10}(P_B/\sigma^2)$ calculated for the UE located at the intersection point of the 3 cells. In both Figs. 6.8 and 6.9, $\Upsilon_N = 20\,\text{dB}$ for the interference-limited scenario is considered.

In Fig. 6.8(a), the system performance of different cooperation and detection schemes considering different significant channels is compared in terms of the cumulative distribution function (CDF) of the output SINRs. Here the system performance of the proposed decentralized CoMP scheme is investigated based on the limiting values of the iterative JD algorithm. It is shown that the conventional intra-cell matched filtering scheme denoted by "MF-cell" gives the worst performance among all the investigated schemes. The reason is that this scheme

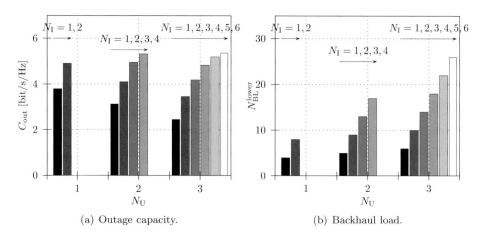

(a) Outage capacity. (b) Backhaul load.

Figure 6.9 Outage capacity and backhaul load vs. numbers of significant channel groups considered in JD with $L = 4$ iterations, $p_{\text{out}} = 0.1$, $\Upsilon_{\text{N}} = 20\,\text{dB}$, $M = K = 2$.

considers only intra-cell useful channels but no inter-cell interfering channel in the signal processing for each UE. The system performance is strongly limited by the inter-cell interference. Applying the proposed decentralized CoMP scheme considering significant CSI, the CDF curve of the SINRs for "JD-$(2,1)$" and that for "JD-$(2,3)$" are plotted. Obviously, the proposed decentralized CoMP scheme considering a few appropriately selected UE-oriented significant channels can strongly improve the system performance as compared to the conventional intra-cell matched filtering scheme. The communication scheme denoted by "JD-$(3,6)$" is nothing else but the decentralized CoMP scheme applying the iterative ZF JD algorithm with full CSI considering all the useful and interfering channels for every UE. All the inter-cell interference is eliminated, and the system performance is only limited by the Gaussian noise.

The SINR performance of the proposed decentralized CoMP scheme considering different numbers of iterations in the iterative JD algorithm is investigated in Fig. 6.8(b). It is shown that after only a few iterations, the SINR performance of the iterative JD converges to that of the corresponding JD with an unlimited number of iterations. The convergence behavior can be well retained even if significant CSI instead of full CSI is considered in JD.

In Fig. 6.9, the influence of different amounts of considered significant CSI on the system performance and on the backhaul load of the proposed decentralized CoMP scheme is investigated. According to the proposed significant channel selection scheme, for each number of significant useful channel groups N_{U}, the number of significant interfering channel groups N_{I} could range from 1 to $(K-1)\,N_{\text{U}}$. In Fig. 6.9, $(K-1)\,N_{\text{U}}$ bars are plotted for each number N_{U} corresponding to all possible numbers of significant useful and interfering channels considered in "JD-$(N_{\text{U}}, N_{\text{I}})$". In Fig. 6.9(a), the outage capacity of UEs in the 3 cells is plotted. In Fig. 6.9(b), the lower bound of the backhaul load as described by (6.39) is plotted. Since i.i.d. Gaussian data symbols and i.i.d. noise signals are considered, it is reasonable to assume that the remaining interfering signals and the noise signals after the linear JD are uncorrelated and Gaussian distributed. With $\gamma^{(k)}$ indicating the SINR for UE k, the corresponding instantaneous capacity is calculated as

$$C_{\text{int}}^{(k)} = \log_2\left(1 + \gamma^{(k)}\right). \tag{6.43}$$

The outage capacity C_{out} is defined w.r.t. its outage probability p_{out} as

$$p_{\text{out}} = \text{Prob}\left\{C_{\text{int}}^{(k)} < C_{\text{out}}\right\}, \quad k = 1, 2, 3, \tag{6.44}$$

where p_{out} describes the probability that the instantaneous capacity $C_{\text{int}}^{(k)}$ of one UE is smaller than the outage capacity C_{out}. For every pair of $(N_{\text{U}}, N_{\text{I}})$, an outage capacity C_{out} can be calculated based on a given p_{out}. The most important results derived from Fig. 6.9 are the following:

- Generally, the more significant channels are considered in the decentralized CoMP scheme, the better system performance can be achieved. The more significant useful channels are considered, the more noise can be suppressed. The more interfering channels are considered in JD, the more interference can be eliminated, but the larger the noise enhancement will be.
- Interestingly, it is shown that for a given number of significant interfering channel groups N_{I}, a larger outage capacity can be achieved when considering a smaller number of significant useful channel groups N_{U}. The reason is that the more significant useful channel groups are considered, the more BSs are involved in JD, and the more interference is included in the matched filtering data estimate for each UE. The system performance of the proposed scheme is mainly limited by the remaining interfering channels which are not considered as significant interfering channels in JD. In fact, the outage capacity increases with the number of significant useful channel groups N_{U} and the ratio

$$\eta = \frac{N_{\text{I}}}{(K-1)\,N_{\text{U}}}, \tag{6.45}$$

rather than directly with the number of significant interfering channel groups N_{I}. Variable η indicates the ratio of the number of considered significant interfering channels to the total number of interfering channels for each UE.
- Considering only a few appropriately selected significant channels, a good system performance with a moderate backhaul load can be achieved by the proposed JD scheme. For example, JD with full CSI, i.e., "JD-(3, 6)", requiring at least a backhaul load of $N_{\text{BL}} = 26$ exchanged messages, and achieves an outage capacity of $C_{\text{out}} = 5.37$ bit/s/Hz, while "JD-(2, 3)" which requires only a backhaul load of $N_{\text{BL}} = 13$ can already achieve an outage capacity of $C_{\text{out}} = 4.96$ bit/s/Hz. Hence, a good compromise between backhaul load and data rate can be made by considering significant CSI in JD.

6.2.3 Summary

A practical uplink decentralized CoMP scheme has been proposed in this section. The decentralized CoMP scheme can be directly implemented at the coordinated BSs in a flexible way without requiring a central unit. Distinguishing the significant useful channels from the significant interfering channels, only the channel state information which plays a significant role in the system performance of each UE is required in joint detection. A good compromise between data rate and backhaul load can be made in the proposed decentralized CoMP scheme considering significant CSI.

6.3 Downlink Distributed CoMP Approaching Centralized Joint Transmission

Lars Thiele, Thomas Haustein, Volker Jungnickel, Wolfgang Zirwas and Federico Boccardi

In this section, we focus on the downlink and consider *centralized joint transmission*. Here, multiple base stations (BSs) cooperate in jointly transmitting precoded data symbols to multiple user equipments (UEs) such that desired signals overlap coherently and the interference is partially canceled out. First, CoMP transmission requires knowledge on the compound channel matrix, i.e. multi-cell channel state information at the transmitter (CSIT), between all UEs and BSs involved in the coherent downlink transmission, in order to obtain the spatial precoding weights. Second, we initially assume that both the multi-cell CSIT as well as the scheduled data bits to be transmitted to the UEs are distributed to all involved BSs, an assumption which will be alleviated later in Section 6.4.

Concerning coherent joint downlink transmission, information theoretical publications typically consider a centralized network architecture, where perfect synchronization, unlimited backhaul and negligible delay are assumed. All BSs in the network are grouped in a huge cluster. Their transmit antennas act as inputs of a generalized multiple-input multiple-output (MIMO) broadcast channel (BC), while antennas from multiple UEs are considered as the outputs. The BS antennas are connected via a fast backhaul link to a CoMP central unit (CCU) [BMWT00, SZ01, WMSL02, GJJV03]. Non-linear signal preprocessing, known as dirty paper coding (DPC) [Cos83], was shown to achieve the BC capacity [CS03, VT03]. By allowing full coordination among the whole network, multi-cell interference can be removed to a certain degree, which depends on the selected precoding strategy as well as on the channel knowledge. Due to additional beamforming gain (referred to in Section 3.5 as *array* gain), the system throughput can exceed the one known from an isolated cell [HV04]. Recently, maximum Eigenvalue transmission (MET) was introduced as a linear transceiver optimization method, which was shown to approach the BC capacity in an isolated cluster [BH07a]. The Eigenmode concept was shown to reduce the peak-

6.3 DL Distributed CoMP Approaching Centralized Joint Transmission

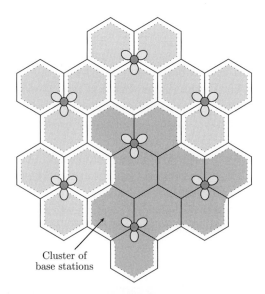

Figure 6.10 Illustration of a cluster of collaborative base stations.

to-average power ratio (PAPR) for linear precoding [JHJvH02] as well as for non-linear Tomlinson-Harashima precoding (THP) [NMK+07]. In general, by limiting each user to report its strongest eigenmodes only, feedback may be reduced.

Recent results applying generalized MIMO techniques in wireless networks show huge gains [FHK+05, KFV06]. In addition, [ZD04] proposes a common framework to study multi-user CoMP downlink transmission, considers practical signal processing issues and emphasizes the advantage of array gain, enhanced channel rank and macro diversity. In [JJT+09], the authors confirm these findings based on channel measurements from a real cellular urban-macro deployment. The work in [VHLV09] promises significant gains obtained from a simulator including realistic operational conditions valid for a WiMAX system operated in an indoor scenario.

However, higher complexity, growing data rates on the backhaul and the additional overhead remain serious challenges for the introduction of CoMP in next generation mobile networks. Note that these costs can be scaled down with the size of the cooperation cluster. In the downlink, backhaul requirements increase at least linearly with the number of BSs belonging to the cluster [HS09] (in a centralized approach). See a detailed discussion on backhaul requirements in both centralized and decentralized downlink CoMP in Section 12.2. Hence, a distributed implementation of CoMP is realistic where the serving BS cooperates with a small subset of BSs (see Fig. 6.10) in its direct vicinity [ZSK+06, MF07a, JTS+08b, NEHA08, PHG09, ZMS+09a, TWH+09].

Sections 13.3 and 13.4 present details on a real-time implementation [JTW+09] of downlink CoMP demonstrating its feasibility.

The subsequent section is organized as follows: In Subsection 6.3.1, we introduce an extended system model which covers the algorithms described in this work. Then we continue with a general description of CoMP joint transmission (JT) obtained in a fully centralized setup and determine the system capacity by use of DPC under a sum power constraint in Subsection 6.3.2. In the next steps, we introduce concepts to alleviate the major drawbacks related to joint transmission, as e.g. its higher complexity, increased backhaul and signaling overhead. Those concepts cover linear precoding techniques, a greedy user selection process and *clustering* solutions in Subsection 6.3.3. The clustering can be carried out statically or dynamically and restrict joint processing techniques to a limited number of base stations. Moreover, the cluster formation may be performed and optimized by a central entity (*network-centric*), or in a per-user way (*user-centric*). In Subsection 6.3.4, we introduce a concept for a unified channel state information (CSI) feedback framework to cope with different vendor specific types of channel feedback provided by mobile devices. This concept is well-aligned with the estimation of an *effective* channel described in Subsection 9.1. Finally, we summarize a system concept where each terminal provides channel feedback to its serving base station only; the base stations in the same cluster exchange the channel feedback and payload data in order to determine the precoding weights and perform the spatial precoding, both in a distributed manner.

6.3.1 System Model

We consider a cellular orthogonal frequency division multiplex (OFDM) downlink where a central site is surrounded by multiple tiers of sites. As in Chapter 3, we assume each site to be partitioned into three $120°$ sectors or cells, i.e. yielding a set $\mathcal{M}$ consisting of $M = |\mathcal{M}|$ sectors in total. In our notation, each sector constitutes a cell which is controlled by one BS, and frequency resources are fully reused in all M cells. In joint transmission, the data to each user is simultaneously transmitted from multiple BSs. In order to mitigate the overhead related to joint transmission techniques, BSs are grouped into C subsets or *clusters*, of which one example is shown in Fig. 6.10. $\mathcal{M}_c$ represents the set of cells included in a cluster c and $M_c = |\mathcal{M}_c|$ denotes its maximum dimension. Joint processing is only allowed between BSs belonging to the same cluster, whereas BSs belonging to different clusters are not coordinated and thus produce residual inter-cluster interference. As an extension, coordinated beamforming techniques may be used to deal with the interference between clusters, i.e. to coordinate the inter-cluster interference, as introduced in Section 5.3. Further, we assume disjoint clusters, i.e. a given BS cannot belong to more than one cluster operated at the same time/frequency resource, as for example created through clustering techniques described in Chapter 7. Let us assume that each BS is equipped with N_bs transmit antennas, and the scheduler of a cluster c has assigned a set of UEs $\mathcal{K}_c$ to the same resource in time and frequency, while $\mathcal{K}$ denotes all UEs in all clusters assigned to this resource. Assuming that UE k is served and scheduled by cluster c, i.e. $k \in \mathcal{K}_c$, the received downlink signal $\mathbf{y}_k$ at UE k is given by

$$\mathbf{y}_k = \underbrace{(\mathbf{H}_k^c)^H \mathbf{s}_k^c}_{\text{Desired signal}} + \underbrace{\sum_{j \in \{\mathcal{K}_c \setminus k\}} (\mathbf{H}_k^c)^H \mathbf{s}_j^c}_{\text{Intra-cluster interference } \zeta_k} + \underbrace{\sum_{j \in \{\mathcal{K} \setminus \mathcal{K}_c\}} (\mathbf{H}_k)^H \mathbf{s}_j + \mathbf{n}}_{\text{Inter-cluster interference } \mathbf{z}_k} \quad (6.46)$$

assuming rank-1 transmission to the scheduled users. Analog to the notation introduced in Section 3.5, $\mathbf{s}_k^c \in \mathbb{C}^{[M_c N_\text{bs} \times 1]}$ denotes the precoded signals targeted towards UE k and to be transmitted from all BS antennas connected to cluster c. We emphasize that the rank-1 transmission assumption is only made to simplify the notation: this assumption can be easily extended to the more general case of sending multiple independent streams to a subset of scheduled users. The desired data stream x_k is distorted by intra-cluster and inter-cluster interference plus noise aggregated in ζ_k and $\mathbf{z}_k$, respectively. Note that $\mathbf{H}_k^c$ spans the $M_c N_\text{bs} \times N_\text{ue}$ channel matrix between all BSs in cluster c and UE k.

6.3.2 Theoretical Limits for Static Clustering and DPC

To study the limits of the system under the assumption of a static clustering, we assume sum-rate maximized transmission within each cluster. The maximum sum-rate in the c-th cluster can be obtained under the assumption of a distributed antenna system, where a centralized unit calculates the optimal transmission strategy for all BSs and users inside the cluster. The optimal transmission strategy can be easily derived from works on the MIMO BC (see for example the seminal works [CS03], [JVG04]), and is given by the so-called DPC scheme. We define as R_c the maximum throughput in the c-th cluster with a common power constraint

$$R_c = \max_{\mathbf{\Phi}_{\text{ss},k} \succeq 0, \sum_{m=1}^{M} \sum_{k=1}^{K} E\{\text{tr}\{\mathbf{\Phi}_{\text{ss},k}^m\}\} \leq M_c P_\text{max}} \log_2 \det \left(\mathbf{I} + \sum_{k=1}^{K} (\mathbf{H}_k^c)^H \mathbf{\Phi}_{\text{ss},k}^c \mathbf{H}_k^c \right), \quad (6.47)$$

where $\mathbf{\Phi}_{\text{ss},k}^c = E\{\mathbf{s}_k^c (\mathbf{s}_k^c)^H\}$ and $\mathbf{\Phi}_{\text{ss},k}^m = E\{\mathbf{s}_k^m (\mathbf{s}_k^m)^H\}$, and P_max is the per-base station power budget. We note that the original MIMO BC capacity [CS03], [JVG04]) was calculated under a sum-power constraint. The problem of finding the sum-capacity region of a downlink system with a per-antenna power constraint was considered in [YL07], where a generalized uplink-downlink duality was established.

Simulation Results for DPC with Sum-Power Constraint

In the sequel, we consider the case where each BS is equipped with $N_\text{bs} = 2$ transmit antennas, and each UE with $N_\text{ue} = 2$ receive antennas. All $M = K = 21$

BSs are considered as a huge distributed antenna system (DAS) (i.e. spanning one huge cluster of cells), which jointly serve a set of users $\mathcal{U}$, of which $\mathcal{K} \subseteq \mathcal{U}$ are served on the same resource in time and frequency. Note, in contrast to the typical Rayleigh fading assumption, the MIMO channels in this evaluation do not have the same average signal-to-noise ratio (SNR). This is caused by the different pathloss coefficient to the different antenna arrays of the BSs in the cellular deployment. The cellular channels are generated by use of the spatial channel model extended (SCME) with a 3D antenna pattern. We are using an iterative water-filling algorithm with a sum-power constraint [JRV+05] to determine the maximum sum-rate of the system as a function of the size of the active set of users U. While in practice we would rather consider a per-BS or per-antenna power constraint than a sum-power constraint, these results should provide an overview on a well-known water-filling algorithm and its achievable sum-rates in a cellular deployment. Note that Section 6.3.4 evaluates a linear precoding scheme with per-antenna power constraint.

For the results shown in Fig. 6.11, we assume that the transmit power per physical resource block (PRB) emitted by each BS is set to $P_i = 400$ mW (equivalent to the full transmit power in LTE systems of 40 W for 20 MHz of bandwidth), $P_i = 40$ mW or $P_i = 4$ mW. As noise, we assume thermal noise given at $20°C$ and an additional receiver noise figure of 9 dB. In particular, the high transmission power of $P_i = 400$ mW yields a very high average SNR of ≈ 38 dB for each user in $\mathcal{U}$ and its specific serving cell in $\mathcal{M}$. We are aware that such high SNRs cannot be achieved in practice due to various impairments in the system hardware, such as resolution of analog to digital conversion (ADC), phase noise etc.

Fig. 6.11 shows the achievable BC capacity by use of an iterative water-filling algorithm with sum power constraint [JRV+05]. The capacity is given for 1 to 5 active users per cell and for the low to high SNR regime. The capacity increases for an increasing set size $\mathcal{U}$ of available users, hence multi-user diversity helps to improve the capacity of the BC. For the low SNR regime, i.e. $P_{\max} = 4$ mW or $P_{\max} = 40$ mW, the capacity increases by 78% when assuming 3 active users instead of 1 user per cell. In contrast, we observe a slightly reduced slope in the high SNRs regime ($P_{\max} = 400$ mW), i.e. for an equivalent gain we need to have 4 active users per cell. Peak capacities, i.e. 90%-ile per cell approach 6.7 bit/s/Hz, 14.6 bit/s/Hz and 23.5 bit/s/Hz for 3 users per cell for low to high SNR regime, respectively. As a reference, we include the BC capacity for Rayleigh fading channels and an average SNR of 38 dB per transmit antenna. It turns out that the typical Rayleigh fading assumption with equivalent average SNR overestimates the capacities by approx. 33%.

6.3 DL Distributed CoMP Approaching Centralized Joint Transmission

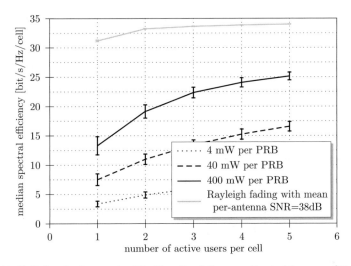

Figure 6.11 Cellular deployment with 21 cells, fully coordinated by use of iterative WF [JRV+05] and sum power constraints. Rayleigh fading CDF is given for an average per-antenna SNR of 38 dB. Error bars indicate the standard deviation of the MIMO BC distributions.

6.3.3 Practical (Linear) Precoding

The study presented in the last section for non-linear precoding can be easily extended to the case of linear precoding, in which the transmitted signal is a spatially multiplexed, linear combination of the users' data signals. Thus, we write $\mathbf{s}_k^c = \mathbf{w}_k^c x_k$, where $\mathbf{w}_k^c \in \mathbb{C}^{[M_c N_{bs} \times 1]}$ is the involved precoding vector. The achievable signal-to-interference-and-noise ratio (SINR) is estimated at each UE, according to

$$\text{SINR}_k = \frac{\left|\mathbf{g}_k^H \left(\mathbf{H}_k^c\right)^H \mathbf{w}_k^c\right|^2}{\sum_{j \in \{\mathcal{K}_c \setminus k\}} \left|\mathbf{g}_k^H \left(\mathbf{H}_k^c\right)^H \mathbf{w}_j^c\right|^2 + \mathbf{g}_k^H \left[\mathbf{z}_k \mathbf{z}_k^H\right] \mathbf{g}_k}, \quad (6.48)$$

with $\mathbf{g}_k$ being the combining weights at the receiver.

One class of linear precoding techniques for the case of a single-antenna receiver is based on zero-forcing [BTC06, YG06b, DS05, PHS05], where each user receives only its desired signal with no interference. Because the number of spatial channels formed using linear beamforming is limited by the number of transmit antennas, the transmitter selects a set of active users for receiving data. This user selection could be done optimally using a brute-force search over all possible combinations of users, but due to the high complexity when the number of users is large, suboptimum techniques based on a greedy algorithm have shown to provide near-optimum performance [BTC06]. This topic is studied in detail in Section 11.1. Extensions of the zero-forcing technique to the case of multiple receive antennas appear in [VVH03, CM04, SSH04], where multiple spatial streams (or eigenmodes) are transmitted to each user with no inter-user interference, resulting in a block diagonal (BD) covariance matrix.

An extension of the BD concept, called MET, was proposed in [BH07a] and uses a linear transmission strategy based on zero-forcing beamforming for maximizing the weighted sum-rate. On a frame-by-frame basis, MET distributes up to $M_c N_{\mathrm{bs}}$ spatially multiplexed streams for one or multiple users.

MET was initially proposed for multi-user MIMO (MU-MIMO) transmissions and its extension to the CoMP case can be summarized as follows. Let's assume that each user multiplies its channel matrix by the Hermitian of the left dominant eigenvector. The effective channel after linear antenna combining is then

$$\mathbf{h}_{k\ \mathrm{MET}}^c = (\mathbf{u}_k^c)^H \mathbf{H}_k^c = (\mathbf{u}_k^c)^H \mathbf{U}_k^c \boldsymbol{\Sigma}_k^c (\mathbf{V}_k^c)^H = \lambda_{k,c} (\mathbf{v}_k^c)^H . \qquad (6.49)$$

As an extension to the MET and according to [TWH+09], we introduce the following concept of an eigenmode-aware optimum combiner (EOC).

$$\mathbf{h}_{k\ \mathrm{EOC}}^c = \left(\alpha \left(\mathbf{Z}_k \right)^{-1} \mathbf{u}_k^c \right)^H \mathbf{H}_k^c. \qquad (6.50)$$

Let us assume the inter-cluster interference aggregated in the covariance matrix $\mathbf{Z}_k$ is known to the receiver. Thus, we can determine an effective channel after linear combining, which considers the projection of residual interference according to the optimum combining (OC) strategy from [Win84]. This filter determines a specific channel direction, based on the dominant eigenmode $\mathbf{u}_k^c$, which provides highest post-equalization SINR for the desired signal and thus leads to an improved system throughput with respect to (6.49).

Based on (6.49) and (6.50), the central unit can schedule a set $|\mathcal{K}| \leq M_c N_{\mathrm{bs}}$ of users, either with a brute force algorithm or with a greedy selection algorithm (see [DS05] and its extension to the case of a per-antenna power constraint [BH06]). Therefore, the effective channels from multiple users are collected in a compound channel matrix in an iterative manner as

$$\mathbf{H}_{\mathrm{virtual}}^c = \left[\mathbf{h}_{MT_1}^c \ldots \mathbf{h}_{MT_K}^c \right]. \qquad (6.51)$$

The linear precoder may be obtained by the Moore-Penrose pseudo-inverse of the compound channel matrix (6.51):

$$\mathbf{W}_i = \mathbf{H}_{\mathrm{virtual}}^c \left((\mathbf{H}_{\mathrm{virtual}}^c)^H \mathbf{H}_{\mathrm{virtual}}^c \right)^{-1}. \qquad (6.52)$$

In practice, a per-BS or an even more restrictive per-antenna power constraint is a mandatory constraint if such algorithms are used in a cellular deployment, where multiple BSs belonging to the same cluster $\mathcal{M}_c$ have to meet their own power constraints while jointly serving the users $\mathcal{K}$. Thus, the maximum available transmit power at each BS is restricted to a P_{max} value and in case of a very strict per-antenna power constraint, P_{max} can be equally divided to all antenna elements, i.e. $P_{\mathrm{max}}/N_{\mathrm{bs}}$. In order to meet this constraint, we use the expression for matrix $\sqrt{\mathbf{P}_c}$ as given in [ZD04]:

$$\sqrt{\mathbf{P}_c} = \left\{ \min_{m=1,\ldots,M} \sqrt{\frac{P_{\max}}{\|\mathbf{W}^m\|_2}} \right\} \cdot \mathbf{I}_{[K \times K]}, \tag{6.53}$$

where $\mathbf{W}^m$ are the rows of matrix $\mathbf{W}$ related to the antennas of BS m. Note that this power allocation is suboptimal and typically results in only one BS antenna transmitting with maximum power, and hence, the remaining $M_c N_{\mathrm{bs}} - 1$ antennas transmit with less than $P_{\max}/N_{\mathrm{bs}}$.

Clustering for Reducing Feedback and Backhaul Data Traffic

From a practical point of view, one of the major drawbacks related to joint processing is its higher complexity, i.e. increasing backhaul and signaling overhead. To reduce these complexity requirements, *clustering* solutions that restrict joint processing techniques to a limited number of BSs have been proposed. In these approaches, the network is statically or dynamically divided into clusters of cells [BH07b, PGH08, TWH+09]. Moreover, the cluster formation may be performed and optimized by a central entity (*network-centric*), or in a per-user way (*user-centric*). As a result of [TBB+10], dynamic BS clustering was found to be key relation for spectrally efficient CoMP transmission while keeping the backhaul traffic at a moderate level.

The work described in [BHA08] considers that BS clusters are created in a dynamic way (see Section 7.2), in other words at each time slot t the sets of coordinated BSs are generated in order to maximize a given objective function. This work demonstrates a significant reduction of signaling overhead in the backhaul due to data sharing between cooperating base stations, while achieving a high fraction of the full coordination performance.

6.3.4 Scheme for Distributed, Centralized Joint Transmission

First of all, let us clarify the usage of the three different terms: *decentralized, centralized* and *distributed*. In case of a *decentralized* system concept, each BS is assumed to have no or only a subset of feedback information (e.g. CSI) from other BS's users, as considered in Section 6.4. In contrast, the *centralized* system concept assumes full CSIT as well as scheduled user data shared among all BSs in the cluster. And finally, the term *distributed* indicates that such a centralized concept can also be implemented in a distributed manner, i.e. without a central unit (CU).

In modern mobile networks, there is a general tendency of using distributed signal processing. The adaptation to the time variation of the wireless channel can be much faster if it is performed directly in the serving BS. For downlink CoMP, we reduce the overall delay in the closed transmitter adaptation loop if the waveforms are generated at the serving BS. Concepts for a distributed implementation of CoMP, where the serving BS cooperates

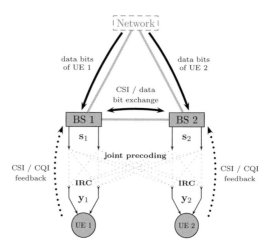

Figure 6.12 Cooperative transmission and CSI/CQI feedback and exchange concept, as illustrated for a toy scenario with $M = K = 2$.

with a small subset of BSs, Fig. 6.10, in its direct vicinity are reported in [ZSK$^+$06, JTS$^+$08b, NEHA08, PHG09, ZMS$^+$09a, TWH$^+$09][1]. Chapter 13.3 reports on a first real-time implementation of downlink CoMP demonstrating its feasibility and [JTW$^+$09, JFJ$^+$10] summarize this work. Terminals are assumed to estimate the multi-cell CSI in the downlink using CSI reference signals (CSI RSs). Subsequently, UEs deliver CSI feedback in combination with channel quality indicator (CQI) values to their serving BS, as illustrated in Fig. 6.12. Next, BSs in the cluster exchange the CSI as well as scheduled user data over a low-latency signaling network denoted as X2 interface [3GP10i]. Precoding weights for the joint beamforming are determined at each BS. The relevant set of weights is applied to the data signals and in this way, the transmitted waveforms are obtained locally. Similar to the centralized approach, the desired signals sum up constructively while the mutual interference inside the cluster is canceled. We emphasize that under the assumption of low Doppler shift, i.e for low mobility or even static users, the backhaul bandwidth required for sharing the user data between cooperating BSs is much higher than the one required for updating the channel estimates within the cluster, as discussed in Section 12.2. Let us assume an average throughput per cell denoted as $\overline{rate}$, hence, each BS has to receive the scheduled user data for its own UEs according to that data rate. Further, we consider that hybrid automatic repeat request (HARQ) processes for each user in the active set of users $\mathcal{M}_k$ are running decentralized at each BS k. Thus,

---

[1] Note, the length of the cyclic prefix (CP) limits the tolerable backhaul latency in the centralized approach. For distributed downlink CoMP, latency is more related to the ongoing aging process of the CSI while it is exchanged over the backhaul. A few ms may be tolerated for slowly moving UEs. Hence, capacity and latency requirements for the backhaul are significantly relaxed compared to the centralized approach.

6.3 DL Distributed CoMP Approaching Centralized Joint Transmission

each BS has to perform the channel coding with a given *code_rate*, according to the CQI feedback provided by the users in $\mathcal{M}_k$. For simplicity, we fix this rate to 1/2 in the sequel. According to (6.54), all remaining $K-1$ BSs in the cluster $\mathcal{K}$ convey their coded user data over the backhaul to the k-th BS. Thus, the backhaul overhead scales linear with the number of BSs exchanging their scheduled data in the cluster.

$$\text{traffic} = \overline{\text{rate}} \left(1 + \frac{K-1}{\text{code\_rate}}\right) \quad (6.54)$$

The process is split into three phases:

- **Phase I: Channel feedback.**
  Each user performs a cluster-wide channel estimation using reference signals (see Section 9.1). Each UE generates multiple-input single-output (MISO)-CSI according to [BH07a, TBH08, TWH+09]

  $$\mathbf{h}_k^c = (\mathbf{v}_k^c)^H \mathbf{H}_k^c, \quad (6.55)$$

  where the Euclidean norm equals $\|\mathbf{v}_k^c\|_2 = 1$. Besides, $\mathbf{v}_k^c$ is always used to denote the linear combining scheme to generate CSI MISO feedback. In Section 6.3.4, we assume the combining metrics defined in (6.49) and (6.50) and denote them as eigenmode-aware receive combining (ERC) and eigenmode-aware optimum combiner (EOC). This channel information is fed back in conjunction with the expected post-equalization SINR

  $$\text{SINR}_k^{(I)} = \frac{|(\mathbf{h}_k^c)^H \widehat{\mathbf{w}}_k^c|^2}{(\mathbf{v}_k^c)^H \left[\mathbf{z}_k \mathbf{z}_k^H\right] \mathbf{v}_k^c}, \quad (6.56)$$

  where each user assumes a precoder according to $\widehat{\mathbf{w}}_k^c = \sqrt{\widehat{p}_k}\mathbf{h}_k^c/\|\mathbf{h}_k^c\|_2$ and no intra-cluster interference, since this interference will be removed by the joint precoder. In particular, the achievable SINR (6.56) together with the CSI (6.55) is then conveyed to the serving BS.

- **Phase II: Distributed precoder calculation.**
  A scheduling instance in the cluster c combines a total number of $M_c N_{\text{bs}} = K N_{\text{bs}}$ MISO channels to a compound MIMO channel matrix[2]. In the following, each BS is responsible for a specific sub-band of the overall bandwidth where CoMP JT is employed. Therefore, BSs partially exchange their collected CSI and combine the channel feedback $\mathbf{h}_k^c$ to a compound virtual MIMO channel matrix of size $M_c N_{\text{bs}} \times K$ according to (6.51). Subsequently, each BS determines the linear precoder for its specified sub-bands but for all $M_c N_{\text{bs}}$ antennas of the cluster according to (6.52).

---

[2] With proper user selection, the full rank condition of the compound channel can be frequently met in the multi-point-to-multi-point case with independent links [ZD04, JJT+09].

Afterwards, BSs exchange their precoding weights $\mathbf{W}^m$ and power allocation $\mathbf{P}^m$, both obtained per sub-band, as well as the complete user (payload) data. Therefore, BSs use logical interconnections, e.g. the X2-interface in LTE-A. Finally, all BSs in the n-th cluster perform the coherently precoded downlink transmission, where each BS is using the weights corresponding to its own transmit antennas.

- **Phase III: "Intra-cluster-interference-free" data reception at the terminal side.**
  In this step, each UE performs its own preferred spatial equalization strategy $\mathbf{g}_k$. Therefore, each user may select the same weights as used in *Phase I* or may perform the equalization using the optimal linear receive combining and introduced as OC [Win84] in (5.4) in Section 5.1. The post-equalization SINR is determined by (6.48) and is used as inputs for the link adaptation.

Simulation Results

The combined transmitter-receiver concepts described in the previous chapters are evaluated in a triple-sectorized hexagonal cellular network with $M = 57$ BSs in total. All cells operate with full frequency reuse. We employ the wrap-around technique, which ensures that performance evaluation can be based on all users in all cells. The different channel matrices are generated by employing the widely used SCME [BHD+05] with urban macro scenario parameters [3GP10d], 3D BS antenna characteristics and an electrical downtilt angle set to achieve a main lobe range of one third of the inter-site distance (ISD).

In particular, we determine the system performance by assuming a dynamic and user-driven clustering method, as in Section 7.2. An active set of users is selected according to the following metric: A set $\mathcal{U}_c$ of active multi-antenna terminals is uniformly distributed in the c-th cluster of the cellular environment. This set contains multiple disjoint user sets $\mathcal{K}_m \subset \mathcal{U}_c$, where the users in $\mathcal{K}_m$ experience highest channel gain to the m-th BS. Thus, each UE is connected to a master BS. Further, we emulate a cluster selection which is user-centric and dynamic over frequency: the M/C strongest channel gains of the users in $\mathcal{K}_m$ are the ones of the M_c BSs within the cluster, i.e. each UE is placed in a perfect cluster in the way that its strongest BS links are all covered by this cluster. A round-robin scheduling policy ensures that only N_{bs} UEs per BS $m \in \mathcal{M}$ are selected for CoMP JT. This enables us to obtain gains from CoMP transmission which are separated from gains due to multi-user diversity. Results are provided for different cluster sizes of $M/C \in \{1, 2, 3, 4, 5, 10\}$. All results in Fig. 6.13 are based on an equal per-beam power constraint with a per-antenna power constraint [ZD04], aligned with LTE assumptions. To determine the data rates which can be achieved in a practical system, we map post-equalization SINRs using 5 bit modulation and coding schemes (MCSs) known from LTE standardization,

6.3 DL Distributed CoMP Approaching Centralized Joint Transmission

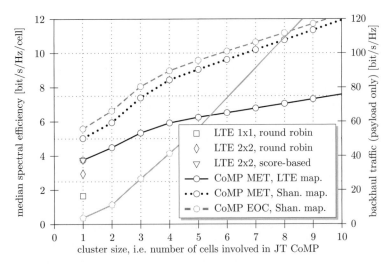

Figure 6.13 Performance results as a function of the cluster size M_c. Channel feedback is assumed according to the ERC (6.49) and EOC (6.50).

and assume 630 bit codeword length and 1% target block error rate (BLER). In addition, some results are provided based on Shannon information rates.

Performance of Reference Cases

For the scheduling in one cell, UEs provide feedback on their SINRs in the form of so-called CQI values for subgroups of sub-carriers denoted as PRBs. These CQIs correspond to a specific spatial transmission mode, which is indicated by the precoding matrix indicator (PMI). As a first extension towards multi-cell processing, adjacent base stations are synchronized and multi-cell demodulation reference signal (DRS) are introduced. They enable interference-aware equalization at the UE and improve the SINR estimation accuracy, leading to a more precise link adaptation at the BS side [TSWJ09].

For reference purposes, we include the performance results for interference-limited single-input single-output (SISO) as well as a MIMO 2×2 transmission from Section 5.1. For $N_\mathrm{bs} = 2$, two active fixed discrete Fourier transform (DFT)-based beams are sent to $K = 2$ different users in a round-robin manner or taking CQI feedback into account. The CQI-aware score-based solution, described in Section 5.1, outperforms both other reference cases with a relative throughput gain of $\Delta_{M_c=1} = 1.27$ and $\Delta_{M_c=1} = 2.2$ compared to round-robin and SISO, respectively. Note, with $K = N_\mathrm{bs}$, the MIMO setup benefits from an additional user in conjunction with an increase of antennas $N_\mathrm{bs} = N_\mathrm{ue} = 2$. All results in Fig. 6.13 are based on an equal per-beam power constraint with a per-antenna power constraint according to LTE assumptions.

Gain from CSI Feedback and CoMP Transmission

CSI-aware precoding within a given cluster reduces the interference experienced at each UE. Therefore, we use Equations (6.49), i.e. MET, and (6.50), i.e. EOC. The case of MET-based CSI feedback and zero-forcing (ZF) beamforming at a single BS, i.e. $M_c = 1$, with subsequent OC, provides a system gain of $\Delta_{M_c=1} = 2.2$ and $\Delta_{M_c=1} = 1.0$ compared to SISO and score-based beam assignment, respectively. In case of EOC-based CSI feedback and ZF beamforming, the data rate increases by approx. 10%. Note that there is no or only a small additional gain from CSI based ZF beamforming in a cluster of size $M_c = 1$. This is mainly caused by the two facts: First, we assume a simplified per-antenna power constraint, which leads to a suboptimal power allocation where only one antenna transmits with full power and all others are scaled accordingly [ZD04]. In contrast, in the case of fixed precoding from Section 5.1 all BS antennas transmit with full power. Second, within the score-based spatial layer assignment (reference case), multi-user diversity is used to assign both users to these fixed beams. In the case of ZF beamforming, both users are directly served on orthogonal beams. In both cases the inter-cell interference is not affected.

In the next step, we increase the cluster size for the CoMP system and the gain from MET-based feedback attributes to $\Delta_{M_c=2} = 2.7$, $\Delta_{M_c=3} = 3.2$, $\Delta_{M_c=4} = 3.6$, $\Delta_{M_c=5} = 3.8$ and $\Delta_{M_c=10} = 4.6$ w.r.t the SISO case. The EOC-based concept provides roughly 10% additional data rate compared to the MET assumption. Note, the gap between Shannon information rates and a practical LTE link adaptation attributes to 4.3 bit/s/Hz/cell for $M_c = 10$. Focusing on the cooperation gain, we observe that the median cell spectral efficiencies are increased compared to the 2×2 round-robin system by 81%, 112% and 157% for coordinating 3, 5 and 10 cells, respectively. However, backhaul requirements per feeder link increase as well, i.e. by 5, 9 and 19 bits per bit-on-air-interface assuming a fixed code rate of 1/2, for the coordination of 3, 5 and 10 cells, respectively. Altogether, significant gains from coordination have already been realized by using small clusters, despite residual interference from non-coordinated cells.

For $K = 10$, the estimated median traffic for user (payload) data exchange per backhaul link will exceed a value of 120 bit/s/Hz. This is only a rough estimate according to (6.54), where $\overline{rate}$ is the median cell data rate in bit/s/Hz/cell for CoMP transmission using the MCSs as defined in LTE. The backhaul traffic consists of the transmitted data over the air interface and the required user data exchange from $K - 1$ other BSs. Since BSs have to coherently transmit the same data, i.e. using the same MCS, we consider the BSs to exchange their coded user payload data and independent mapping to identical QAM symbols. For sake of simplicity, we assume here a code rate of 1/2. Further, we do not consider additional overhead due to CSI exchange, as this would be only a small part of the overall traffic requirement [HS09], as emphasized in Section 12.2.

6.3.5 Summary

In this section, we investigated centralized joint transmission in the context of CoMP transmission in the downlink of next generation mobile networks. Starting from a general system model, we first determined the MIMO broadcast capacity in a cellular system for different SNR regimes. Second, we introduced the concept of linear joint precoding for a subset of BSs, i.e. a cluster, in the system, whereas BSs belonging to different clusters are not coordinated. Hence, each cluster is surrounded by multiple non-coordinated cells. For removing the interference inside the cluster the common multi-user eigenmode transmission has been further developed towards optimum combining. The gains from receive antenna combining have been included in the overall optimization. The performance has been studied in detail in a triple-sectored multi-cell scenario covering 57 cells. At first, we observed that median data rates per cell can be increased by 81%, 112% and 157% assuming a cluster size of 3, 5 and 10 cells, respectively, compared to a non-cooperative system with the same 2×2 antenna configuration. However, backhaul requirements per feeder link increase as well, i.e. by 5, 9 and 19 bits per bit-on-air-interface assuming a fixed code rate of 1/2, for the coordination of 3, 5 and 10 cells, respectively. Second, as a function of the cluster size ranging from 1 to 5 and up to 10, the linear eigenmode-aware optimum combiner scheme achieves 28%, 34%, 41%, 46%, 49% up to 62% of the capacities provided by system-wide dirty paper coding. Altogether, significant gains from coordination have already been realized by using small clusters.

Acknowledgements

The authors are grateful for financial support from the German Ministry of Education and Research (BMBF) in the national collaborative project EASY-C under contract No. 01BU0631.

6.4 Downlink Decentralized Multi-User Transmission

David Gesbert and Randa Zakhour

In this section, we study *decentralized* downlink CoMP schemes that differ from the schemes introduced in Section 6.3 in such a way that compound channel state information (CSI) and data bits corresponding to the different user equipments (UEs) are *not* shared among all cooperating base stations (BSs). Instead, different cooperating BSs may have different extents of knowledge on subsets of the compound channel matrix, and on subsets of user data. This means that globally optimal precoding vectors cannot be found, but decentralized algorithms can aim at finding sub-optimal solutions.

We consider two forms of (partial) decentralized processing. In the first, we assume that user data is fully shared at the cooperating BSs, whereas the amount

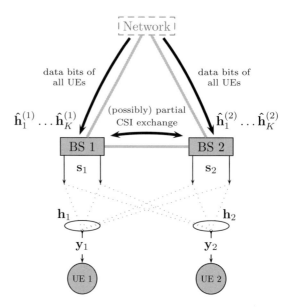

Figure 6.14 Setup for decentralized beamforming with limited CSIT, for a toy example with $M = K = 2$.

of shared CSI is limited [ZG10b]. In the second, the CSI is shared ideally across the cooperating cells, while the user data is only partially shared to lift the burden off the backhaul [ZG10a]. In the latter case, we use the notion of *superposition coding*, as already introduced in an uplink context in Section 4.3, assuming now that each terminal receives a superposition of conventionally and cooperatively transmitted signals. We will see that the optimal ratio of these signals varies with both the interference strength statistics and the backhaul capacity constraint.

6.4.1 Decentralized Beamforming with Limited CSIT

Let us first consider a multi-cell scenario where the cooperating BSs share the user data and aim at joint multiple-input multiple-output (MIMO) precoding (beamforming (BF)) to serve the mobile users in the downlink. Thus, each user receives a transmission from all cooperating BSs, as illustrated in Fig. 6.14.

For perfect downlink beamforming, channel state information at the transmitter (CSIT) must be acquired and shared by the transmitters through a combination of user feedback and backhaul signaling [KFVY06, SSBN+06]. In practice, it is convenient to derive robust beamforming schemes able to cope with imperfect and, importantly, *different* CSI available at each BS, as this will allow us to deal with various scenarios for decentralized knowledge such as

- Each BS obtaining an independent estimate of the same global CSI, or

6.4 Downlink Decentralized Multi-User Transmission

- Each BS having precise knowledge about a different subset of the user channels (e.g. a BS may not wish or be able to decode the CSI feedback from a very distant user).

Note that this is the essential difference to Section 6.3, where the same extent of CSI is assumed to be fully distributed among all cooperating BSs. In this section, we introduce a feedback model using the concept of *hierarchical codebooks*, which allows us to incorporate additional structure into this problem and as a result facilitate robust beamforming design.

Despite possible differences in their acquired CSIT, the different transmitters wish to conciliate their views so as to design a consistent set of precoding vectors that maximizes the user rate. This problem can be categorized as a so-called *team-decision problem* or a decentralized statistical decision making problem [Ho80, Rad62].

System Model

Consider a set of M BSs communicating with K UEs. Each BS has $N_{\text{bs}} \geq 1$ antennas, whereas each UE has a single antenna. $\mathbf{h}_k^m \in \mathbb{C}^{N_{\text{bs}} \times 1}$ is the channel from BS m to UE k and $\mathbf{h}_k = [(\mathbf{h}_k^1)^T, \ldots, (\mathbf{h}_k^M)^T]^T \in \mathbb{C}^{N_{\text{BS}} \times 1}$ is UE k's whole channel: $\mathbf{h}_k^m \sim \mathcal{N}_{\mathbb{C}}(\mathbf{0}, \sigma_{k,m}^2 \mathbf{I}_{[N_{\text{bs}}]})$ and different $\mathbf{h}_k^m$ are independent of each other. The overall unquantized channel matrix $\mathbf{H}$ groups the channels to all users, i.e. $\mathbf{H} = [\mathbf{h}_1 \ldots \mathbf{h}_K]$. Similarly, we let $\hat{\mathbf{h}}_k^{(m)}$ denote UE k's quantized channel as perceived by BS m and group the whole of BS m's channel knowledge into $\hat{\mathbf{H}}^{(m)} = [\hat{\mathbf{h}}_1^{(m)} \ldots \hat{\mathbf{h}}_K^{(m)}]$. The signal received by UE k is given by:

$$y_k = \mathbf{h}_k^H \mathbf{s} + n_k, \qquad (6.57)$$

where $\mathbf{s} \in \mathbb{C}^{N_{\text{BS}} \times 1}$ is the concatenated transmit signal sent by all transmitters and $n_k \sim \mathcal{N}_{\mathbb{C}}(0, \sigma^2)$ is the noise at receiver k. Cooperative transmit processing in the form of joint linear precoding with per-transmitter power constraint P is adopted. Thus, $\mathbf{s}$ can be expressed as:

$$\mathbf{s} = \mathbf{W}\mathbf{x} = \sum_{k=1}^{K} \mathbf{w}_k x_k, \qquad (6.58)$$

where $\mathbf{x} \in \mathbb{C}^{K \times 1}$ is the vector of transmit symbols, its entries being independent and with $\mathbf{x} \sim \mathcal{N}_{\mathbb{C}}(\mathbf{0}, \mathbf{I})$. The overall beamforming matrix $\mathbf{W}$ groups the precoding vectors $\mathbf{w}_k$ carrying the different users' symbols, so that $\mathbf{W} = [\mathbf{w}_1 \ldots \mathbf{w}_K] \in \mathbb{C}^{N_{\text{BS}} \times K}$, where precoding vector $\mathbf{w}_k = [(\mathbf{w}_k^1)^T \ldots (\mathbf{w}_k^M)^T]^T$ carries user k's symbols, and $\mathbf{w}_k^m \in \mathbb{C}^{N_{\text{bs}} \times 1}$ is BS m's precoding contribution towards UE k. The rate achievable for user k is equal to

$$R_k = \log_2(1 + \gamma_k), \qquad (6.59)$$

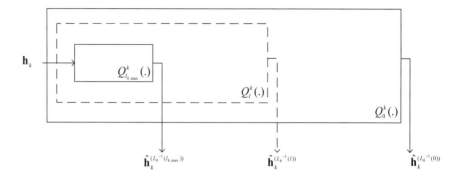

Figure 6.15 Distributed hierarchical CSI model: the quantization codebooks are designed to be hierarchical to offer additional structure. $Q_l^k(.)$ denotes the l-level (specifying the accuracy) quantization function of user k's channel.

where the signal-to-interference-and-noise ratio (SINR) γ_k is equal to

$$\gamma_k = \frac{\left| \sum_{m=1}^{M} (\mathbf{h}_k^m)^H \mathbf{w}_k^m \right|^2}{\sigma^2 + \sum_{j \neq k} \left| \sum_{m=1}^{M} (\mathbf{h}_j^m)^H \mathbf{w}_j^m \right|^2}. \quad (6.60)$$

In the following, it will also be useful to focus on a given BS's beamforming decisions jointly. We denote this $\mathbf{W}^m$, so that $\mathbf{W}^m = [\mathbf{w}_1^m \ldots \mathbf{w}_k^m]$, the overall precoding matrix at BS m. We now consider how these will be designed.

Decentralized Beamforming using Team Decision

In a decentralized processing scenario, each BS computes its beamforming matrix based on its local CSI. In addition to its own local CSI, statistical CSI (slow varying) concerning other transmitters' local CSI may be available. The M transmitters may be viewed as members of a team who need to take decisions in order to attain a common payoff, but who do not have access to the same information. Thus, transmitter m chooses $\mathbf{W}^m$ based on its local CSI, the quantized channel matrix $\hat{\mathbf{H}}^{(m)}$, and the extra statistical information it has.

Hierarchical CSI Structure

We now show how structure can be exploited into the CSI model so as to facilitate team decision making at the BSs. In the setup below, the channel $\mathbf{h}_k$ corresponding to user k is quantized using a hierarchical codebook, such that different BSs acquire a given channel vector information up to one particular level of quantization error. The most precise codebook, composed of a series of embedded less precise sub-codebooks, is assumed to be known at all BSs. Each BS is also assumed to be aware of the level of feedback accuracy decodable at other BSs (typically a function of the distance to the user).

We define $L_k(m)$ as the number of quantization bits successfully decodable by BS m for the channel vector $\mathbf{h}_k$ corresponding to user k. Thus, BS m acquires a quantized version of this channel denoted by $\hat{\mathbf{h}}_k^{(m)}$. An interesting feature of hierarchical codebooks is the strong information structure: Assuming without loss of generality that $L_k(m_1) > L_k(m_2)$, then BS m_1 knows exactly what is known by BS m_2 in addition to its own estimate. Thus BS m_1 also knows $\hat{\mathbf{h}}_k^{(m_2)}$ while BS m_2 only knows that $\hat{\mathbf{h}}_k^{(m_1)}$ belongs to a subset of codewords located in the Voronoi region centered at $\hat{\mathbf{h}}_k^{(m_2)}$. This is shown in Fig. 6.15, where $L_k^{-1}(l)$ provides the set of BSs (if any) which decode the feedback information of user k up to accuracy level l.

Bayesian Optimization
As perfect CSI is neither available nor shared, we consider the maximization of a global average utility $\mathcal{U}$ which may be decoupled as a sum of utilities over the users:

$$\mathcal{U} = E\left\{U\left(\mathbf{H}, \mathbf{W}^1\left(\hat{\mathbf{H}}^{(1)}\right), \ldots, \mathbf{W}^M\left(\hat{\mathbf{H}}^{(M)}\right)\right)\right\}$$
$$= \sum_{k=1}^{K} E\left\{U_k\left(\mathbf{h}_k, \mathbf{W}^1\left(\hat{\mathbf{H}}^{(1)}\right), \ldots, \mathbf{W}^M\left(\hat{\mathbf{H}}^{(M)}\right)\right)\right\}. \quad (6.61)$$

For instance, an average weighted sum-rate, where U_k denotes the (possibly weighted) rate of user k, fits into this model. Restricting ourselves to deterministic decisions, in the sense that there will be a single $\mathbf{W}^m$ corresponding to each state of channel knowledge at transmitter m, $\hat{\mathbf{H}}^{(m)}$, $\mathcal{U}$ can be expanded into:

$$\mathcal{U} = \int \ldots \int d\mathbf{H} f_\mathbf{H}(\mathbf{H}) U\left(\mathbf{H}, \tilde{\mathbf{W}}^1(\mathbf{H}), \ldots, \tilde{\mathbf{W}}^M(\mathbf{H})\right), \quad (6.62)$$

where

$$\tilde{\mathbf{W}}^m(\mathbf{H}) \triangleq \mathbf{W}^m\left(\hat{\mathbf{H}}^{(m)} \big| \mathbf{H}\right)$$

is the beamforming strategy at transmitter m given the local knowledge at that transmitter corresponding to a true channel $\mathbf{H}$. $f_\mathbf{H}$ denotes the probability distribution function of the overall channel matrix $\mathbf{H}$.

Global Optimization
A globally optimal set of beamforming decisions maximizing the above expected utility $\mathcal{U}$ consists of sets of beamforming matrices $\mathbf{W}^m$, $m \in \{1, \ldots, M\}$, (one set per user, consisting of as many matrices as there are possible states of knowledge at that user), which jointly maximize $\mathcal{U}$. As stated in [Ho80, TA84], for example, it is often intractable to find the globally optimal strategies at the different team members. In such cases, a suboptimal solution may be obtained by finding strategies that are *person-by-person optimal*, as specified next. We note the differences with conventional game-theoretic approaches for beamforming design in

that here all transmitters agree on maximizing a common utility (despite their lack of shared CSI knowledge), as opposed to optimizing a selfish utility.

Person-by-Person Optimization

Person-by-person optimal strategies[3] are such that for each team member, his strategy is optimal given the other team members' strategies. Clearly, the globally optimal strategies are person-by-person optimal, but the converse is in general not true. In our particular setup of distributed CSIT, an optimal strategy for transmitter m, given that the other transmitters' strategies are fixed, may be characterized, for a local channel knowledge equal to $\hat{\mathbf{H}}^{(m)}$, as follows:

$$\mathbf{W}^{m*}\left(\hat{\mathbf{H}}^{(m)}\right)$$
$$= \arg \max_{\|\mathbf{W}^m\|_F^2 \leq P} \int \cdots \int d\mathbf{H} f_{\mathbf{H}|\hat{\mathbf{H}}^{(m)}}\left(\mathbf{H}|\hat{\mathbf{H}}^{(m)}\right) \tilde{U}\left(\mathbf{H}, \mathbf{W}^m\right) \qquad (6.63)$$

where

$$\tilde{U}\left(\mathbf{H}, \mathbf{W}^m\right) = U\left(\mathbf{H}, \tilde{\mathbf{W}}^1\left(\mathbf{H}\right), \ldots, \mathbf{W}^m, \ldots, \tilde{\mathbf{W}}^M\left(\mathbf{H}\right)\right), \qquad (6.64)$$

and $f_{\mathbf{H}|\hat{\mathbf{H}}^{(m)}}$ denotes the probability distribution function of the overall channel matrix $\mathbf{H}$ conditioned on $\hat{\mathbf{H}}^{(m)}$, the quantized overall channel matrix at transmitter m. This is equivalent to

$$\mathbf{W}^{m*}\left(\hat{\mathbf{H}}^{(m)}\right) = \arg \max_{\|\mathbf{W}^m\|_F^2 \leq P} \int \cdots \int_{\mathcal{R}(\hat{\mathbf{H}}^{(m)})} d\mathbf{H} f_{\mathbf{H}}\left(\mathbf{H}\right) \tilde{U}\left(\mathbf{H}, \mathbf{W}^m\right), \qquad (6.65)$$

where we define $\mathbf{R}\left(\hat{\mathbf{H}}^{(m)}\right)$ as the Voronoi region corresponding to this state of knowledge at transmitter m. The Voronoi region indicates the set of all possible values for the actual channel $\mathbf{H}$ given that channel estimate $\hat{\mathbf{H}}^{(m)}$ has been observed at transmitter m.

A Decentralized Beamforming Example for $M = K = 2$

To simplify the presentation of the solution to the problem, we focus on the $M = K = 2$ case. The hierarchy in the knowledge at the two transmitters, and as a result the beamforming strategies to follow, fall into one of three cases, which may be characterized as follows:

Common Knowledge

In this case, $L_1(1) = L_1(2)$ and $L_2(1) = L_2(2)$. It corresponds to the *traditional* assumption under limited CSIT, where both transmitters have the same knowledge. This is the case if the BSs mutually exchange their CSIT, as considered in Sections 6.3 and 13.3, or if the CSI feedback is designed such that it can be

---

[3] Note that in game theory, Nash equilibria, which correspond to strategies from which no user has any incentive to deviate, are also person-by-person optimal, but there users in general do not share a common objective and are often competing for resources.

decoded by both BSs individually, which is the approach pursued in Section 13.4. Considering a hierarchical CSI structure, the assumption of common knowledge can be regarded reasonable if both users are located at the cell-edge. In terms of performance, having the same global CSI available at each BS is equivalent to having centralized beamforming decisions being made.

Degraded Knowledge
In this case, $L_1(1) \geq L_1(2)$ and $L_2(1) \geq L_2(2)$, or $L_1(1) \leq L_1(2)$ and $L_2(1) \leq L_2(2)$. In other words, one of the transmitters has a better representation of both channels, and will adapt its beamforming on a finer scale than the other transmitter. This is typical, for example, of when both users lie in the same cell.

Symmetric Knowledge
Here, $L_1(1) > L_1(2)$ and $L_2(1) < L_2(2)$, or $L_1(1) < L_1(2)$ and $L_2(1) > L_2(2)$. Hence, one of the transmitters has a better representation of the channel of a given user, and a worse one for the other, with opposite knowledge at the other transmitter. This corresponds, for instance, to the BSs serving users each within their own cell.

We now focus on the symmetric case where $L_1(1) > L_1(2)$ and $L_2(1) < L_2(2)$: this represents the more common setup among the ones described and is also the more challenging to formulate; the remaining cases can be dealt with in a similar manner. We characterize each user's quantized CSI by a pair $\mathbf{i}_1 = (i_{1,2}, i_{1,1})$ for user 1, and another $\mathbf{i}_2 = (i_{2,1}, i_{2,2})$ for user 2. The first index in each pair corresponds to the coarse knowledge (hence is shared by both users), i.e. the index of the codeword in the coarsest codebook, to which the channel is quantized, $Q_{L_k^{-1}(\min_m L_k(m))}(\mathbf{h}_k)$ (see Fig. 6.15), and the second index provides the missing bits to locate the finer codeword around the coarsest one, $Q_{L_k^{-1}(\max_m L_k(m))}(\mathbf{h}_k)$. Given the structure of the distributed CSI, the beamforming matrix decisions may be parameterized in terms of these indices, so that $\mathbf{W}^1$ varies with $(\mathbf{i}_1, i_{2,1})$, whereas $\mathbf{W}^2$ is a function of $(i_{1,2}, \mathbf{i}_2)$. Taking this into consideration, we expand (6.62) to

$$\sum_{i_{1,2}=1}^{2^{L_1(2)}} \sum_{i_{2,1}=1}^{2^{L_2(1)}} S(i_{1,2}, i_{2,1}) \tag{6.66}$$

where $S(i_{1,2}, i_{2,1})$ is given by

$$\sum_{i_{1,1}=1}^{I_1} \sum_{i_{2,2}=1}^{I_2} \int_{\mathbf{R}_1(\mathbf{i}_1)} \int_{\mathbf{R}_2(\mathbf{i}_2)} d\mathbf{h}_1 d\mathbf{h}_2 f_\mathbf{H}(\mathbf{H}) U\left(\mathbf{H}, \mathbf{W}^1(\mathbf{i}_1, i_{2,1}), \mathbf{W}^2(i_{1,2}, \mathbf{i}_2)\right), \tag{6.67}$$

where $I_1 = 2^{L_1(1)-L_1(2)}$, $I_2 = 2^{L_2(2)-L_2(1)}$, $\mathbf{R}_1(\mathbf{i}_1)$ and $\mathbf{R}_2(\mathbf{i}_2)$ correspond to the Voronoi regions associated with the indexed codewords.

It is easy to verify that the beamforming decisions for each $S\left(i_{1,2}, i_{2,1}\right)$ term may be optimized separately. For given $i_{1,2}$ and $i_{2,1}$, we optimize the corresponding $S\left(i_{1,2}, i_{2,1}\right)$. To simplify notation, we remove the dependence on $i_{1,2}$ and $i_{2,1}$ from the expressions. The problem is thus:

$$\max. \sum_{i_{1,1}=1}^{I_1} \sum_{i_{2,2}=1}^{I_2} \int_{\mathbf{R}_1(i_{1,1})} \int_{\mathbf{R}_2(i_{2,2})} d\mathbf{h}_1 d\mathbf{h}_2$$

$$\left[f_{\mathbf{H}}(\mathbf{H}) U\left(\mathbf{H}, \mathbf{W}^1\left(i_{1,1}\right), \mathbf{W}^2\left(i_{2,2}\right)\right)\right] \quad (6.68)$$

$$\text{s.t. } \|\mathbf{W}^1\left(i_{1,1}\right)\|_F^2 \leq P, i_{1,1}=1, \ldots, I_1 \quad (6.69)$$

$$\|\mathbf{W}^2\left(i_{2,2}\right)\|_F^2 \leq P, i_{2,2}=1, \ldots, I_2. \quad (6.70)$$

Recalling the separable nature of our utility function (refer to (6.61)), this can be reformulated as:

$$\max. \sum_{i_{1,1}=1}^{I_1} \sum_{i_{2,2}=1}^{I_2} \sum_{k=1}^{2} \Pr\left[\mathbf{R}_{\bar{k}}(i_{\bar{k},\bar{k}})\right] \int_{\mathbf{R}_k(i_{k,k})} d\mathbf{h}_k$$

$$\left[f_{\mathbf{h_k}}\left(\mathbf{h}_k\right) U_k\left(\mathbf{h}_k, \mathbf{W}^1\left(i_{1,1}\right), \mathbf{W}^2\left(i_{2,2}\right)\right)\right]$$

$$\text{s.t. } \|\mathbf{W}^1\left(i_{1,1}\right)\|_F^2 \leq P, i_{1,1}=1, \ldots, I_1$$

$$\|\mathbf{W}^2\left(i_{2,2}\right)\|_F^2 \leq P, i_{2,2}=1, \ldots, I_2, \quad (6.71)$$

where $\bar{k} = \mod(k, 2) + 1$ and

$$\Pr\left[\mathbf{R}_{\bar{k}}(i_{\bar{k},\bar{k}})\right] = \int_{\mathbf{R}_{\bar{k}}(i_{\bar{k},\bar{k}})} d\mathbf{h}_{\bar{k}} f_{\mathbf{h}_{\bar{k}}}\left(\mathbf{h}_{\bar{k}}\right), \quad (6.72)$$

is the probability of user $\bar{k}$'s channel being quantized to the codeword indexed by the pair $(i_{\bar{k},k}, i_{\bar{k},\bar{k}})$.

Application to Sum-Rate Maximization

The above problem may be approximately solved via a projected gradient ascent method. Moreover, to avoid integration, we resort to approximations. As we deal with sum-rate maximization in our illustrative examples, the following approximation is plugged into problem formulation (6.71) above:

$$\int_{\mathbf{R}_k(i_{k,k})} d\mathbf{h}_k U_k\left(\mathbf{h}_k, \mathbf{W}^1\left(i_{1,1}\right), \mathbf{W}^2\left(i_{2,2}\right)\right)$$

$$= \int_{\mathbf{R}_k(i_{k,k})} d\mathbf{h}_k^H \log_2\left(1 + \frac{|\mathbf{h}_k^H \mathbf{w}_k\left(i_{1,1}, i_{2,2}\right)|^2}{\sigma^2 + |\mathbf{h}_k^H \mathbf{w}_{\bar{k}}\left(i_{1,1}, i_{2,2}\right)|^2}\right)$$

$$\approx \log_2\left(1 + \frac{\mathbf{w}_k\left(i_{1,1}, i_{2,2}\right)^H \mathbf{C}_k^{(i_{k,k})} \mathbf{w}_k\left(i_{1,1}, i_{2,2}\right)}{\sigma^2 + \mathbf{w}_{\bar{k}}\left(i_{1,1}, i_{2,2}\right)^H \mathbf{C}_k^{(i_{k,k})} \mathbf{w}_{\bar{k}}\left(i_{1,1}, i_{2,2}\right)}\right), \quad (6.73)$$

where $\mathbf{C}_k^{(i_{k,k})} = \mathbb{E}\left\{\mathbf{h}_k \mathbf{h}_k^H \middle| \mathbf{h}_k \in \mathbf{R}_k(i_{k,k})\right\}$, and $\mathbf{w}_k\left(i_{1,1}, i_{2,2}\right)$, $k = 1, 2$ is obtained from $\mathbf{W}^1\left(i_{1,1}\right)$ and $\mathbf{W}^2\left(i_{2,2}\right)$ by extracting the appropriate entries as defined in our system model. A similar approximation was used in [KC07, HBO08] for

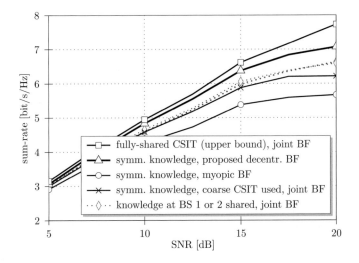

Figure 6.16 Sum-rates for $L_1(2) = L_2(1) = 2$, $L_1(1) = L_2(2) = 6$ bits and $\beta = 0.1$, from [ZG10b]. © 2010 IEEE.

example. The quality of this approximation increases and becomes asymptotically optimal with the size of the codebook.

Reference Schemes
Simple upper and lower bounds to the proposed schemes correspond to joint beamforming based on the more accurate CSIT (unachievable in a distributed system) and the least accurate (achievable) CSIT, respectively. In another decentralized scheme which uses the local channel knowledge, each BS designs its transmission assuming all the other BSs share the same knowledge as itself. This is simpler than the proposed decentralized scheme, and has similar complexity to joint beamforming design based on the coarse CSIT.

Numerical Results
To show the gains from this decentralized scheme, we plot average rates achieved for a symmetric $M = K = 2$, $N_{\text{bs}} = 1$ channel, where $\mathbf{h}_{1,1}$, $\mathbf{h}_{2,2}$ are $\mathcal{N}_\mathbb{C}(\mathbf{0}, 1)$, and $\mathbf{h}_{1,2}$, $\mathbf{h}_{2,1}$ are $\mathcal{N}_\mathbb{C}(\mathbf{0}, \beta)$, β modeling the strength of the interference links. The hierarchical codebooks are designed using Lloyd's algorithm: first the coarse codebook, then for each codeword in it, the corresponding finer codebook.

Fig. 6.16 compares the proposed decentralized scheme to the upper and lower bounds stated before for $L_1(2) = L_2(1) = 2$ and $L_1(1) = L_2(2) = 6$. We label the scheme which attempts to use local channel knowledge as if it were shared 'myopic' BF, in the sense that each BSs ignores some of the information it could be using. Thus, the upper bound scheme would require $2(L_1(1) + L_2(2)) = 24$ bits of CSIT being shared, whereas the schemes based on distributed, symmetric CSIT would require $L_1(1) + L_2(2) + L_1(2) + L_2(1) = 16$ bits. The benefit of the second layer of CSI over the more coarse shared representation of the channel depends on the signal-to-noise ratio (SNR) and on the value of β. At low SNR and for β low, there is little use for the extra information. The performance of myopic BF, even though it relies on more information than the joint beamforming relying on coarse CSI, is significantly worse, highlighting the importance of coordinated action. For reference, we also plot the performance that would be obtained if the knowledge at transmitter i, $i = 1, 2$ were indeed common to both transmitters and joint beamforming would result; clearly this yields more gain than joint beamforming based on coarse CSI.

6.4.2 Multi-cell Beamforming with Limited Data Sharing

We now assume the CSI is shared perfectly across the cooperating cells, and consider a limited backhaul dedicated to user data sharing. Imposing finite capacity constraints on the backhaul links brings with it a set of interesting research questions, in particular:

- Given the backhaul constraints, what kind of rates can we expect to achieve? What is the capacity region of the resulting multi-cell channel? In fact, the multi-cell MIMO channel under finite backhaul no longer corresponds to a MIMO broadcast channel (BC), nor to an interference channel (IC).
- How useful is data sharing when backhaul constraints are present? In other words, how do the rates achieved with a data sharing-, and therefore joint transmission enabling scheme compare to those achieved without data sharing, when the backhaul is limited?

These questions have lead to a number of recent interesting research efforts. To cite a few, in [SSSP08] and [SSPS09b], joint encoding for the cellular downlink is studied under the assumption that the BSs are connected to a CoMP central unit (CCU) via finite-capacity links. One of their main conclusions is that "central encoding with oblivious cells", whereby quantized versions of the signals to be transmitted from each BS and computed at a CCU, are sent over the backhaul links, is shown to be a very attractive option for both ease of implementation and performance, unless high data rates are required. If this is the case, the BSs need to be involved in the encoding, i.e. at least part of the backhaul link should be used for sending the messages themselves not the corresponding codewords. In [MF08b, MF09a], an optimization framework is proposed for an adopted backhaul usage scheme within clusters of cooperating cells. Here, the BSs can either be provided by the CCU with already precoded signals, hence being oblivious to the used codewords as stated before, or with (possibly quantized) representations of the messages to be transmitted to the terminals, which are then encoded and precoded locally.

Here, we consider a setup in which the backhaul is between the network and each of the BSs, and focus on how to use this given backhaul to serve the users

6.4 Downlink Decentralized Multi-User Transmission

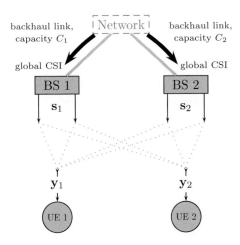

Figure 6.17 Setup for multi-cell beamforming with limited data sharing, for a toy setup with $M = K = 2$.

in the system. We focus on the two-cell problem. We specify a transmission scheme whereby superposition coding is used to transmit signals to each user, as also suggested in [MF08b]: this allows us to formulate a continuum between full message sharing between BSs (resembling a BC) and the conventional network with a single serving BS per user (an interference channel (IC)). To do so, the data rate is in fact split between two distinct forms of data to be received by the users, a private form to be sent by the serving BS alone, and a common form to be transmitted via multiple bases. The intuition is that the stronger the interference the most interesting it becomes to share data. The interesting problem is therefore to determine how much of the traffic should be shared vs. non shared in general cases.

System Model
The system considered is shown in Fig. 6.17. In this preliminary study, we focus on a two-transmitter, two-receiver setup. We assume a noiseless backhaul link of capacity C_m between the network and transmitter m, for $m \in \{1, 2\}$: it will be used to transmit the messages for each user. We distinguish between different types of messages:

- private messages are sent from the network to only one of the BSs, and
- shared or common messages, which are sent from the network to both transmitters, and are consequently jointly transmitted. Note that this notion of a common message is different from that commonly used in the context of interference channels (the Han-Kobayashi scheme and schemes derived from it) for example, as they do not correspond to messages to be decoded by both receivers, but rather to messages to be sent by both transmitters.

Thus user k's message rate r_k is split into common and private rates, $r_{k,c}$ and $r_{k,p}$, respectively:

$$r_k = r_{k,p} + r_{k,c}. \tag{6.74}$$

In the sequel, we assume full CSITs at both BSs, since we focus on the cost of sharing data, and denote as $\bar{k}$ the *other* BS or UE, depending on the context.

Proposed Backhaul Usage

Backhaul link k with finite capacity C_k serves to carry both private and common messages for user k as well as the common messages for user $\bar{k}$, so that:

$$C_k \geq r_{k,p} + r_{k,c} + r_{\bar{k},c} = r_k + r_{\bar{k}} - r_{\bar{k},p}, \quad k = 1, 2. \tag{6.75}$$

Since $r_{k,p} \geq r_k$, we have that $r_1 + r_2 \leq C_1 + C_2$.

Over the Air Transmission

The transmitted signal is the superposition of the precoded symbols corresponding to both categories of messages described above: 'private symbols' only arrive from the transmitter that knows them, whereas 'common symbols' are jointly precoded by both transmitters [MF08b]. Recall again that, unlike in the context of the IC where common messages arrive from a certain transmitter but are to be decoded by both receivers, 'common' messages here mean messages dedicated to a given user but shared across both transmitters, enabling them to jointly encode them. The transmitted signal may be written as follows:

$$\mathbf{s} = \begin{bmatrix} \mathbf{w}_{1,c} & \mathbf{w}_{2,c} \end{bmatrix} \begin{bmatrix} x_{1,c} \\ x_{2,c} \end{bmatrix} + \begin{bmatrix} \mathbf{w}_{1,p}^1 \\ 0 \end{bmatrix} x_{1,p} + \begin{bmatrix} 0 \\ \mathbf{w}_{2,p}^2 \end{bmatrix} x_{2,p}, \tag{6.76}$$

where $\mathbf{s} \in \mathbb{C}^{N_{\text{BS}}}$ is the transmitted signal, such that the first N_{bs} elements are transmitter 1's transmit signal, the remaining N_{bs} elements are transmitter 2's signal. $\mathbf{w}_{k,c} \in \mathbb{C}^{N_{\text{BS}}}$ are the beamforming vectors carrying user k's common message symbols $x_{k,c}$, and $\mathbf{w}_{k,p} \in \mathbb{C}^{N_{\text{bs}}}$ are those carrying his private message symbols $x_{k,p}$. Gaussian signaling is assumed: $x_{1,p}, x_{1,c}, x_{2,p}, x_{2,c}$ are all $\mathcal{N}_{\mathbb{C}}(0,1)$.

Similarly to the previous subsection, we split $\mathbf{w}_{k,c}$ into the contributions of the different transmitters, so that $\mathbf{w}_{k,c} = \left[(\mathbf{w}_{k,c}^1)^T \ (\mathbf{w}_{k,c}^2)^T \right]^T$. Using this notation, per base station power constraints P imply that:

$$\|\mathbf{w}_{1,c}^m\|^2 + \|\mathbf{w}_{2,c}^m\|^2 + \|\mathbf{w}_{m,p}^m\|^2 \leq P, \ m \in \{1, 2\}. \tag{6.77}$$

Background: MAC with Common Message

Given the assumption that each UE decodes only desired signals and not interference, the channel between the two transmitters and user k can be viewed as a multiple access channel (MAC) with a common message [SW73b], where the receiver noise is replaced by receiver noise plus interference due to transmission to user $\bar{k}$. Denoting by $\sigma_{\bar{k}}^2$ this power:

$$\sigma_{\bar{k}}^2 = \sigma^2 + \left|\mathbf{h}_{\bar{k}}^H \mathbf{w}_{\bar{k},p}^{\bar{k}}\right|^2 + \left|\mathbf{h}_{\bar{k}}^H \mathbf{w}_{\bar{k},c}\right|^2, \tag{6.78}$$

the following rate region is achievable:

$$r_{k,p} \leq \log_2 \left(1 + \frac{\left|(\mathbf{h}_k^k)^H \mathbf{w}_{k,p}^k\right|^2}{\sigma_k^2}\right),$$

$$r_k = r_{k,p} + r_{k,c} \leq \log_2 \left(1 + \frac{\left|(\mathbf{h}_k^k)^H \mathbf{w}_{k,p}^k\right|^2 + \left|\mathbf{h}_k^H \mathbf{w}_{k,c}\right|^2}{\sigma_k^2}\right). \tag{6.79}$$

Proof. This follows from results obtained for the two-user MAC with a common message by [SW73b]. □

Particular Cases
The transmission scheme introduced here covers the two particular cases of:

- no message sharing (IC), obtained by forcing $r_{k,p} \equiv r_k, k = 1, 2$, and
- full message sharing (BC), obtained by forcing $r_{k,p} \equiv 0, k = 1, 2$.

Achievable Rate Region
An achievable rate region $\mathcal{R}$ is the set of $(r_1, r_{1,p}, r_2, r_{2,p})$, as specified above, that satisfies the specified backhaul and power constraints. One way to get its boundary is to use the rate profile notion from [MZC06]. Points along the boundary are thus obtained by solving the following optimization problem for $\alpha \in [0, 1]$: thus, α specifies how the sum-rate achieved, r, is split between the two users.

max. r

s.t. $r_1 \geq \alpha r, \quad r_2 \geq (1-\alpha)r$

$r_1 + r_2 - r_{2,p} \leq C_1, \quad r_1 + r_2 - r_{1,p} \leq C_2$

$$r_k \leq \log_2 \left(1 + \frac{\left|(\mathbf{h}_k^k)^H \mathbf{w}_{k,p}^k\right|^2 + \left|\mathbf{h}_k^H \mathbf{w}_{k,c}\right|^2}{\sigma^2 + \left|(\mathbf{h}_k^{\bar{k}})^H \mathbf{w}_{\bar{k},p}^{\bar{k}}\right|^2 + \left|\mathbf{h}_k^H \mathbf{w}_{\bar{k},c}\right|^2}\right), k = 1, 2, \tag{6.80}$$

$$r_{k,p} \leq \log_2 \left(1 + \frac{\left|(\mathbf{h}_k^k)^H \mathbf{w}_{k,p}^k\right|^2}{\sigma^2 + \left|(\mathbf{h}_k^{\bar{k}})^H \mathbf{w}_{\bar{k},p}^{\bar{k}}\right|^2 + \left|\mathbf{h}_k^H \mathbf{w}_{\bar{k},c}\right|^2}\right), k = 1, 2, \tag{6.81}$$

$$\|\mathbf{w}_{k,p}^k\|^2 + \|\mathbf{w}_{k,c}^k\|^2 + \|\mathbf{w}_{\bar{k},c}^k\|^2 + \leq P, k = 1, 2. \tag{6.82}$$

The above optimization may be solved using a bisection over the sum-rate r: an essential part of the solution consists of establishing feasibility of a given rate.

Establishing Feasibility of a Given Rate

Assume sum-rate r and α to be fixed. Thus, $r_1 = \alpha r$, $r_2 = (1-\alpha)r$. Establishing feasibility of a given rate pair hinges on the following two remarks:

- For r_k to be supported, it cannot possibly exceed C_k, and
- Sharing information whenever possible outperforms not doing so. Thus a rate pair is not achievable unless it is so for the minimum possible private message rates $r_{k,p}$, $k = 1, 2$. Given the backhaul constraints, these are:

$$(r_{k,p})_{min} = \max(0, r_1 + r_2 - C_{\bar{k}}), k = 1, 2. \quad (6.83)$$

Thus the rate pair is only feasible if $r_1 \leq C_1$, $r_2 \leq C_2$, and rate tuple $\left(r_1, (r_{1,p})_{min}, r_2, (r_{2,p})_{min}\right) \in \mathbb{R}^4$ is achievable.

Feasibility of $(r_1, r_{1,p}, r_2, r_{2,p})$

Assume r_1, r_2, $r_{1,p}$ and $r_{2,p}$ are fixed. Establishing their feasibility and obtaining beamforming vectors to achieve them may be done by solving the total transmit power minimization problem subject to constraints (6.80), (6.81) and (6.82): feasibility of this problem implies feasibility of the set of rates, and its optimal solution yields the most power efficient beamforming strategies to attain it. This problem can be formulated as:

$$\min. \sum_{k=1}^{2} [\|\mathbf{w}_{k,c}^k\|^2 + \|\mathbf{w}_{k,p}^k\|^2]$$

$$\text{s.t. } 2^{r_k} - 1 \leq \frac{\left|(\mathbf{h}_k^k)^H \mathbf{w}_{k,p}^k\right|^2 + \left|\mathbf{h}_k^H \mathbf{w}_{k,c}\right|^2}{\sigma^2 + \left|(\mathbf{h}_k^{\bar{k}})^H \mathbf{w}_{\bar{k},p}^{\bar{k}}\right|^2 + \left|\mathbf{h}_k^H \mathbf{w}_{\bar{k},c}\right|^2}, k = 1, 2,$$

$$2^{r_{k,p}} - 1 \leq \frac{\left|(\mathbf{h}_k^k)^H \mathbf{w}_{k,p}^k\right|^2}{\sigma^2 + \left|(\mathbf{h}_k^{\bar{k}})^H \mathbf{w}_{\bar{k},p}\right|^2 + \left|\mathbf{h}_k^H \mathbf{w}_{\bar{k},c}\right|^2}, k = 1, 2,$$

$$\|\mathbf{w}_{k,c}^k\|^2 + \|\mathbf{w}_{\bar{k},c}^k\|^2 + \|\mathbf{w}_{k,p}^k\|^2 \leq P, k = 1, 2.$$

We can transform the above problem into an equivalent convex optimization.

- If $r_{k,p} \equiv 0$ or $r_k \equiv r_{k,p}$, we can reduce the problem as follows:
  - If $r_{k,p} \equiv 0$, the corresponding constraint becomes redundant, and $\mathbf{w}_{k,p} = \mathbf{0}$.
  - If $r_k \equiv r_{k,p}$, the optimum $\mathbf{w}_{k,c} = \mathbf{0}$ and the constraint on r_k is removed.

  In both cases, the remaining constraint can be transformed into a second-order cone constraint as in [VGL03, WES06, DY08].
- Otherwise, both constraints on UE k's rates will be tight at the optimum, and

$$\frac{2^{r_{k,p}} - 1}{2^{r_k} - 2^{r_{k,p}}} \left|\mathbf{h}_k^H \mathbf{w}_{k,c}\right|^2 = \left|(\mathbf{h}_k^k)^H \mathbf{w}_{k,p}^k\right|^2. \quad (6.84)$$

Further noting that $\mathbf{h}_k^H \mathbf{w}_{k,c}$ and $\mathbf{h}_{k,k}^H \mathbf{w}_{k,p}^k$ being real does not restrict the solution, we obtain the following equivalent convex problem:

$$\min. \sum_{k=1}^{2} [\|\mathbf{w}_{k,c}\|^2 + \|\mathbf{w}_{k,p}^k\|^2]$$

$$\text{s.t. } \sqrt{\frac{2^{r_k} - 2^{r_{k,p}}}{2^{r_{k,p}} - 1}} \left(\mathbf{h}_k^k\right)^H \mathbf{w}_{k,p}^k = \mathbf{h}_k^H \mathbf{w}_{k,c}, k = 1, 2$$

$$\sqrt{2^{r_{k,p}} - 1} \left\| \left[\sigma \; \left(\mathbf{h}_k^{\bar{k}}\right)^H \mathbf{w}_{\bar{k},p}^k \; \mathbf{h}_k^H \mathbf{w}_{\bar{k},c} \right] \right\| \leq \left(\mathbf{h}_k^k\right)^H \mathbf{w}_{k,p}^k, k = 1, 2$$

$$\|\mathbf{w}_{k,c}^k\|^2 + \|\mathbf{w}_{\bar{k},c}^k\|^2 + \|\mathbf{w}_{k,p}\|^2 \leq P, k = 1, 2.$$

Comparison with a Quantized Backhaul

As noted in the introduction, the transmission may alternatively be designed centrally and the backhaul used to provide each BSs with the signals they are to transmit [SSPS09b]. Since the finite capacity of the backhaul restricts the rate of these signals, one can interpret them as quantized versions of those that would be transmitted in the absence of this limitation. We compare such an approach to the proposed rate splitting by modifying the scheme in [SSPS09b] which was published in the form of a high complexity non-linear dirty paper coding (DPC)-based scheme using the so-called Wyner simplified channel model to our situation which deals with lower complexity linear precoding as well as more realistic fading channels. For $N_{\text{bs}} = 1$, the power minimization problem subject to SINR constraints, central to finding the boundary of the achievable rate region, can be transformed into a convex optimization, thus solved efficiently.

Numerical Results

Figs. 6.18(a)-6.18(c) show the rate regions corresponding to the proposed scheme, which we label *hybrid IC/BC*, the IC ($r_{k,c} = 0, k = 1, 2$), the BC scheme ($r_{k,p} = 0, k = 1, 2$), and the quantized use of backhaul for different values of backhaul capacity (we let $C_1 = C_2 = C$), for a particular channel instance with $N_{\text{bs}} = 1$. For low backhaul capacity, the rate regions corresponding to the hybrid scheme and the IC almost overlap and are both larger than the BC region. As the backhaul capacity increases, all 3 regions become larger (up to the point where the system is no longer backhaul-constrained), the BC region becomes larger than the IC region and closer to the hybrid scheme's region, until eventually these two regions overlap. Moreover, depending on the strength of the interfering links and on the backhaul constraints, one or the other scheme will be better.

Fig. 6.18(d) illustrates the average common to total rate ratio as a function of C, when $\alpha = 0.5$, for a Rayleigh block-fading channel such that $\mathbf{h}_k^k \sim \mathcal{N}_{\mathbb{C}}(0, \mathbf{I}_{N_{\text{bs}}})$ and $\mathbf{h}_k^{\bar{k}} \sim \mathcal{N}_{\mathbb{C}}(0, \beta \mathbf{I}_{N_{\text{bs}}})$, when $N_{\text{bs}} = 1$: As for earlier simulation results, parameter β controls the strength of the interference links. In general, the maximum cannot be achieved without sharing some data. How much depends on β.

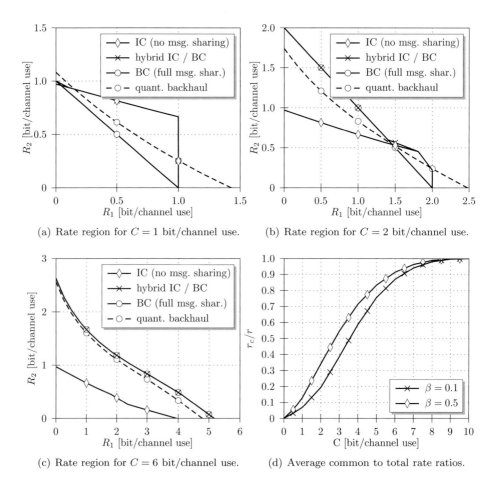

Figure 6.18 a)-c) Sample rate regions for 10 dB SNR, $N_{\text{bs}} = 1$,
$\mathbf{H} = [-0.8152 + \mathrm{i}0.8872 \quad 0.3489 - \mathrm{i}0.2163; -0.2150 - \mathrm{i}0.6359 \quad -0.2714 + \mathrm{i}0.1499]$,
d) average common to total rate ratios vs. backhaul for Rayleigh fading channels with $\mathbf{h}_k^k \sim \mathcal{N}_\mathbb{C}(0, \mathbf{I}_{N_{\text{bs}}})$ and $\mathbf{h}_k^{\bar{k}} \sim \mathcal{N}_\mathbb{C}(0, \beta \mathbf{I}_{N_{\text{bs}}})$.

6.4.3 Summary

In this section, decentralized downlink CoMP strategies suitable for reducing the backhaul required for information exchange between cooperating BSs were introduced. The backhaul related to channel state information exchange can be mitigated by the use of transmit beamforming schemes which explicitly account for the lack of CSI accuracy and the difference in CSI estimates at the involved base stations. The backhaul related to user data exchange can be reduced by limiting the exchange to only a fraction of the total traffic, and adjusting the ratio of shared versus non shared traffic with respect to key system parameters, such as the number of antennas, the interference strength and the backhaul capacity limits. Numerical evaluation showed the benefit of such approaches.

Part III

Challenges Connected to CoMP

7 Clustering

Patrick Marsch, Stefan Brück, Andrea Garavaglia,
Matthias Schulist, Ralf Weber and Armin Dekorsy

As mentioned in previous chapters, CoMP has the capability to significantly enhance spectral efficiency and cell-edge throughput. However, CoMP may require additional signaling overhead on the air interface and the backhaul, in particular joint signal processing CoMP as introduced in Chapter 6. Therefore, in practice only a limited number of base stations can cooperate in order to keep the overhead manageable. This raises the question which base stations should form cooperation clusters in order to exploit the advantages of CoMP efficiently at limited complexity.

In general, one can distinguish between static and dynamic clustering algorithms. Static clusters are kept constant over time and designed based on geographical criteria as the positions of base stations and the morphology of the surroundings. In the case of dynamic clustering, the system continuously adapts the clustering strategy to changing parameters such as user equipment (UE) locations and radio frequency (RF) conditions. Here, the central question is on which information the adaptation of clusters shall be based, and where in the system clustering decisions are made.

To illustrate concrete clustering results and their corresponding performance, we use two different setups in this chapter. On one hand, we consider an idealistic setup, i.e. a hexagonal layout of up to $M = 111$ cells, grouped into sites of 3 cells each, with an inter-site distance (ISD) of 500 m. We here calculate pathlosses according to a flat-plane pathloss model with

$$\text{PL} = 130.5 + 37.6 \cdot \log_{10}(d/\text{km}) \text{ [dB]}, \quad (7.1)$$

where d is the distance between transmitter and receiver, and take into account the impact of directive base station (BS) antennas with an azimuth-dependent attenuation of

$$\text{AL} = \min\left(12 \cdot \left|\frac{\theta}{70°}\right|^2, 20\right) \text{ [dB]} \quad (7.2)$$

and an antenna gain of 14 dBi. On the other hand, we observe a real-world setup with $M = 54$ BSs, as it exists in downtown Dresden, Germany, and again calculate pathlosses based on (7.1) and (7.2), but now also considering signal reflection, diffraction and obstruction based on ray-tracing using a 3D-model of

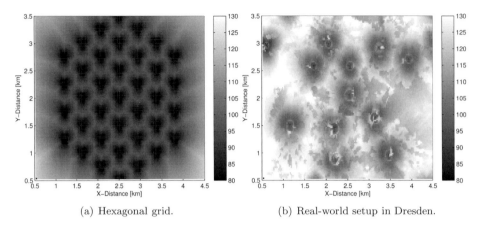

Figure 7.1 Setups considered in this chapter (pathloss to best serving cell in dB).

the city. The two setups are illustrated in Fig. 7.1, where the pathloss to the best serving cell is shown as a function of potential UE location. Clearly, the real-world setup has a larger average ISD than the hexagonal grid, but we use the former as it corresponds to the setup of the test bed discussed in Sections 13.2, 13.4 and 13.5, and the latter as it corresponds to standard next generation mobile networks (NGMN) simulation assumptions.

Before going into the details of static and dynamic clustering, let us introduce the concept of *ideal clustering* as the case where each potential UE location is served by exactly the set of cells to which it has the strongest links. This is clearly infeasible in practice, as it will be unlikely to find other UEs that can be jointly served through exactly the same set of optimal cells, and as this would involve a substantial signalling overhead between BSs. However, this concept serves as a good upper performance bound for any concrete clustering scheme usable in practice. Assuming for simplification that we have only $N_{bs} = 1$ BS antenna per cell, and that downlink joint signal processing CoMP is performed such that interference between jointly transmitted streams is completely removed and the maximum array gain is obtained (i.e. an idealistic assumption), we can state the downlink signal-to-interference-and-noise ratio (SINR) obtained by a UE j on an exemplary orthogonal frequency division multiplex (OFDM) sub-carrier if it is served by a cluster of cells $\mathcal{M}'$ as

$$\text{SINR}_j^{\mathcal{M}'} = \frac{P \cdot \sum_{m \in \mathcal{M}'} \lambda_j^m}{P \cdot \sum_{m \in \{\mathcal{M} \backslash \mathcal{M}'\}} \lambda_j^m + \sigma^2}, \tag{7.3}$$

where $\mathcal{M}$ is the set of all cells in the system, and λ_j^m is the path gain (linear and in terms of power) from UE j to the BS serving cell m. P is the transmit power, assumed to be equal at each BS, and σ^2 the noise variance. If each UE is now served by the *best* M' cells in terms of signal strength, cumulative distribution

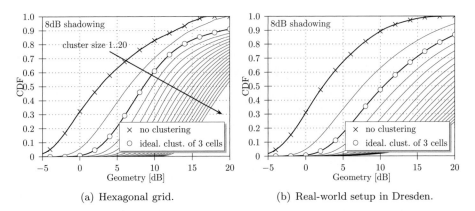

Figure 7.2 SINRs achievable under perfect interference cancelation in ideal clusters of different sizes.

functions (CDFs) of the resulting downlink SINRs can be calculated as shown in Fig. 7.2. This is typically referred to as the interference *geometry*, which in this case varies as different extents of (idealized) CoMP are applied. We can see that serving each UE by an ideal cluster of 3 cells offers a median SINR improvement of about 6 dB or 8 dB as compared to a cluster size 1 (i.e. without BS cooperation), for the idealistic and real-world setup, respectively.

7.1 Static Clustering Concepts

Patrick Marsch

Let us first focus on the case of *static clustering*, i.e. where clusters are created based on time-invariant information such as base station (BS) locations and signal propagation properties, considering *any potential* user equipment (UE) locations. Such schemes have been proposed in, e.g., [MF07a, BH07b, Ven07]. As mentioned before, it is unfeasible to serve each UE by the best set of cells in practice. Due to various practical constraints that will be revealed in the next chapters, we furthermore assume that cooperation takes place in clusters of no more than $|\mathcal{M}'| = 3$ cells. In the next subsections, we will hence explore how *fixed* clusters of 3 cells each can be defined so that most potential UEs locations can experience a moderate gain through CoMP, even though it may be served by a sub-optimal set of cells. We will use the lines marked with crosses and circles in Fig. 7.2 as lower and upper performance bounds, i.e. representing the non-cooperative case and idealized clusters of 3 cells, respectively, and denote all potential UE locations as a set $\mathcal{J}$. We generally differentiate between *non-overlapping* and *overlapping* clusters, which are treated in Subsections 7.1.1 and 7.1.2, respectively, before the impact of the introduced clustering schemes on the interference geometry is discussed in Subsection 7.1.3.

7.1.1 Non-Overlapping Clusters

Let us first consider the case where clusters may not overlap, i.e. where clusters are *disjunct* w.r.t. the cells involved. The question is now how such a set of clusters can be found in accordance to some performance metric at reasonable complexity.

For this, we consider two different optimization criteria. On one hand, it can be desirable to maximize the *mean* signal-to-interference-and-noise ratio (SINR) that the points in $\mathcal{J}$ can achieve under a particular fixed clustering. On the other hand, we can consider maximizing a certain *outage* measure, i.e. the number of locations in $\mathcal{J}$ for which a certain *minimum* SINR can be achieved. For both cases, let us assume that a set of C potential clusters has already been chosen heuristically (for example based on a ranking of the most frequently desired clusters for the locations in $\mathcal{J}$), and a matrix

$$\mathbf{A} \in \{0,1\}^{[M \times C]} \tag{7.4}$$

is given, where each non-zero element $a_{m,c}$ means that cell m is involved in cluster c. We now state the SINR achievable at a location j if served by cluster c similarly as in (7.3) as

$$\mathrm{SINR}_j^c = \frac{P \cdot \sum_{m \in \mathcal{M}, a_{m,c}=1} \lambda_j^m}{P \cdot \sum_{m \in \mathcal{M}, a_{m,c}=0} \lambda_j^m + \sigma^2}. \tag{7.5}$$

The *mean* SINR that could be achieved over all locations in $\mathcal{J}$ for a particular potential cluster c can then be stated as

$$f_c = \frac{1}{|\mathcal{J}|} \sum_{j \in \mathcal{J}} \mathrm{SINR}_j^c. \tag{7.6}$$

If this mean SINR is to be optimized, we have the optimization problem

$$\max\ \mathbf{f}^T \mathbf{x} \tag{7.7}$$
$$\mathrm{s.t.}\ \mathbf{A}\mathbf{x} \leq \mathbf{1}_{[M \times 1]} \tag{7.8}$$
$$\mathrm{and}\ \mathbf{x} \in \{0,1\}^{[C \times 1]}\ \mathrm{binary}, \tag{7.9}$$

where $\mathbf{x}$ is a vector of binary values stating whether a particular cluster has finally been selected or not, and the (element-wise) linear inequality constraint in (7.8) assures that each cell is involved in at most one cluster. This is a standard *binary optimization problem* with linear constraints, for which standard solvers can be used [BV04]. Note that averaging SINRs in linear domain does not reflect the fact that capacity is a concave function of SINR, but this has nevertheless shown to lead to a numerically stable optimization yielding good results.

If an *outage* criterion is to be optimized, we introduce a certain target SINR γ, and a matrix $\mathbf{B} \in \{0,1\}^{[J \times C]}$ where each element $b_{j,c}$ denotes whether a terminal in location j and served by cluster c can achieve the target, i.e.

$$\forall j, c: b_{j,c} = \begin{cases} 1 & \text{SINR}_j^c \geq \gamma \\ 0 & \text{otherwise} \end{cases}. \tag{7.10}$$

We then maximize the number of locations that can achieve the SINR target through a particular clustering concept by re-stating the optimization problem from (7.7) as

$$\max \mathbf{1}^T \mathbf{z} \tag{7.11}$$
$$\text{s.t. } \mathbf{B}\mathbf{x} \geq \mathbf{z}, \tag{7.12}$$
$$\text{and } \mathbf{A}\mathbf{x} \leq \mathbf{1} \tag{7.13}$$
$$\text{and } \mathbf{x} \in \{0,1\}^{[C \times 1]}, \mathbf{z} \in \{0,1\}^{[J \times 1]} \text{ binary}, \tag{7.14}$$

where $\mathbf{z}$ is a binary auxiliary variable indicating whether the chosen clustering approach fulfills the SINR targets at different locations or not. This is similar to an approach used for determining optimal BS locations in [MN00]. The problem can be re-written into the standard notation for binary linear optimization problems by jointly optimizing over both $\mathbf{x}$ and the auxiliary variable $\mathbf{z}$, i.e.

$$\max \begin{bmatrix} \mathbf{0}^T \mathbf{1}^T \end{bmatrix} \begin{bmatrix} \mathbf{x} \\ \mathbf{z} \end{bmatrix} \tag{7.15}$$

$$\text{s.t. } \begin{bmatrix} \mathbf{A} & \mathbf{0} \\ -\mathbf{B} & \mathbf{I}_{[J]} \end{bmatrix} \begin{bmatrix} \mathbf{x} \\ \mathbf{z} \end{bmatrix} \leq \begin{bmatrix} \mathbf{1}_{[M \times 1]} \\ \mathbf{0}_{[J \times 1]} \end{bmatrix} \tag{7.16}$$

$$\text{and } \mathbf{x} \in \{0,1\}^{[C \times 1]}, \mathbf{z} \in \{0,1\}^{[J \times 1]} \text{ binary}. \tag{7.17}$$

Typically, a standard problem solver would first treat $\mathbf{x}$ and $\mathbf{z}$ as arbitrary, positive real-valued variables (i.e. solve a linear programm), and then introduce additional linear constraints and perform branching to reflect the fact that $\mathbf{x}$ and $\mathbf{z}$ are constrained to be binary (often referred to as *branch-and-cut*). As the number of inequality constraints grows linearly in M and J, and the complexity of solving the problem can grow exponentially in this number of constraints, it is essential to choose reasonable values for C, M and J. Optimization complexity can for example be reduced significantly by optimizing smaller parts of cellular systems separately (hence equally scaling down C, M and J), performing a better heuristical pre-selection of clusters (reducing C), or reducing the granularity of potential UE locations (hence reducing the number of SINR constraints J).

In this work, for the setups stated before, we perform a pre-selection of potential clusters by observing which cluster of 3 cells would be preferred by which potential UE location, and then rank the clusters by their frequency of occurrence. We then further reduce the granularity of potential UE locations to obtain $J = 300$ SINR constraints. After solving the problems in (7.7) or (7.15), respectively, we then assign all potential UE locations at the initial granularity to the selected clusters according to the largest geometry. For both the pathloss data based on a flat-plane model and a hexagonal grid of BSs, as well as the

144   **Clustering**

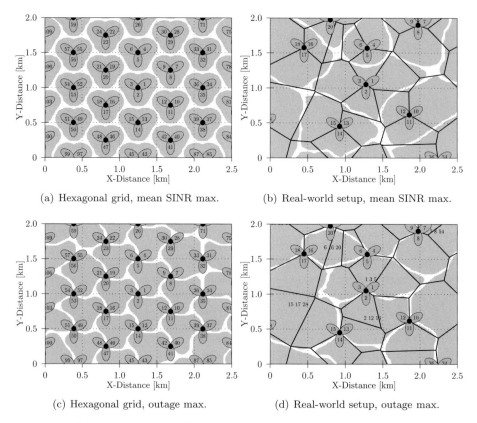

(a) Hexagonal grid, mean SINR max.

(b) Real-world setup, mean SINR max.

(c) Hexagonal grid, outage max.

(d) Real-world setup, outage max.

Figure 7.3 Optimization results for non-overlapping clusters.

data based on ray-tracing, the clustering results are shown in Fig. 7.3. We can here see the sets of locations that are assigned to particular clusters, where the indices of the involved cells are given (except in cases where this is obvious). For the hexagonal grid, the result is rather intuitive. In the case of a mean SINR optimization, complete sites are declared as clusters, as users at sector borders can benefit most from CoMP due to a large signal-to-noise ratio (SNR), while for the outage optimization, cells of different sites are grouped to clusters, as this can mainly increase the performance of the weak users. For real-world BS locations, however, clusters may of course span co-located as well as distributed cells. Clearly, the actual assignment of potential UE locations to the best serving cluster here leads to a more scattered result due to shadowing, but we have here averaged over these effects for illustration purposes. The performance obtainable with these clustering schemes is shown in Fig. 7.5 and will be discussed later.

7.1.2 Overlapping Clusters

Clearly, the clustering schemes introduced in the last subsection inherit the problem that UEs at the border between clusters will always experience a low SINR. Hence, one may consider using spatially *overlapping* clusters using different subsets of system resources [MF07a, Mar10]. This can be seen as a kind of *fractional frequency reuse*, but employed in such a way that the overall reuse factor is 1.

Let us assume, for example, that the system resources are split into $R = 3$ equally-sized resource blocks (RBs). Each cell can then be involved in up to 3 clusters with different partnering cells. We can solve both the problem of finding the optimal choice of overlapping clusters as well as the optimal assignment of resources to clusters by a simple extension of the problem in (7.7) to

$$\max \; [\mathbf{f}^T \mathbf{f}^T \mathbf{f}^T] \, \mathbf{x} \tag{7.18}$$

$$\text{s.t.} \; \begin{bmatrix} \mathbf{A} & \mathbf{0} & \mathbf{0} \\ \mathbf{0} & \mathbf{A} & \mathbf{0} \\ \mathbf{0} & \mathbf{0} & \mathbf{A} \\ \mathbf{I}_{[C]} & \mathbf{I}_{[C]} & \mathbf{I}_{[C]} \end{bmatrix} \mathbf{x} \leq \mathbf{1}_{[R \cdot M + C \times 1]} \tag{7.19}$$

$$\text{and} \; \mathbf{x} \in \{0,1\}^{[R \cdot C \times 1]} \; \text{binary}, \tag{7.20}$$

in the case that the mean SINR is to be optimized. We here simply observe $R \cdot C$ *virtual clusters* connected to one of the three resource blocks. The constraint in (7.19) assures that each cell is only involved in one cluster for each RB, and that each potential cluster is only chosen on one RB. Equivalently, if an outage measure is to be optimized, we can change (7.15) to

$$\max \; [\mathbf{0}^T \mathbf{1}^T] \begin{bmatrix} \mathbf{x} \\ \mathbf{z} \end{bmatrix} \tag{7.21}$$

$$\text{s.t.} \; \begin{bmatrix} \mathbf{A} & \mathbf{0} & \mathbf{0} & \mathbf{0} \\ \mathbf{0} & \mathbf{A} & \mathbf{0} & \mathbf{0} \\ \mathbf{0} & \mathbf{0} & \mathbf{A} & \mathbf{0} \\ -\mathbf{B} & -\mathbf{B} & -\mathbf{B} & \mathbf{I}_{[J]} \end{bmatrix} \begin{bmatrix} \mathbf{x} \\ \mathbf{z} \end{bmatrix} \leq \begin{bmatrix} \mathbf{1}_{[R \cdot M \times 1]} \\ \mathbf{0}_{[J \times 1]} \end{bmatrix} \tag{7.22}$$

$$\text{and} \; \mathbf{x} \in \{0,1\}^{[R \cdot C \times 1]}, \mathbf{z} \in \{J \times 1\} \; \text{binary}. \tag{7.23}$$

The clustering results for before mentioned setups and based on overlapping clusters are shown in Fig. 7.4. For a hexagonal setup, the result is interestingly almost the same, independent of whether mean SINR or outage is optimized. We can see that both co-located cells and those belonging to 3 different sites are grouped to clusters, and that each cell is now involved in exactly one *intra*-site and two *inter*-site clusters. The resulting clustering approach is similar to those proposed intuitively in [MF07a, Mar10]. For the real-world setup, the chosen clustering differs strongly depending on the optimization criterion, and a cell must not necessarily be involved in 3 clusters.

146   Clustering

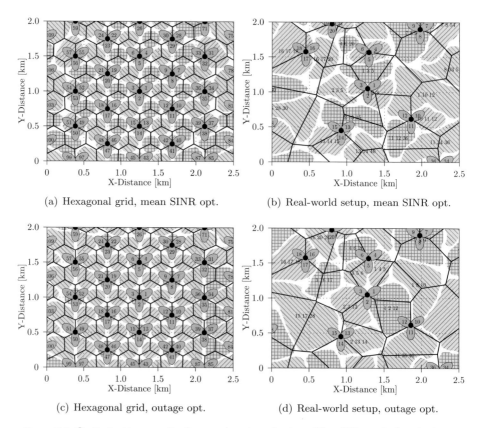

Figure 7.4 Optimization results for overlapping clusters. The differently hatched areas represent clusters connected to the three different system resource partitions.

7.1.3 Resulting Geometries

Fig. 7.5 finally shows the geometries that can be achieved with the different proposed static clustering strategies. As before, the upper two plots refer to the case where the mean SINR is optimized, and the lower two to the case of outage optimization. We generally consider fixed clusters, but calculate geometries based on many shadow fading realizations with a standard pathloss deviation of 2 dB or 8 dB, respectively, where the shadowing from one UE to multiple co-located cells is assumed fully correlated, and that to arbitrary cells has a correlation coefficient of 0.5. While a standard deviation of 2 dB is mostly used to model channels for indoor UE positions, a value of 8 dB reflects outdoor locations. As mentioned before, the case of no cooperation at all and ideal clusters of size 3 are considered as upper and lower bounds. In Plots 7.5(a) and 7.5(c), reflecting a hexagonal setup, we can see that static, non-overlapping clusters can help to improve either the strong users (mean SINR optimization) or the outage by about 3-5 dB. Significantly improved geometries can be obtained for overlapping clusters, where for the hexagonal setup there is no difference between mean SINR

7.1 Static Clustering Concepts

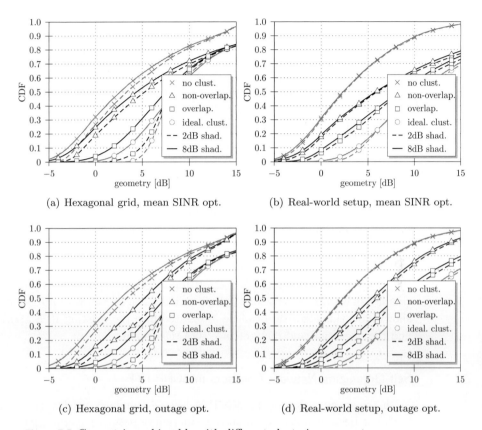

Figure 7.5 Geometries achievable with different clustering concepts.

and outage optimization. Performance then approaches that of ideal clustering within less than 2 dB. These results are also paralleled in Plots 7.5(b) and 7.5(d) for a real-world setup. Here, however, we can see a stronger benefit of non-overlapping clusters, as interference is more localized due to morphology than in a plat-plane model. Again, employing overlapping clusters allows to approach the performance of ideal clustering by less than 2 dB.

While previous results strongly suggest the usage of overlapping clusters, one must mention that these come at a certain price. As the system resources are split into $R = 3$ blocks, one looses multi-user diversity in each cluster, and also a certain extent of diversity in frequency for each UE [Mar10].

While the clustering techniques introduced in this section were based on optimizing the performance for *all potential* UE locations (as clustering is calculated once and then fixed), it is intuitively clear that better clustering results can be obtained if this is performed *dynamically*, i.e. for a set of *actual* UE locations, and changing over time. This will be addressed in the following section, along with the question of how clustering can be based on UE-centric measurement data available in legacy networks rather than on ray-tracing data.

7.2 Self-Organizing Clustering Concepts

Stefan Brück, Andrea Garavaglia, Matthias Schulist, Ralf Weber and Armin Dekorsy

Initial work on dynamic clustering based on radio frequency (RF) channel measurements of the mobile stations can be found in [PGH08] and [PA10]. A key requirement for any dynamic clustering algorithm is that it fits into the architecture of the radio access and/or the core network of LTE as described in [3GP10i]. The 3GPP standard already offers a framework for self-organizing concepts to support automatic configuration and optimization of the network [3GP10i]. The aim of this section is to introduce an adaptive terminal-aware clustering concept and to show how it can be integrated into the existing network architecture and the self-organizing network (SON) concept of LTE.

The remaining part of this section is organized as follows: First, we briefly provide an overview of the SON framework in LTE in Subsection 7.2.1. Then, the proposed adaptive clustering algorithm is introduced in Subsection 7.2.2 and simulation results are provided in Subsection 7.2.3. In Subsection 7.2.4, it is then shown how this algorithm can be integrated into the existing LTE architecture and the operation and maintenance (OAM) system.

7.2.1 Self-Organizing Network Concepts in 3GPP LTE

With the specification of the 3GPP Release 8 standard of LTE, the concept of SON has been included to pursue a reduction of network operational costs. The basic principle is to include appropriate instruments into the standards such that typically time-consuming tasks related to network operations can be automated as much as possible. With this principle in mind, different areas where such savings could be applied have been identified, e.g.:

- **Self-Configuration**: The ability of (a group of) network elements to configure themselves automatically, e.g. at power-up or after a major failure or change propagated by the OAM.
- **Self-Optimization**: The ability of the system to automatically adjust system parameters (could be per single site or cell, as well as per cluster) in order to maximize a certain predefined performance objective, according to the operator's strategy.
- **Self-Healing**: The ability of the system to recover from a major failure (for example a site gets out of service), by temporarily reconfiguring the surrounding elements or nodes in order to guarantee a minimum service quality.

The LTE system has been designed from the beginning to include these features by identifying what tasks could be automated both from a practical and feasibility point of view in a real system and specifying the means to achieve such automation (see [3GP10e], [3GP10i]).

Some functions have been specified in Release 8, like the automatic neighbor relation (ANR) function, the physical cell identifier (PCI) selection function and self-configuration functionalities, while others have been added in Release 9 and later refined. Among them are mobility robustness optimization, mobility load balance - see [3GP10e], [3GP10i] for latest updates. The specification work relates to both detailed interfaces within the network, which is mainly done in the radio access network (RAN) working groups, as well as details related to the management of the system (OAM), mainly done in the SA5 group (see www.3gpp.org). Both centralized and distributed functions are considered from an architecture point of view, where the particular choice is made case-by-case, based on trade-offs between complexity to implement the function and overall benefits for the operations.

The approach that is followed in this work for CoMP clustering also considers the possible impact on the standardization, and how the functionality could be integrated in the future 3GPP specifications. By looking at existing functions, it turns out that the ANR function, which collects radio quality information to automatically create and adjust neighbor relation tables (NRTs) in the systems, provides a good framework for the integration of a clustering algorithm with limited complexity. In fact, CoMP clustering could be regarded from this perspective as an extension of the information that is included in the neighbor relations, by considering what cells are suited to cooperate.

7.2.2 Adaptive Clustering Algorithms

The idea of adaptive clustering for CoMP is to provide the system with the ability to capture variations of the perceived radio environment and user locations, to achieve better CoMP performance. In fact, due to the time-variant characteristics of the wireless channel, the variations of system loading and the mobility of the users, it is expected that a clustering algorithm able to adapt to such conditions will enable CoMP to perform better from a system point of view than a static clustering approach as considered in Section 7.1, where all cooperative sets are pre-defined based on proximity information and on network planning predictions.

Let's consider a group of CoMP-enabled user equipments (UEs) served by a cell $m \in \mathcal{M}$ in the macro area where CoMP functionalities are principally available (which we denote as the CoMP *top cluster*). These UEs will report radio quality measurements to the serving cell $m \in \mathcal{M}$, which can thus collect statistics of RF measurements and exploit them for CoMP clustering purposes. One simple option to collect UE inputs is to make use of existing measurements, for example extracting data from measurement report messages (MRMs) in terms of averaged reference signal received power (RSRP) of the measured cells. Let us assume that a UE reports a set of cells $\mathcal{M}_c \subseteq \mathcal{M}$, where for example only the cells stronger than a certain configurable threshold are considered forming the set, reported in any arbitrary order. The serving cell m collects several of

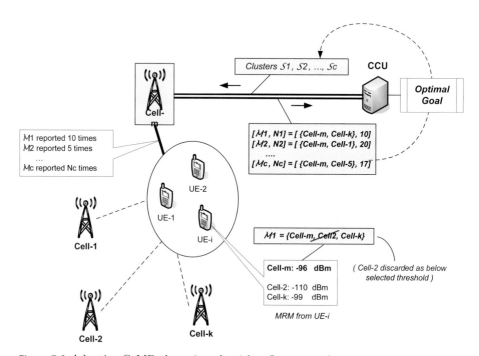

Figure 7.6 Adaptive CoMP clustering algorithm flow example.

such sets from different terminals over the observation period T and computes relevant statistical properties. In each serving cell, the reported information can be summarized with a list of pairs $[\mathcal{M}_c, N_c]$, whereby $N_c \geq 1$ is the number of occurrences the set $\mathcal{M}_c$ has been reported by all UEs to the serving cell during the period T. The idea behind $[\mathcal{M}_c, N_c]$ is that cell combinations that have been observed very often offer a higher potential to improve the system performance for several users when a CoMP scheme is adopted.

This information is eventually collected in a CoMP central unit (CCU) associated with the considered top cluster, which computes the cell clusters in an adaptive manner by optimizing selected objectives. Fig. 7.6 illustrates the steps of the entire process and the involved logical entities: At each period T, information is collected at the CCU and passed to an optimization algorithm that adapts the cell clustering and redistributes back the new sets to all base stations in the top cluster.

As in Section 7.1, the optimization problem to be performed by the CCU can be written in a classical linear programming notation. Let $\mathcal{C}$ denote a set of C potential clusters chosen based on the before mentioned UE reports and according to some heuristic. As before, binary matrix $\mathbf{A} \in \{0,1\}^{[M \times C]}$ states which cell is involved in which potential cluster. Different from Section 7.1, however, clusters may consist of varying numbers of cells within a heuristically chosen range

7.2 Self-Organizing Clustering Concepts

$\forall\, c:\ K_{\min} \leq |\mathcal{M}_c| \leq K_{\max}$. To each potential cluster c there is an associated cost σ_c and a cardinality $|\mathcal{M}_c|$, i.e. the number of cells belonging to the set. The choice of clusters from all potential clusters in $\mathcal{C}$ can now be stated as a *cost minimization* problem, i.e.

$$\min \sum_{c \in \mathcal{C}} \sigma_c x_c \qquad (7.24)$$

$$\text{s.t.} \ \ \mathbf{A}\mathbf{x} \geq 1 \qquad (7.25)$$

$$\text{and} \ \ \mathbf{x} \in \{0,1\}^{[C \times 1]} \ \ \text{binary.} \qquad (7.26)$$

where $\mathbf{x} \in \{0,1\}^{[C \times 1]}$ is a binary vector stating which clusters have finally been selected, and (7.25) assures that each cell is involved in at least one cluster. If this latter constraint is changed to an equality, disjoint clusters are enforced, as in Section 7.1.1. Furthermore, equally-sized clusters can be obtained by choosing $K_{\min} = K_{\max}$. One of the key factors to define an appropriate optimization problem is the selection of the cost function σ_c. Looking at the CoMP functionality, a trade-off between system complexity and performance could for example be obtained by making the cost proportional to the cluster cardinality $|\mathcal{M}_c|$ or to the number of required X2 interfaces (as large clusters increase system complexity) and inversely proportional to the combined radio conditions of the cluster cells (better radio conditions means higher performance). In order to account for the number of UEs that would benefit from a certain cluster c, the cost can included a term inversely proportional to N_c, i.e. to the number of UEs that have reported the cluster, leading to

$$\sigma_c \propto \frac{|\mathcal{M}_c|}{\sum_{m \in \mathcal{M}_c} \text{RSRP}_m \times 10^{N_c}}, \qquad (7.27)$$

where RSRP_m represents the average of the UE measurements for cell m expressed in linear scale and collected over all received measurement reports. Equation (7.27) captures the combined radio conditions of the set of cells $\mathcal{M}_c$ as an estimate of the CoMP performance potential of that cluster. This cost function has been selected heuristically after comparing different similar options and will be used later on for a simulative analysis of the algorithm. As mentioned in Section 7.1, obtaining the optimum of the problem described by equations (7.24) to (7.27) may be of prohibitive computational complexity, depending on the number of cells and potential clusters. Hence, we describe a simplified, heuristic algorithm leading to a good sub-optimal solution in the sequel:

1. Generate a set $\mathcal{C}$ of potential clusters according to cardinality constraints in an exhaustive way
   - This is possible as the computational complexity, which grows exponentially with the number of cells and neighbor relations, remains small in practical cases, considering a top cluster includes around 30-40 cells in realistic cases and $K_{\max}$ is practically constrained to a few cells, comparable

to the active set size in wideband code division multiple access (WCDMA) systems.
2. Associate to each potential cluster $c \in \mathcal{C}$ its cost σ_c according to (7.27)
3. Create an initial optimization solution by adding the sets in increasing cost order, till all cells are included in the final solution or there are no more potential clusters available
   – A modified version is needed in case of disjoint sets (set partitioning), as at each step the sets overlapping with the ones already put in the solution shall be removed from the candidate list - the process stops when all cells are covered or the candidate list is empty.
4. Improve the solution by step-wise replacing two (or more) sets with one set not yet included, whose cost is lower than the sum of the costs of the replaced sets.
   – This is in fact the only way to decrease the cost, as the initial solution was built by selecting sets in cost-increasing order.

Cluster computations and simulation results following this scheme are detailed in the next section.

7.2.3 Simulation Results

In order to evaluate the performance of the adaptive clustering principle, system level simulations were run employing the hexagonal setup shown in Fig. 7.7 and the simulation assumptions according to [3GP10d]. A 3GPP reference network layout was configured with 19 3-sector sites of 500 m inter-site distance (ISD). Each of the 57 cells was equipped with $N_{bs} = 2$ antennas of 15 degrees downtilt. Herewith, the typical 3GPP urban macro spatial channel model extended (SCME) in the 2 GHz band was used. A number of 100 UEs were placed at random locations within each of the 4 hotspot areas indicated in Fig. 7.7. UEs were simulated with $N_{ue} = 2$ antennas moving at a speed of 3 km/h. For each UE, the 8 strongest interfering sectors were simulated as spatially correlated. A signal bandwidth of 5 MHz was used, and the maximum transmit power per sector was set to 20 W (43 dBm) per 5 MHz.

Fig. 7.7 shows the result of the applied clustering algorithm which was configured to obtain the optimal solution for a disjoint set of clusters with up to 3 cells using a shadow fading standard deviation of 2 dB. The clustering algorithm took all RSRP measurements from UEs into account that were greater than -120 dBm. It is apparent that for the two circular-type UE hotspots on the upper right and lower left, the closest 3 sectors $(4, 6, 26)$ and $(14, 16, 42)$ from three different sites were selected, respectively. For the other two line-type hotspots on the upper left and lower right, the three geographically closest sectors $(10, 33, 35)$ and $(18, 20, 52)$ in the middle of the area as well as the adjacent cells each belonging to a different site were selected.

7.2 Self-Organizing Clustering Concepts

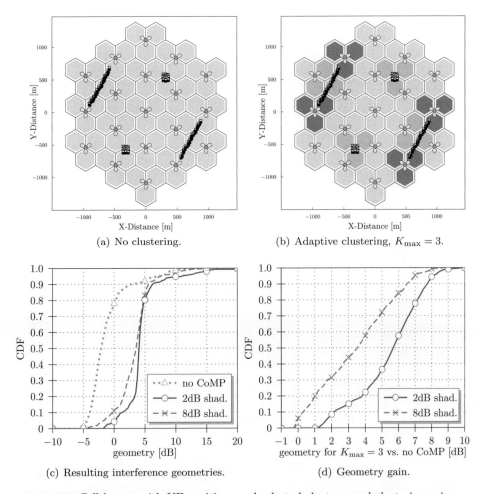

Figure 7.7 Cell layout with UE positions and selected clusters, and clustering gains.

In order to assess the performance of the adaptive clustering algorithm, network simulations were run with the calculated clusters obtained in Fig. 7.7(b). As in the previous section, performance is measured via the *interference geometry* introduced in (7.3). The cumulative distribution function (CDF) of UE geometries obtained for the calculated CoMP clusters using 2 and 8 dB shadow fading standard deviation are depicted in Fig. 7.7 c) and compared to the corresponding geometries if the UE is served by one cell only[1]. While a shadow fading with 8 dB standard deviation represents the default value for outdoor scenarios, a standard deviation of 2 dB is selected for indoor scenarios. It is seen in the figure

---

[1] Simulations without CoMP resulted in similar geometry curves for the selected UE distributions in Fig. 7.7 a) for different shadow fading standard deviations due to the underlying 3GPP cross-correlation coefficient which is set to 0.54 for interfering cells from other sites and to 1.0 for interfering cells belonging to the same site.

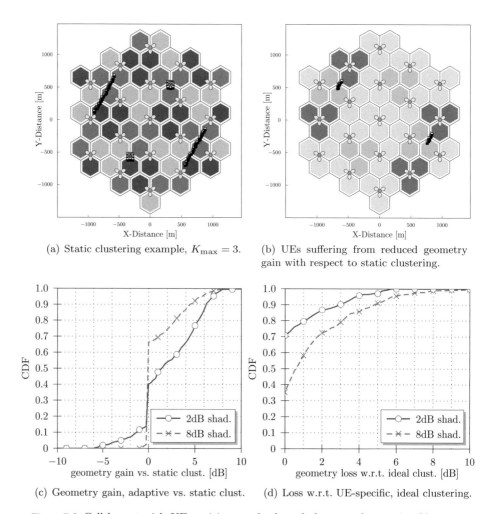

Figure 7.8 Cell layout with UE positions and selected clusters, cluster size $K_{\max} = 3$.

that for 50% of the observed geometries, the CoMP clustering algorithm results in a 6 dB better geometry environment. Though the CDFs in Fig. 7.7 c) are showing the geometry statistics of all UEs, the curves do not reflect the effective improvement experienced by individual UEs at the very same position in the network. These are instead represented in Fig. 7.7(d).

For the outdoor case with 8 dB shadow fading standard deviation, the CoMP cluster in Fig. 7.7(b) achieves a median geometry improvement of 3.5 dB for individual UEs, whereas for the indoor case, the reduced standard deviation of 2 dB leads to an even higher median gain of 5.7 dB.

One drawback of the adaptive clustering algorithm compared to pre-defined (static) clusters is the need for an additional control entity in the network and the increased signalling overhead to estimate a good set of clusters. Such addi-

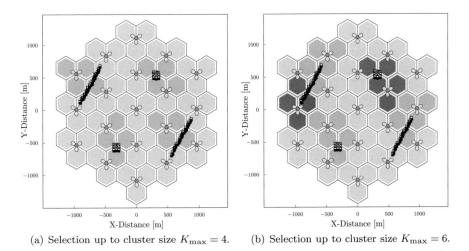

Figure 7.9 Clustering results for different maximum cluster sizes $K_{\max}$.

tional complexity is only justified if gains compared to static clustering can be achieved. In order to evaluate the performance improvement of the adaptive algorithm, simulations were run and compared with a static cluster as defined in Fig. 7.8(a). The selection of the static cluster was done based on empirical proximity layout considerations, where each cell of a cluster belongs to a different site. A wrap-around mechanism was used to assign sectors at the borders of the macro cluster. As can be seen from Fig. 7.8(c), using a radio channel aware adaptive clustering, 32% (60%) of the UEs experience a performance improvement, 66% (30%) experience the same performance in terms of geometry gain for 8 dB (2 dB) shadow fading standard deviation, respectively.

The optimization algorithm for adaptive CoMP clustering considers all UEs and selects a set of clusters under the given constraints (i.e. number of cells per clusters, disjoint sets etc.) in order to optimize the performance for the majority of UEs. However, some UEs may not experience an improvement if compared with a different cluster set selection depending on their location and the standard deviation of the shadow fading. Herewith, UEs can be affected that are located in areas between sectors which do not belong to the same CoMP cluster. This can be seen from Fig. 7.8(c), where around 14% of the UEs perceived a lower geometry than the static cluster set of Fig. 7.7(c) in the 2 dB shadow fading case. While in the static cluster case, these UEs were located between sectors belonging to the same CoMP cluster, they were located at the border of two CoMP clusters selected by the adaptive clustering algorithm. In case of the higher shadow fading standard deviation of 8 dB, this effect is reduced and affects only 2% of the UEs. Fig. 7.8(b) depicts those UEs that experience a lower geometry with adaptive than with static clustering.

As stated at the beginning of this chapter, the case of UE-specific clustering, where each UE is assigned to its set of best-serving cells, can be seen as a

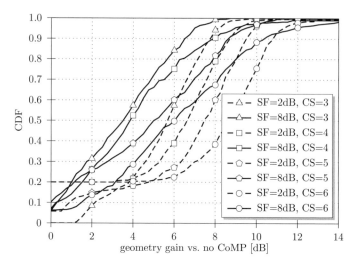

Figure 7.10 Interference geometry obtained with different cluster sizes.

performance benchmark. The performance gap between the proposed UE-aware adaptive clustering algorithm leading to Fig. 7.7(b) and UE-specific clustering with cluster size 3 is shown in Fig. 7.8(d). In the 2 dB shadow fading case, 30% of the UEs would experience an improvement, where 10% perceive more than 3 dB geometry gain. For 8 dB shadow fading standard deviation, 64% of the UEs enjoy better performance and 10% even benefit from a 5 dB geometry gain.

In order to improve the performance gain obtained through adaptive clustering, higher cluster set sizes and/or overlapping clustering can be applied, as in Section 7.1.2. Fig. 7.9 shows the cluster selection and performance for increased cluster sizes of up to 4 and 6 cells per cluster. It is apparent from Fig. 7.10 that further geometry gains can be achieved when higher number of sectors per cluster can be accomplished by the network. For example, in case clusters of up to 6 cells can be selected, the median geometry gain compared to no clustering reaches 6 and 9 dB for shadow fading standard deviations of 2 and 8 dB, respectively. Compared with a maximum cluster size of $K_{max} = 3$, this gives an additional performance improvement of $2.5 - 3$ dB. Nevertheless, the additional price to pay with the use of larger cluster sizes is again an increased complexity and signalling amount needed to coordinate the cluster scheduling.

At the end of this section, some aspects that deserve further investigations shall be briefly mentioned. The presented results have been achieved based on the cost function from (7.27). During the investigations it turned out that the performance is very sensitive to changes in the cost function. Further investigations of reliable cost functions are therefore an interesting topic for future investigation. It should further be noted that the antenna pattern model that has been chosen according to [3GP10d] has a strong backlobe component, which has a significant impact on the footprint of the sector and therefore impacts clustering. Considering more realistic antenna patterns is therefore of interest for future evaluations. Another aspect to be mentioned is the modeling of shadow fading in the 3GPP spatial channel model (SCM). Shadow fading is modeled as being spatially uncorrelated in the SCM. A more realistic model, however, would take a correlation of the shadow fading over distance into account. This missing correlation impacts the presented results as well, since UEs being located next to each other can measure very different RSRP values from the same cells. Taking a spatial correlation of shadow fading into account, it becomes more likely that closely located UEs report similar sets of cells S_j, which should improve the reliability of the adaptive clustering algorithm.

7.2.4 Signaling and Control Procedures

In this subsection, it is illustrated how self-organizing clustering concepts can be included in 3GPP LTE-A as described in [3GP10i]. In particular, it is shown how the SON functionality named ANR can be enhanced to enable clustering of cells. The purpose of the ANR function is to relieve the operator from the burden of manually managing neighbor relations (NRs), which are needed to e.g. execute handovers from one cell to the other.

The ANR function resides in the enhanced Node B (eNB) and manages the conceptual NRT. One main sub-function is the neighbor detection function that finds new neighbors and adds them to the NRT. Moreover, to remove outdated neighbor relations, the ANR function also contains the neighbor removal function. Both subfunctions are implementation specific. Further, an NR in the context of the ANR functionality is defined between a source and a target cell. The NR relates the source and the target cell such that the source cell

- knows the enhanced UMTS terrestrial radio access network (E-UTRAN) cell global identifier (CGI) and the PCI of the target cell.
- has an entry in its NRT in order to identify the target cell.
- has set the attributes in the NR for the target cell. Either the attributes are configured by default or set via OAM.

The NRT contains an entry for each NR with a target cell, which is addressed by the target cell identifier (TCI).

The base station instructs its mobile stations to perform measurements in order to identify the TCI of neighbor cells. Both the type of the measurement as well as the periodicity of the reports is determined by the base station. A typical example of those measurements are average received power levels obtained from the LTE reference signals (RSRP). Since the NRT contains mid to long term measurements obtained from specific mobile stations, this knowledge of the base station about neighbor cells can be exploited for adaptive cell clustering.

In the following, a framework for control signaling and the network architecture is presented that allows using the ANR and the NRT for cell clustering.

158 Clustering

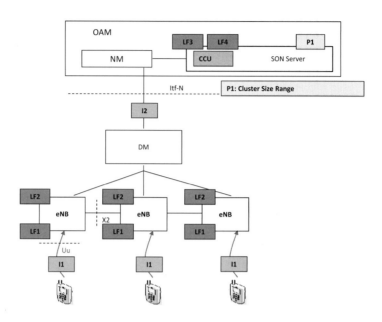

Figure 7.11 Architecture for adaptive cell clustering.

To manage clustering of an entire network with a huge number of cells with acceptable complexity, we firstly group a set of cells in top-clusters that are pre-computed and static. We assume that a CCU controls such a top-cluster, such that the clustering method introduced here is a central approach as well. Distributed approaches are not covered by the presented architectural framework.

The main task of the CCU is to determine sets of cells within a top-cluster that form a CoMP cluster. For simplicity of the signaling, it is assumed that all cells of a base station belong to the same top-cluster. Fig. 7.11 depicts the architecture for the adaptive cell clustering. It is assumed that the CCU is part of the SON server of the OAM functionality. The OAM system is connected by the Itf-N interface to the domain manager device manager (DM), which again is connected to several eNBs. Fig. 7.11 further illustrates the inputs (Ix), the parameters (Py) as well as the logical functions (Lz) being required for adaptive cell clustering. In this architecture, the mobile stations report over the LTE-A air interface MRMs including e.g. RSRP measurements and PCIs of those cells that are strongly received to their serving base stations (I1). It should be noted that for the algorithm described above only existing LTE Release 8 messages are required. The base station extracts the RF measurements and set of cells contained in the MRM (LF1). These measurements are used to update the NRT and to set extension attributes in the NRT required for CoMP clustering (I2). This includes the RF measurement values themselves as well as the frequency of the measurements. Once the NRT is updated, the base stations transmit the extended NRTs to the CCU via the DM and the Itf-N interface (LF2). The CCU in the SON server receives the updated and extended NRT (LF3) and computes the updated clusters (LF4) within a top-cluster as a result of the optimization algorithm described in the previous subsections. Finally, the CCU transmits the updated cluster information (LF3), which is then received by the base stations (LF2). If a master/slave concept is applied for CoMP, the message may also contain information which cells act as master and slaves. Since in this architectural framework the CCU is part of the OAM, the Itf-N interface is impacted and needs to be enhanced to support sending the extended NRT tables to the CCU and the updated cluster information back to the base stations. Principally, the CCU could also be part of the serving gateway (S-GW) or the mobility management entity (MME). In this case the S1-U and the S1-C interfaces could be extended, respectively. In case the CCU is a completely separate entity, introducing a new interface is required.

7.3 Summary

In this chapter, cell clustering techniques were observed, which are an essential prerequisite of using CoMP in practical cellular systems. One can here mainly differentiate between *static* clustering concepts, where certain sets of cooperation-enabled base stations are defined once and then fixed, and *adaptive*, or *self-organizing* clustering concepts, where clusters are adapted over time to user locations. Interference geometries have been introduced as a performance metric for clustering, where all results can then be compared to the theoretical benchmark of *UE-specific, ideal* clustering, where each terminal is served by the best possible set of cells. We have seen that static clustering can already obtain a geometry within a few dB of that of UE-specific clustering if *overlapping* clusters are defined, hence if each cell is involved in multiple clusters connected to different portions of the system resources. Performance can be further improved at the expense of signaling overhead through adaptive clustering, where a concrete algorithm based on terminal-side radio channel measurements already supported in LTE Release 8 was described. With adaptive clustering, up to 70% of all terminal stations experienced geometry gains compared to the case of static, non-overlapping clusters. It has to be noted, however, that the performance of adaptive clustering is very sensitive to the choice of the cost function, which is hence an important topic for future investigations.

Finally, it was shown that the proposed adaptive clustering scheme fits well into the SON framework of LTE with only slight extensions of the system architecture and of the already existing SON ANR concept. In this case, the clustering algorithm is run on the SON server of the OAM system.

8 Synchronization

This chapter deals with another major challenge connected to CoMP, namely the synchronization of cooperating and cooperatively served devices in time and frequency. On one hand, there are different local oscillators in each base station and mobile terminal that lead to deviations in the carrier frequency according to its nominal value. On the other hand, there are variations in the symbol timing between each transmitter and receiver station. Both effects need to be compensated by synchronization techniques.

In cellular networks, we can distinguish between a network synchronization among all involved base stations and the alignment of the user equipments to that time and frequency reference. The basic definitions of the synchronization terms as well as procedures for the reference network synchronization are described in Section 8.1. The impact of symbol timing mismatches on CoMP is then treated in Section 8.2, before Section 8.3 concludes this chapter with the analysis of the impact of residual carrier frequency offsets on CoMP performance.

8.1 Synchronization Concepts

D. Richard Brown III and Andrew G. Klein

Synchronization is the process of establishing a common notion of time among two or more entities. In the context of wired and wireless communication networks, synchronization enables coordination among the nodes in the network and can facilitate applications such as distributed sensing. Precise synchronization can also facilitate scheduling of communication resources as well as interference avoidance in multi-access networks. This section provides an overview of some of the synchronization concepts and techniques used in coordinated communication networks.

8.1.1 Synchronization Terminology

In the context of wireless communication networks, each node in the network keeps a local notion of time, i.e. a clock, by counting cycles of a local oscillator (LO). Among other parameters, all oscillators are characterized in terms of their nominal frequency and accuracy. The accuracy of an oscillator is typically specified in parts per million (ppm) with respect to the nominal frequency. For example, a 10 MHz LO with ±10 ppm accuracy oscillates with a frequency between ±100 Hz of the LO's nominal 10 MHz frequency. Low-cost oscillators typically provide ±100 ppm accuracy [FMd09], and higher-cost temperature-compensated or oven-controlled oscillators can provide accuracies better than ±1 ppm [Rak09]. A clock derived from an unsynchronized low-cost ±100 ppm LO can gain or lose, with respect to a perfect reference clock, up to 8.64 seconds in a day.

Two clocks are said to be perfectly *syntonized* if they agree exactly on the duration of an interval between two events. In other words, syntonized clocks share the same rate or frequency, but there is no requirement for the clocks to agree on the time of a single event. Syntonized clocks are sometimes said to be frequency synchronized. A clock can be said to be syntonized to a specified level of uncertainty if the frequency difference with respect to a reference clock (often normalized and specified in ppm) is no more than the uncertainty. This frequency difference is commonly called *frequency offset* or *skew* and is typically specified in statistical terms, e.g. standard deviation or maximum frequency offset. For example, the LTE specification requires user equipments (UEs) to have a maximum frequency offset of ±0.1 ppm [STB09].

Two clocks are said to be perfectly *synchronized* if they agree exactly on the time of occurrence of an event at an arbitrary time. Note that synchronized clocks must also be syntonized since unsyntonized clocks can only agree on the occurrence of an event at a particular time. Clocks can also be said to be synchronized to a specified level of uncertainty if they do not agree precisely on the time of occurrence of an event, but the difference in the measured event times between the clocks is no more than the uncertainty. The time difference between two clocks is commonly called *clock offset* or *phase offset* and is typically specified in statistical terms, e.g. standard deviation or maximum clock offset.

Since each node in a wireless network keeps time with its own LO, syntonation and synchronization is necessary to establish a common time base among the nodes in the network to a desired level of precision. In fact, if the nodes in the network have a maximum clock offset requirement, periodic frequency and phase re-synchronization is necessary to correct for unavoidable phase drift between any pair of nodes caused by frequency offset and oscillator instabilities. Several factors can affect the accuracy of phase and frequency synchronization among nodes in communication networks. These factors include local oscillator stability, network stability, and the re-synchronization interval, i.e. how often synchronization messages are exchanged among the nodes in the network.

Figure 8.1 shows an example of clock and frequency offset between two clocks labeled "clock A" and "clock B". The re-synchronization interval in this example is denoted as T. Prior to the first synchronization attempt at time t_1, the phase of clock B is ahead of clock A and the phases of the clocks are drifting apart due to the frequency offset. At time t_1, clock B adjusts its frequency and phase

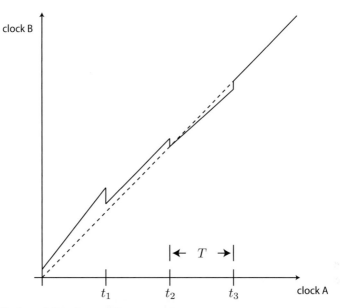

Figure 8.1 Clock and frequency offset example.

in response to a synchronization attempt with clock A. In this example, the clocks become syntonized at t_1, but they remain unsynchronized. At time t_2, the clocks again attempt to synchronize. In this example, clock B and clock A are temporarily synchronized at time t_2, but are now unsyntonized. The nodes become synchronized and syntonized after the synchronization attempt at t_3.

In the following subsections, we provide an overview of several synchronization techniques suitable for distributing a common notion of time to nodes in wired and wireless communication networks. We begin with a discussion of two *network synchronization* techniques: network time protocol (NTP) and precision time protocol (PTP), i.e. IEEE 1588. We then describe *satellite-based synchronization* techniques with a focus on global positioning system (GPS). We conclude with a discussion of *endogenous wireless distributed synchronization* techniques suitable for the precisely synchronizing and syntonizing the carriers of wireless transmitters for distributed phase coherent communication.

8.1.2 Network Synchronization

The Network Time Protocol

The NTP is a protocol for synchronizing the clocks of nodes that are connected through variable-latency networks [Mil91]. NTP is an application-layer protocol that operates over the Internet protocol (IP), and can therefore be implemented completely in software. The protocol has been in use since the 1980's, and today it is responsible for synchronizing the clocks of the majority of computers connected to the Internet. Nodes in the network are assigned to a class or *stratum*,

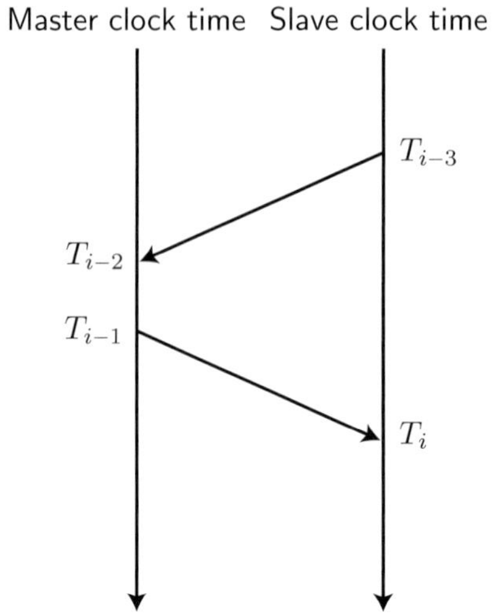

Figure 8.2 NTP message exchange.

and those with the lowest stratum number are assumed to be perfectly synchronized with Coordinated Universal Time (UTC). Nodes with higher stratum numbers synchronize their clocks with nodes having lower stratum numbers. This hierarchical structure of NTP results in it being highly scalable.

To estimate clock offsets, a master and slave exchange timestamps which are 64-bit descriptions their current local clock time. Figure 8.2 demonstrates the exchange of timestamps between a master and slave. If T_i, T_{i-1}, T_i, and T_{i-1} are the four most recent timestamps, then the clock offset of the slave relative to the master at time T_i can be calculated via

$$\Delta_i = \frac{T_{i-2} + T_{i-1} - T_{i-3} - T_i}{2}. \tag{8.1}$$

Since each NTP message contains the last three timestamps $T_{i-1}, T_{i-2}, T_{i-3}$, and the final timestamp T_i is estimated upon arrival of the message, the clock offset can be estimated from a single message exchange between slave and master.

Equation (8.1) implicitly assumes that the two transmission paths are symmetric and have equal delay. In practice, however, network delays are stochastic quantities. Consequently, NTP performs multiple offset estimates in combination with a filtering and selection scheme to obtain a more accurate estimate of the clock offset. The estimated clock offsets are fed to a Type-II adaptive parameter phase-locked loop (PLL), which corrects the LO phase and frequency. An adaptive Type-II PLL has one integrator in the loop filter (or two poles in the open-loop transfer function) and continuously adjusts the phase and frequency [Smi86].

The accuracy of the protocol depends on a variety of factors, including the update interval and network topology. Several studies (e.g. [Mil03, MTH97,

KZM07, Min99]) have investigated the performance of NTP under typical use, showing that clock offsets have a standard deviation on the order of several milliseconds, and residual frequency offsets on the order of ±0.1 ppm.

The Precision Time Protocol
Also known as IEEE 1588 [IEE08a], the PTP attains sub-microsecond accuracy which is necessary in applications such as networked control systems and precision machinery in factories. The phase and frequency correction in PTP are quite similar in principle to NTP: after a sequence of messages are exchanged between slave and master, the clock offsets are estimated through filtering and selection, and are used to adjust a PLL which corrects the LO phase and frequency. There are, however, several fundamental differences between PTP and NTP. The primary difference is that PTP is implemented in hardware rather than software. By moving the clock synchronization as close to the physical layer as possible, sources of jitter and processing delay introduced in network layers higher up the stack can be mitigated. In addition, PTP is primarily intended to be used in a local area network (LAN) setting as opposed to NTP, which may synchronize to an Internet clock reference located some far distance away. While PTP can achieve a higher accuracy than NTP, it does require the use of dedicated hardware. The performance of PTP will again depend on a variety of factors, including the quality of the LO, as well as the network topology. Products already available on the market today [Sem10] claim clock offsets within 1 μs and frequency offsets better than ±0.01 ppm. Similar results were achieved [Ton05] in a test of PTP over a metropolitan area network.

The most recent version of the standard, referred to as IEEE 1588-2008, offers a *transparent clock* mode which requires dedicated network switches that support the standard. Such switches employ a transparent clock that further minimizes delay by providing an alternate local clock for network nodes so that they need not rely on the master clock. This mode permits maximum clock offset errors on the order of tens of nanoseconds [HJ10].

8.1.3 Satellite-Based Synchronization

A Global Navigation Satellite System (GNSS) permits nodes to determine their location to within a few meters using time signals received line-of-sight from satellites. While the primary intent of a GNSS is for determining position information, such systems are also very useful as an accurate, common clock reference. In contrast to NTP and PTP, clock synchronization using a GNSS is done wirelessly using one-way communication links (i.e. by receiving signals broadcast from the satellites). In order for a terrestrial node to be able to receive the relatively weak signals from distant satellites, however, a line-of-sight link is typically necessary. In the absence of precise location information, a node must be able to receive signals from four satellites since there are four unknowns: latitude, longitude,

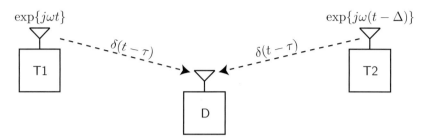

Figure 8.3 Two-transmitter one-destination distributed beamforming scenario.

altitude, and time. If precise location information is available, only one satellite is needed for clock synchronization since propagation delay is known.

Examples of GNSS's include the United States' GPS, the Russian GLObal NAvigation Satellite System (GLONASS), and the European Galileo system. As stated in [LAK99], GPS provides clock synchronization "to better than 100 ns in time and 10^{-13} in frequency." Other satellite systems are expected to give synchronization accuracy of a similar order, as they share many of the same parameters as GPS [HP05a].

8.1.4   Endogenous Distributed Wireless Carrier Synchronization

Coherent downlink CoMP techniques have recently been proposed in which the base stations (BSs) transmit with phase-aligned carriers such that the *bandpass signals* are aligned at the intended destination. These coherent transmission techniques require the BSs to accurately pre-compensate for the downlink channel phases and maintain close synchronization. One approach to coherent downlink CoMP, as discussed in Sections 13.3 and 13.4, is to closely syntonize the BSs' carriers using, for example, highly stable GPS-referenced local oscillators. Coherent downlink transmission can then be achieved by having the mobiles estimate the downlink channel state information (CSI) and feeding this back to the BSs for carrier phase pre-compensation. However, periodic downlink CSI re-estimation and low-latency feedback on the order of a few milliseconds is necessary to maintain phase coherence at the mobile, as pointed out in Sections 13.3 and 13.4, and the used oscillators may be considered a cost issue in a large-scale deployment.

A different approach to coherent downlink CoMP is to have the BSs endogenously synchronize their carriers without the aid of GPS and without CSI feedback from the mobiles. The "two-way" downlink beamforming technique proposed in [PD10] is one example of this approach. Two-way downlink beamforming is a retrodirective transmission technique in which the BSs closely synchronize their carriers (in both phase and frequency) to emulate a conventional retrodirective antenna array [Pon64] and achieve coherent downlink transmission through uplink channel conjugation. Note that this approach requires uplink/downlink channel reciprocity and also precise synchronization of the BSs.

8.1 Synchronization Concepts

To understand just how accurately the BSs must be synchronized to facilitate retrodirective transmission, consider the two-transmitter distributed beamforming scenario shown in Fig. 8.3 where both BSs simultaneously transmit unmodulated carriers at radian frequency ω with the goal of having the carriers arrive with identical phase, i.e. "coherently combine", at the destination, i.e. the mobile. Note that the BSs are implicitly syntonized in this scenario, but they are unsynchronized such that transmitter 2 has a clock offset of Δ with respect to transmitter 1. After propagation through the unit-gain single-path channels, the received signal at the destination can be written as

$$y(t) = \exp\{j\omega(t-\tau)\} + \exp\{j\omega(t-\Delta-\tau)\},$$

where the baseband signals modulated by each carrier are omitted for clarity. The received power can be computed as $|y(t)|^2 = 2 + 2\cos(\omega\Delta) \leq 4$. When the transmitters are perfectly synchronized, i.e. $\Delta = 0$, the carriers combine coherently at the destination and the received power $|y(t)|^2 = 4$. This corresponds to the "ideal coherent" case in distributed beamforming. When the transmitters are not synchronized, the received power will be less than in the ideal coherent case. To illustrate the effect of unsynchronized transmitters, the clock offset Δ can be modeled as a zero-mean Gaussian distributed random variable with standard deviation σ_Δ. Fig. 8.4 shows the received power at the destination as a function of σ_Δ and carrier frequency $f_0 = \frac{\omega}{2\pi}$. This example shows that, even at the lowest carrier frequency of 800 MHz, the standard deviation of the transmitter clock offset must be smaller than approximately 130 picoseconds in order to achieve, on average, 90% or better of the ideal coherent received power. Please note that this level of synchronization accuracy is not required by the downlink CoMP techniques described in the remainder of this book, which use GPS and CSI feedback to achieve baseband signal alignment at the mobile. The approach described here is based on *distributed beamforming* in which the *radio frequency* signals of the base stations are aligned at the mobile.

Since conventional synchronization techniques like GPS and PTP are unable to provide the accuracy required for retrodirective downlink transmission at typical radio frequencies, several recent studies have focused on the development of precise endogenous distributed wireless carrier synchronization techniques including full-feedback closed-loop [TP02], one-bit closed-loop [MHMB05, MWMR06, MHMB10], master-slave open-loop [MBM07], round-trip open-loop carrier synchronization [DPM05, DH08], and two-way open-loop carrier synchronization [PD10]. Each of these techniques has advantages and disadvantages in particular applications, as discussed in the survey article [MDMH09].

Many of the distributed wireless carrier synchronization techniques described in [MDMH09] operate on the principle of exchanging beacons between the BSs and synchronizing carriers based on estimates of the phase and frequency of these beacons. For example, in the two-way carrier synchronization protocol [PD10], a series of "forward beacons" are exchanged from node 1 to node 2 and so on to

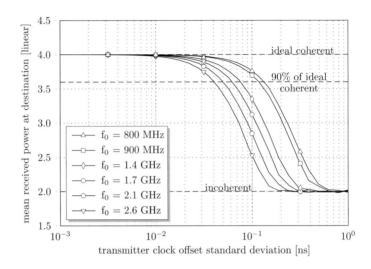

Figure 8.4 The effect of transmitter clock offset on distributed beamforming power for several common cellular carrier frequencies.

node M where node m transmits a periodic extension of the beacon it received from node $m - 1$. A series of "backward beacons" are also exchanged in the same way from node M to node $M - 1$ and so on to node 1. Each node then sums its phase and frequency estimates obtained from the forward and backward beacons to derive a synchronized local oscillator with frequency and phase identical (in the absence of estimation error) to the other nodes in the system. These synchronized local oscillators can then be used to enable retrodirective downlink transmission from two or more BSs to a mobile in the network.

While the various carrier synchronization techniques described in [MDMH09] differ in terms of how beacons are exchanged, the overall performance of each of these techniques tends to be limited by the accuracy of the beacon phase and frequency estimators. A common technique for the estimation of the phase and frequency of a single tone in additive white Gaussian noise is the maximum likelihood estimator (MLE) [RB74]. Under mild regularity conditions [H.V94], the MLE is known to asymptotically achieve the Cramér-Rao lower bound (CRLB). Given a beacon of amplitude a in complex Gaussian white noise with power spectral density $\frac{N_0}{2}$, the CRLB is given as [RB74]

$$\mathrm{cov}\left\{\begin{bmatrix}\tilde{\omega}, \tilde{\theta}\end{bmatrix}^\top\right\} \succeq \frac{N_0}{a^2} \begin{bmatrix} \frac{12}{T^3} & \frac{-6}{T^2} \\ \frac{-6}{T^2} & \frac{4}{T} \end{bmatrix},$$

where T is the duration of the observation, and the notation $\mathbf{A} \succeq \mathbf{B}$ means that $\mathbf{A} - \mathbf{B}$ is positive semi-definite.

As an example, Fig. 8.5 shows the clock and frequency offset standard deviations for the two-way carrier synchronization protocol developed in [PD10]. This example assumes seven transmitters serially exchange 1 GHz wireless beacons

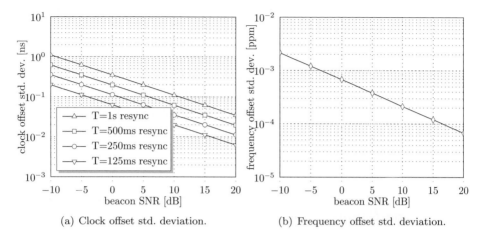

Figure 8.5 Clock and frequency offset standard deviations as a function of beacon SNR and re-synchronization interval for seven transmitters synchronized via the two-way carrier synchronization protocol.

according to the two-way synchronization protocol. Each beacon has a duration $T = 1$ ms and a total of 12 beacons are exchanged to synchronize the transmitters. The transmitters use MLE to form local phase and frequency estimates from the noisy observations. After synchronization is complete in this example, clock offset standard deviations better than 100 ps and frequency offset standard deviations better than 0.4 parts per billion (ppb) can be obtained when the beacon signal-to-noise ratio (SNR) is greater than 5 dB and the re-synchronization interval $T \leq 500$ ms.

Distributed wireless carrier synchronization techniques achieve high accuracy by exploiting the timing information contained in the phase and frequency of the bandpass beacons exchanged during synchronization. While distributed wireless carrier synchronization can, in principle, achieve much more precise clock and frequency offset than PTP and GPS, the use of unmodulated beacons can lead to periodic ambiguities in the phase estimates. Hence, distributed wireless carrier synchronization techniques may be used in conjunction with conventional lower-precision synchronization techniques for to provide appropriate synchronization at different timescales. For example, GPS can be used with two-way carrier synchronization to provide symbol synchronization and also to stabilize the frequencies of the carriers at the BSs.

8.1.5 Summary

In this section, we have introduced the concept of synchronization and described several approaches to the problem of synchronizing nodes in a coordinated communication network. These techniques can be used separately or in conjunction to facilitate the establishment of a common notion of both frequency and time to a desired level of accuracy. Existing synchronization techniques are in fact sufficient to enable synchronization-sensitive CoMP schemes such as downlink multi-cell joint transmission in the context of orthogonal frequency division multiple access (OFDMA), if *baseband* signals are to overlap coherently. However, these techniques require low-latency channel feedback, and current solutions may be considered a cost issue in large-scale deployments, rendering further research on alternative techniques interesting. The following sections will now analyze the effect of residual time and frequency errors on the performance of coordinated communication networks.

8.2 Imperfect Synchronization in Time: Performance Degradation and Compensation

Vincent Kotzsch and Gerhard Fettweis

In this section, we investigate the effect of unavoidable time delay of arrival (TDOA) between the transmitter and receiver stations in orthogonal frequency division multiplex (OFDM) based CoMP systems. As we can see on the left side of Fig. 8.6, they originate from arbitrary distances d between each user equipment and base station that lead to different path delays $\tau(d)$ with

$$\tau(d) = \frac{d}{c} \qquad (8.2)$$

on each link, where c is the speed of light. In common cellular wireless communication systems, where one base station serves a certain number of mobiles, synchronization procedures are used to compensate those path delays (e.g. [MKP07]). In the uplink, this can be done by adjusting the timing of the mobiles to the base station network reference via timing advance commands that are sent to the mobiles via control channels. In the context of CoMP, however, in particular for joint signal processing concepts as introduced in Chapter 6, we are interested in exploiting the signal propagation between base stations (BSs) and user equipments (UEs) across multiple cells. In these scenarios, we have a delay coupling matrix $\mathbf{T}_d$ containing the resulting non-compensable TDOAs of the different links between each transmitter and receiver. In OFDM systems, a cyclic extension also known as cyclic prefix (CP) of length T_{CP} is used to overcome synchronization mismatches as well as to avoid the inter-symbol interference (ISI) between two consecutive OFDM symbols. Given this constraint, the remaining TDOAs $\Delta\tau(d) = \tau(d_{\max}) - \tau(d_{\min})$ in $\mathbf{T}_d$ need to be less than the cyclic prefix minus the maximum channel excess delay $\tau(L)$, i.e.

$$\Delta\tau(d) \leq T_{\text{CP}} - \tau(L). \qquad (8.3)$$

A possible timing scenario is shown on the right side of Fig. 8.6, where the signals of three transmitters are received by one receiver. The desired transmitter (Tx#1) is synchronized to the receiver, while the others are delayed such that

8.2 Imperfect Sync in Time: Perf. Degradation and Compensation

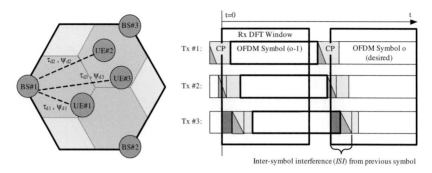

Figure 8.6 Hexagonal cell structure and possible timing scenarios in CoMP systems.

the TDOA of transmitter 2 (Tx#2) lies within the CP, but the ISI that is caused by the channel decay already leaks into the discrete Fourier transform (DFT) window. The third transmitter (Tx#3) even violates the CP limit, such that a portion of the previous OFDM symbol leaks into the DFT span.

As we will see in Subsection 8.2.1, ISI is introduced in the system on top of multi-user interference (MUI) if the maximum TDOAs are not limited to be within the CP (see e.g. [WG00], [WXBD09], [ZMM+08], [Ham10]). The amount of additional interference depends on the grade of timing mismatch. Fig. 8.7(a) depicts the joint distribution of occurring TDOAs after synchronization for different inter-site distances (ISDs), for the case that three users are uniformly distributed within a hexagonal cooperative cell that is served by three base stations. As an example, we also indicate the bounds for the short and the long CP length that is used in 3GPP LTE systems ($T_{CP} = 4.7 / 16.7\mu s$) by vertical lines. Besides the path delays, another important effect that needs to be considered is the pathloss $\psi(d)$ with

$$\psi(d) = \Xi \left(\frac{d}{d_0}\right)^{\eta}, \tag{8.4}$$

which also depends on the distance between the transmitters and receivers. Note that η is used as pathloss exponent, d_0 as the reference distance and Ξ as an attenuation factor that depends on the environment here. If we consider the pathloss attenuations from all links, we can also form a coupling matrix $\boldsymbol{\Psi}_d$ similar to $\mathbf{T}_d$. As it is known from literature, the attenuation of each link due to pathloss leads to special structures of the channel matrix, e.g. diagonal or row dominated matrices (depending on the relative position of the users). Likewise, also the possible interference power is attenuated. To compensate the pathloss, transmit power can be controlled, which is however limited to the maximum transmit power $P_{\max}$. Therefore, especially in large cells we may not be able to achieve a required target signal power level throughout the whole serving area. A convenient metric to assess the decoupling of two links is the separation factor (SF) that defines the ratio between maximum and minimum receive power

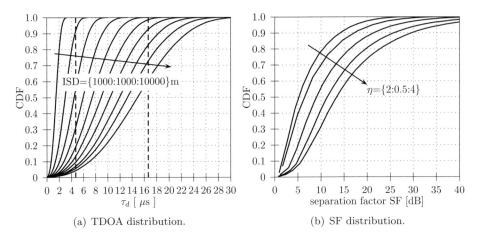

Figure 8.7 TDOA and SF distributions in a hexagonal cell.

among all transmitters at one receiver station, given as

$$\Delta\psi(d) = \frac{P_{R,\max}}{P_{R,\min}} = \frac{P_{\max}\,\psi(d_{\max})}{\psi(d_{\min})\,P_{\max}} = \left(\frac{d_{\max}}{d_{\min}}\right)^\eta. \quad (8.5)$$

A distribution of occurring link separations is shown in Fig. 8.7(b) for different values of the pathloss exponent η in a system with 3 BSs and 3 UEs, as it is already used in Fig. 8.7(a). As expected, the lower the pathloss exponent is, the higher the probability is that all three users are within a certain cooperation range (CR) Ω. To simplify the analysis of the occurring timing delays, we define a *symmetric* user configuration scenario, where three users directly move from the cell corner towards their primary serving base stations on the border of a circle with radius r_C. In those scenarios, we only have to consider the direct distance from each mobile to the primary serving base station $d_1 = D - r_C$ and the distance between the mobile and the secondary serving base stations $d_2 = D(1 + \frac{r_C}{D} + \frac{r_C^2}{D^2})^{1/2}$. We use $D = d_{\text{ISD}}/\sqrt{3}$ as cell diameter here. In order to fulfill the cyclic prefix restriction, we can re-write (8.3) as

$$\frac{d_2 - d_1}{c} \leq T_{\text{CP}} - \tau(L). \quad (8.6)$$

If we reorder (8.6), we obtain an expression for the *cooperation radius* of the circle in which ISI-free CoMP is possible (see [KJRF10] for details), i.e.

$$r_{C,\text{TDOA}} \leq \frac{2c(T_{\text{CP}} - \tau(L))D + (c(T_{\text{CP}} - \tau(L)))^2}{3D + 2c(T_{\text{CP}} - \tau(L))}. \quad (8.7)$$

If we only allow a joint signal processing for users who are within a certain cooperation range Ω, we can define the second constraint as

$$\left(\frac{d_2}{d_1}\right)^\eta \leq \Omega. \quad (8.8)$$

8.2 Imperfect Sync in Time: Perf. Degradation and Compensation

Then, a cooperation radius fulfilling the maximum link separation is given as

$$r_{\text{C,CR}} \leq \left(-\frac{a}{2} - \sqrt{\frac{a^2}{4} - 1}\right) D \quad \text{with} \quad a = \frac{1 + 2\Omega^{2/\eta}}{1 - \Omega^{2/\eta}}. \tag{8.9}$$

However, in systems with limited transmit power we also have to ensure that a minimum link signal-to-noise ratio (SNR) can be achieved although a low link separation is available. The synchronization effects that need to be characterized for CoMP systems can be summarized as follows:

- The TDOAs must not exceed the CP limitation, otherwise ISI is induced.
- The channel length increases above the CP constraint.
- The level of ISI power depends on the TDOA and the link separation.

8.2.1 MIMO OFDM Transmission with Asynchronous Interference

While the notation used throughout the book so far focussed on the transmission over one sub-carrier of a perfectly synchronized orthogonal frequency division multiple access (OFDMA) system, we now have to extend our model to also capture the time and frequency dimensions of our signals. All variables from Chapter 3 are used in the same way as before, but are followed by parentheses containing these new indices:

- OFDM symbol indices o and o'.
- Sample indices i and i' (for *time domain* representation).
- Sub-carrier indices q and q' (for *frequency domain* representation).

While all transmitted and received signals, channel matrices etc. were so far given in *frequency domain*, we now additionally introduce variables in *time domain*, which are the same but with a tilde on top. As an example, the frequency domain channel from transmitter k to receiver m in OFDM symbol o and on sub-carrier q is given as $H_k^m(o,q)$, and the corresponding time domain representation (now using a sample index i instead of a sub-carrier index q) is $\tilde{H}_k^m(o,i)$.

Let us assume that the transmit symbols in frequency domain of a transmitter k in OFDM symbol o are given as $x_k(o,q)$, where only a subset of sub-carriers $q \in \mathcal{Q}$ is used. After a DFT operation of size N and the insertion of a cyclic prefix of length N_{CP} samples, the corresponding time domain signals of the user are given as

$$\tilde{x}_k(o,i) = \frac{1}{\sqrt{N}} \sum_{\forall q \in \mathcal{Q}} x_k(o,q)\, e^{\frac{j2\pi qi}{N}}, \quad -N_{\text{CP}} \leq i \leq N - 1. \tag{8.10}$$

After transmission over a channel specified by the taps $h_k^m(o, i')$ for the link between transmitter k and receiver m, the time domain signal is given as

$$\tilde{y}_m(o, i) = \sum_{k=1}^{K} \sum_{i'=1}^{L} \left(\tilde{h}_k^m(o, i') \tilde{x}_k(o, i - i' - \mu_k^m) \right) + \tilde{n}_m(o, i). \tag{8.11}$$

Here, $\tilde{n} \sim \mathcal{N}_\mathbb{C}(0, \sigma_n^2 \mathbf{I})$ is the receive noise in time domain. The channel taps of the link specific channel impulse responses are modeled as $h_k^m(o, i') \sim \mathcal{N}_\mathbb{C}(0, \sigma_{h_k^m(o,i')}^2 \psi_k^m(d))$, where $L = \lfloor \tau(L)/T_S \rfloor$ represents the discrete channel length and $\sigma_{h_k^m(o,i')}^2$ the tap variance given by the corresponding power delay profile. The parameter $\mu_k^m = \lfloor \Delta \tau_k^m(d)/T_S \rfloor$ expresses a timing offset (given in samples). The received signal in frequency domain is obtained by the DFT operation applied to the received samples $\tilde{y}_m(o, i)$ as

$$y_m(o, q) = \frac{1}{\sqrt{N}} \sum_{i=0}^{N-1} \tilde{y}_m(o, i) \, e^{\frac{-j 2 \pi q i}{N}}, \quad q \in \mathcal{Q}. \tag{8.12}$$

In [KF10], a closed-form solution of (8.12) is derived that gives us an expression for the frequency domain transmission with arbitrary symbol timing offsets μ. The transmission can then be summarized for the received signal at the m-th receiver branch with

$$y_m(o, q) = \sum_{k=1}^{K} \sum_{\forall q' \in \mathcal{D}} \left(E_k^m(o, q, q') H_k^m(o, q', q') x_k(o, q') e^{-j \frac{2\pi}{N} \mu_k^m q'} \right)$$

$$- \sum_{k=1}^{K} \sum_{\forall q' \in \mathcal{D}} \left(\sum_{q''=0}^{N-1} E_k^m(o, q, q'') C_k^m(o, q'', q') \right) x_k(o, q')$$

$$+ \sum_{k=1}^{K} \sum_{\forall q' \in \mathcal{D}} \left(E_k^m(o-1, q, q') H_k^m(o, q', q') x_k(o-1, q') e^{-o \frac{2\pi}{N} (\mu_k^m - N_{\text{CP}}) q'} \right)$$

$$+ \sum_{k=1}^{K} \sum_{\forall q' \in \mathcal{D}} \left(\sum_{q''=0}^{N-1} E_k^m(o, q, q'') C_k^m(o-1, q'', q') \right) x_k(o-1, q') + n_m(o, q), \tag{8.13}$$

where E_k^m are elements of the matrices $\mathbf{E}_k^m(o)$, $\mathbf{E}_k^m(o-1) \in \mathbb{C}^{[N \times N]}$ which include the inter-carrier interference (ICI) due to the windowing of the current and previous OFDM symbols in time domain in the case that the CP limit is

8.2 Imperfect Sync in Time: Perf. Degradation and Compensation

exceeded. These matrix elements can be denoted as

$$E_k^m(o, q, q') = \begin{cases} \frac{N_B - \mu_k^m}{N} & \kappa_k^m = 0 \\ \frac{1}{N} e^{j\frac{\pi}{N}\theta_o} \frac{\sin(\frac{\pi \kappa_k^m}{N}(N_B - \mu_k^m))}{\sin\left(\frac{\pi \kappa_k^m}{N}\right)} & \text{otherwise} \end{cases}$$

$$E_k^m(o-1, q, q') = \begin{cases} \frac{\mu_k^m - N_{CP}}{N} & \kappa_k^m = 0 \\ \frac{1}{N} e^{j\frac{\pi}{N}\theta_{o-1}} \frac{\sin(\frac{\pi \kappa_k^m}{N}(\mu_k^m - N_{CP}))}{\sin\left(\frac{\pi \kappa_k^m}{N}\right)} & \text{otherwise} \end{cases}$$

with $\kappa_k^m = q' - q \ \forall \ q, q' = 1...N$ and

$$\theta_o = \kappa_k^m (\mu_k^m + N_B - 2N_{CP} - 1)$$
$$\theta_{o-1} = \kappa_k^m (\mu_k^m - N_{CP} - 1).$$

In (8.13), H_k^m are elements of the diagonal matrix with the channel transfer function (CTF) in frequency domain on a certain link $\mathbf{H}_k^m \in \mathbb{C}^{[N \times N]}$. By using the Fourier transform matrix $\mathbf{F} \in \mathbb{C}^{[N \times N]}$ with the elements $F(q, q') = e^{-j\frac{2\pi}{N} q q'}/\sqrt{N}$, this channel matrix can also be written as $\mathbf{H} = \mathbf{F}\tilde{\mathbf{H}}\mathbf{F}^H$, where $\tilde{\mathbf{H}} \in \mathbb{C}^{[N_B \times N_B]}$ includes the time domain channel impulse response (CIR) matrix in Toeplitz structure with the first column $[h_1 \cdots h_L \ 0 \cdots 0]^T$. $N_B = N + N_{CP}$ is used as OFDM block length here. The elements $C_k^m(o)$, $C_k^m(o-1)$ of the matrices $\mathbf{C}(o), \mathbf{C}(o-1) \in \mathbb{C}^{[N \times N]}$ contain the inter-symbol interference that is induced by the $(L-1)$-tap channel decay from the previous OFDM symbol. They are given as

$$C_k^m(o, q, q') = \frac{1}{N} \sum_{a=0}^{N-1} \sum_{b=0}^{N-1} \tilde{C}_k^m(o, a, b) e^{j\frac{2\pi}{N}(bq' - aq)} \tag{8.14}$$

$$C_k^m(o-1, q, q') = \frac{1}{N} \sum_{a=0}^{N-1} \sum_{b=0}^{N-1} \tilde{C}_k^m(o-1, a, b) e^{j\frac{2\pi}{N}(bq' - aq)}, \tag{8.15}$$

where $\tilde{C}_k^m(o)$ and $\tilde{C}_k^m(o-1)$ are elements of special Toeplitz matrices in time domain that are explained in more detail in [KF10]. It should be noted that for $\mu_k^m \leq N_{CP}$, $\mathbf{E}_k^m(o)$ becomes an identity matrix and $\mathbf{E}_k^m(o-1)$ changes to a zero matrix. Within the range of $N_{CP} - L + 1 \leq \mu_k^m \leq N_{CP}$, we define an effective cyclic extension $N_{CP,eff} = N_{CP} - L + 1$ for the following which gives us the interference free range within the CP. For the case that $\mu_k^m \leq N_{CP,eff}$, $\mathbf{C}(o)$ and $\mathbf{C}(o-1)$ also become zero matrices and we get the well known asynchronous interference free transmission equation in frequency domain with

$$y^m(o, q) = \sum_{k=1}^{K} \left(H_k^m(o, q, q) x_k(o, q) e^{-j\frac{2\pi}{N} \mu_k^m q} \right), \tag{8.16}$$

where we only have a phase slope caused by μ over all sub-carriers within one OFDM symbol.

If we reorder (8.13), we can define $\mathbf{Z}(o,q) \in \mathbb{C}^{[M \times K]}$ as the representation of the interference channel on the q-th sub-carrier in frequency domain. In a next step, we can summarize the transmission equation of the user symbols $\mathbf{x}(o,q) \in \mathbb{C}^{[K \times 1]}$ to the receivers $\mathbf{y}(o,q) \in \mathbb{C}^{[M \times 1]}$ as follows:

$$\mathbf{y}(o,q) = \underbrace{\mathbf{Z}(o,q)\mathbf{x}(o,q)}_{\text{MUI}} + \underbrace{\sum_{\forall q' \in \mathcal{D} \setminus q} \mathbf{Z}(o,q')\mathbf{x}(o,q')}_{\text{ICI}} + \underbrace{\sum_{\forall q' \in \mathcal{D}} \mathbf{Z}(o-1,q')\mathbf{x}(o-1,q')}_{\text{ISI}} + \mathbf{n}. \tag{8.17}$$

$$\underbrace{\phantom{\sum_{\forall q' \in \mathcal{D} \setminus q} \mathbf{Z}(o,q')\mathbf{x}(o,q') + \sum_{\forall q' \in \mathcal{D}} \mathbf{Z}(o-1,q')\mathbf{x}(o-1,q')}}_{\mathbf{u}(o,q)}$$

As we can observe in this expression, we now have a coupling between adjacent sub-carriers (ICI) and consecutive OFDM symbols (ISI) in addition to the coupling between multi-user interference (MUI). A characterization of the ISI and ICI is done for example in [SDAD02], [SM03], [MC06] and [NK02].

8.2.2 Interf.-Aware Multi-User Joint Detection and Transmission

If we assume a signal transmission with a transmit filter $\mathbf{W} \in \mathbb{C}^{[K \times K]}$ and receive filter $\mathbf{G} \in \mathbb{C}^{[K \times M]}$, we can re-write (8.17) in order to get a generalized expression for the signal transmission in frequency domain:

$$\begin{aligned}
\widehat{\mathbf{x}}(o,q) = & \; \mathbf{G}(o,q)\mathbf{Z}(o,q)\mathbf{W}(o,q)\sqrt{\mathbf{P}(o,q)}\mathbf{x}(o,q) \\
& + \sum_{\forall q' \in \mathcal{D} \setminus q} \mathbf{G}(o,q)\mathbf{Z}(o,q')\mathbf{W}(o,q')\sqrt{\mathbf{P}(o,q')}\mathbf{x}(o,q') \\
& + \sum_{\forall q' \in \mathcal{D}} \mathbf{G}(o,q)\mathbf{Z}(o-1,q')\mathbf{W}(o-1,q')\sqrt{\mathbf{P}(o-1,q')}\mathbf{x}(o-1,q') \\
& + \mathbf{G}(o,q)\mathbf{n}(o,q),
\end{aligned} \tag{8.18}$$

where we also introduced a transmit power scaling through the diagonal matrices $\mathbf{P} \in \mathbb{C}^{[K \times K]}$. In that case we assume $E\left\{\mathbf{x}\mathbf{x}^H\right\} = \mathbf{\Phi}_{\mathbf{xx}} = \mathbf{I}_K$ for all sub-carriers q and OFDM symbols o. Unless the sub-carrier and OFDM symbol indices are explicitly given, we assume the q-th sub-carrier and the o-th OFDM symbol for our equations in the following. The error covariance matrix of the mean square error (MSE) between the received and transmitted symbols $\mathbf{\Phi}_{\text{ee}} = E\left\{|\mathbf{x} - \widehat{\mathbf{x}}|^2\right\}$ is given by

$$\mathbf{\Phi}_{\text{ee}} = (\mathbf{GZW} - \mathbf{I})\mathbf{P}\left(\mathbf{W}^H\mathbf{Z}^H\mathbf{G}^H - \mathbf{I}\right) + \mathbf{G}\left(\mathbf{\Phi}_{\text{uu}} + \mathbf{\Phi}_{\text{nn}}\right)\mathbf{G}^H \tag{8.19}$$

with

$$\Phi_{uu} = \sum_{\forall q' \in \mathcal{D} \setminus q} \mathbf{Z}(o,q')\mathbf{W}(o,q')\mathbf{P}(o,q')\mathbf{W}(o,q')^H \mathbf{Z}(o,q')^H$$
$$+ \sum_{\forall q' \in \mathcal{D}} \mathbf{Z}(o-1,q')\mathbf{W}(o-1,q')\mathbf{P}(o-1,q')\mathbf{W}(o-1,q')^H \mathbf{Z}(o-1,q')^H. \tag{8.20}$$

It should be noted that for frequency-selective channels we have to include a separate precoding filter and power allocation vector for all adjacent sub-carriers in order to get an exact expression for the MSE.

The interference aware receive filter can then be obtained by minimizing the sum mean squared error (SMSE), i.e.

$$\mathbf{G} = \underset{\mathbf{G}}{\operatorname{argmin}} \left\{ \operatorname{tr}\left\{\Phi_{ee}\right\}\right\} = \underset{\mathbf{G}}{\operatorname{argmin}} \left\{ E\left\{\|\mathbf{x} - \widehat{\mathbf{x}}\|_2^2\right\}\right\}. \tag{8.21}$$

The minimum argument is obtained by setting the derivative of the SMSE to zero (e.g. [Say08]). In this way we obtain

$$\mathbf{G} = \mathbf{P}\mathbf{W}^H \mathbf{Z}^H \Phi_{yy}^{-1} \tag{8.22}$$

with $\Phi_{yy} = E\{\mathbf{y}\mathbf{y}^H\}$. The interference-aware transmit filter can be derived by solving the optimization criterion of (8.21) with respect to $\mathbf{W}$ such that

$$\mathbf{W} = \underset{\{\mathbf{W},\beta\}}{\operatorname{argmin}} \left\{ E\left\{\|\mathbf{x} - \beta^{-1}\widehat{\mathbf{x}}\|_2^2\right\} \mid E\left\{\|\mathbf{W}\sqrt{\mathbf{P}}\mathbf{x}\|_2^2\right\} = P_{\max}\right\}, \tag{8.23}$$

where we included an additional sum-power constraint such that the maximum transmit power does not exceed $P_{\max}$. This restriction is fulfilled by introducing the transmit power scaling parameter β that has to be reversed at the receiver input. By following the steps in [JJT+09] and setting the derivative of the resulting Lagrangian function w.r.t $\mathbf{W}$ and β to zero, we get

$$\mathbf{W} = \beta\, \mathbf{Z}^H \left(\mathbf{Z}\mathbf{Z}^H + \underbrace{\frac{\operatorname{tr}(\Phi_{nn} + \Phi_{uu})}{P_{\max}} \mathbf{I}_K}_{\Phi'_{nn}} \right)^{-1} \tag{8.24}$$

$$E\left\{\|\mathbf{W}\sqrt{\mathbf{P}}\mathbf{x}\|_2^2\right\} = \beta^2 \operatorname{tr}\left\{\mathbf{W}\mathbf{P}\mathbf{W}^H\right\} = P_{\max} \tag{8.25}$$

$$\beta = \sqrt{\frac{P_{\max}}{\operatorname{tr}\left\{\mathbf{Z}\mathbf{Z}^H \left(\mathbf{Z}\mathbf{Z}^H + \Phi'_{nn}\right)^{-2} \mathbf{P}\right\}}}. \tag{8.26}$$

If we assume an uplink transmission where the users have only one transmit antenna and are not able to communicate to each other the transmit filter becomes an identity matrix $\mathbf{W} = \mathbf{I}_K$. Under this constraint, we can derive an expression for the post equalization signal-to-interference-and-noise ratio (SINR) of the k-th user after the joint BS signal processing as the ratio of the desired signal power and the portion of MUI, ISI and ICI plus noise as

$$\text{SINR}_k = \frac{p_k \mathbf{g}_k^H \mathbf{z}_k \mathbf{z}_k^H \mathbf{g}_k}{\mathbf{g}_k^H \left(\sum_{r=1, r \neq k}^{K} p_r \mathbf{z}_r \mathbf{z}_r^H + \mathbf{\Phi}_{uu} + \mathbf{\Phi}_{vv} \right) \mathbf{g}_k}. \tag{8.27}$$

It is worth mentioning that the used filter matrix only aims at canceling the multi-user interference for the desired sub-carrier with the knowledge of the colored noise $\mathbf{u}$ (see (8.17)). If we use the filter defined in (8.22), the post equalization SINR for the k-th user yields

$$\text{SINR}_k = \frac{(\mathbf{z}_k)^H \mathbf{\Phi}_{yy}^{-1} \mathbf{z}_k}{1 - (\mathbf{z}_k)^H \mathbf{\Phi}_{yy}^{-1} \mathbf{z}_k}. \tag{8.28}$$

For a downlink transmission to M non-cooperative receiver stations with one receiver antenna each, the receive filter becomes an identity matrix $\mathbf{G} = \beta^{-1} \mathbf{I}_M$. In that case, the SINR for the k-th user stream can be expressed by

$$\text{SINR}_k = \frac{p_k \mathbf{z}_k^H \mathbf{w}_k \mathbf{w}_k^H \mathbf{z}_k}{\sum_{r=1, r \neq k}^{K} p_r \mathbf{z}_k^H \mathbf{w}_r \mathbf{w}_r^H \mathbf{z}_k + \beta^{-2} \left(\sigma_u^2 + \sigma_v^2 \right)} \tag{8.29}$$

with

$$\sigma_u^2 = \sum_{\forall q' \in \mathcal{D} \setminus q} \mathbf{z}_k(o, q')^H \mathbf{W}(o, q') \mathbf{P}(o, q') \mathbf{W}(o, q')^H \mathbf{z}_k(o, q')$$

$$+ \sum_{\forall q' \in \mathcal{D}} \mathbf{z}_k(o-1, q')^H \mathbf{W}(o-1, q') \mathbf{P}(o-1, q') \mathbf{W}(o-1, q')^H \mathbf{z}_k(o-1, q').$$

$$\tag{8.30}$$

In [AA09], a joint optimization of the receive and transmit filter as well as power control in systems with asynchronous interference has been presented based on the results in [SB04] and [ZMM+08]. In order to simplify the analysis, we introduce a simple power control scheme here. The transmit power of each transmitter is controlled in order to achieve a target SNR γ_k on the strongest link within one column of the channel matrix. The uplink transmit power values in that case can be obtained by

$$p_k = \frac{\gamma_k \, \sigma_v^2}{\max_m \left\{ |z_k^m|^2 \right\}}. \tag{8.31}$$

8.2.3 System Level SINR Analysis

After the derivation of the OFDM CoMP transmission model, it is used for the numerical analysis of different channel effects in a system model with hexagonal cells as shown in Fig. 8.6. Unless otherwise stated, the parameter set of Table 8.1 is used for all simulations, which is based on a 3GPP LTE setup.

We used the non line-of-sight (NLOS) rural macro model for the pathloss calculations. Therefore, we assumed an average building height of 5 m and a BS

8.2 Imperfect Sync in Time: Perf. Degradation and Compensation

Table 8.1. Simulation Parameters

Parameter	Value
DFT size N	256
Used sub-carriers N_{SC}	120
CP length $N_{CP} = T_{CP}/T_S$	18
System bandwidth $B_S = 1/T_S$	3.84 MHz
Sub-carrier bandwidth B_{SC}	15 kHz
Carrier frequency f_C	800 MHz
Target SNR γ	20 dB
UE maximum Tx power $p_{k,\max}$	23 dBm
UE noise figure $\Psi_{UE,NF}$	7 dB
UE antenna gain $\Psi_{UE,AG}$	-1 dBi
BS maximum Tx power $p_{k,\max}$	43 dBm
BS noise figure $\Psi_{BS,NF}$	4 dB
BS antenna gain $\Psi_{BS,AG}$	15 dBi
Noise power per sub-carrier σ_n^2	$10\log_{10}(1.38^{-23}\frac{J}{K} \cdot 290K \cdot B_{SC}) = -132$ dBm
Fast fading margin Ψ_{FFM}	2 dB
Inter-cell interf. margin Ψ_{IIM}	3 dB
RM prop. env. factor $10\lg(\Psi)$	1.69 dB @800 MHz
RM pathloss coefficient η	3.86
Power delay profile of length L	$\sum_{\lambda=1}^{L} \frac{1}{\sigma_\lambda} e^{\frac{-(\lambda-1)}{\sigma_\lambda}} = 1$, $\sigma_\lambda = -\frac{L-1}{\ln(0.1)}$

antenna height of 35 m. The height for a mobile terminal is assumed to be 1.5 m and a street width is chosen as 20 m. For the numerical simulations we use the already mentioned *symmetric* user positioning model with three users and three base stations in the uplink direction in order analyze the average SINR in the case that the user circle radius r_C is increased. In that case, we have only the link channels as random numbers for averaging. The advantage of this model is that we have the same propagation conditions for all users at one fixed point r_C. This is helpful to understand the basic limitations within CoMP.

The experiment that we introduce here is intended to get an idea about the impact of the ISI. The result in terms of the average SINR loss according to target SNR is shown in Fig. 8.8(a), where we assume a single link power control as denoted in (8.31). We normalized the user radius r_C to the cell diameter D. Thus, the point 0 marks the cell center, and at point 1 the users are very close to their primary serving base stations.

As one reference, we include performance values for a scenario where no pathloss is considered ($\eta = 0$). This is usually done in investigations on link level where we mainly look at results in which we only increase the timing offset. Here, we use two channel lengths $\tau(L)$. In one case, we use a flat channel ($\tau(L) = 0$), and in the other we set the channel length to be in the order of half of the CP length ($\tau(L) = 2.3\mu s$). If we look at the performance curves, one can clearly see from the points where the interference-free range of the CP is exceeded we have strong SINR degradations. In Fig. 8.8(b), we depicted the occurring TDOAs of that experiment where we marked the two CP limits. As another reference,

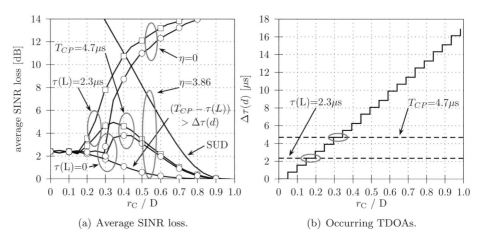

Figure 8.8 Average SINR loss and occurring TDOAs in a symmetric user scenario.

we show results where we assume that we have a CP length that can cover all possible TDOAs $((T_{\text{CP}} - \tau(L)) > \Delta\tau(d))$. For that simulation we used now the pathloss model as described above. As it can be observed, the signal attenuations due to the pathloss lead to a decoupling of the channel matrix between the diagonal and off-diagonal entries with increasing r_C. In the cell-center, the single-user power control scheme is not able to control the transmit powers to achieve the target SNR, since the multi-user interference is not considered. That is the reason why we can observe an initial SINR loss of ≈ 2 dB. In a next scenario, we limit the CP as is denoted in Table 8.1 with $T_{\text{CP}} = 4.7$ µs. One can see an SINR degradation from the point where the CP limit is exceeded again, but we can also observe that the ISI power is attenuated due to pathloss with increasing r_C. The single user detection (SUD) performance is included as a general bound where we assume that each base station only wants to detect the closest user without the knowledge of the others. A key observation of the presented results is that, as expected, only in a small region the asynchronous interference is the dominant degrading effect. On the one hand, the TDOAs must exceed the ISI-free range within the CP. It should be noted that in real systems, due to timing estimation errors, the ISI-free range can be reduced in addition to the reduction caused by the channel decay. On the other hand, the ISI power from the interfering users which is also attenuated by the pathloss needs to exceed a certain threshold such that it leads to additional interference. Among others, this threshold mainly depends on inter-site distance, carrier frequency, pathloss exponent, transmit power etc. Furthermore it should be mentioned that for the downlink case we can achieve similar results since the problem of asynchronous interference is equivalent in both directions.

8.2.4 Summary

In this section, we investigated CoMP systems in the case that the time differences of arrival after single link synchronization exceed the limitation which is given by the cyclic prefix in OFDM systems. We analytically described the impact on the OFDM transmission model as well as on multi-user joint detection and transmission. In numerical simulations, we analyzed how the additional asynchronous interference affects the SINR performance in hexagonal cells within a simple symmetric user configuration setup. We could show that the SINR loss increases if the residual symbol timing offsets violate the cyclic prefix limitation until the pathloss leads to a decoupling of the users and consequently to an attenuation of the asynchronous interference.

8.3 Imperfect Synchronization in Frequency: Performance Degradation and Compensation

Malte Schellmann

In a CoMP system, frequency errors may be introduced if the simultaneously transmitting entities are not perfectly synchronized in their carrier frequency. These frequency errors impose a time-continuous phase rotation on the signals transmitted from the single antennas, giving the effective channels seen at the receivers a time-variant nature. The time variance of the channel may degrade the system performance severely, as the channel-dependent matrices used for the spatial pre- or post-processing may no longer match the true channel, and correspondingly additional interference between the simultaneously transmitted spatial streams will arise. In orthogonal frequency division multiplex (OFDM) systems, channel time variances may also destroy the orthogonality between the sub-carrier signals, giving rise to inter-carrier interference (ICI). In this section, we will focus on these two performance degrading effects, namely the inter-stream interference and the ICI, separately for the downlink and the uplink. We will derive analytical expressions for the resulting signal-to-interference ratios (SIRs), allowing to reveal their direct dependence on the system parameters and the frequency errors. From these results, we draw some conclusions on the required synchronization accuracy of the transmitting nodes. Furthermore, we give a brief overview on existing techniques to compensate for frequency errors from the literature.

In the literature, only few studies exist on the effect of synchronization errors in CoMP systems, especially for the precoded downlink. The most comprehensive study available in this field can be found in [Zar08]. This study is based on pure numerical evaluations of high-dimensional CoMP systems with 32 transmit antennas in total. Analytical investigation on this topic are barely available; the reason can certainly be found in the fact that analysis of CoMP systems with more than two transmit antennas is hardly tractable. Nevertheless, we have decided to thoroughly analyze the simplest case of a CoMP system, with 2 cooperating single-antenna base stations (BSs) serving two single-antennas user equipments (UEs). The obtained closed-form expressions give us clear insights into the exact relations between the frequency errors and the measures of interest. For systems of higher dimension, it can be expected that these relations are basically sustained, and performance simply scales (to some extent) with the number of antennas.

8.3.1 Downlink Analysis

Inter-stream Interference of Spatially Precoded Streams

We consider the downlink of an exemplary CoMP scheme, where M single-antenna BSs transmit to K single-antenna UEs. The K UEs are all assumed to be assigned to the same physical resource block (PRB). In the sequel, we focus on the signal conditions observed at a single sub-carrier of this PRB. According to the notation introduced in Section 3.5, the transmission equation for this sub-carrier can be given as

$$\mathbf{y} = \mathbf{H}^H \mathbf{W} \mathbf{x} + \mathbf{n}. \tag{8.32}$$

If the cooperating BSs are not properly synchronized, then an independent, time-continuous phase rotation is imposed on the signal transmitted from each BS antenna, which results from the carrier frequency offset (CFO) between the transmitting BS and a common reference. This phase rotation is represented by the diagonal matrix

$$\mathbf{\Omega}(t) = \mathrm{diag}\left(\exp(j2\pi\Delta f_1 t), \ldots, \exp(j2\pi\Delta f_M t)\right), \tag{8.33}$$

where Δf_m is the CFO between BS m and the common reference. For notational convenience, we introduce the angular frequencies of the CFOs $\omega_m = 2\pi\Delta f_m$. The matrix $\mathbf{\Omega}(t)$ is incorporated into the transmission equation (8.32) according to

$$\mathbf{y} = \mathbf{H}^H \mathbf{\Omega}(t) \mathbf{W} \mathbf{x} + \mathbf{n}. \tag{8.34}$$

In general, there are also CFOs observed at the side of the receiver, representing the offsets of the receivers' oscillators. These could be captured by another CFO matrix, as it has been done in the equation for the uplink in (8.54). However, those CFOs have been neglected in (8.34), as they can easily be tracked continuously and compensated by standard synchronization techniques for the downlink [MKP07].

We now draw our attention to the joint precoding matrix $\mathbf{W}$. This matrix can be separated into the product of two matrices

$$\mathbf{W} = \mathbf{C} \cdot \mathbf{P} = [p_1 \mathbf{c}_1 \ \ldots \ p_K \mathbf{c}_K] \tag{8.35}$$

$$\mathbf{C} = [\mathbf{c}_1 \ \ldots \ \mathbf{c}_K]$$

$$\mathbf{P} = \mathrm{diag}\left(p_1, \ldots, p_K\right).$$

8.3 Imperfect Sync in Frequency: Perf. Degradation and Compensation

The matrix $\mathbf{C}$ represents the algebraic function used to diagonalize the effective transmission channel $\mathbf{H}^H\mathbf{W}$, while $\mathbf{P}$ is a diagonal matrix, whose diagonal elements p_k represent the scaling of the column vectors $\mathbf{c}_k$ in $\mathbf{C}$ (i.e. the precoding beams for the transmit symbols x_k in vector $\mathbf{x}$) according to the power allocation. As we do not assume the columns $\mathbf{c}_k$ to be normalized, the relation between p_k and the transmit power P_k allocated to beam k can be characterized by $p_k = \|\mathbf{c}_k\|^{-1}\sqrt{P_k}$.

To analyze the impact of CFOs on the signal conditions at the receivers, we focus on the simplest case of a CoMP system, where two BSs each with one single antenna cooperate to simultaneously serve two single-antenna terminals. Then the channel matrix is given as

$$\mathbf{H}^H = \begin{bmatrix} h_{11} & h_{12} \\ h_{21} & h_{22} \end{bmatrix} = \begin{bmatrix} \mathbf{h}_1^T \\ \mathbf{h}_2^T \end{bmatrix}, \quad (8.36)$$

where $\mathbf{h}_k$ specify the channel components connected to UE k. To diagonalize the effective channel $\mathbf{H}^H\mathbf{W}$, we use zero-forcing (ZF) precoding. Assuming that both BS have ideal channel knowledge at time instant $t = 0$, the ZF matrix $\mathbf{C}$ is calculated according to

$$\mathbf{C} = (\mathbf{H}^H)^{-1} = \underbrace{\frac{1}{h_{11}h_{22} - h_{12}h_{21}}}_{\alpha} \begin{bmatrix} h_{22} & -h_{12} \\ -h_{21} & h_{11} \end{bmatrix} = \begin{bmatrix} c_{11} & c_{12} \\ c_{21} & c_{22} \end{bmatrix}. \quad (8.37)$$

The constant scaling factor in front of the matrix will in the following be denoted as α. The effective channel matrix for time instant $t \neq 0$ then yields

$$\mathbf{H}^H \mathbf{\Omega}(t) \mathbf{W} = \begin{bmatrix} \mathbf{h}_1^T\mathbf{\Omega}(t)\mathbf{c}_1 p_1 & \mathbf{h}_1^T\mathbf{\Omega}(t)\mathbf{c}_2 p_2 \\ \mathbf{h}_2^T\mathbf{\Omega}(t)\mathbf{c}_1 p_1 & \mathbf{h}_2^T\mathbf{\Omega}(t)\mathbf{c}_2 p_2 \end{bmatrix}. \quad (8.38)$$

Here, the k-th diagonal element of the matrix represents the effective channel of the transmit symbol x_k intended for k-th receiver, while the off-diagonal elements in row k at position m represent the interference from symbol x_m on the signal seen at k-th receiver. As we assume symmetric conditions for all receivers, we focus exemplarily on the the signal received at the first receiver, i.e. $y_1 = \mathbf{h}_1^T\mathbf{\Omega}(t)\mathbf{W}\mathbf{x}$. The signal y_1 will be separated into a desired and an interference part, $y_1 = y_{1,d} + y_{1,i}$, which will be analyzed separately in the following. The desired signal $y_{1,d}$ contains the contribution from x_1; with (8.33), (8.36) and (8.37), it amounts to

$$y_{1,d} = \mathbf{h}_1^T\mathbf{\Omega}(t)\mathbf{c}_1 p_1 x_1$$

$$= p_1 \alpha \exp(j\omega_1 t) \left[h_{11}h_{22} - \exp(j\underbrace{(\omega_2 - \omega_1)}_{\Delta\omega_{21}} t) h_{12} h_{21} \right] x_1$$

$$= p_1 \exp(j\omega_1 t)\left[1 - (1 - \exp(j\Delta\omega_{21} t)) h_{12} c_{21}\right] x_1. \quad (8.39)$$

In the above equation, we observe that the ZF pre-compensated channel is distorted by the complex term $(1 - \exp(j\Delta\omega_{21}t))h_{12}c_{21}$. It is reasonable to assume that BSs and UEs are spaced sufficiently apart from each other, and hence the single channel coefficients h_{ij} can be considered to be mutually independent. Consequently, the product $h_{12}c_{21}$ represents a random complex number with zero mean. The amplitude of the ZF pre-compensated channel may therefore increase or decrease with the same probability.

In a similar manner, we can calculate the interference signal $y_{1,i}$, which reflects the inter-stream interference from the unintended signal x_2:

$$y_{1,i} = \mathbf{h}_1^T \mathbf{\Omega}(t) \mathbf{c}_2 p_2 x_2$$
$$= p_2 \exp(j\omega_1 t) \alpha \left[(\exp(j\Delta\omega_{21} t) - 1) h_{11} h_{12} \right] x_2$$
$$= p_2 \exp(j\omega_1 t) \left[0 - (1 - \exp(j\Delta\omega_{21} t)) h_{12} c_{22} \right] x_2. \tag{8.40}$$

Note the similar structure in (8.40) compared to (8.39), which both exhibit the same weighting factor $(1 - \exp(j\Delta\omega_{21} t))$ scaling the complex distortion. The complex distortion is formed here by the product of h_{12} and c_{22}, which is independent of the value altering the useful signal part in (8.39), as c_{21} is independent of c_{22}. We can therefore conclude that the inter-stream interference introduced by the CFOs has a magnitude that is similar to that of the amplitude change of the desired signal, as long as c_{21} and c_{22} can be assumed to have similar mean power.

From the above results, we will now derive an expression for the SIR between desired and interference signal. From (8.39) and (8.40), the instantaneous SIR_i can be given as

$$\text{SIR}_i = \frac{|y_{1,d}|^2}{|y_{1,i}|^2} = \left| \frac{p_1 (h_{11} h_{22} - \exp(j\Delta\omega_{21} t) h_{12} h_{21})}{p_2 (1 - \exp(j\Delta\omega_{21} t)) h_{11} h_{12}} \right|^2. \tag{8.41}$$

To obtain an estimate for the mean SIR, we use Jensen's inequality, allowing us to determine the mean value for enumerator and denominator separately:

$$\text{SIR} \geq \frac{E\{|y_{1,d}|^2\}}{E\{|y_{1,i}|^2\}}$$
$$= \frac{P_1}{P_2(|1 - \exp(j\Delta\omega_{21} t)|^2)} \cdot \frac{E\{\|\mathbf{c}_2\|^2 (|h_{11} h_{22} - \exp(j\Delta\omega_{21} t) h_{12} h_{21}|^2)\}}{E\{\|\mathbf{c}_1\|^2 |h_{11} h_{12}|^2\}}. \tag{8.42}$$

Additional to the mutual independence, we assume for further simplification that the channel coefficients h_{ij} have the same mean power and zero mean. This then yields for the mean SIR

$$\text{SIR} \geq \frac{P_1}{P_2(|1 - \exp(j\Delta\omega_{21} t)|^2)} \cdot \frac{E\{\|\mathbf{c}_2\|^2 (|h_{11} h_{22}|^2 + |h_{12} h_{21}|^2)\}}{E\{\|\mathbf{c}_1\|^2 |h_{11} h_{12}|^2\}}$$
$$= \frac{P_1}{P_2(1 - \cos(\Delta\omega_{21} t))} \approx \frac{2}{(\Delta\omega_{21} \cdot t)^2} \cdot \frac{P_1}{P_2}, \tag{8.43}$$

8.3 Imperfect Sync in Frequency: Perf. Degradation and Compensation

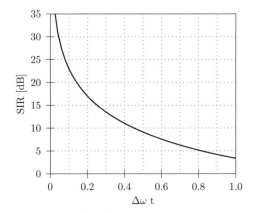

Figure 8.9 SIR degradation due to CFO-induced inter-stream interference of two spatially precoded streams according to (8.43) ($P_1 = P_2$).

where the Taylor expansion of cosine for small angles, $\cos(\alpha) \approx 1 - 0.5\alpha^2$, was used to obtain the approximation on the right-hand side. An illustration of this SIR relation for $P_1 = P_2$ is given in Fig. 8.9.

Eq. (8.43) reveals that the mean SIR resulting from the CFO-induced interstream interference between the spatially precoded streams strongly depends on the difference of the CFOs observed at the transmitters, $\Delta\omega_{21} = \omega_2 - \omega_1$, as well as ratio of the powers P_m allocated to the single transmission streams. As long as the precoding matrix $\mathbf{C}$ is not updated, the interference grows continuously over time with the factor $(\Delta\omega_{21} \cdot t)^2$. From the above result, we can thus deduce a requirement for the minimum update interval of the precoding matrix $\mathbf{C}$ to achieve a desired SIR constraint γ:

$$t = (\Delta\omega_{21})^{-1}\sqrt{2\gamma^{-1}}. \qquad (8.44)$$

The CFO difference $\Delta\omega_{21}$ can be related to the accuracy a achieved for the oscillators used at the transmitters after synchronization. With the carrier frequency f_c, the relation $\Delta\omega_{21} = 2\pi a f_c$ holds. To give an example, assume that a CoMP system operates at a carrier frequency $f_c = 2$ GHz, synchronization of the transmitters has achieved an accuracy of $a = 1$ parts per billion (ppb) $= 10^{-3}$ parts per million (ppm), and the desired SIR target the system should not fall below is $\gamma = 20$ dB. Then the update interval should be around 11 ms, which lies in the dimension of the duration of only a few radio frames in modern wireless communication systems. This example clearly indicates that synchronization requirements for downlink CoMP precoding are very strict; however, we have seen in Section 8.1 that synchronization methods are readily available that are capable of establishing these.

Inter-Carrier Interference in an OFDM system

To get insights into the general characteristics of the ICI firstly, we will derive an analytical expression for a single-user single-input single-output (SISO) channel, which is distorted by a single CFO Δf. Afterwards we will use the obtained results to characterize the ICI in the downlink CoMP system.

We assume a (static) frequency-selective channel is characterized by the discrete channel impulse response $\tilde{h}(l)$, $l \in 1, \ldots, L$, where the discrete positions in time, l, are based on the sampling period T_s. Given a CFO of Δf, the time-varying channel seen at the receiver can be given in the time/delay domain according to

$$\tilde{h}(t, \tau) = \sum_{l=0}^{L-1} \tilde{h}(l) \exp(j2\pi \Delta f t) \delta(\tau - lT_s), \qquad (8.45)$$

where δ denotes the Dirac-function. As derived in detail in [STJ08, Sch09], the corresponding OFDM channel with Q sub-carriers can be given according to

$$h(q, \Delta q) = \underbrace{\sum_{l=0}^{L-1} \tilde{h}(l) \exp\left(-j2\pi l \frac{q}{Q}\right)}_{h(q)} \exp(j\pi \Delta f Q T_s) \frac{\sin(\pi \Delta f Q T_s)}{\pi(\Delta f Q T_s - \Delta q)}, \qquad (8.46)$$

with

$$\begin{aligned} q &\in \mathcal{Q} := \{0, \ldots, Q-1\} \\ \Delta q &\in \mathcal{D} := \{-Q/2+1, \ldots, Q/2\}. \end{aligned} \qquad (8.47)$$

According to this definition, $h(q, \Delta q)$ for $\Delta q \neq 0$ represents the interference that is imposed from the sub-carrier signal at position q on the sub-carrier that is spaced Δq sub-carriers apart, whereby the sign of Δq specifies the direction. For $\Delta q = 0$, however, $h(q, \Delta q)$ delivers the useful channel seen at sub-carrier q.

The first part of the expression in (8.46) represents the sub-carrier channel $h(q)$, which is seen at sub-carrier q if no CFO distortions are present. We can clearly see that the CFO reduces the amplitude of this useful channel ($\Delta q = 0$) and introduces an additional phase rotation according to the expression on the right hand side. Accordingly, fractions of the signal power of the channel coefficient $h(q)$ are spread as ICI on the other sub-carriers in the system ($\Delta q \neq 0$). Based on the time-variant OFDM channel $h(q, \Delta q)$, the transmission equation for the received signal $y(q)$ at a fixed sub-carrier q can be given as

$$y(q) = h(q, 0)x(q) + \sum_{\Delta q \in \mathcal{D} \setminus 0} h(q - \Delta q, \Delta q) x(q - \Delta q) + n(q). \qquad (8.48)$$

The first term in this expression represents the useful signal, while the sum term represents the distortions from ICI.

8.3 Imperfect Sync in Frequency: Perf. Degradation and Compensation

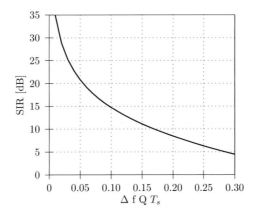

Figure 8.10 SIR degradation due to CFO-induced inter-carrier interference in a SISO-OFDM system according to (8.51).

As derived in detail in [Sch09], it can be shown that the mean power of the useful signal, P_u, and the mean power of the ICI, P_{ICI}, amount to

$$P_u = P_s \sigma_h^2 \cdot \text{si}^2(\pi \Delta f Q T_s) \qquad (8.49)$$
$$P_{ICI} = P_s \sigma_h^2 \cdot (1 - \text{si}^2(\pi \Delta f Q T_s)), \qquad (8.50)$$

where P_s is the mean power of the transmit signals, σ_h^2 is the total mean power of the channel and $\text{si}(x) = \sin(x)/x$ is the si-function. These results are based on the wide sense stationary uncorrelated scattering (WSSUS) assumption for the channel coefficients; further it has been assumed that a signal with mean power P_s has been transmitted from all available sub-carriers. The SIR resulting from the ICI can therefore be given as

$$\text{SIR} = \frac{P_u}{P_{ICI}} = \frac{\text{si}^2(\pi \Delta f Q T_s)}{1 - \text{si}^2(\pi \Delta f Q T_s)}. \qquad (8.51)$$

This SIR degradation caused by ICI is illustrated in Fig. 8.10.

ICI for CoMP Transmission in Downlink

To ease analytical derivations and allow for simple analytical expressions, we assume here that on each of the available OFDM sub-carriers, the same set of users is served by the same set of base stations.

In (8.46) we have seen that the ICI from sub-carrier q on a sub-carrier at a distance Δq depends on the channel coefficient $h(q)$ at that sub-carrier. Therefore, we can conclude that these relations hold similarly also for the *effective* channels in a CoMP system that include the spatial precoding weights. According to (8.34), the effective channel at sub-carrier q reads $\mathbf{H}(q)^H \mathbf{\Omega} \mathbf{W}(q)$. As synchronization has to ensure that the inter-stream interference does not grow too large, the majority of the power of this effective channel can be found in the diagonal elements of this channel, see (8.38). Thus, for the ICI consideration, the inter-stream interference may be neglected and the effective channel may be understood as a set of independent, orthogonal SISO channels. The sub-carrier channel $h(q)$ seen by the first receiving UE is thus the effective SISO channel

$$h(q) = \mathbf{h}_1^T(q)\mathbf{\Omega}\mathbf{c}_1(q)p_1(q) \approx p_1(q). \qquad (8.52)$$

Before inserting this effective channel into (8.46) to obtain useful channel and ICI channels seen by the first receiving UE, some considerations on the CFOs Δf_m are necessary: As both CFOs Δf_m have an influence on the exact value of the ICI, we can use their maximum $\Delta f = \max_m \Delta f_m$ to upper bound it. Based on this, (8.46) can be directly translated for the effective channel seen by the first UE to

$$h(q, \Delta q) = p_1(q) \exp\left(j\pi \Delta f Q T_s\right) \operatorname{si}\left(\pi \left(\Delta f Q T_s - \Delta q\right)\right). \qquad (8.53)$$

If we assume that the mean power of $p_1(q)$ is identical for all sub-carriers, the upper bound for the ICI from (8.51) can be found to be valid also for the CoMP downlink.

By comparing the SIR for the inter-stream interference from (8.43) and for the ICI from (8.51), we immediately see that the former expression grows with the time t, while the latter only depends on the absolute value of the CFO Δf normalized to $(QT_s)^{-1}$, which is, in fact, the sub-carrier spacing. Therefore, the requirement on the synchronization accuracy derived from the SIR for the inter-stream interference is orders of magnitude larger than that derived from the SIR for the ICI. Resorting to our example from the preceding subsection, the CFO Δf to achieve a SIR of 20 dB should be below 5 % of the sub-carrier spacing. If the sub-carrier spacing is at 15 kHz like in current OFDM-based mobile radio systems, the maximum allowed CFO would be 750 Hz. Compared to the maximum allowed CFO of only a few Hertz that is required to enable updates of the precoding matrices within a few milliseconds, this difference amounts to a factor larger than 100. From these considerations, it may be concluded that the ICI in OFDM systems does not play a significant role for the synchronization of the CoMP downlink, and therefore its influence may be neglected.

Techniques for the Compensation of Degradation Effects
Unfortunately, there are no methods available to compensate for the degradation effects due to inter-stream interference at the side of the receiver. The reason for that is related to the fact that the dimension of the spatial receive space in the downlink is usually much smaller than the dimension of the spatial transmit space. For an illustration, consider the following: If M BSs with N_{bs} transmit antennas each cooperate, they are able to form up to MN_{bs} orthogonal transmit beams. If the BSs are not properly synchronized, the beams will loose their orthogonality and will thus interfere at each UE. Even if the UE has multiple antennas, it will not be able to resolve the interfering beams, disabling suitable approaches for proper compensation.

The only remaining solution is to compensate for CFO distortions at the side where they appear, that is the transmitter. However, this can only be accomplished if knowledge on the CFOs is available there. In [Zar08], it has been proposed to estimate the CFOs by all UEs that are involved in the CoMP transmission, who then feed back their estimates to the BSs. Techniques to estimate these CFOs with high accuracy have also been proposed there. However, this approach has the drawback that it requires computationally complex estimators at the UEs and a continuous feedback of the estimates from all UEs. Moreover, it is questionable whether the proposed technique can achieve a better synchronization accuracy in practice than techniques for the synchronization between BSs do (see Section 8.1). To summarize, we can conclude here that tight synchronization between the BSs is of fundamental importance to enable reliable CoMP transmission in the downlink, which calls for inter-BS synchronization techniques achieving the best accuracy possible.

8.3.2 Uplink Analysis

Inter-Stream Interference of Spatially Precoded Streams

The transmission equation for the uplink can be given as

$$\mathbf{y} = \mathbf{\Omega}_r(t)\mathbf{H}\mathbf{\Omega}_t(t)\mathbf{P}\mathbf{x} + \mathbf{n}. \tag{8.54}$$

To point out the duality to the downlink, we consider the CFO distortion matrices for both the transmitter and receiver, $\mathbf{\Omega}_t(t)$ and $\mathbf{\Omega}_r(t)$, here. The diagonal matrix $\mathbf{P}$ now is constituted of the elements $p_k = \sqrt{P_j}$ only, so it reduces to a simple power allocation matrix here. To obtain the transmitted signals, the cooperating BSs jointly equalize the received signal $\mathbf{y}$

$$\tilde{\mathbf{x}} = \mathbf{W}^H \mathbf{y} = \mathbf{W}^H (\underbrace{\mathbf{\Omega}_r(t)\mathbf{H}\mathbf{\Omega}_t(t)}_{\bar{\mathbf{H}}(t)}\mathbf{P}\mathbf{x} + \mathbf{n}). \tag{8.55}$$

As all CFO distortions can be continuously tracked at the receiving BSs, the effective matrix $\bar{\mathbf{H}}(t)$ can be compensated completely if we choose the equalization matrix at time instant t according to

$$\mathbf{W}^H(t) = \mathbf{\Omega}_t(-t)\mathbf{H}^{-1}\mathbf{\Omega}_r(-t). \tag{8.56}$$

However, note that this requires a continuous update of the equalization matrix on a per-symbol base. If we assume, for some reason, that such a continuous update cannot be conducted and equalization is performed with the matrix $\mathbf{W}^H(t) = \mathbf{H}^{-1}$ instead, then the situation is very similar to the dual downlink case considered in (8.34), and correspondingly we obtain the same SIR expression as derived in (8.43) for the two-user case. (Note that the CFOs in matrix $\mathbf{\Omega}_t(t)$ do not have any effect on the SIR, as they only impose a phase rotation on desired and interfering signal. This phase rotation, however, can always be compensated at the receiver by phase tracking after channel equalization).

In any case, the constraints on the synchronization requirements will by far be not as strict as in the downlink case, as the compensation matrix $\mathbf{W}^H$ can be updated much more frequently, say at least every sub-frame. If a continuous update of the compensation matrix is conducted on a per-symbol basis, then the requirement for the synchronization accuracy comes rather from the ICI conditions, which are addressed in the next paragraph.

Inter-Carrier Interference in an OFDM System

Similarly as for the downlink analysis, we assume here that on each of the available OFDM sub-carriers the same set of users transmits to the same set of base stations. Within this subsection, we refer again to the exemplary case of 2 BSs and two UEs, all being equipped with one antenna each.

For the ICI investigations, it is reasonable to assume that the CFO distortion effects occurring at the side of the UEs dominate those at the side of the BSs, as it is much easier to establish a tight synchronization between BSs rather than between UEs. Therefore, we neglect matrix $\mathbf{\Omega}_r(t)$ in (8.54) for our considerations here. The sub-carrier channel at sub-carrier q then reads $\mathbf{H}(q)\mathbf{\Omega}_t(t)$. Correspondingly, the channel function for the OFDM system in (8.46) can be updated for the CoMP uplink as follows:

$$\mathbf{H}(q, \Delta q) = [\mathbf{h}_1(q) \quad \mathbf{0}] \exp(j\pi \Delta f_1 Q T_s) \mathrm{si}\left(\pi \left(\Delta f_1 Q T_s - \Delta q\right)\right) \\ + [\mathbf{0} \quad \mathbf{h}_2(q)] \exp(j\pi \Delta f_2 Q T_s) \mathrm{si}\left(\pi \left(\Delta f_2 Q T_s - \Delta q\right)\right). \quad (8.57)$$

We observe here that the two different CFOs Δf_1 and Δf_2 generate independent ICI from the channel vectors $\mathbf{h}_1(q)$ and $\mathbf{h}_2(q)$, respectively, which are simply superimposed. The transmission equation given for the SISO case in (8.48) is modified to match the CoMP uplink according to

$$\mathbf{y}(q) = \mathbf{H}(q, 0)\mathbf{x}(q) + \underbrace{\sum_{\Delta q \in \mathcal{D} \setminus 0} \mathbf{H}(q - \Delta q, \Delta q)\mathbf{x}(q - \Delta q)}_{\text{ICI}} + \mathbf{n}(q). \quad (8.58)$$

If we jointly equalize the received signal $\mathbf{y}(q)$ at sub-carrier q now by using the equalizer matched to the corresponding sub-carrier channel, i.e. $\mathbf{W}^H(q) = \mathbf{\Omega}_t(-t)\mathbf{H}^{-1}(q)$, we observe that this equalizer will not be able to suppress any of the ICI in (8.58), because the equalizer is not properly matched to any of the interference channels $\mathbf{H}(q - \Delta q, \Delta q)$. (This relation holds as long the transmission channels are of frequency-selective nature and the channels between different antenna pairs are mutually independent). As a result, the ICI from all simultaneously transmitted data streams will affect the useful signal $\tilde{x}_k(q)$ after equalization. Hence, the mean ICI power affecting the useful signal at any sub-carrier q will for the CoMP uplink amount to

$$\bar{P}_{ICI} = P_1 \left(1 - \mathrm{si}^2(\pi \Delta f_1 Q T_s)\right) + P_2 \left(1 - \mathrm{si}^2(\pi \Delta f_2 Q T_s)\right). \quad (8.59)$$

8.3 Imperfect Sync in Frequency: Perf. Degradation and Compensation

For the power of the useful signal, conditions remain the same as in the SISO case, i.e. we achieve according to (8.51) for the SIR of the useful signal $\tilde{x}_k(q)$

$$\mathrm{SIR}\,(\tilde{x}_k) = \frac{P_k \mathrm{si}^2(\pi \Delta f_k Q T_s)}{P_1 \left(1 - \mathrm{si}^2(\pi \Delta f_1 Q T_s)\right) + P_2 \left(1 - \mathrm{si}^2(\pi \Delta f_2 Q T_s)\right)}. \tag{8.60}$$

Compared to the SIR conditions valid for the CoMP downlink in (8.51), we clearly see that the major difference for the uplink lies in the fact that the ICI from all simultaneously transmitted data streams affect the SIR of each desired signal. The ICI power in the uplink thus scales with the number K of simultaneously transmitting UEs. Considering that it is more difficult to establish a tight synchronization between different UEs, compensation techniques for ICI may become an issue in the uplink.

Remark: The brief analysis presented here points out the most important aspects of the ICI effects in the OFDM-based CoMP uplink only. For a more detailed investigation, the interested reader is referred to [SJ09].

Techniques for Compensation of Degradation Effects

As mentioned already in the corresponding subsection, compensation of the degradation effects due to inter-stream interference is simple, as we only have to update the equalization matrix (8.56) continuously. However, we have pointed out above that in the uplink the ICI may be a more serious problem, as the CFOs caused by the low-cost oscillators at the UEs may become much larger than those caused by the oscillators of the BSs. Although compensation of the ICI distortions on the side of the receiver is possible, it turns out to be a complex task. For an overview on existing compensation techniques for CFO-induced ICI, refer to [MKP07, SJ09].

In general, the OFDM channel including the CFO distortions can be represented as a large square matrix of dimension $Q \cdot N_{\mathrm{bs}} \times Q \cdot N_{\mathrm{bs}}$, which has a band structure with all elements within this band being non-zero. By zero-forcing this huge channel matrix, the channel including all CFO-distortions can be fully compensated. However, it is obvious that this would imply an enormous computational effort. To cut this effort down, several approaches have been proposed that exploit the specific structure of the large channel matrix to divide it into a set of smaller sub-matrices. The computational effort remaining is still considerable, though. Also some solutions have been proposed that try to maintain the sub-carrier-wise processing OFDM systems are favored for. Although they are not capable of removing the CFO-induced ICI completely, they can compensate for a large amount of it, improving the SIR conditions significantly. One such approach was presented in [SJ09]; the signal processing for the compensation process is depicted in Fig. 8.11: The received signals r_m at antenna m are individually transformed to the frequency domain, where the sub-carrier-wise channel equalization with matrix $\mathbf{G}^H = \mathbf{H}^{-1}$ is applied that separates the simultaneously transmitted signal streams. After signal separation, the equalized symbols of each stream are convoluted in frequency domain with a func-

192 Synchronization

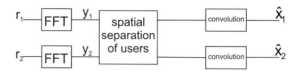

Figure 8.11 Receiver processing for simplified signal reconstruction with CFO compensation in uplink.

tion derived from (8.46) to compensate for the CFO distortions. This method fully compensates for the CFO-induced ICI of the transmit symbols, however, as shown analytically in [SJ09], it cannot compensate for the ICI that results from the CFO-induced violation of the periodic property of the cyclic prefix, which is used for OFDM transmission.

8.3.3 Summary

In this section, we analyzed the inter-stream interference and the ICI induced by CFO distortions for the downlink and uplink of an OFDM-based CoMP system, consisting of 2 BSs and 2 UEs, each equipped with a single antenna. For the downlink, it has been shown that the SIR conditions derived for the inter-stream interference set strict requirements for the synchronization accuracy, while the effect of ICI can be neglected. As compensation for CFO distortions at the receivers is not possible, tight synchronization between the simultaneously transmitting BSs is mandatory. In the uplink, we can fully compensate for the inter-stream interference if the CFO distortions are continuously tracked. Therefore, requirements for the synchronization accuracy are rather deduced from the SIR conditions resulting from the ICI. The ICI distortion may still be compensated at the receiver; however, the additional computational complexity required for this purpose is considerable.

9 Channel Knowledge

In this chapter, we address the issue how channel knowledge - referring to both desired channels and the channels towards interferers - needed for various CoMP schemes can be made available where it is needed. We first investigate channel estimation techniques at the receiver side in Section 9.1, and then discuss how the obtained channel knowledge can be efficiently fed back to the transmitter side in Section 9.2, which is for example a crucial requirement for the downlink CoMP schemes investigated in Sections 6.3 and 6.4. The chapter shows that standard channel estimation and feedback concepts can principally be extended to enable CoMP in general. However, it also becomes apparent that large CoMP cooperation sizes may be considered questionable in practice, due to the fact that weak links cannot be estimated accurately, and the involved pilot and channel state information (CSI) feedback overhead may become prohibitive.

9.1 Channel Estimation for CoMP

Wolfgang Zirwas, Lars Thiele, Tobias Weber, Nico Palleit and Volker Jungnickel

One of the main challenges for CoMP schemes like joint transmission (JT) is to obtain accurate channel information in a multi-cell mobile radio environment with acceptable overhead for pilot signals.

The section is structured as follows. In Subsection 9.1.1, main characteristics of the mobile radio channel and state-of-the-art estimation and interpolation techniques like Wiener filtering will be introduced, with a special focus on channel prediction. For CoMP, the analysis then has to be extended to multiple channel components and multi-cell scenarios, which will be done in Subsections 9.1.2 and 9.1.3, respectively. While most of the section focusses on a *downlink* transmission and hence channel estimation at the terminal side, Subsection 9.1.4 finally looks into specific aspects of *uplink* channel estimation.

In this section, we observe an orthogonal frequency division multiplex (OFDM) system, as considered in LTE Release 8 and beyond, having the main benefit of a very effective orthogonalization of resources in time and frequency. This simplifies the implementation of complex precoding or equalization schemes in the context of multiple-input multiple-output (MIMO) or CoMP, as proved by first CoMP test beds or demonstrators described in Chapter 13, which are all OFDM implementations.

9.1.1 Channel Estimation - Single Link

CoMP setups are basically distributed multi-user MIMO (MU-MIMO) systems involving a large number of radio channel components. However, before analyzing CoMP including many channel components, one has to understand fundamental aspects of channel estimation like optimal filtering, sampling or interpolation for a single channel component, being subject of this first subsection.

Note that the structure of this first part is motivated by a proper understanding of the physical aspects of channel estimation based on the time domain channel impulse response (CIR). Starting point is the multi-path channel itself, eventually leading to the well-known reference signal design and receiver processing blocks for OFDM systems. The description does not follow exactly the transmit and receive chain of an OFDM system, but rather gives the reader the underlying reasoning e.g. for standardized LTE Release 8 parameters as well as the potential for further optimizations. Channel prediction - as one such possible optimization - will be introduced in more detail, as it is currently a hot CoMP research topic promising to overcome the problem of outdated time-variant channel information.

The Channel Transfer Function

Fig. 9.1 illustrates a typical simplified mobile radio channel comprising a direct and several reflected multi-path components (MPCs). According to Fig. 9.1(b), $\tilde{h}(t,\tau) \in \mathbb{R}$ is the resulting CIR varying over time t, given as

$$\tilde{h}(t,\tau) = \sum_n \alpha_n(t)\delta(\tau - \tau_n(t)), \quad \text{with} \quad \tau_n(t) = d_n(t) - d_0(t), \qquad (9.1)$$

where $n \in \{1, \ldots, N_\mathrm{m}\}$ are the relevant MPC indices, $\alpha_n(t)$ the amplitude of the n-th MPC, and $d_0(t)$ is the delay connected to the shortest MPC. Note that we here use $\tilde{h}$ with a tilde, as we are observing a channel in *time domain*, as opposed to h, which always refers to a channel coefficient in *frequency domain* throughout this book. The length of the CIR $\tilde{h}(t,\tau)$ is $T_{\mathrm{CIR}} = \tau_{N_\mathrm{m}} - \tau_1$, which is directly related to the path length difference Δs of the shortest and longest MPC. For example, a value of T_{CIR} of 1 μs corresponds to $\Delta s = 1~\mu\mathrm{s} \cdot c = 300$ m, with c being the velocity of light of $3 \cdot 10^8$ m/s. Fig. 9.1(c) illustrates the ideally infinite[1] channel transfer function (CTF) $h(t,f) \in \mathbb{C}$, i.e. the *frequency domain*

---

[1] In reality, one can observe a wideband similarity of the radio channel over several 100 MHz, allowing downlink beamforming based on uplink covariance estimation in frequency division duplex (FDD) systems.

9.1 Channel Estimation for CoMP

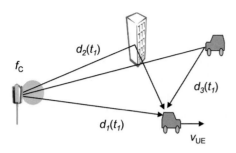
(a) Illustration of multi-path components.

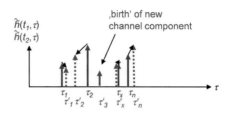
(b) Development of the CIR over time.

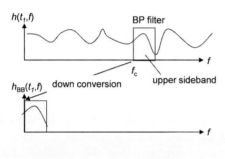
(c) Illustration of down-conversion.

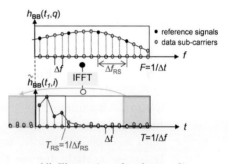
(d) Illustration of undersampling.

Figure 9.1 Typical multi-path radio channel.

representation of $\tilde{h}(t,\tau)$ over frequency f:

$$h(t,f) = \sum_n \alpha_n(t)\delta(\tau - \tau_n(t))e^{-j2\pi f \tau_n(t)}. \quad (9.2)$$

Any real-valued time-domain signal $\tilde{s}(t)$ requires that its frequency-domain representation fulfills $s(-f) = s^*(f)$, leading to positive and negative complex conjugate frequency components. Direct conversion will generally shift the frequency bands with positive and negative frequencies together from base- to radio frequency (RF)-band, resulting in the so called lower and upper sidebands plus additionally the carrier signal at RF frequency f_c. Upper and lower sidebands contain basically the same information twice and therefore for single side band (SSB) transmission the upper or lower sideband is filtered out, thereby doubling the spectral efficiency. Standard transmitters and receivers according to the LTE Release 8 specification are based on in- and quadrature phase modulators/demodulators, which generate/demodulate the SSB signal directly without filtering and even avoiding the carrier signal. The CTF of the baseband signal $h_{\rm BB}(t,f)$ is a down-converted copy of the generally complex-valued CTF of the SSB RF-band:

$$h_{\rm BB}(t,f) = {\rm BPF}(f) \cdot h(t,f)e^{-j2\pi f_c t}. \quad (9.3)$$

For simplicity, we will assume in the following an ideal rectangular bandpass filter $\text{BPF}(f) = \text{rect}(\frac{f-(f_c+B/2)}{B})$ of bandwidth B corresponding in time domain to the well-known SI-function representing $\sin(x)/x$.

Analog to digital conversion (ADC) leads to $h_{\text{BB}}(t,q) \in \mathbb{C}$, the sampled CTF of $h_{\text{BB}}(t,f)$ at frequency bins $q \cdot \Delta f$, $q \in \{1,...,Q\}$ with $Q \cdot \Delta f = B$, the bandwidth of the baseband signal. In the sequel, we will write the sampled CTF in vector notation as $\mathbf{h}_{\text{BB}}(t) = [h_{\text{BB}}(t,1) \ldots h_{\text{BB}}(t,Q)]^T \in \mathbb{C}^{[Q \times 1]}$. Note that this is different from the notation used in most parts of the book, where channel coefficient vectors reflect the *spatial* signal domain, but now we use these to reflect a frequency dimension (i.e. sub-carrier indices).

The equidistantly sampled CIR $\tilde{h}(t,i) \in \mathbb{C}$ with $i \in \{1,\ldots,Q\}$ - or $\tilde{\mathbf{h}}(t) \in \mathbb{C}^{[Q \times 1]}$ in vector notation where the elements represent sample indices - is obtained by an inverse fast Fourier transform (IFFT) of $h_{\text{BB}}(t,q)$ with the fast Fourier transform (FFT) matrix $\mathbf{F} \in \mathbb{C}^{[Q \times Q]}$:

$$\tilde{\mathbf{h}}(t) = \frac{1}{Q}\mathbf{F}\mathbf{h}_{\text{BB}}(t), \text{ with } f_{i,q} = e^{j2\pi i q/Q},\ i,q \in \{1,\ldots,Q\}, \qquad (9.4)$$

where the maximum time duration of the CIR $\tilde{\mathbf{h}}(t)$, i.e. the time duration of one FFT window, is $T = 1/\Delta f = Q \cdot \Delta t$ with sampling time $\Delta t = T/Q = 1/B$. Note that, as explained before, we do not strictly follow the structure of an OFDM transmission chain, but use the time domain CIR as basis for description of some main aspects of channel estimation.

For OFDM systems, a guard interval is inserted in the form of a cyclic prefix (CP), where the first receive samples within the CP of length $T_g = L \cdot \Delta t$ from indices $-L+1$ to 0 are a copy of the samples from $Q-L+1$ to Q and are not evaluated at the receiver side. OFDM in combination with a CP results in a circular convolution of the signal with the channel. In case of a properly designed CP with a length longer than the maximum expected length of the CIR, i.e. $T_{\text{CIR, max}} = \tau_{N_m} \le T_g$, inter-symbol interference (ISI) between consecutive OFDM symbols and inter-carrier interference (ICI) can be avoided completely.

As a direct consequence, the samples $i \in \{L+1,\ldots,Q\}$ of a CIR $\tilde{h}(t,i)$ fulfilling the above design criteria for T_g can be assumed to be zero or at least below a certain threshold. The relevant part of the shortened CIR is $\tilde{\mathbf{h}}_s(t) = [h(t,0),\ldots,h(t,L-1)]^T \in \mathbb{C}^{[L \times 1]}$. Note that in the case of LTE, the length of an OFDM symbol is $T = T_{\text{symbol}} = 66{,}7\ \mu\text{s}$, and that of the normal CP is $4.69\ \mu\text{s}$. This corresponds to a maximum MPC path length difference between longest and shortest path of $\Delta l = 4.69 \cdot 300\text{m} \approx 1400$ m for ISI-free transmission.

Undersampling in Frequency Domain

Undersampling of a CTF by a factor of $\Delta f_{\text{RS}}/\Delta f$ reduces the length of the IFFT converted time domain signal from $T = Q \cdot \Delta t = 1/\Delta f$ to $T_{\text{RS}} = L_{\text{RS}} \cdot \Delta t = 1/\Delta f_{\text{RS}}$, meaning that only the first L_{RS} samples of the CIR are reconstructed, while the sampling rate $1/\Delta t$ is unaffected (see Fig. 9.1(d)). As a consequence, a CIR $\tilde{\mathbf{h}}_s(t)$ of length $T_{\text{CIR}} \leq T_g$ can be fully reconstructed even for the above mentioned undersampled frequency domain signal $h(t,q)$. This can and - in case of LTE Release 8 - is being used for an efficient design of so-called pilots or reference signals (RSs), as will be explained later.

Receive Filter

In case of a rectangular bandpass RF- (or baseband) filter BPF(f), the Dirac functions $\delta(\tau - \tau_n(t))$ of the MPCs will be convoluted with the SI-function sinc($t/\Delta t$), so that each tap of the CIR $\tilde{h}(t,i)$ is a superposition of different MPCs. Note that in the general case, the MPC delays $\tau_n(t)$ will not coincide with the sampling timing $i \cdot \Delta t$. Hence, from the measured, quantized and potentially shortened CIR $\tilde{\mathbf{h}}_s(t)$, one cannot directly derive the real channel MPC delays τ_n and amplitudes α_n. Accurate knowledge would be desirable - e.g. for channel prediction - as the superposition of MPCs affects the further evolution of a tap.

Time-varying Radio Channel

Mobile radio channels are time-variant due to movements of the user equipments (UEs) themselves, moving objects in the environment, or time-variant scatterers. From Eq. (9.1) and Fig. 9.1(b), the main effects on the CIR for a moving UE are clearly visible, i.e. at time $t + \Delta t$, the delays of the MPCs will have changed from $\tau_n(t)$ to $\tau_n(t + \Delta t)$. The delay values $\tau_n(t + \Delta t)$ compared to $\tau_n(t)$ are determined by the variation of the corresponding path lengths between transmit and receive antennas, which increase or decrease dependent on the relative - for downlink (DL) - incident angles at the moving UE for particular MPCs.

Estimation in Time and Frequency - Two-Dimensional Wiener Filter

Mobile radio OFDM or single carrier frequency domain multiple access (SC-FDMA) systems like LTE Release 8 have been designed for training based channel estimation, i.e. they rely on predefined and standardized pilots or RSs. To save overhead, RSs are placed only on every n_{RS}-th sub-carrier, where $n_{\text{RS}} = \Delta f_{\text{RS}}/\Delta f$ as introduced above for undersampling of the radio CTF. In general, RSs may be allocated to any sub-carrier, but in practical systems regular sampling of the CTF $h(t,q)$ is most common.

Considering the transmission of a signal vector $\mathbf{s} \in \mathbb{C}^{[Q \times 1]}$ having zero mean over a mobile radio channel $\mathbf{h}_{\text{BB}}(t)$ in frequency domain can be written as

$$\mathbf{y} = \mathbf{h}_{\text{BB}}(t) \odot \mathbf{s} + \mathbf{n}, \tag{9.5}$$

where $\odot$ denotes element-wise multiplication, and $\mathbf{n} \in \mathbb{C}^{[Q \times 1]}$ is a zero mean additive i.i.d. Gaussian noise vector with $E\{\mathbf{nn}^H\} = \sigma^2 \mathbf{I}$. As mentioned above, every n_{RS}-th sub-carrier in $\mathbf{s}$ carries a known RS. All these RSs are collected in the reduced length reference signal vector $\mathbf{s}_{\text{RS}} \in \mathbb{C}^{[(Q/n_{\text{RS}}) \times 1]}$. Undersampling - at least in the noise-free case - allows full reconstruction of sampled CIR $\tilde{\mathbf{h}}_s(t,i)$ of a length smaller or equal to L, while the overhead for RSs is limited to Q/n_{RS}.

Frequency-spaced RSs with undersampling factor $1/n_{\mathrm{RS}}$ reduce processing overhead for channel estimation, and one can introduce

$$[\mathbf{F}_{\mathrm{RS}}]_{i',q'} = e^{j2\pi i'q'/Q}, \quad i' \in \{1,\ldots,L,\ldots Q/n_{\mathrm{RS}}\}, \; q' \in \{1,\ldots,Q/n_{\mathrm{RS}}\}. \quad (9.6)$$

Here, $\mathbf{F}_{\mathrm{RS}} \in \mathbb{C}^{[Q/n_{\mathrm{RS}} \times Q/n_{\mathrm{RS}}]}$ is the row- and column-reduced inverse discrete Fourier transform (IDFT) matrix $\mathbf{F}$, and $\mathbf{y}_{\mathrm{RS}}(t)$ is the row-reduced receive vector for the Q/n_{RS} sub-carriers carrying RSs:

$$\widehat{\tilde{\mathbf{h}}}(t) = \mathbf{F}_{\mathrm{RS}} \mathbf{y}_{\mathrm{RS}}(t), \quad (9.7)$$

where $\widehat{\tilde{\mathbf{h}}}(t) \in \mathbb{C}^{[Q/n_{\mathrm{RS}} \times 1]}$ is the noisy estimate of the CIR $\tilde{\mathbf{h}}(t)$. For $\widehat{\tilde{\mathbf{h}}}_s(t)$, the elements of $\widehat{\tilde{\mathbf{h}}}$ with indices $L+1$ to Q/n_{RS} can be set to zero, based on the assumption that the CIR is limited to L taps and the taps $L+1,\ldots,Q/n_{\mathrm{RS}}$ carry only noise or at least very low power taps. This zero setting of elements or taps is called *denoising* and improves the signal-to-interference ratio (SIR) of the estimated CTF after conversion of the CIR into frequency domain.

$$\widehat{\tilde{\mathbf{h}}}_s(t) = \left[\underbrace{\widehat{\tilde{h}}(t,1),\ldots,\widehat{\tilde{h}}(t,L)}_{\text{excess delay}}, 0, \ldots, 0 \right]^T. \quad (9.8)$$

The shortened CIR $\widehat{\tilde{\mathbf{h}}}_s(t)$ can be calculated based on the row-reduced matrix $\mathbf{F}'_{\mathrm{RS}} = \mathbf{F}_{\mathrm{RS}}|_{u=1\ldots L, q=1\ldots Q}$. We finally calculate $\hat{\mathbf{h}}(t)$ as the estimated and interpolated CTF for all Q sub-carriers from the noisy time-domain estimate $\widehat{\tilde{\mathbf{h}}}_s$:

$$\hat{\mathbf{h}}(t) = \mathbf{F}^H \cdot \widehat{\tilde{\mathbf{h}}}_s(t). \quad (9.9)$$

It is also possible to generate $\hat{\mathbf{h}}(t)$ directly from $\tilde{\mathbf{y}}_{\mathrm{RS}}$ by applying the interpolation matrix $\mathbf{F}_{\mathrm{int}} \in \mathbb{C}^{[Q \times Q/n_{\mathrm{RS}}]}$, which may be pre-calculated for real systems and known RS positions and signals, i.e.

$$\hat{\mathbf{h}}(t) = \mathbf{F}^H \widehat{\tilde{\mathbf{h}}}_s(t) = \mathbf{F}^H \cdot \mathbf{F}'_{\mathrm{RS}} \cdot \mathbf{y}_{\mathrm{RS}}(t) = \mathbf{F}_{\mathrm{int}} \cdot \mathbf{y}_{\mathrm{RS}}(t). \quad (9.10)$$

Interpolation Gain

Estimation accuracy and overhead for RSs in terms of resources and power are important design parameters for an OFDM system. For a given signal-to-interference-and-noise ratio (SINR), the achievable channel estimation accuracy depends on the number of RSs or, equivalently, the undersampling factor of the CTF. By doubling the number of RSs, the so-called *interpolation* or *processing* gain is increased by 3 dB at the cost of two-fold pilot overhead. Note that the interpolation gain is the improvement of channel estimation accuracy due to the above described denoising effect, compared to a baseline channel estimation performed individually for each sub-carrier and OFDM symbol.

The length of the guard interval has been designed for expected *worst case scenarios*. In scenarios where τ_{N_m} is significantly smaller than $L \cdot \Delta t$, further interpolation gains may be obtained by setting further samples of the CIR within the GI to zero. For this purpose, the length of the CIR $\widehat{\mathbf{h}}(t)$ has to be estimated requiring additionally an estimate of the signal-to-noise ratio (SNR) of $\tilde{\mathbf{y}}(t)$ to find those taps being below the noise level.

Wiener Filter
A well-known solution for exploiting potential interpolation gains is to apply *Wiener filtering* [Hay02], which finds the optimum filter $\mathbf{F}_{\text{int,opt}}$ with respect to a minimum mean square error (MMSE) criterion, i.e. minimizing $\sigma_{\Delta H}^2 = 1/Q \cdot E\{||\widehat{\mathbf{h}}(t) - \mathbf{h}(t)||^2\}$. Wiener filters generally exploit estimated or known statistical properties of the signals, i.e. the auto-covariance matrix $\mathbf{\Phi}_{\mathrm{yy}} = E\{\mathbf{yy}^H\} \in \mathbb{C}^{[Q \times Q]}$ and cross-covariance matrix $\mathbf{\Phi}_{\mathrm{hy}} = E\{\mathbf{hy}^H\} \in \mathbb{C}^{[Q \times Q]}$ to calculate

$$\mathbf{F}_{\text{int,opt}} = \mathbf{\Phi}_{\mathrm{hy}} \mathbf{\Phi}_{\mathrm{yy}}^{-1}. \tag{9.11}$$

Note that (9.11) states the general solution, while in the case of sub-sampled RSs as explained above the dimensionality of $\mathbf{\Phi}_{\mathrm{yy}}$ will have to be changed accordingly. The equation leads to the optimal solution under the assumption of fully uncorrelated channel $\mathbf{h}$ and observation noise $\mathbf{n}$. The interpolation gain of Wiener filtering is due to noise suppression for radio channels with large coherence bandwidth. Assume as an illustrative example a fully correlated frequency-flat radio channel, where one single complex value can be estimated from Q/n_{RS} observations. $\mathbf{\Phi}_{\mathrm{hy}}$ is a matrix carrying the SNR values on the diagonal elements. With decreasing SNR, the estimated covariance values from $\mathbf{\Phi}_{\mathrm{yy}}$ will be scaled down according to their reliability.

The interpolation matrix $\mathbf{F}_{\text{int}}$ as already introduced in (9.10) targets the same interpolation gain as the Wiener filter. Note there is an inverse relation between the coherence bandwidth in frequency domain affecting $\mathbf{\Phi}_{\mathrm{yy}}$ and the length of the CIR $\tilde{\mathbf{h}}(t)$. Instead of estimation of $\mathbf{\Phi}_{\mathrm{hy}}$, the estimated noise level σ_n^2 is used in (9.10) for direct estimation of relevant taps of the CIR above noise level.

In [HSJ+05, SJ06], it has been found out that sensitivity to estimation errors regarding the number of valid taps of $\tilde{\mathbf{h}}(t)$ seems to be relatively small, and a good compromise with respect to complexity was in the envisaged scenarios to assume $L/2$ as the standard length of the channel.

The general way of finding the discrete Wiener finite impulse response (FIR) is by solving the Wiener Hopf equations [Hay02], while the methodology based on $\mathbf{F}_{\text{int,opt}}$ is an implementation-friendly solution as it makes use of easily available FFT processing blocks. The scheme has been implemented also for the demonstration system in [HKR+97b].

Two-Dimensional Wiener Filter
For OFDM systems - and specifically for LTE Release 8 - it is possible to extend the one-dimensional filtering in frequency into the time domain leading to the two-dimensional Wiener filter. LTE aims - at least for UEs with low to moderate

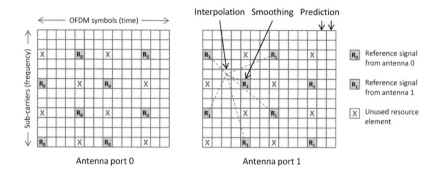

Figure 9.2 Pattern of LTE Release 8 reference signals (CRS) for 2 antenna ports.

mobility - at exploiting opportunistic multi-user scheduling gains. Supporting this in LTE, so-called physical resource blocks (PRBs) were defined consisting of 12 sub-carriers $q = 1, \ldots, 12$ and 14 OFDM symbols $o = 1, \ldots, 14$ forming a sub-frame length of $t_{\text{SF}} = 14 \cdot 70\ \mu\text{s} = 1$ ms. Note that in LTE one frame of length 10 ms consists of 10 subframes.

The PRBs are the smallest entity addressed by the multiple access channel (MAC) layer - i.e. the scheduler - and might be rescheduled in each transmit time interval (TTI). For that reason, PRBs have to be decodeable independently. At least in case of dedicated user specific RSs, channel estimation has to be possible individually per PRB. In the sequel, one single PRB is assumed.

Fig. 9.2 illustrates the characteristic pilot grid $R_0(q, o) = \{1, 1; 7, 1; \ldots; 10, 12\}$ for antenna port 0 and $R_1(q, o) = \{3, 1; 9, 1; \ldots 7, 12\}$ for antenna port 1 as standardized for LTE Release 8 for a single PRB (so-called *common reference signal (CRS)*). R_0 and R_1 have a regular structure as far as possible for the given number of sub-carriers and OFDM symbols. 2 antenna ports is the baseline assumption for LTE Release 8, and for optimum channel estimation accuracy, full orthogonality between the sets R_0 and R_1 is being achieved by mutual muting (crosses of unused resource elements (REs) in the figure) of REs carrying RSs for the other port. The resulting overhead for 2 ports is $16/(12 \cdot 14) = 9.5\%$.

From Fig. 9.2, we can deduce the effect of two dimensional filtering, where channel estimation for any RE is the result of an interpolation in time and frequency. In case of optimal Wiener filtering and for low mobility UEs, the staggered allocation of RSs in different OFDM symbols has the special benefit of almost doubling the frequency resolution.

Fig. 9.2 illustrates a further important aspect of Wiener filtering, i.e. that dependent on the REs carrying RSs or not, the Wiener filter will have the effect of *smoothing*, *interpolation* or even channel *prediction*.

Generally the *smoothing* effect is interesting for those REs carrying RSs, which are in a notch of the CTF, so that without smoothing, channel estimation accuracy for these REs would be poor.

Interpolation allows - as long as the sampling theorem is fulfilled - adapting of RS overhead to the intended estimation accuracy. With respect to CoMP, channel *prediction* might be the most interesting aspect, as it seems to be a viable option to overcome the issue of channel state information (CSI) outdating. The current goal is to extend the prediction range for a single PRB for the last 2 OFDM symbols of 70 μs up to at least several milliseconds, as explained later.

The two-dimensional Wiener filtering solution is an extension of the one-dimensional case. In the two-dimensional case it is necessary to stack all $O = 14$ channel vectors within a PRB into one channel vector $\mathbf{h}_2(t)$. The same has to be applied to the two-dimensional receive signal, which results in the vector $\mathbf{y}_2$. With these matrices, it is possible to compute the auto-covariance matrix $\mathbf{\Phi}_{2,\mathrm{yy}} = E\{\mathbf{y}_2\mathbf{y}_2^H\}$ and the cross-covariance matrix $\mathbf{\Phi}_{2,\mathrm{hy}} = E\{\mathbf{h}_2\mathbf{y}_2^H\}$ to calculate the optimum filter $\mathbf{F}_{2,\mathrm{int,opt}}$. More details can be found in [HKR+97b].

Subspace Concept - Channel Prediction

The so-called subspace concept as proposed in [WMZ05b] is closely related to the optimum Wiener filter solution, i.e. it exploits long-term channel statistics to improve channel estimation quality. The subspace is spanned by the relevant MPCs within the excess delay according to (9.8) being unequal to zero or - more precisely - above a certain threshold. Hence the subspace dimension might vary between 1 and the maximum number of taps L of the CIR. Most easily it is explained from Fig. 9.1(b), where the CIR $\tilde{\mathbf{h}}(t,\tau)$ is depicted for t_1 and $t_2 = t_1 + \Delta t$. For small Δt, the MPCs τ_n will change only marginally, i.e. will mainly change their phases $\varphi(\tau_n)$, while the amplitudes $\alpha(\tau_n)$ remain almost constant. In other words, the large scale fading is almost stable, and only the small-scale fading varies. If e.g. the main MPCs defining the subspace have been properly identified by according long-term channel observation of the auto covariance matrix $\mathbf{\Phi}_{\mathrm{yy}}$, it will be sufficient to limit estimation (and reporting) to the short term variations of the main MPCs $\varphi(\tau_n)$ or, equivalently, of the subspace defined by all elements in $\widehat{\mathbf{h}}_s(t)$ that exceed a certain power threshold.

In the case of low sub-space dimensions of the radio channel, it could be shown in [WMZ05a] that significant interpolation gains can be possible. As an extreme example, for one single relevant MPC, an interpolation gain of about 15 dB has been reported, where some of the gain is due to proper identification of additional irrelevant taps within the CIR, and not only of the maximum length of $\tilde{\mathbf{h}}(t)$.

For low mobility below 3 km/h and a prediction horizon of less than 5 ms, even simple linear prediction over two previous CSI estimates of the interpolated CTF might yield a mean square error (MSE) of ≤ -20 dB, simulated based on the spatial channel model extended (SCME). For LTE, a useful target is a prediction horizon of $10 - 20$ ms under real world radio channel conditions. An interesting candidate is *model-based CSI prediction* [PW09, PW10]. The idea is to estimate from long-term observations the evolution of main MPCs under the assumption of a linear movement of the UE, and to generate a corresponding artificial channel model reproducing the time-variant CIR. Future channel conditions are predicted by further linear movement of the UE within the model. This type of prediction works fine as long as the large scale parameters of the radio channel remain unchanged. In reality, radio channels are more volatile due to birth and death of radio channel components, e.g. at street crossings, or due to varying characteristics of the multi-path reflections.

9.1.2 Channel Estimation for CoMP

Up to now, a single channel component $h(t)$ and its corresponding time domain representation $\tilde{h}(t)$ have been investigated. For advanced CoMP schemes like JT, there are specific challenges not sufficiently addressed yet in LTE Release 8. In the following, the main requirements as well as some status regarding discussion in LTE-A will be given.

Typical CoMP Scenarios

In general, the channel estimation requirements depend on the used CoMP scheme. In the sequel, we will focus on downlink multi-cell JT, as introduced in Section 6.3, which is most sensitive to imperfect CSI.

Compared to single cell MIMO, for CoMP the number of channel components is multiplied with the number of cooperating cells. Hence, the number of channel components might explode, e.g. under assumption of $N_{bs} = 4$ or $N_{bs} = 8$ antennas per base station (BS) and 3 cooperating cells, there are already 12 or 24 channel coefficients to be estimated at each UE receive antenna. This is a serious challenge keeping in mind that in LTE Release 8 pilot overhead is already 9.5% for the estimation of only two channel components!

In addition, interference cancelation for cell-edge UEs between uncorrelated sites is highly sensitive to channel estimation errors, potentially requiring high frequency granularity e.g. per PRB or even per half PRB.

Multi-cell JT is generally seen as a very low mobility solution, e.g for nomadic users with speeds below 3 km/h, which is also emphasized in Chapter 13. Nonetheless, many simulation results as well as test bed implementations [JFJ+10] have clearly indicated that despite this low mobility, CSI outdating of only a few ms degrades JT performance significantly, due to a time delay between CSI estimation and precoding. CSI prediction might be a smart way to overcome CSI outdating, motivating a lot of research in this area lately.

Effective Channel Concept

As mentioned above CoMP has specific new challenges. The so-called *effective channel* concept is a high level and flexible means to handle the issue of high number of channel components and potentially different numbers of antennas at different BSs. In 3GPP, similar concepts are discussed under the label of *implicit* versus *explicit* channel estimation and feedback, where the latter is more or less the direct feedback of the quantized channel components, while implicit includes pre- or postcoders like e.g. beamformers from an LTE Release 8 codebook. In case of implicit feedback, the UEs might be aware or not of the pre-/postcoders, leading to transparent versus non-transparent solutions, as discussed in more detail in Section 9.2. As noted before, there might be easily 10-50 explicit channel components motivating implicit solutions. The most effective approach is to reduce the number of channel components to the number of transmitted spatial layers within the cluster, leading to the stated effective channel concept [ZMS+09b]. For more details, see (6.49) to (6.52), where the effective channel concept has already been introduced. The main idea is to form so-called *virtual antennas* by linear combination of antenna elements at transmitter and receiver side.

Typically, antennas of a cell will be co-located and correlated, while correlation between different cells will be small. Therefore, the currently most promising scheme uses virtual antennas per cell (implicit CSI) in combination with explicit inter-cell channel estimation, including phase- and amplitude information per virtual cell-specific effective channel component.

Assume as a simple example one spatial layer per UE and one active UE with $N_{ue} = 2$ antennas, then for a cooperation cluster of size $M_c = 3$ with $N_{bs} = 4$ antenna ports per cell each UE has $3 \times 4 \times 2 = 24$ explicit estimates and the overall channel matrix $\mathbf{H}$ would be of size $(3 \times 4) \times (2 \times 3) = 72$. For the effective channel concept, each UE just reports one single virtual channel component per cell, leading to the effective channel matrix $\mathbf{H}_{\text{eff}}$ of size $3 \times 3 = 9$ with

$$\mathbf{H}_{\text{eff}} = \mathbf{V}_{\text{UE}}^H \mathbf{H}^H \mathbf{V}_{\text{BS}}. \qquad (9.12)$$

$\mathbf{V}_{\text{BS}} \in \mathbb{C}^{[M_c N_{bs} \times M_c]}$ contains the precoding vectors per BS forming the virtual transmit antenna ports and $\mathbf{V}_{\text{UE}} \in \mathbb{C}^{[M_c \times M_c N_{ue}]}$ the UE specific postcoders for all UE. Note that these postcoders might be cell-specific, i.e. in contrast to (6.49), the UE calculates the left dominant eigenvectors with respect to their serving cells and not all cooperating cells.

This simplifies channel estimation, and can be motivated by the additional precoders for cancelation of inter-cell interference within the cooperation area. For $\mathbf{H}_{\text{eff}} \in \mathbb{C}^{[M_c \times M_c]}$ this linear - potentially zero-forcing (ZF) - precoder is obtained by the Moore-Penrose pseudo-inverse

$$\mathbf{W}_{\text{eff}} = \mathbf{H}_{\text{eff}} (\mathbf{H}_{\text{eff}}^H \mathbf{H}_{\text{eff}})^{-1}. \qquad (9.13)$$

Note that the cell-specific pre- and postcoders lead to an implicit *per cell* channel estimation, while the estimation of $\mathbf{H}_{\text{eff}}$ is done explicitly per *effective* channel component. This scheme achieved already some consensus in 3GPP, as it takes care of the different levels of correlation of antenna elements within one and between different cells.

Precoder Selection Schemes

The effective channel concept is a powerful means to limit channel estimation overhead, but requires as an additional step the selection of precoders per BS. There are basically two procedures possible.

One option is to start with the explicit estimation of all channel components of **H** of the serving cell based on an accordingly high number of orthogonal CSI reference signals (CSI RSs) (for which in this particular case of per-cell channel estimation the CRSs defined in LTE Release 8 may be used). For the estimated channels, the UE selects the most suitable wide-band serving cell precoders $\mathbf{V}_{\mathrm{BS}}$ and e.g. left dominant eigenvectors as UE-specific postcoders. This explicit channel estimation is done semi-statically, as spatial correlation can be assumed for antennas of one cell, limiting overall overhead. In a step two, the BSs apply their wide-band precoders for broadcasting of dedicated demodulation reference signals (DRSs), which allows the UEs to estimate the resulting implicit channel components, which will be estimated continuously.

A second approach is to combine JT with coordinated beamforming, i.e. there is a set of precoders $\mathbf{V}_{\mathrm{BS}}$ varying over frequency sub-bands and/or time slots in a predictable manner so that each UE can estimate individually the best fitting wide-band beamformers (sub-band/slot) and the resulting effective channel components. The BS then schedules the UE onto subsequent sub-bands/slots fitting to the reported beamformers and calculates the precoder for the corresponding $\mathbf{H}_{\mathrm{eff}}$. The benefit is that there is no extra phase for explicit estimation of **H**, but this has to be paid by some scheduling restrictions and possibly some extra delay as UEs have to be scheduled onto specific sub-bands/slots.

Demodulation Reference Signals

Specifically for transparent solutions, where UEs are not fully aware about the chosen precoders, precoded or dedicated RS, so-called DRS have to be inserted for demodulation of each PRB carrying user data over the so-called physical downlink shared channel (PDSCH).

Dedicated RS means that the DRSs use the same precoders as the PDSCH. CSI estimation accuracy has to be high as highest modulation and coding schemes (MCSs) up to 64-QAM have to be decodeable, generating an overhead per UE of about 15%. Making things worse, in CoMP mode the overhead scales with the cooperation size due to required inter-UE orthogonality. Research is ongoing for improved and integrated CSI RS and DRS solutions trading performance versus overall overhead.

9.1.3 Multi-Cell Channel Estimation

Up to now, single-cell channel estimation has been analyzed, where the CRSs defined in LTE Release 8 were used as CSI RS. In case of mobile cellular radio systems, there is a high number of interfering radio cells, generating a significant interference floor. Even in case of the effective channel concept, where the number of relevant channel components is significantly reduced, channel estimation will suffer from inter-RS interference. Helpful is a localization of interference as far as possible by applying strong antenna tilting [TWB+09, TWS+09]. In addition, LTE Release 8 foresees different levels of inter-RS orthogonality:

- full orthogonality for antenna ports of the same cell (mutual muting of REs),
- cell-specific frequency shifting of antenna patterns (with and without muting in other cells) and
- so-called quasi orthogonality between cells based on cell-specific Zadoff-Chu sequences s_{ZC} [HT09].

Note that these sequences s_{ZC} are a variant of the well-known constant amplitude zero autocorrelation codes (CAZAC) sequences. While they provide zero auto- and cross-correlation, the amplitude is not really constant, but at least the value of the cubic metric is comparable to that of QPSK modulation. Zadoff-Chu sequences s_{ZC} run over all CRS of one OFDM symbol and provide good wide-band orthogonality. For CoMP, the challenge is that in case of shortened sequence lengths - as required for frequency-selective CSI - the performance degrades significantly. For example, the cross-correlation for a sequence length of 50 PRBs achieves an MSE of about $\sigma^2_{CRS} = -16$ dB for all sequence shifts, while for a length of 10 PRBs this degrades to about $\sigma^2_{CRS} = -10$ dB. For one PRB, the effect will be more detrimental, making improvements in the RS design mandatory for advanced CoMP.

Due to the required backward compatibility to LTE Release 8, where UEs rely on a constant transmission of the *CRS* grid as illustrated in Fig. 9.2, it is obvious that these existing reference signals have to be kept as they are. A possible way forward is hence to design new CSI RS, specifically intended for channel estimation for up to 8 antenna ports in a multi-cell environment with sufficiently high accuracy. Based on the maximum MCS of 64-QAM with code rate 5/6-th, a σ^2_{CSI} in the range of -20 dB seems to be a reasonable, but challenging, target.

For given mobile radio channels, the MSE of channel estimation σ^2_{CSI} will be affected by the design parameters according to (9.14), i.e. the number of channel components N_{CC}, the number of RSs Q/n_{RS}, the length of the pilot sequence L_{RS} and its relative power P_{RS}, where the last point is known as power boosting.

$$\sigma^2_{CSI} = \frac{N_{CC} \cdot n_{RS}}{L_{RS} \cdot P_{RS}} \cdot \text{const.} \tag{9.14}$$

The main challenge of (9.14) is to find a good trade-off between overhead and channel estimation accuracy. For LTE-A for example - based on the assumption of low mobility for MU-MIMO or CoMP UEs - the current agreement is to make CSI RS sparse in frequency and time.

Making n_{RS} large reduces pilot overhead, but beside limited interpolation gain, also the frequency selectivity of the radio channel might not be fully captured. Increasing the length of the pilot sequence L_{RS} means to allocate more mutually orthogonal RSs for different antenna ports as well as different cell IDs. This can be done by frequency division multiplex (FDM), time division multiplex (TDM), code division multiplex (CDM), or any hybrid solution, as has been intensively investigated in 3GPP. Performance-wise there are only minor differences, but there might be side effects like backward compatibility to LTE Release 8, muting and corresponding power offset issues etc.

The parameter N_{CC} - the number of channel components to be estimated - can be minimized by applying the effective channel concept as explained in the previous subsection. In [TSS+08], a specific solution allows to increase L_{RS} without adding extra overhead. For that purpose, CSI RS are multiplied in time domain with cell-specific Hadamard sequences s_{H} (or so called orthogonal *cover codes*) and a suitable regular allocation of s_{H} to cells ensures that the required length of s_{H} for full orthogonality increases with the increasing inter-cell distance.

Depending on UE mobility, the longer sequences might more or less violate the coherence time of the radio channel, leading to corresponding inter-code interference. Statistically, the more distant cells will contribute less interference, so that higher sensitivity to mobility due to long sequences is easier acceptable. The additional orthogonality comes basically for free, as the Hadamard sequences are applied to the already available CSI RS. In Fig. 9.3, the normalized MSE of the channel estimator for correlation over several TTIs for the top 5 strongest cells is compared with Hadamard or random sequences on top of the reference signals. With increasing length of the correlation time, Hadamard sequences provide significantly lower MSE than the random sequences.

In Fig. 9.3 we can observe a further important aspect: the MSE of channel estimation degrades significantly with decreasing receive power of the estimated channel components, as already seen in Section 4.2. Fortunately, lower receive power relates to lower interference and for JT, corresponding simulations verified a 'self scaling' effect, i.e. that precoding sensitivity to channel estimation errors decreases with decreasing receive power.

9.1.4 Uplink Channel Estimation

In theory, uplink (UL) and DL channels are reciprocal to each other (see Section 3.5), but in real systems there are some substantial differences. In the DL, RSs are continuously broadcasted, allowing all UEs to perform channel estimation simultaneously and over longer periods of time, while for the UL, CSI has to be estimated based on UE-specific RSs. In LTE Release 8, UEs use SC-FDMA to transmit so-called sounding RSs as well as DRSs for coherent detection and demodulation. Sounding RSs are wideband and intended for scheduling, while DRS are UEs-specific, dedicated and precoded RSs for data demodulation, having the same bandwidth as UL user data [HT09]. Orthogonality between simultaneously transmitting UEs is achieved by selecting one out of 30 Zadoff-Chu sequences [HT09] per cell plus one out of 8 cyclic shifts. In case of different

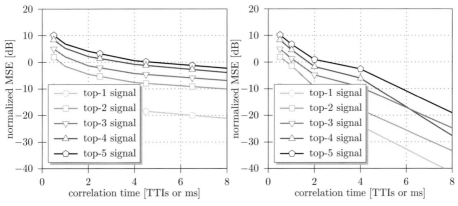

(a) Conventional interference estimation. (b) Hadamard sequences on top of reference signals.

Figure 9.3 Normalized MSE of the correlation estimator for the five strongest cells [TSS+08]. © 2008 IEEE.

bandwidth of the Zadoff-Chu sequences in adjacent cells, cross-correlation will be poor, requiring either enhancements of the overall RSs scheme or synchronization of the bandwidths between cells, which would lead to scheduling restrictions.

Note that Zadoff-Chu sequences provide only wideband orthogonality as discussed already for CSI RS. For powerful joint detection (JD) schemes, this might be a serious limitation in case of frequency-selective radio channels as being expected for typical cell-edge users. For UL CoMP, one, two or several UEs per cell transmit in UL their physical uplink shared channel (PUSCH) data. BSs will do JD and channel estimation based on the received UE-specific DRSs of all jointly processed UEs. Compared to the DL, there is the advantage that DRSs are embedded into user data and are therefore not subject to being outdated at all. In principal, JD can be done after any delay without performance degradation (unless hybrid automatic repeat request (HARQ) with limited round-trip time (RTT) is used, as addressed in Section 11.2, but this constraint is not connected to channel estimation).

LTE applies open and closed loop power control for the PUSCH, specified as $P_{\text{PUSCH}} = min(P_{\text{max}}, 10\log_{10}(\text{no. PRBs}) + P_0 + \alpha \cdot \text{PL} + ...)$, with P_{max} as the maximum allowed transmit power and α between 0.4 and 1 as cell-specific pathloss compensation factor. For $\alpha \neq 1$, the pathloss will not be fully compensated, leading to a pathloss-dependent and UE-specific receive power per UE, rendering channel estimation and JD more challenging. For UL CoMP, it might be useful to set α to values similar to 1. Even then, DRSs of cooperating UEs will arrive with different receive power due to the pathloss differences leading to accordingly different channel estimation quality. However, the same self-scaling effect as in the DL can be observed, i.e. low power channel components will have accordingly lower influence on JD performance.

Similar as CSI RS suffer in the DL from multi-cell interference, channel estimation in the UL will be degraded due to simultaneously transmitting UEs, specifically from cells reusing the same Zadoff-Chu sequences for the DRSs. Therefore, it is important to localize interference by strong BS antenna tilting and to minimize inter-cluster interference by proper clustering of cells and user grouping [TSS+08], still being a field of extensive research.

9.1.5 Summary

Accurate channel estimation is the basis for any CoMP scheme like coordinated beamforming / scheduling or - more importantly - joint precoding or detection. Channel estimation affects precoding accuracy and defines an upper limit for possible performance gains.

Specifically *joint transmission* faces several further challenges compared to more conventional single link systems, such as

- a high number of channel components to be estimated,
- strong multi-cell interference due to frequency reuse one, and
- a high sensitivity of joint precoding with respect to estimation errors combined with the need for frequency-selective channel information.

A sound understanding of the time-variant radio channel is important, motivating well-known channel estimation techniques like two dimensional Wiener filtering or the so-called *sub-space* concept. Discussed enhancements are the *effective channel* concept, which limits the number of effective channel components to be estimated to the number of supported data streams - and reference signals carrying orthogonal time domain sequences reducing inter-cell interference for low mobility users without extra overhead.

9.2 Channel State Information Feedback to the Transmitter

Guido Dietl and Wolfgang Utschick

While the last section was concerned with the estimation and prediction of channel state information (CSI) at the receiver side, we now want to look into the problem of feeding back this information to the transmitter side, which is for example required for the multi-cell joint transmission (JT) schemes discussed in Sections 6.3 and 6.4. Although we restrict the feedback considerations in this section to single-cell multi-user MIMO (MU-MIMO) systems, these can be easily extended to CoMP schemes where base stations are cooperating. In fact, MU-MIMO is seen as one of the main ingredients for CoMP.

In this section, we consider downlink MU-MIMO transmission (cf. Section 6.3) in frequency division duplex (FDD) systems where CSI is fed back from the user equipments (UEs) to a single base station (BS). Due to feedback channels of limited rate, mobiles can only provide imperfect CSI, i.e., a small finite number of bits is used for the information fed back to BSs (cf., e.g., [MRF10]). The consideration of systems with limited feedback assuming nonlinear transmit processing can be found in, e.g., [DLZ06, CJCU07]. However, we are especially interested in linear precoders because of their implementation advantages over non-linear schemes, like being in general computationally more efficient, having smaller processing delays by avoiding successive encoding, and inducing less requirements on hardware like the dynamic range of amplifiers or analog-to-digital converters. Compared to the investigation of linear precoders in [MSEA03, LHSH04] (see also references therein) for a single-user system, or in [Jin06] (see also references therein) for a multi-user system, we consider a multi-user system with *user scheduling* in order to fully exploit *multi-user diversity*.

Note that in the standardization of 3GPP LTE-A, i.e., where non-cooperative single-user MIMO (SU-MIMO) is considered, two major feedback schemes have been discussed, viz., *implicit* and *explicit feedback* of CSI. Here, implicit feedback means the feedback of a precoder index from each mobile to its assigned BS. The corresponding codebook entry is used by the BS as the precoder in the downlink. Although the index directly represents the precoder, the CSI is still included in an implicit manner. Contrary to that, explicit feedback denotes the direct feedback of CSI in terms of an index which represents the codebook entry which is closest to the exact CSI with respect to some distance criterion. In this section, we focus solely on explicit CSI feedback because it is the most promising feedback scheme in the context of MU-MIMO and CoMP (cf., e.g., [DB07]).

Please note that, while we were concerned with a most efficient estimation and representation of channel characteristics in the time- and frequency domain in Section 9.1, we are now exploiting *spatial* dimensions of channel matrices in order to make CSI feedback as efficient as possible. In this section, we mainly focus on channel vector quantization (CVQ) schemes for efficiently capturing these spatial dimensions. Most of the explicit CSI feedback schemes are based on CVQ using a finite channel codebook as proposed in [3GP06a, 3GP06b, DB07, DLU09]. Precisely speaking, each user quantizes a product of his channel matrix and an estimation of its receive filter, in the following denoted as the composite channel vector, and feeds back the corresponding codebook index together with an approximate signal-to-interference-and-noise ratio (SINR) value. Note that users need to estimate their receivers because the finally chosen receive filters depend on the precoder at the BSs, which is determined after quantization. In fact, the BSs use the quantized composite channel vectors to compute the precoder based on the zero-forcing (ZF) criterion, and use the available SINR values to schedule the users by maximizing the sum-rate.

Usually, CVQ is based on choosing the codebook entry with minimum Euclidean distance to the composite channel vector. However, minimizing the Euclidean distance is not necessarily related to the final goal of designing a communications system, i.e., maximizing the sum-rate. Therefore, we propose

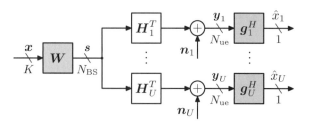

Figure 9.4 Downlink of a multi-user MIMO system on one OFDMA sub-carrier, from [DLU09]. © 2010 IEEE.

to estimate the receive filter and quantize the corresponding composite channel vector by maximizing the approximate SINR which is directly related to the achievable rate of the corresponding user [DLU09]. Note that this idea is strongly related to the method presented in [WSJ+10], however, the latter approach is based on two codebooks and a different type of sum-rate approximation.

Before reviewing the state-of-the-art CVQ methods and deriving the proposed CVQ approaches in Subsections 9.2.3-9.2.6, we introduce our transmission model in the next two subsections. Finally, we investigate the performance of the proposed schemes when applied to a MU-MIMO system with linear ZF precoding in Subsections 9.2.7-9.2.9.

9.2.1 Transmission Model

Here, we consider the downlink transmission from one BS with N_{bs} antennas to U UEs with N_{ue} receive antennas each, out of which $K \leq U$ UEs have been scheduled to be served on the same resources in time and frequency. Note that in this section, $N_{\text{BS}} = N_{\text{bs}}$ since the number of BSs is assumed to be $M = 1$. In the sequel, we will capture all K selected users in the set $\mathcal{K} \subseteq \{1, \ldots, U\}$, $|\mathcal{K}| = K$. As introduced in Chapter 3, the transmission taking place on one exemplary orthogonal frequency division multiplex (OFDM) sub-carrier of the system can be stated as

$$\mathbf{y}_k = \mathbf{H}_k^T \mathbf{s} + \mathbf{n}_k, \tag{9.15}$$

where $\mathbf{y}_k \in \mathbb{C}^{[N_{\text{ue}} \times 1]}$ are the signals received by UE k, $\mathbf{H}_k \in \mathbb{C}^{[N_{\text{BS}} \times N_{\text{ue}}]}$ is the channel matrix connected to UE k, $\mathbf{s} \in \mathbb{C}^{[N_{\text{BS}} \times 1]}$ are the signals transmitted from the BS antennas, and $\mathbf{n}_k \in \mathbb{C}^{[N_{\text{ue}} \times 1]}$ is additive Gaussian noise at receiver k, i.e., with $\mathbf{n}_k \sim \mathcal{N}_{\mathbb{C}}(\mathbf{0}, \sigma^2 \mathbf{I})$. In the following, we assume that maximal one data stream is assigned to each user, regardless of the number N_{ue} of antennas each UE has. The transmit symbols $\mathbf{s}$ are the result of linear precoding, i.e.,

$$\mathbf{s} = \sum_{k \in \mathcal{K}} \mathbf{w}_k x_k = \mathbf{W}\mathbf{x}, \tag{9.16}$$

where $\mathbf{W} \in \mathbb{C}^{[N_{\mathrm{BS}} \times K]}$ is the precoding matrix, and $\mathbf{x} \in \mathbb{C}^{[K \times 1]}$ are the symbols of the K scheduled UEs before precoding. As in Section 3.5, we assume that these symbols have unit power, i.e., $\mathbf{x} \sim \mathcal{N}_{\mathbb{C}}(\mathbf{0}, \mathbf{I})$, and that the transmit power assigned to different streams is inherently contained in the precoding matrix $\mathbf{W}$.

Next, we describe the receivers at the mobile stations. We assume that each user is applying a linear filter $\mathbf{g}_k \in \mathbb{C}^{[N_{\mathrm{ue}} \times 1]}$ to the receive vector $\mathbf{y}_k$ to get the estimate

$$\hat{x}_k = \mathbf{g}_k^H \mathbf{y}_k \in \mathbb{C} \quad (9.17)$$

of the symbol x_k. In particular, we consider the linear minimum mean square error (LMMSE) filter obtained via the minimization of the mean square error (MSE) between x_k and $\hat{x}_k$, where the solution computes as (e.g., [Ver98])

$$\mathbf{g}_k = \left(\mathbf{H}_k^T \mathbf{W} \mathbf{W}^H \mathbf{H}_k^* + \sigma^2 \mathbf{I} \right)^{-1} \mathbf{H}_k^T \mathbf{w}_k. \quad (9.18)$$

Note that we assume throughout this section that each receiver has perfect CSI connected to its own channel $\mathbf{H}_k$, while it has no CSI about the channels of the other users. We refer the reader to Section 9.1 for details on receiver-side channel estimation.

9.2.2 Sum-Rate Performance Measure

A typical measure for the downlink transmission performance of a MU-MIMO system is the sum-rate over all users. With the assumptions made in Section 9.2.1, the SINR at the receive filter output $\mathbf{g}_k$ of the kth user can be written as [DLU09]

$$\gamma_k = \frac{\left| \mathbf{g}_k^H \mathbf{H}_k^T \mathbf{w}_k \right|^2}{\|\mathbf{g}_k\|_2^2 \sigma^2 + \sum_{j \neq k} \left| \mathbf{g}_k^H \mathbf{H}_k^T \mathbf{w}_j \right|^2}, \quad (9.19)$$

and the sum-rate computes as

$$R_{\mathrm{sum}} = \sum_{k \in \mathcal{K}} \log_2 (1 + \gamma_k). \quad (9.20)$$

9.2.3 Channel Vector Quantization (CVQ)

In order to compute the precoder and schedule the users for transmission, the base station requires information about the channel matrices $\mathbf{H}_k$ for all $k \in \{1, \ldots, U\}$. This CSI is fed back from the terminals to the base station. Precisely speaking, each user quantizes its channel based on a channel codebook and feeds back the corresponding codebook index as the channel direction indicator (CDI) together with an SINR value as the channel quality indicator (CQI) which includes a rough estimate of the interference caused by the quantization error [TBT08, 3GP06a, 3GP06b, Jin06]. The base station then computes a ZF precoder based on the CDIs, and allocates resources and chooses the proper modulation and coding scheme (MCS) using the CQIs of the different users. Again, since we are especially interested in schemes of strongly constrained feedback, we restrict the maximum number of transmitted data symbols per user to be one.

Assume for the first that the precoder $\mathbf{W}$ is known at the mobile receivers such that the LMMSE filters $\mathbf{g}_k$ can be computed according to (9.18). In order to compute the CDI, each user k quantizes the composite channel vector $\mathbf{c}_k = \mathbf{H}_k \mathbf{g}_k \in \mathbb{C}^{[N_{\mathrm{BS}} \times 1]}$, being a combination of the LMMSE filter and the physical channel matrix, by applying CVQ based on the channel codebook

$$\mathcal{C} = \{\mathbf{u}_1, \ldots, \mathbf{u}_{2^B}\}, \tag{9.21}$$

where B denotes the number of necessary bits for indexing the 2^B normalized codebook vectors $\mathbf{u}_q \in \mathbb{C}^{[N_{\mathrm{BS}} \times 1]}$, $q \in \{1, \ldots, 2^B\}$. By doing so, only N_{BS} entries in $\mathbf{c}_k$ instead of $N_{\mathrm{BS}} \cdot N_{\mathrm{ue}}$ entries in $\mathbf{H}_k$ need to be quantized at each UE, leading to a smaller quantization error if one keeps the feedback amount constant [DLU09].

However, in a real system, the finally chosen precoder $\mathbf{W}$, and therefore the resulting receive filter $\mathbf{g}_k$ is unknown to UE k at the time CVQ is applied. This is because each user has no knowledge about channels of other users due to the non-cooperative nature of the downlink channel. As a consequence, the quantizer $Q_\mathcal{C}$ needs to compute the quantized composite channel vector $\hat{\mathbf{c}}_k \in \mathbb{C}^{[N_{\mathrm{BS}} \times 1]}$ based on an estimate of the receive filter, in the following denoted as $\hat{\mathbf{g}}_k \in \mathbb{C}^{[N_{\mathrm{ue}} \times 1]}$, whose estimation quality compared to the finally chosen LMMSE filter $\mathbf{g}_k$ depends mainly on the chosen quantization method. In the following, we define the quantizer output as

$$\mathrm{CDI}:\ (\hat{\mathbf{c}}_k, \hat{\mathbf{g}}_k) = Q_\mathcal{C}\left(\mathbf{H}_k\right). \tag{9.22}$$

Moreover, due to the fact that the channels of other users and the finally chosen precoder are not known when the feedback information is computed at the mobile, CQI must be approximated as well. This is usually done by taking into account a rough estimate of the multi-user interference caused by the imperfect CSI at the base station due to quantization. As derived in [TBT08, 3GP06a], the CQI of user k, which is here a scaled version of the SINR at the kth mobile receiver, is approximated via

$$\mathrm{CQI}:\ \hat{\gamma}_k\left(\hat{\mathbf{c}}_k, \hat{\mathbf{g}}_k, \mathbf{H}_k\right) = \frac{\|\mathbf{H}_k \hat{\mathbf{g}}_k\|_2^2 \cos^2 \theta_k}{\sigma^2 + K \|\mathbf{H}_k \hat{\mathbf{g}}_k\|_2^2 \sin^2 \theta_k / N_{\mathrm{BS}}}, \quad \cos \theta_k = \frac{|\hat{\mathbf{c}}_k^H \mathbf{H}_k \hat{\mathbf{g}}_k|}{\|\mathbf{H}_k \hat{\mathbf{g}}_k\|_2}, \tag{9.23}$$

where $\theta_k \in [0, \pi]$ denotes the angle between the normalized composite channel vector and the quantized version thereof (quantization angle), and where, without loss of generality, we set $\|\hat{\mathbf{g}}_k\|_2 = 1$. In the following, we present two CVQ methods based on two different quantization criteria.

9.2.4 Minimum Euclidean Distance Based CVQ

Remember that one problem of CVQ is the fact that the finally chosen LMMSE filter is not known when the user computes the feedback information due to its dependency on the finally chosen precoder, and this precoder cannot be computed at the mobile because of the lack of knowledge about the CSI at other terminals. To overcome this obstacle, we first assume arbitrary receive filters $\mathbf{g}_k$. Since the resulting composite channel vector is then an arbitrary linear combination of the columns of $\mathbf{H}_k$, it lies in the column space of $\mathbf{H}_k$. This fact can be utilized for CVQ in the sense that the codebook entry is chosen such that its Euclidean distance to the row space of $\mathbf{H}_k$ is minimized (minimum quantization angle or error, cf. Fig. 9.5).

Mathematically speaking, with the QR factorization of $\mathbf{H}_k = \mathbf{Q}_k \mathbf{R}_k$ where $\mathbf{Q}_k \in \mathbb{C}^{[N_{\mathrm{BS}} \times N_{\mathrm{ue}}]}$ is an orthonormal basis of the column space of $\mathbf{H}_k$, and $\mathbf{R}_k \in \mathbb{C}^{[N_{\mathrm{ue}} \times N_{\mathrm{ue}}]}$ is upper triangular, the quantized composite channel vector reads as

$$\hat{\mathbf{c}}_k^{\mathrm{Euclid}} := \mathbf{u}_\ell, \quad \ell = \arg\max_{q \in \{1,\ldots,2^B\}} \left\| \mathbf{Q}_k^H \mathbf{u}_q \right\|_2^2. \tag{9.24}$$

Here, ℓ denotes the codebook index which is fed back to the base station as the CDI using B bits. To compute the CQI, we need not only the quantized composite channel vector but also an estimate of the receive filter. In the case of minimum Euclidean distance based CVQ, the receive filter is also chosen such that the resulting composite channel vector in the column space of $\mathbf{H}_k$ has the minimum Euclidean distance to the quantized composite channel vector.

In other words, the estimate of the receive filter is obtained by projecting $\hat{\mathbf{c}}_k^{\mathrm{Euclid}}$ back into the column space of $\mathbf{H}_k$ according to

$$\tilde{\mathbf{c}}_k = \frac{\mathbf{Q}_k \mathbf{Q}_k^H \hat{\mathbf{c}}_k^{\mathrm{Euclid}}}{\left\| \mathbf{Q}_k \mathbf{Q}_k^H \hat{\mathbf{c}}_k^{\mathrm{Euclid}} \right\|_2}, \tag{9.25}$$

and applying the left-hand side pseudo-inverse of the matrix $\mathbf{H}_k$, i.e., $\mathbf{H}_k^\dagger = (\mathbf{H}_k^H \mathbf{H}_k)^{-1} \mathbf{H}_k^H$, as well as normalization in order to finally get

$$\hat{\mathbf{g}}_k^{\mathrm{Euclid}} := \frac{\mathbf{H}_k^\dagger \tilde{\mathbf{c}}_k}{\left\| \mathbf{H}_k^\dagger \tilde{\mathbf{c}}_k \right\|_2}. \tag{9.26}$$

Note that the optimization criterion of the resulting receive filter estimate is no longer the MSE such as in the finally applied LMMSE receiver but the Euclidean distance. This leads to a mismatch between the true SINR and the one fed back as CQI. Finally, minimum Euclidean distance based CVQ can be summarized as

$$Q_{\mathcal{C}}^{\mathrm{Euclid}} : \mathbf{H}_k \mapsto (\hat{\mathbf{c}}_k^{\mathrm{Euclid}}, \hat{\mathbf{g}}_k^{\mathrm{Euclid}}), \quad \begin{cases} \hat{\mathbf{c}}_k^{\mathrm{Euclid}} = \arg\max_{\mathbf{u} \in \mathcal{C}} \left\| \mathbf{Q}_k^H \mathbf{u} \right\|_2^2, \quad \mathbf{H}_k = \mathbf{Q}_k \mathbf{R}_k, \\ \hat{\mathbf{g}}_k^{\mathrm{Euclid}} = \dfrac{\mathbf{H}_k^\dagger \mathbf{Q}_k \mathbf{Q}_k^H \hat{\mathbf{c}}_k^{\mathrm{Euclid}}}{\left\| \mathbf{H}_k^\dagger \mathbf{Q}_k \mathbf{Q}_k^H \hat{\mathbf{c}}_k^{\mathrm{Euclid}} \right\|_2}, \end{cases} \tag{9.27}$$

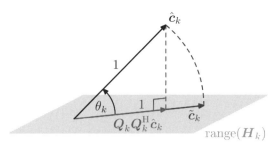

Figure 9.5 Quantization of the composite channel vector, from [DLU09]. © 2010 IEEE.

and the corresponding CQI computes as $\hat{\gamma}_k(\hat{\mathbf{c}}_k^{\text{Euclid}}, \hat{\mathbf{g}}_k^{\text{Euclid}}, \mathbf{H}_k)$ according to (9.23).

9.2.5 Maximum SINR Based CVQ

In this section, we propose an alternative CVQ method which is based on the maximization of the SINR expression [DLU09]. The original CVQ approach from [Jin06] is only concerned with minimizing the quantization error represented by the quantization angle θ_k. However, the ultimate objective in our system should be to target the highest possible sum-rate. Due to (9.19) and (9.20), this is achieved by maximizing the SINR, which can be approximated via expressions such as Eq. (9.23).

Due to the above reasons, we argue that an approach based on the maximization of an SINR expression may yield superior performance in a sum-rate sense than simply minimizing the quantization angle. Our method is solely concerned with finding the best receive filter estimate $\hat{\mathbf{g}}_k$ and the quantized composite channel vector $\hat{\mathbf{c}}_k$. In that sense, it provides an alternative to Section 9.2.3.

Let us have a look again at the approximate expression for the scaled SINR in (9.23). Note that this SINR expression will approach $\cotan^2 \theta_k$ as the noise variance σ^2 goes to zero. Therefore, at large signal-to-noise ratio (SNR), the SINR is almost fully determined by the quantization error, which justifies the minimization of θ_k in such conditions. However, the situation might be different at lower SNRs.

Thus, our proposed approach is to maximize (9.23) over all possible codebook entries $\mathbf{u} \in \mathcal{C}$ and receiver weights $\mathbf{v} \in \mathbb{C}^{[N_{\text{ue}} \times 1]}$ of unit norm, i.e.,

$$(\hat{\mathbf{c}}_k^{\text{SINR}}, \hat{\mathbf{g}}_k^{\text{SINR}}) = \underset{(\mathbf{u},\mathbf{v}) \in \mathcal{C} \times \{\mathbb{C}^{[N_{\text{ue}} \times 1]} : \|\mathbf{v}\|_2 = 1\}}{\arg \max} \hat{\gamma}_k(\mathbf{u}, \mathbf{v}, \mathbf{H}_k), \qquad (9.28)$$

yielding an optimal $\hat{\mathbf{c}}_k^{\text{SINR}} \in \mathcal{C}$ and $\hat{\mathbf{g}}_k^{\text{SINR}} \in \mathbb{C}^{[N_{\text{ue}} \times 1]}$. Finally, the corresponding CQI fed back to the transmitter side is given by $\hat{\gamma}_k(\hat{\mathbf{c}}_k^{\text{SINR}}, \hat{\mathbf{g}}_k^{\text{SINR}}, \mathbf{H}_k)$.

In order to perform the above maximization, let us re-write $\hat{\gamma}_k(\mathbf{u}, \mathbf{v}, \mathbf{H}_k)$ according to (9.23) by substituting $\|\mathbf{H}_k \mathbf{v}\|^2 \cos^2 \theta_k = \mathbf{v}^H \mathbf{H}_k^H \mathbf{u} \mathbf{u}^H \mathbf{H}_k \mathbf{v}$ and using the identities $\sin^2 \theta_k = 1 - \cos^2 \theta_k$ and $1 = \mathbf{g}_k^H \mathbf{g}_k$ (unit norm constraint of receive filters):

$$\hat{\gamma}_k(\mathbf{u},\mathbf{v},\mathbf{H}_k) = \frac{\mathbf{v}^H\left(\mathbf{H}_k^H\mathbf{u}\mathbf{u}^H\mathbf{H}_k\right)\mathbf{v}}{\mathbf{v}^H\left(\sigma^2\mathbf{I}+\mathbf{H}_k^H\left(\mathbf{I}-\mathbf{u}\mathbf{u}^H\right)\mathbf{H}_k\right)\mathbf{v}}$$
$$= \frac{\mathbf{v}^H\mathbf{A}(\mathbf{u})\mathbf{v}}{\mathbf{v}^H\mathbf{B}(\mathbf{u})\mathbf{v}}. \qquad (9.29)$$

It is well known that expressions in the form of (9.29) are maximized by setting $\mathbf{v}$ to the eigenvector corresponding to the largest eigenvalue μ_i solving the generalized eigenvalue problem $\mathbf{A}(\mathbf{u})\mathbf{v}_i = \mu_i \mathbf{B}(\mathbf{u})\mathbf{v}_i$ (e.g., [BX05]). Moreover, if $\mathbf{B}(\mathbf{u})$ is invertible, the eigenvalues and eigenvectors are the same as for the regular eigenvalue decomposition of $\mathbf{B}^{-1}(\mathbf{u})\mathbf{A}(\mathbf{u})$.

Note that this maximization finds the best $\mathbf{v}$ given a specific codebook entry $\mathbf{u}$. The optimal $\hat{\mathbf{c}}_k^{\text{SINR}}$ is the one yielding the largest SINR value over all codebook entries, and $\hat{\mathbf{g}}_k^{\text{SINR}}$ is the corresponding optimal weight vector for this $\hat{\mathbf{c}}_k^{\text{SINR}}$, i.e.,

$$Q_C^{\text{SINR}} : \mathbf{H}_k \mapsto (\hat{\mathbf{c}}_k^{\text{SINR}}, \hat{\mathbf{g}}_k^{\text{SINR}}), \begin{cases} \hat{\mathbf{c}}_k^{\text{SINR}} = \arg\max_{\mathbf{u}\in\mathcal{C}} \max_{\mathbf{v}\in\{\mathbb{C}^{[N_{\text{ue}}\times 1]}:\|\mathbf{v}\|_2=1\}} \hat{\gamma}_k(\mathbf{u},\mathbf{v},\mathbf{H}_k), \\ \hat{\mathbf{g}}_k^{\text{SINR}} = \arg\max_{\mathbf{v}\in\{\mathbb{C}^{[N_{\text{ue}}\times 1]}:\|\mathbf{v}\|_2=1\}} \hat{\gamma}_k(\hat{\mathbf{c}}_k^{\text{SINR}},\mathbf{v},\mathbf{H}_k). \end{cases}$$
(9.30)

A drawback of this method applied directly is the computational complexity, since the maximization over $\mathbf{v}$ needs to be performed for all entries of the channel codebook, and thus, requires 2^B generalized eigenvalue decompositions per subcarrier on which the optimization is performed.

9.2.6 Pseudo-Maximum SINR based CVQ

To overcome the high computational burden of the maximum SINR solution, we propose a pseudo-maximization algorithm as an alternative to the exact maximization in (9.28) [DLU09]. We note that $\hat{\gamma}_k$ is only a function of the composite channel magnitude (CCM) $\|\mathbf{H}_k\mathbf{v}\|_2^2$ and the quantization error θ_k for a given transmit power. The only way to increase $\hat{\gamma}_k$ is by increasing $\|\mathbf{H}_k\mathbf{v}\|_2^2$ or decreasing θ_k. The critical assumption we will make here is that $\hat{\gamma}_k$ is close to its maximum when either $\|\mathbf{H}_k\mathbf{v}\|_2^2$ is maximized or θ_k is minimized. Therefore, we evaluate $\hat{\gamma}_k$ at two specific points, denoted as $(\hat{\mathbf{c}}_k^{\text{CCM}}, \hat{\mathbf{g}}_k^{\text{CCM}})$ and $(\hat{\mathbf{c}}_k^{\text{Euclid}}, \hat{\mathbf{g}}_k^{\text{Euclid}})$, where we define

$$\hat{\mathbf{g}}_k^{\text{CCM}} = \arg\max_{\mathbf{v}\in\{\mathbb{C}^{[N_{\text{ue}}\times 1]}:\|\mathbf{v}\|_2=1\}} \mathbf{v}^H\mathbf{H}_k^H\mathbf{H}_k\mathbf{v},$$
$$\hat{\mathbf{c}}_k^{\text{CCM}} = \arg\max_{\mathbf{u}\in\mathcal{C}} \left|\mathbf{u}^H\mathbf{H}_k\hat{\mathbf{g}}_k^{\text{CCM}}\right|, \qquad (9.31)$$

and the point $(\hat{\mathbf{c}}_k^{\text{Euclid}}, \hat{\mathbf{g}}_k^{\text{Euclid}})$ is the minimum Euclidean distance solution that we already presented in Section 9.2.3. Note that $\hat{\mathbf{g}}_k^{\text{CCM}}$ is in the direction of the eigenvector corresponding to the largest eigenvalue of $\mathbf{H}_k^H\mathbf{H}_k$. The increase in computational complexity of the pseudo-maximization solution with respect to the minimum Euclidean distance method therefore includes one (regular) eigenvalue decomposition and one search for the closest quantization vector $\hat{\mathbf{c}}_k^{\text{CCM}}$. We finally summarize the Pseudo-Maximum (PM) solution to the SINR maximization problem as

$$Q_{\mathcal{C}}^{\text{PM}} : \mathbf{H}_k \mapsto \left(\hat{\mathbf{c}}_k^{\text{PM}}, \hat{\mathbf{g}}_k^{\text{PM}}\right) := \underset{(\mathbf{u},\mathbf{v}) \in \left\{\left(\hat{\mathbf{c}}_k^{\text{CCM}}, \hat{\mathbf{g}}_k^{\text{CCM}}\right), \left(\hat{\mathbf{c}}_k^{\text{Euclid}}, \hat{\mathbf{g}}_k^{\text{Euclid}}\right)\right\}}{\arg\max} \hat{\gamma}_k(\mathbf{u}, \mathbf{v}, \mathbf{H}_k). \tag{9.32}$$

9.2.7 Application to Zero-Forcing (ZF) Precoding

With the quantized composite channel matrix $\hat{\mathbf{C}} \in \mathbb{C}^{[N_{\text{BS}} \times K]}$ composed by the columns $\hat{\mathbf{c}}_k$, $k \in \mathcal{K}$, the ZF precoder at the base station computes as [Ver98, DLU09]

$$\mathbf{W} = \mathbf{W}' \mathbf{\Lambda}^{1/2}, \quad \mathbf{W}' = \left(\hat{\mathbf{C}}^T\right)^\dagger = \hat{\mathbf{C}}^* \left(\hat{\mathbf{C}}^T \hat{\mathbf{C}}^*\right)^{-1}, \tag{9.33}$$

where the diagonal matrix $\mathbf{\Lambda} \in \mathbb{C}^{K \times K}$ represents power loading. For the simulations in Section 9.2.9, we assume equal power loading according to

$$\mathbf{\Lambda} = \text{diag}\left(\frac{1}{\|\mathbf{W}' \mathbf{e}_k\|_2^2}\right)_{k=1}^K, \tag{9.34}$$

where $\mathbf{e}_k \in \{0,1\}^{[K \times 1]}$ contains a one in element k, and otherwise zero.

9.2.8 Resource Allocation

With the codebook indices and the scaled SINR values of all users, the base station schedules the users and computes the ZF precoder as described in the previous subsection. To do so, it calculates the SINR approximations based on the scaled versions thereof. It holds [TBT08, 3GP06a]

$$\mathbf{\Gamma} = \mathbf{\Lambda}_{\mathcal{K}} \text{diag}\left(\gamma'_k\right)_{k \in \mathcal{K}}. \tag{9.35}$$

Then, it uses these SINR approximations in order to schedule the users according to a greedy algorithm as described in [DLU09, TBT08, 3GP06a], see also Section 9.1. Finally, the set $\mathcal{K}$ of scheduled users is used to compute the ZF precoder according to (9.33) and (9.34).

9.2.9 Simulation Results

We investigate the proposed schemes in a MIMO OFDM system with the parameters as given in Table 9.1 and assuming the typical urban macro-cell channel model of the WINNER project [HKK+07].

Fig. 9.6 illustrates the performance difference between the CVQ schemes with Pseudo-Maximization (PM) and full maximization of the CQI indicator (approx-

Table 9.1. Simulation parameters, see [DLU09]. © 2010 IEEE.

Parameter	Variable	Value
Num. of BSs	M	1
Num. of Tx ant. per BS	N_{bs}	4
Num. of overall users	U	10
Num. of users sched. to same res.	K	4
Num. of Rx ant.	N_{ue}	2
Speed of users		1 m/s
Carrier frequency		2.0 GHz
Bandwidth		18 MHz
FFT size		2048
Num. of sub-carriers		1200
Num. of feedback bits	B	4
Feedback period		1.0 ms
SINR quantization		No
Channel model		Typical urban macro
Path loss used		No
Ant. spacing at Tx		0.5 wavelength
Ant. spacing at Rx		0.5 wavelength

imate SINR), and compares them with the minimum Euclidean Distance (minimum quantization error) approach. Here, we assume a random codebook with $B = 4$, where the elements of $\mathcal{C}$ are chosen from an isotropic distribution on the N_{BS}-dimensional unit sphere, i.e., normalized versions of vectors with random entries that correspond to $\sim \mathcal{N}_{\mathbb{C}}(0, 1)$.

The maximal gain of the pseudo and full maximization schemes over the minimum Euclidean distance method seems to be about 1.2 bit/s/Hz at 0 dB SNR. Moreover, one can see that the pseudo-maximization scheme acceptably approaches the performance of the full maximization scheme. Indeed, the performance gap between the two is never more than about 0.7 bit/s/Hz. Surprisingly, pseudo-maximization and Euclidean distance minimization are both slightly superior to full maximization in the mid and high SNR regions. This is possible due to the fact that the SINR measure used as the cost criterion for maximization does not represent the exact SINR, but is rather an approximation of this quantity. It appears that Euclidean distance minimization is the best option at mid and high SNR, and that in that range the SINR pseudo-maximization achieves the same performance. As pseudo-maximization makes a choice between the codebook entry that maximizes channel magnitude and the one that minimizes the quantization angle (Euclidean distance), one can suspect that at high SNR, the minimum quantization angle solution is chosen most of the time, such that no performance difference is noticeable with the scheme that only chooses the entry with the minimum Euclidean distance.

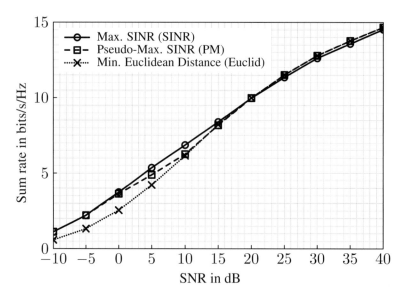

Figure 9.6 Performance comparison in case of random codebook and typical urban macro-cell, from [DLU09]. © 2010 IEEE.

Remember that the pseudo-maximization scheme requires much less computational complexity than full maximization (see the discussion on that effect in Section 9.2.5). Due to the above results, it also appears that pseudo-maximization always performs better or equivalently to quantization angle minimization, and that not much is gained by using full maximization (it can even cause slight performance degradation in certain SNR ranges). Therefore, the pseudo-maximization scheme seems like the preferred alternative to quantization angle minimization and produce significant sum-rate gains with respect to this scheme in the low SNR region. In [KDA$^+$10], the same conclusion has been drawn on the basis of system-level simulations.

9.2.10 Summary

In this section, we presented explicit CSI feedback schemes based on CVQ. Whereas the minimum Euclidean distance method achieves a good performance in the high SNR region, it degrades enormously in cases of small SNRs. In these cases, CVQ based on maximizing an estimate of the SINR outperforms Euclidean distance based schemes. However, the maximum SINR based CVQ method suffers from a very high computational complexity. Here, the suboptimal solutions presented in this section provide a good trade-off between complexity and performance. While this section focussed on a single-cell case, all discussed CSI feedback techniques can principally also be applied to multi-cell CSI (possibly capturing both desired and interfering channels), and hence be used in the context of CoMP.

10 Efficient and Robust Algorithm Implementation

In this chapter, we now look into algorithm implementation aspects connected to CoMP. In Section 10.1, the issue of numerically robust and flexible multi-cell precoding is addressed, while Section 10.2 looks into the performance of interference rejection combining filters under practical conditions.

10.1 Robust and Flexible Base Station Precoding Implementation

Jörg Holfeld and Gerhard Fettweis

As we have seen in Sections 6.3 and 6.4, spectral efficiency can be significantly increased if downlink CoMP schemes based on multi-cell joint signal processing are applied. Here, multiple base stations (BSs) perform a joint and coherent transmission towards multiple user equipments (UEs). Especially cell-edge users with symmetric links gain from the joint transmission (JT) with minimized inter-cell and inter-user interference compared to conventional cellular mobile networks.

In a cellular system, a spatial downlink resource allocation algorithm may switch between a single link transmission comprising one BS and one UE, a single BS multi-user scenario or finally a multi-cell joint transmission. The precoder must be able to cope with different system setups or matrix dimensions respectively. Such an implementation requires a higher signal processing flexibility as compared to non-cooperative setups. Physical link parameters like line-of-sight (LOS), non-LOS, or correlated transmit and receive antenna patterns influence the CoMP multi-user MIMO (MU-MIMO) eigenvalue spread and finally the spatial diversity and multiplexing gains.

Furthermore, in real-time hardware, the link parameters are related to finite precision effects. This must also be considered, since hardware architectures are constrained by a limited number of processing units and by limited numerical precision. The implementation of precoding for CoMP must be intended to preserve high numerical stability and low error propagation between the processing stages in conjunction with the heterogeneous CoMP setups.

This section compares several precoder architectures for downlink JT and focuses on a generalized matrix inverse with assessable link parameters and controllable error propagation. Initially, the transmission model is stated in Subsection 10.1.1, after which precoding architectures are introduced step by step in Subsections 10.1.2 to 10.1.4. The section concludes with a numerical example and a summary in Subsections 10.1.5 and 10.1.6, respectively.

10.1.1 System Model

The considered multi-user downlink system employs an orthogonal frequency division multiple access (OFDMA) scheme, where M BSs with all in all N_{BS} transmit antennas jointly transmit data to K UEs with N_{UE} receive antennas in total. The notation from Section 3.5 is reused for the baseband model of a frequency flat channel in frequency domain per sub-carrier as

$$\hat{\mathbf{x}} = \mathbf{G}^H \left(\mathbf{H}^H \mathbf{W} d(\mathbf{x}) + \mathbf{n} \right) = \mathbf{G}^H \left(\bar{\mathbf{H}} \mathbf{x} + \mathbf{n} \right), \quad (10.1)$$

where $\mathbf{x}, \hat{\mathbf{x}} \in \mathbb{C}^{[N_{\mathrm{UE}} \times 1]}$ are the stacked symbols which are transmitted to the UEs and estimated after receive processing, respectively.

On one hand, the statistical quantities of this model are the transmit symbols, whose transmit covariance matrix is given with $\mathbf{\Phi}_{\mathrm{xx}} = E\left\{\mathbf{x}\mathbf{x}^H\right\} = \sigma_x^2 \mathbf{I}_{N_{\mathrm{UE}}}$. On the other hand, the vector $\mathbf{n} \in \mathbb{C}^{[N_{\mathrm{UE}} \times 1]}$ is assumed to be additive white Gaussian noise (AWGN) with covariance matrix $\mathbf{\Phi}_{\mathrm{nn}} = E\left\{\mathbf{n}\mathbf{n}^H\right\}$.

The transmit symbols are processed by the linear spatial precoder $\mathbf{W} \in \mathbb{C}^{[N_{\mathrm{BS}} \times N_{\mathrm{UE}}]}$ which forms together with the physical channel $\mathbf{H} \in \mathbb{C}^{[N_{\mathrm{BS}} \times N_{\mathrm{UE}}]}$ an effective channel $\bar{\mathbf{H}} \in \mathbb{C}^{[N_{\mathrm{UE}} \times N_{\mathrm{UE}}]}$. Additionally, the transmit power is limited by the gain $\beta \in \mathbb{R}$ according to the sum-power constraint

$$E\left\{\|\mathbf{W}\mathbf{x}\|^2\right\} = \beta^2 \cdot \mathrm{tr}\left\{\mathbf{W}\mathbf{\Phi}_{\mathrm{xx}}\mathbf{W}^H\right\} = E_{\mathrm{Tx}}. \quad (10.2)$$

The aim of the linear precoder is to decouple the received data symbols from each other already at BS side. Each of the K non-cooperative UEs spatially filters its received signals independently. Consequently, the receive filter matrix $\mathbf{G} \in \mathbb{C}^{[N_{\mathrm{UE}} \times N_{\mathrm{UE}}]}$ is a block matrix of the compound UEs receive filters. The equalized data signals at each UE k can be decomposed into

$$\hat{\mathbf{x}}_k = \mathbf{G}_k^H \cdot \left(\mathbf{H}_k^H \mathbf{W}_k \mathbf{x}_k + \sum_{j=1, j \neq k}^{K} \mathbf{H}_k^H \mathbf{W}_j \mathbf{x}_j + \mathbf{n}_k \right). \quad (10.3)$$

Here, $\mathbf{H}_k^H \mathbf{W}_k \mathbf{x}_k$ describes the desired symbol part. The second term $\sum_{j=1, j \neq k}^{K} \mathbf{H}_k^H \mathbf{W}_j \mathbf{x}_j$ represents the spatial interference. The AWGN vector $\mathbf{n}_k \in \mathbb{C}^{[N_{\mathrm{ue}} \times k]}$ per UE is defined with $E\left\{\mathbf{n}_k \mathbf{n}_k^H\right\} = \sigma_{n_k}^2 \mathbf{I}_{N_{\mathrm{ue}}}$ as local covariance matrix.

For the sake of simplicity, the compound receive filter is reduced to a scaled diagonal matrix $\mathbf{G} = \beta^{-1} \mathbf{I}_{N_{\mathrm{UE}}}$. This means that each user stream is handled independently from other streams like single antenna UEs. Each UE should estimate this scalar β based on the effective channel. Admittedly, real handset implementations have to employ more advanced synchronization and channel estimation techniques which are not within the scope of this section [SBM+04, KHF09]. The topic of spatial interference rejection combining (IRC) and minimum mean square error (MMSE) filtering will be discussed in detail in Section 10.2.

Based on the previously mentioned decomposition, the receiver-side signal-to-interference-and-noise ratio (SINR) of one single *stream* $1 \leq l \leq N_{\mathrm{UE}}$ is chosen as performance metric

$$\mathrm{SINR}_l = \frac{\mathbf{h}_l^H \mathbf{w}_l \mathbf{w}_l^H \mathbf{h}_l}{\sum_{l' \neq l}^{N_{\mathrm{UE}}} \mathbf{h}_l^H \mathbf{w}_{l'} \mathbf{w}_{l'}^H \mathbf{h}_l + \sigma_n^2} \Bigg|_{\mathbf{G} = \beta^{-1} \mathbf{I}_{N_{\mathrm{UE}}}} . \tag{10.4}$$

This expresses a power ratio based on the effective channel $\bar{\mathbf{H}} = \mathbf{H}^H \mathbf{W}$ which describes the relation of the useful signal part l to interference terms.

10.1.2 Transmit Filter Eigendecomposition

Linear precoding algorithms can be classified into transmit zero-forcing (ZF) filter and Wiener filter (WF) [JUN05] and, not considered here, block diagonalization [SH02] and iterative methods [SB04]. The WF was derived for the MMSE criterion as

$$\mathbf{W} = \beta_{\mathrm{WF}} \cdot \mathbf{H} \cdot \Bigg(\mathbf{H}^H \mathbf{H} + \underbrace{\frac{\mathrm{tr}\{\boldsymbol{\Phi}_{\mathrm{nn}}\}}{E_{\mathrm{Tx}}} \mathbf{I}_{N_{\mathrm{UE}}}}_{\boldsymbol{\Phi}'} \Bigg)^{-1} \tag{10.5}$$

$$\text{with} \quad \beta_{\mathrm{WF}} = \sqrt{\frac{E_{\mathrm{Tx}} / \sigma_x^2}{\mathrm{tr}\left\{ \mathbf{H}^H \mathbf{H} \left(\mathbf{H}^H \mathbf{H} + \boldsymbol{\Phi}' \right)^{-2} \right\}}} . \tag{10.6}$$

The term $\mathbf{H}^H \mathbf{H}$ is regularized with the noise covariance matrix in (10.5). In interference (and not noise) limited environments with dense BS deployments this term can be neglected. Hence, only the eigenvalue decomposition (EVD) can be considered, i.e.

$$\mathbf{H}^H \mathbf{H} = \mathbf{U} \boldsymbol{\Lambda} \mathbf{U}^H = \sum_{k=1}^{n_{\min}} \lambda_k^2 \, \mathbf{u}_k \, \mathbf{u}_k^H , \tag{10.7}$$

with the unitary matrix $\mathbf{U} \in \mathbb{U}^{[N_{\mathrm{UE}} \times N_{\mathrm{UE}}]}$ and $\boldsymbol{\Lambda} \in \mathbb{R}^{[N_{\mathrm{UE}} \times N_{\mathrm{UE}}]}$. This decomposition bounds the system performance.

Spatial conditions are determined, e.g., through antenna correlations, path loss or shadowing effects, so that reduced rank situations can occur. Consequently, in a multiple-input multiple-output (MIMO) multi-point scenario there exist at least $\lambda_1^2 \geq \lambda_2^2 \geq \cdots \geq \lambda_{n_{\min}}^2$ dominant eigenvalues with $\mathbf{u}_k$ as the corresponding eigenvector to eigenvalue λ_k^2 and $n_{\min} = \min(M, K)$. In the full rank case, there exist at most $n_{\min} = \min(N_{\text{BS}}, N_{\text{UE}})$ non-zero eigenvalues. Related to the EVD, each user symbol is transmitted on a different eigenmode with the power allocated according to corresponding eigenvalues. With the EVD, the transmit filter can be written as

$$\mathbf{W} = \beta_{\text{WF}} \cdot \mathbf{H} \cdot \mathbf{U} \underbrace{\left(\mathbf{\Lambda} + \mathbf{U}^H \mathbf{\Phi}' \mathbf{U} \right)^{-1}}_{\mathbf{\Lambda} + \frac{\text{tr}\{\mathbf{\Phi}_{\text{nn}}\}}{E_{\text{Tx}}} \mathbf{I}_{N_{\text{UE}}}} \mathbf{U}^H, \qquad (10.8)$$

and the scalar β_{WF} as

$$\beta_{\text{WF}} = \sqrt{\frac{E_{\text{Tx}}/\sigma_x^2}{\text{tr}\left\{ \mathbf{\Lambda} \left(\mathbf{\Lambda} + \mathbf{U}^H \mathbf{\Phi}' \mathbf{U} \right)^{-2} \right\}}} \qquad (10.9)$$

with $\displaystyle\lim_{\sigma_n^2 \to 0} \beta_{\text{WF}} = \sqrt{\dfrac{E_{\text{Tx}}/\sigma_x^2}{\sum_{i=1}^{n_{\min}} \frac{1}{\lambda_i^2}}}$ and $i \leq n_{\min}$. \qquad (10.10)

The interpretation of (10.10) leads to following basic results: Under a fixed number of transmit antennas in CoMP and with a fixed β_{WF}, each additional user stream increases the transmit power. For a fixed ratio E_{Tx}/σ_x^2, each additional user stream reduces the SINR of the effective channel. The small eigenvalues affect the transmit power or the SINR at most. They also determine the performance from a numerical point of view as condition number

$$\kappa(\mathbf{H}) = \sqrt{\lambda_1^2 / \lambda_{n_{\min}}^2} \geq 1. \qquad (10.11)$$

If eigenvalues are rounded towards zero due to limited precision, then the observed numerical rank, which is denoted as $\text{rank}_{\text{num}}(\mathbf{H})$, is smaller than the rank of $\mathbf{H}$. The threshold ε_{num} depends on the algebraic computations.

$$\text{rank}(\mathbf{H}) = \sum_{i=1}^{\min(N_{\text{BS}}, N_{\text{UE}})} \Gamma_i \quad \text{with } \Gamma_i = \begin{cases} 0 & \text{for } \lambda_i^2 = 0 \\ 1 & \text{otherwise} \end{cases} \qquad (10.12)$$

$$\text{rank}_{\text{num}}(\mathbf{H}) = \sum_{i=1}^{\min(N_{\text{BS}}, N_{\text{UE}})} \Gamma_{i,\text{num}} \quad \text{with } \Gamma_{i,\text{num}} = \begin{cases} 0 & \text{for } |\lambda_i^2| < \varepsilon_{\text{num}} \\ 1 & \text{otherwise} \end{cases}$$
$$(10.13)$$

Finally, the number of user streams cannot exceed the number of eigenmodes. Hence, a spatial resource allocation may not include an additional stream which would reduce the SINR of the already precoded streams.

10.1.3 Transmit Filter Computations

MU-MIMO algorithms for CoMP must be designed for an upper number of cooperative BSs that serve an upper number of UEs. To realize the linear filter (10.5),

10.1 Robust and Flexible Base Station Precoding Implementation

one has to choose an appropriate algebraic decomposition [GVL96]. In this section, three basic approaches will be compared before the idea of order-recursions will be presented.

The **Cholesky decomposition (CD)**, which is similar to Gaussian elimination, separates symmetric, positive-definite matrices into a triangular matrix $\mathbf{R}$. By calculating the inverse of $\mathbf{R}$ with backward recursions and pivoting, CD is a low complex ($\mathcal{O}(1/3 N_{UE}^3)$) and numerical stable algorithm.

$$\text{Substitution:} \quad \mathbf{R}^H \mathbf{R} = \left(\mathbf{H}^H \mathbf{H} + \mathbf{\Phi}' \mathbf{I} \right) \tag{10.14}$$

$$\text{Computation:} \quad \mathbf{W} = \mathbf{H} \cdot \mathbf{R}^{-1} \cdot \left(\mathbf{R}^{-1} \right)^H \tag{10.15}$$

On the other hand, the ability to handle a multitude of MIMO setups requires a further organizational overhead, because the assignment of sub-matrices within $\mathbf{H}$ necessarily leads to an advanced exception handling within the algorithm.

Another possibility is to use the class of **QR decomposition** with its realizations namely the modified Gram-Schmidt orthogonalization, Householder transformation or the Givens rotation. A compound matrix is decomposed into a unitary and a triangular matrix with

$$\text{Substitution:} \quad \begin{bmatrix} \mathbf{H}^H \\ \mathbf{\Phi}'^{-1/2} \end{bmatrix} = \begin{bmatrix} \mathbf{Q}_1 \\ \mathbf{Q}_2 \end{bmatrix} \cdot \begin{bmatrix} \mathbf{R} \\ \mathbf{0} \end{bmatrix} \quad \text{and} \tag{10.16}$$

$$\text{Computation:} \quad \mathbf{W} = \mathbf{R}^{-1} \cdot \mathbf{Q}_1^H . \tag{10.17}$$

Compared to the CD, the orthogonalization step increases the complexity to $\mathcal{O}(4 N_{BS} N_{UE}^2 - 1/3 N_{UE}^3)$. The usability of these algorithms in various spatial setups is given, because the propagation of null elements is inherently realized.

The third method is the **Schur complement**. Submatrices of small dimensions are selected and calculated iteratively. These iterations can be realized very efficiently in pipelined architectures.

$$\text{Substitution:} \quad \begin{bmatrix} \mathbf{A} & \mathbf{U} \\ \mathbf{V} & \mathbf{C} \end{bmatrix}^{-1} = \left(\mathbf{H}^H \mathbf{H} + \mathbf{\Phi}' \right) \tag{10.18}$$

$$\text{Computation:} \quad \mathbf{W} = \mathbf{H} \cdot \begin{bmatrix} \mathbf{A} & \mathbf{U} \\ \mathbf{V} & \mathbf{C} \end{bmatrix}^{-1} \tag{10.19}$$

$$\begin{bmatrix} \mathbf{A} & \mathbf{U} \\ \mathbf{V} & \mathbf{C} \end{bmatrix}^{-1} = \begin{bmatrix} \underbrace{\mathbf{P}}_{} & \underbrace{-\mathbf{P}\mathbf{K}}_{} \\ -\underbrace{\mathbf{S}\mathbf{P}}_{S} \underbrace{\mathbf{C}^{-1}\mathbf{V}}_{P} \underbrace{(\mathbf{A}-\mathbf{K}\mathbf{V})^{-1}}_{} & \underbrace{\mathbf{U}\mathbf{C}^{-1}}_{K} + \mathbf{C}^{-1} \end{bmatrix}$$
$$\tag{10.20}$$

The drawback is the high error propagation in limited precision arithmetics. Numerical errors are accumulated with each processing stage from $\mathbf{C}^{-1}$ to $\mathbf{K}$ until the full matrix has been processed.

The Idea of Order-Recursive Precoding

The idea of order-recursive precoding is based on *changing the view onto the matrix elements of the Schur complement*. Instead of sub-matrices, a column wise or row wise consideration of a matrix is chosen with

$$\mathbf{H} = \left[\begin{array}{c|c} \mathbf{A} & \mathbf{U} \\ \hline \mathbf{V} & \mathbf{C} \end{array}\right] \Rightarrow \mathbf{H} = \underbrace{[\,\mathbf{A}\,|\,\mathbf{u}\,]}_{\mathbf{u} \in \mathbb{C}^{[N_{\mathrm{BS}} \times 1]}} = \left[\begin{array}{c} \mathbf{A} \\ \hline \underbrace{\mathbf{v}}_{\mathbf{v} \in \mathbb{C}^{[1 \times N_{\mathrm{UE}}]}} \end{array}\right]. \tag{10.21}$$

This principle denotes the transmit filter as a series of column- or row-wise matrix extensions. Here, each column in $\mathbf{H}$ is the algebraic mapping of antenna links which relate to one column in $\mathbf{W}$ for each user data stream.

$$\mathbf{w}_1 = \beta_{\mathrm{TxWF}} \cdot \mathbf{h}_1 \cdot \left(\mathbf{h}_1^H \mathbf{h}_1 + \frac{\mathrm{tr}\{\boldsymbol{\Phi}_{\mathrm{nn}}\}}{E_{\mathrm{Tx}}}\right)^{-1} \tag{10.22}$$

$$[\mathbf{w}_1\ \mathbf{w}_2] = \beta_{\mathrm{TxWF}} \cdot [\mathbf{h}_1\ \mathbf{h}_2] \cdot \left([\mathbf{h}_1\ \mathbf{h}_2]^H [\mathbf{h}_1\ \mathbf{h}_2] + \frac{\mathrm{tr}\{\boldsymbol{\Phi}_{\mathrm{nn}}\}}{E_{\mathrm{Tx}} \mathbf{I}_2}\right)^{-1} \tag{10.23}$$

$$\vdots \tag{10.24}$$

$$\mathbf{W} = \beta_{\mathrm{TxWF}} \cdot \mathbf{H} \cdot \left(\mathbf{H}^H \mathbf{H} + \boldsymbol{\Phi}'\right)^{-1}. \tag{10.25}$$

In the following, the algorithm is discussed in detail to explain the filter updates by forward and backward recursion.

10.1.4 The Order-Recursive Filter in Details

As preliminary steps, an ordered set of stream indices $\mathcal{I} \subseteq \{1, \ldots, N_{\mathrm{UE}}\}$ must be introduced, in order to pick out the columns of $\mathbf{H}$ and $\mathbf{W}$ which correspond to the potential spatial streams that can be transmitted. The notation $\mathcal{I}_{[a:b]}$ shall denote the subset of the a-th to b-th elements in $\mathcal{I}$. Further, for any matrix $\mathbf{A}$, the notation $\mathbf{A}_{[\mathcal{I},\mathcal{J}]}$ yields the sub-matrix of $\mathbf{A}$ which correspond to the rows and columns indexed in $\mathcal{I}$ and $\mathcal{J}$, respectively. Then, for any set $\mathcal{I}$ of selected spatial streams, the transmit WF can be expressed as

$$\mathbf{W}_{[:,\mathcal{I}]} = \mathbf{H}_{[:,\mathcal{I}]} \cdot \left(\mathbf{H}_{[:,\mathcal{I}]}^H \mathbf{H}_{[:,\mathcal{I}]} + \boldsymbol{\Phi}'_{[\mathcal{I},\mathcal{I}]}\right)^{-1}. \tag{10.26}$$

The idea of order-recursive precoding is based on a recursive filter structure where in each iteration l, a previously computed precoding matrix $\mathbf{W}_{[:,\mathcal{I}_{[1:l-1]}]}$ connected to stream indices 1 to $l-1$ is extended by a column for the l-th stream index in $\mathcal{I}$. Note that the order in which streams are processed (hence the order of the elements in $\mathcal{I}$) can be arbitrary. Furthermore, a spatial resource allocator can switch BSs and UEs on or off based on a selection criterion $\mathcal{F}$, as it will be detailed later.

In each iteration, the new precoding coefficients are determined by an additive matrix update and the concatenated precoding vector $\mathbf{b} \in \mathbb{C}^{[N_{\mathrm{BS}} \times 1]}$ for user

10.1 Robust and Flexible Base Station Precoding Implementation

stream $\mathcal{I}_{[l]}$ with

$$\mathbf{W}_{[:,\mathcal{I}_{[1:l]}]} = \left[\mathbf{W}_{[:,\mathcal{I}_{[1:l-1]}]} - \mathbf{b}\cdot\mathbf{d}_{\mathcal{I}_{[1:l-1]}}^T \mid \mathbf{b}\right]. \tag{10.27}$$

In the first step, the vector $\mathbf{d} \in \mathbb{C}^{[N_{\text{BS}}\times 1]}$ is calculated as

$$\mathbf{d}_{[\mathcal{I}_{[1:l]}]} = \left[\begin{pmatrix}\mathbf{H}_{[:,\mathcal{I}_{[l]}]}^H \cdot \mathbf{W}_{[:,\mathcal{I}_{[1:l-1]}]}\end{pmatrix}^T\right], \tag{10.28}$$

and is followed by the projection vector $\mathbf{e} \in \mathbb{C}^{[N_{\text{BS}}\times 1]}$ as

$$\mathbf{e} = \mathbf{H}_{[:,\mathcal{I}_{[l]}]}^H - \mathbf{H}_{[:,\mathcal{I}_{[1:l-1]}]}^H \cdot \mathbf{d}_{[\mathcal{I}_{[l-1]}]}. \tag{10.29}$$

Finally, $\mathbf{b}$ is determined dependent on an update criterion F for order extension

$$\mathbf{b} = \begin{cases} \mathbf{0}, \mathbf{H}_{[:,\mathcal{I}_{[l]}]} := 0 & \text{if } F \text{ is not fulfilled} \\ \tilde{\mathbf{b}} & \text{otherwise} \end{cases} \tag{10.30}$$

with $\tilde{\mathbf{b}} = \mathbf{e}^*/\left(\|\mathbf{e}\|^2 + \|\mathbf{\Phi}_{[\mathcal{I}_{[1:l]},\mathcal{I}_{[1:l]}]}^{\prime 1/2} \cdot \mathbf{d}_{[\mathcal{I}_{[1:l]}]}\|^2\right). \tag{10.31}$

The updates are performed with subspace projections (see (10.28) and (10.29)). Thus

$$\mathbf{e} = \underbrace{\left(\mathbf{I}_{N_{\text{BS}}} - \left(\mathbf{W}_{[:,\mathcal{I}_{[1:l-1]}]}\cdot\mathbf{H}_{[:,\mathcal{I}_{[1:l-1]}]}^H\right)^T\right)}_{P_{\mathcal{C}^\perp}(\mathbf{H}_{[:,\mathcal{I}_{[1:l-1]}]}^H)} \cdot \mathbf{H}_{[:,\mathcal{I}_{[l]}]}^* \tag{10.32}$$

can be expressed with the orthogonal projector $P_{\mathcal{C}^\perp}(\mathbf{H}_{[:,\mathcal{I}_{[1:l-1]}]}^*)$ onto the complementary space which is spanned by the set of the already considered row space of the channel matrix. With this approach, the precoding matrix is additively extended with rank-one updates of regularized projections through the channel column vector $\mathbf{H}_{[:,\mathcal{I}_{[l]}]}$ into the existing precoding solution.

The quantity of linear independence between the user stream $\mathcal{I}_{[l]}$ and the existing precoder $\mathbf{W}_{[:,\mathcal{I}_{[1:l-1]}]}$, corresponding to the EVD, is the angle

$$\cos\theta_{\mathcal{I}_{[l]}} = \frac{\mathbf{H}_{[:,\mathcal{I}_{[l]}]}^T \cdot \mathbf{e}}{\|\mathbf{H}_{[:,\mathcal{I}_{[l]}]}\| \cdot \|\mathbf{e}\|}, \quad 0 \le \theta_{\mathcal{I}_{[l]}} \le \frac{\pi}{2} \tag{10.33}$$

as a function of $\mathbf{H}$ and the user stream order in $\mathcal{I}$. If $\mathcal{I}_{[l]}$ is completely dependent of the already spanned row space the angle is zero. Vice versa, a fully independent $\mathcal{I}_{[l]}$ is orthogonal. In this case the eigenvalue is equal to $\|\mathbf{H}_{[:,\mathcal{I}_{[l]}]}\|^2$. The eigenvalues in (10.8) bound the observed norms of $\mathbf{e}$

$$\lambda_1^2 \ge \underbrace{\mathbf{H}_{[:,\mathcal{I}_{[l]}]}^T P_{\mathcal{C}^\perp}(\mathbf{H}_{[:,\mathcal{I}_{[1:l-1]}]}^*)\mathbf{H}_{[:,\mathcal{I}_{[l]}]}^*}_{\|\mathbf{e}\|^2} \ge \lambda_{n_{\min}}^2, \quad \|\mathbf{e}\| > 0. \tag{10.34}$$

The main advantage of subspace projections is the ability to propagate null elements in $\mathbf{H}_{[:,\mathcal{I}_{[l]}]}$ to the corresponding position in $\mathbf{W}_{[:,\mathcal{I}_{[l]}]}$. Hence, it is inherently

Table 10.1. Operation count comparison of matrix inversion algorithms [GVL96, HKF10].

	Cholesky dec.	QR dec.	Order-rec.
$\mathcal{O}$-notation	$1/3 N_{UE}^3$	$4 N_{BS} N_{UE}^2 - 1/3 N_{UE}^3$	$3/2 N_{BS} N_{UE}^2$
Divisions	N_{UE}	N_{UE}	N_{UE}
Sqr. roots	0	N_{UE}	0

possible to handle multiple transmission scenarios such as single-input single-output (SISO), multiple-input single-output (MISO) and MIMO. Reorganizing the matrix elements is not necessary with this approach. The full pseudo code as reference for an implementation in a real-time system and a complexity estimation is given in [HKF10]. Table 10.1 shows the algorithmic complexity which is made up of a complex operation count of the required multiplications, divisions and square root computations.

Update Sequence and Criteria

The user streams are decoupled from the iteration numbering through the choice of $\mathcal{I}$. With each order-recursive iteration, the sequence of each processed user stream can be computed based on update criterion F as already given in (10.30).

Furthermore, quality of service (QoS) oriented precoding criteria like transmit power, SINRs or user stream priorities beside the condition number can be used as F to decide how an additional user affects the system performance. Based on this, stopping thresholds can be defined. With this minor extension, the proposed solution can be extended towards a simple spatial scheduling or resource allocation methodology similar to the tree search of [FDGH07].

For example, in ill-conditioned CoMP channels weak eigenmodes can be skipped to preserve the numerical stability. Here from the base station point of view, the column vector $\mathbf{H}_{[\mathcal{I}_{[i]},:]}$ is filled with zeros. This means that either a UE will not be served by CoMP or a particular stream to this UE is not included. More details on this topic can be found in Section 11.1.

10.1.5 Example: SINR as Function of the Condition Number

This example illustrates the SINR (10.4) as function of the condition number for a physical channel matrix $\mathbf{H}$ with independently and identically distributed (i.i.d.) circular symmetric complex normal coefficients. In this metric, the mean over the desired signal power and the mean over the interference plus AWGN noise power is given as ratio in decibel. Fig. 10.1 compares the SINR under various noise power levels as well as floating and fixed-point precision. The implemented fixed-point number format is the 2's complement with one sign bit, 5 integer bits and 10 fractional bits. The loss of SINR compared to the floating point computations is negligible for good-conditioned channels. In case of ill-conditioning the previously described effects of rounding and numerical propagation errors become

10.2 Low-Complexity Terminal-Side Receiver Implementation

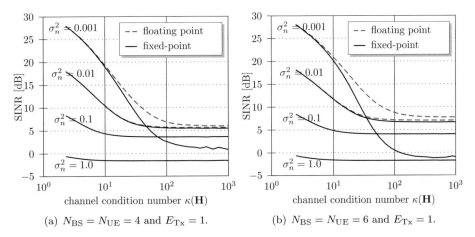

Figure 10.1 Floating and fixed-point SINR for i.i.d. Gaussian MIMO systems as function of the channel condition number.

prominent. If eigenvalues (10.8) are smaller than the machine accuracy ε_{num} or the projection norms (10.34) are observed as zeros then a low desired signal and high spatial interference power results. With an increasing noise, the regularization through $\mathbf{\Phi}'$ affects the division operations to be in a defined range, prevents numerical instabilities and improves the conditioning of the matrix inversion. Anyway, the regularized (or approximated) division results in an imperfect interference suppression.

10.1.6 Summary

This section described an order-recursive algorithm to calculate the Wiener transmit filter for coherent CoMP transmissions between several decentralized base stations to several decentralized mobile terminals in cellular radio networks. State of the art implementations do not offer the flexibility to handle multiple transmission setups as required for multi-user precoding in conjunction with a proposed reduced complexity. The discussion of all signal processing steps shows how algebraic observations are mapped to physical parameters and how they reflect the spatial eigenmodes. These observations can be integrated into spatial resource allocation to extend the precoding coefficients recursively with additional spatial streams.

10.2 Low-Complexity Terminal-Side Receiver Implementation

Udo Wachsmann, Rainer Bachl and Stefan Müller-Weinfurtner

Cellular systems adopting CoMP techniques to increase system capacity impose new requirements on the terminal-side receiver. One particular aspect being presented in this section is the treatment of intra-cell and inter-cell interference in the downlink. In orthogonal frequency division multiplex (OFDM)-based wireless systems like LTE, intra-cell interference is mainly due to spatial multiplexing of multiple users. For inter-cell interference, we have to distinguish the two cases where serving and interfering cell(s) are co-located or located at different sites. In both cases, it is assumed that CoMP techniques lead to dedicated beams towards the user equipments (UEs) in order to increase spectral efficiency. This shall apply for the desired link as well as for the interfering link(s).

A good overview on space-time receivers in general and means to mitigate co-channel interference can be found in [PNG03]. One particular method in a terminal-side receiver to exploit the characteristics of the interference signal at affordable cost is the so-called interference rejection combining (IRC), also known as optimum combining [Win84]. In this chapter, the potential gains of IRC on link level are analyzed. Starting from a simple system model with ideal assumptions on channel and interference estimation, the impact of estimation errors is elaborated in more detail. Particular focus is spent on complexity aspects for interference estimation when taking the LTE Release 8 signal structure into consideration [3GP07b].

10.2.1 Introduction to Interference Rejection Combining (IRC)

The aspects of interference handling in the UE are presented based on a simplified downlink system model where each base station (BS) transmits to exactly one UE. For the considerations in this chapter, we imply further on that only one transmission layer is used per UE, i.e. no single-user spatial multiplexing is applied. However, the study can easily be extended to accommodate single-user spatial multiplexing without loss of generality. These assumptions enable the simple model of having one dedicated beam per BS towards one UE. In terms of the system model introduced in Chapter 3, we have $K = M$ and investigate the transmission on an exemplary frequency-flat sub-carrier of an orthogonal frequency division multiple access (OFDMA) system.

The pair BS 1 and UE 1 represents the desired link, while the remaining (BS, UE) pairs are treated as interference at UE 1. For the development of a terminal-side receiver implementation, we are only interested in the receive signal of the desired link. Hence, the simplified system model being adapted from (3.6) can be written as follows

$$\tilde{x}_1 = \mathbf{g}^H \mathbf{y}_1 = \mathbf{g}^H (\underbrace{[\mathbf{H}_1^1]^H \mathbf{w}_1 x_1}_{\text{desired}} + \underbrace{\sum_{k=2}^{K} [\mathbf{H}_1^k]^H \mathbf{w}_k x_k}_{\text{interference}} + \underbrace{\mathbf{n}_1}_{\text{noise}}), \qquad (10.35)$$

where $\mathbf{g}, \mathbf{y}_1, \mathbf{n}_1 \in \mathbb{C}^{[N_{\text{ue}} \times 1]}$, $\forall k: \mathbf{w}_k \in \mathbb{C}^{[N_{\text{bs}} \times 1]}$, and $\forall k: \mathbf{H}_1^k \in \mathbb{C}^{[N_{\text{bs}} \times N_{\text{ue}}]}$. In (10.35), we further restricted ourselves to linear and unitary preprocessing at the BS side and replaced the term $d(x)$ in (3.6) by x. Furthermore, the index

10.2 Low-Complexity Terminal-Side Receiver Implementation

for the UE receive filter weights $\mathbf{g}$ has been skipped for brevity since we deal exclusively with UE 1 in this section. Please note that due to the single-layer assumption introduced above, the transmit symbols towards the UEs, x_k, are denoted as scalars. For simplified notation in the sequel, we define the effective channel for single-layer transmission from BS k to UE 1 as follows:

$$\overline{\mathbf{h}}_k = [\mathbf{H}_1^k]^H \mathbf{w}_k. \tag{10.36}$$

Then, the interference-and-noise part from (10.35) being effective at the receiver can be formulated as

$$\mathbf{z}_1 = \sum_{k=2}^{K} \overline{\mathbf{h}}_k x_k + \mathbf{n}_1 \tag{10.37}$$

and the resulting simplified system model reads

$$\tilde{x}_1 = \mathbf{g}^H \mathbf{y}_1 = \mathbf{g}^H \left(\overline{\mathbf{h}}_1 x_1 + \mathbf{z}_1 \right). \tag{10.38}$$

The covariance matrix of the effective noise $\mathbf{z}_1$ is given by

$$\mathbf{\Phi}_{z_1 z_1} = E_{\mathbf{x},\mathbf{n}} \{\mathbf{z}_1 \mathbf{z}_1^H\} = \sigma^2 \mathbf{I} + \sum_{k=2}^{K} \overline{\mathbf{h}}_k \overline{\mathbf{h}}_k^H E\{|x_k|^2\}. \tag{10.39}$$

The averaging is performed over interfering transmit symbols $\mathbf{x}$ and noise $\mathbf{n}$, so that only the dependency on the instantaneous effective channels $\overline{\mathbf{h}}_k$ remains. At this stage, we did not average over effective channel realizations, since we assume that they can be estimated with reasonable accuracy in order to exploit spatial correlations for interference rejection.

The concept of IRC in light of the above-mentioned system model can be formally described as a cascade of pre-whitening and maximum ratio combining (MRC). The pre-whitening step transforms the colored noise consisting of interference and white noise into the domain where we have only white noise with unit power on each receive branch. In case of white noise, the MRC receiver is well-known to be optimal [Win84]. Hence, the IRC operation in terms of the UE-side receiver filter can be formulated as

$$\mathbf{g}_{\text{IRC}}^H = \underbrace{\overline{\mathbf{h}}_1^H \mathbf{\Phi}_{z_1 z_1}^{-1/2}}_{\text{MRC}} \cdot \underbrace{\mathbf{\Phi}_{z_1 z_1}^{-1/2}}_{\text{pre-whitening}} = \overline{\mathbf{h}}_1^H \mathbf{\Phi}_{z_1 z_1}^{-1}. \tag{10.40}$$

This conceptual approach of pre-whitening and MRC is also known as whitened matched filter (WMF), see e.g. [Lar07]. Please note that the MRC part in (10.40) includes $\mathbf{\Phi}_{z_1 z_1}^{-1/2}$ since the channel after pre-whitening is given by $\mathbf{\Phi}_{z_1 z_1}^{-1/2} \overline{\mathbf{h}}_1$. Please note further that the cascade of pre-whitening and MRC is only one way to introduce and motivate IRC. However, the actual computation of filter weights is done in a single step.

The performance metric being used for evaluation of numerical results is the residual signal-to-interference-and-noise ratio (SINR) after the UE receive filter. The motivation for this choice is that the SINR is appropriate to represent the effective signal-to-noise ratio (SNR) on an additive white Gaussian noise (AWGN) channel. With the effective SNR approach further performance metrics like error probability or throughput can easily be derived from known AWGN figures. As starting point, the generic expression for the instantaneous SINR with a general UE receive filter is given by (see e.g. [HSP01])

$$\text{SINR}(\mathbf{g}) = \frac{|\mathbf{g}^H \overline{\mathbf{h}}_1|^2}{\mathbf{g}^H \mathbf{\Phi}_{z_1 z_1} \mathbf{g}} E\{|x_1|^2\}. \tag{10.41}$$

The term *instantaneous SINR* relates to the fact that only one specific channel realization of the desired and interfering links is considered. The generic expression in (10.41) is used later on when mismatched UE receive filters due to estimation errors are evaluated. The next step is to insert UE receive filter weights for IRC from (10.40) into (10.41) resulting in the simplified SINR expression

$$\text{SINR}(\mathbf{g}_{\text{IRC}}) = \overline{\mathbf{h}}_1^H \mathbf{\Phi}_{z_1 z_1}^{-1} \overline{\mathbf{h}}_1 \cdot E\{|x_1|^2\} = \gamma \cdot E\{|x_1|^2\}. \tag{10.42}$$

Here, the IRC parameter

$$\gamma = \overline{\mathbf{h}}_1^H \cdot \mathbf{\Phi}_{z_1 z_1}^{-1} \cdot \overline{\mathbf{h}}_1 \tag{10.43}$$

is introduced which can be interpreted as SINR in case of ideal IRC for unit-power transmit symbols. In order to complete the framework on IRC, the UE receive filter weights according to the minimum mean square error (MMSE) approach [Ver98] are derived and compared to the IRC ones. The MMSE solution works on the covariance of the receive signal, which reads

$$\mathbf{\Phi}_{y_1 y_1} = E_{\mathbf{x}, \mathbf{n}}\{\mathbf{y}_1 \mathbf{y}_1^H\} = \sigma^2 \mathbf{I} + \sum_{k=1}^{K} \overline{\mathbf{h}}_k \overline{\mathbf{h}}_k^H E\{|x_k|^2\}. \tag{10.44}$$

The relation between the covariance of the receive signal $\mathbf{y}_1$ and the covariance of the interference-and-noise signal $\mathbf{z}_1$ can be observed as

$$\mathbf{\Phi}_{y_1 y_1} = \mathbf{\Phi}_{z_1 z_1} + \overline{\mathbf{h}}_1 \overline{\mathbf{h}}_1^H E\{|x_1|^2\}. \tag{10.45}$$

Conceptually, the MMSE weights are obtained by whitening the receive signal and applying a matched filter afterwards. This results in the following weight expression

$$\mathbf{g}_{\text{MMSE}}^H = \overline{\mathbf{h}}_1^H \mathbf{\Phi}_{y_1 y_1}^{-1}$$
$$= \overline{\mathbf{h}}_1^H \left[\mathbf{\Phi}_{z_1 z_1} + \overline{\mathbf{h}}_1 \overline{\mathbf{h}}_1^H E\{|x_1|^2\}\right]^{-1}. \tag{10.46}$$

Applying the Sherman-Morrison-Woodbury formula [GVL96], also known as matrix inversion lemma, the MMSE weights can then be re-formulated as follows

$$\mathbf{g}_{\text{MMSE}}^H = \overline{\mathbf{h}}_1^H \left(\mathbf{\Phi}_{z_1 z_1}^{-1} - \frac{\mathbf{\Phi}_{z_1 z_1}^{-1} \overline{\mathbf{h}}_1 E\{|x_1|^2\} \overline{\mathbf{h}}_1^H \mathbf{\Phi}_{z_1 z_1}^{-1}}{1 + \overline{\mathbf{h}}_1^H \mathbf{\Phi}_{z_1 z_1}^{-1} \overline{\mathbf{h}}_1 E\{|x_1|^2\}} \right). \tag{10.47}$$

Inserting the IRC expressions for weights from (10.40) as well as for SINR from (10.42), the MMSE weights read

$$\mathbf{g}_{\text{MMSE}}^H = \left(1 - \frac{\text{SINR}(\mathbf{g}_{\text{IRC}})}{1 + \text{SINR}(\mathbf{g}_{\text{IRC}})}\right) \cdot \mathbf{g}_{\text{IRC}}^H = \frac{1}{1 + \text{SINR}(\mathbf{g}_{\text{IRC}})} \cdot \mathbf{g}_{\text{IRC}}^H. \quad (10.48)$$

This interesting result shows that MMSE weights $\mathbf{g}_{\text{MMSE}}$ are simply a scaled version of IRC weights $\mathbf{g}_{\text{IRC}}$, hence yielding the same SINR

$$\text{SINR}(\mathbf{g}_{\text{MMSE}}) = \text{SINR}(\mathbf{g}_{\text{IRC}}). \quad (10.49)$$

Moreover, from an implementation point of view, the MMSE weights are preferable since they provide inherently higher numerical stability. In order to illustrate this, we consider e.g. the case of 1 interferer ($K = 2$) and high SNR, i.e. $\sigma^2 \ll E\{|x_1|^2\}$. Then, the matrix $\mathbf{\Phi}_{z_1 z_1}$ being used for the IRC weights $\mathbf{g}_{\text{IRC}}$ is almost singular and the inversion tends to get numerically instable if no manual counter measures like e.g. adding a constant are taken. On the other hand, the matrix $\mathbf{\Phi}_{y_1 y_1}$ being used for the MMSE weights $\mathbf{g}_{\text{MMSE}}$ is typically well conditioned (at least for not too bad-conditioned channels $\mathbf{H}_1$) and the inversion is inherently numerically stable. For the numerical results further below in this chapter, we further consider the MRC receiver as a reference [Win84]. The UE receive filter weights for MRC (without pre-whitening) can be formulated as

$$\mathbf{g}_{\text{MRC}}^H = \sigma^{-2} \overline{\mathbf{h}}_1^H. \quad (10.50)$$

So far, the UE receive filter weights have been derived assuming ideal knowledge of channel and effective noise covariance, which is also the basis for the first performance study in Section 10.2.2. In later subsections, we will incorporate different kinds of estimation errors by introducing the following weights:

$$\hat{\mathbf{g}}_{\text{IRC},a}^H = \hat{\mathbf{g}}_{\text{IRC}}^H(\hat{\mathbf{h}}, \mathbf{\Phi}) = \widehat{\overline{\mathbf{h}}}_1^H \mathbf{\Phi}_{z_1 z_1}^{-1} \quad \text{with estimation of } \mathbf{h} \quad (10.51)$$

$$\hat{\mathbf{g}}_{\text{IRC},b}^H = \hat{\mathbf{g}}_{\text{IRC}}^H(\mathbf{h}, \hat{\mathbf{\Phi}}) = \overline{\mathbf{h}}_1^H \hat{\mathbf{\Phi}}_{z_1 z_1}^{-1} \quad \text{with estimation of } \mathbf{\Phi} \quad (10.52)$$

$$\hat{\mathbf{g}}_{\text{IRC},c}^H = \hat{\mathbf{g}}_{\text{IRC}}^H(\hat{\mathbf{h}}, \hat{\mathbf{\Phi}}) = \widehat{\overline{\mathbf{h}}}_1^H \hat{\mathbf{\Phi}}_{z_1 z_1}^{-1} \quad \text{with estimation of } \mathbf{h} \text{ and } \mathbf{\Phi}, \quad (10.53)$$

where $\hat{(\cdot)}$ denotes estimates. Here, the letters a, b, c are used to distinguish different degrees of estimation errors.

10.2.2 Performance Study on IRC with Known Channel and Interference Covariance

In this subsection, numerical results on SINR performance for IRC reception are presented. Hereby, ideal weight calculation is assumed, i.e. ideal estimation of channel and interference covariance. The multiple-input multiple-output (MIMO) channel used for the study is modeled by means of a correlation channel model (CCM) with independent transmit and receive antenna correlation. The adopted channel model is taken from the one agreed upon in 3GPP [3GP10h] with correlation parameters α for transmit correlation and β for receive correlation.

As described in the introduction of this section, the study focuses on scenarios where desired and interferer downlink are generated as dedicated beams. For an appropriate modelling of this beamforming aspect, high correlation at BS side is assumed. Therefore, the value $\alpha = 0.9$ is chosen since it also models in 3GPP [3GP10h] the high correlation case. At the UE side, we assume some polarization diversity and, therefore, low to medium correlation can be achieved. Here, $\beta = 0.3$ is selected matching the medium correlation case in 3GPP [3GP10h].

We consider one BS site with $N_{\text{bs}} = 2$ physical antennas and one UE with $N_{\text{ue}} = 2$ physical antennas. Furthermore, we take $K = 2$, i.e. one desired beam (layer) and one interferer beam (layer). Please note that a beam in this context is equivalent to one layer in the LTE sense. For simplicity, no particular pre-coding is adopted meaning that one BS antenna serves one UE. Such a model gives direct insight into the performance of multi-user spatial multiplexing.

The performance measure for the study is the mean SINR where the expectation is taken over all channel samples. Study parameters are the SNR and signal-to-interference ratio (SIR), respectively. They are defined as

$$\text{SNR} = \frac{E\{|x_1|^2\}}{\sigma^2} \tag{10.54}$$

$$\text{SIR} = \frac{E\{|x_1|^2\}}{E\{|x_2|^2\}}. \tag{10.55}$$

Starting from (10.42), this results in the following SINR expression ($K = 2$) for IRC:

$$\text{SINR}(\mathbf{g}_{\text{IRC}}) = \overline{\mathbf{h}}_1^H \left[\frac{\mathbf{\Phi}_{z_1 z_1}}{E\{|x_1|^2\}} \right]^{-1} \overline{\mathbf{h}}_1$$

$$= \overline{\mathbf{h}}_1^H \left[\text{SNR}^{-1} \mathbf{I} + \overline{\mathbf{h}}_2 \overline{\mathbf{h}}_2^H \text{SIR}^{-1} \right]^{-1} \overline{\mathbf{h}}_1. \tag{10.56}$$

The curves are sketched as mean SINR versus SNR with SIR as additional parameter. Numerical results for IRC are depicted in Fig. 10.2 for the parameter SIR $\in \{-10, 0, 10\}$ dB. Additionally, for reference, the performance of an MRC receiver is shown.

It can be seen that the potential gains of IRC versus MRC strongly depend on the SNR operating point. As soon as the SNR is larger than the SIR, the IRC gain versus MRC gets substantial. This behavior is further affected by transmit and receive correlation. For higher receive correlation values, the point where IRC pays off is moved to slightly higher SNR values.

As a conclusion for interference-limited scenarios (SIR < SNR) with one dominating interferer, we see substantial gains for the discussed IRC approach. In the following sections, it is further elaborated on the losses being implied when estimation errors for channel and interference covariance are taken into account.

10.2 Low-Complexity Terminal-Side Receiver Implementation

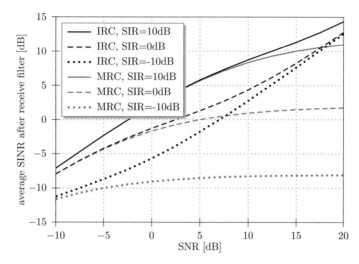

Figure 10.2 Mean SINR versus SNR for ideal IRC and MRC receive filter weights. Correlation channel model from [3GP10h] with $\alpha = 0.9, \beta = 0.3$.

Please note that the underlying channel model assumes simply one antenna array at BS site which is used by desired as well as interfering link. In the strict sense, this model is only applicable for the intra-cell interference case. However, the basic results on algorithm aspects and potential IRC gains can be easily extended to the cases of inter-cell interference.

10.2.3 Implementation Losses from Imperfect Channel Estimation

Model for Channel Estimation Error and Resulting Weight Mismatch
We assume reference signal (RS)-based non-data-aided channel estimation so that interference and noise in reference signal processing is independent from the perturbation in data processing. For the error in the channel estimate for the desired link, we assume a simple model with an additive perturbation [Ahn08], which follows the same spatial correlation like interference and noise in data processing. This assumption is based on the fact that in many mobile communications systems like LTE, the reference signals will suffer from the same spatial interference as data. In LTE, reference signals from directly neighboring cells are differently shifted in frequency so that the characteristics of interference on reference signals is close to that on data symbols (see also Section 9.1). With sufficient accuracy, this can be assumed to hold for all transmission modes in LTE. Channel estimation filtering in time and/or frequency direction usually does not remove the spatial correlation. We assume that precoded demodulation reference signal (DRS) are used as introduced in Section 9.1, enabling each UE to directly estimate the effective channel after precoding, yielding an estimate

$$\widehat{\overline{\mathbf{h}}}_1 = \overline{\mathbf{h}}_1 + \mathbf{e}, \qquad (10.57)$$

where the vector **e** consists of complex-valued Gaussian noise with zero mean and covariance

$$\Phi_{ee} = E\{ee^H\} = \frac{1}{G}\Phi_{z_1 z_1}, \tag{10.58}$$

and where we introduced the *interpolation* or *processing* gain G as a linear power ratio (see Section 9.1). The amount of corresponding de-noising by means of channel estimation filtering in time and/or frequency direction is given by $10\log_{10}(G)$ [dB].

Let us first assume the interference-and-noise covariance estimate to be ideal and deal with estimation in Section 10.2.4. With the additive error model, the non-ideal weight vector suffering from channel estimation errors reads

$$\hat{g}_{\text{IRC},a}^H = \overline{\mathbf{h}}_1^H \Phi_{z_1 z_1}^{-1} + \mathbf{e}^H \Phi_{z_1 z_1}^{-1}. \tag{10.59}$$

Analysis of SINR with Weight Mismatch

The instantaneous SINR suffering from combining-weight mismatch can be obtained by using (10.59) to tailor (10.41) to our needs. We average separately over the perturbations in nominator and denominator to obtain

$$\text{SINR}(\hat{g}_{\text{IRC},a}) = \frac{E_{\mathbf{e}}\left\{\left|\hat{\mathbf{h}}_1^H \Phi_{z_1 z_1}^{-1} \overline{\mathbf{h}}_1\right|^2\right\}}{E_{\mathbf{e}}\left\{\hat{\mathbf{h}}_1^H \Phi_{z_1 z_1}^{-1} \hat{\mathbf{h}}_1\right\}} E\{|x_1|^2\}. \tag{10.60}$$

Including the additive error model for channel estimation from (10.57) into this formula yields

$$\text{SINR}(\hat{g}_{\text{IRC},a}) = \frac{E_{\mathbf{e}}\left\{\left|\left(\overline{\mathbf{h}}_1^H + \mathbf{e}^H\right)\Phi_{z_1 z_1}^{-1}\overline{\mathbf{h}}_1\right|^2\right\}}{E_{\mathbf{e}}\left\{\left(\overline{\mathbf{h}}_1^H + \mathbf{e}^H\right)\Phi_{z_1 z_1}^{-1}\left(\overline{\mathbf{h}}_1 + \mathbf{e}\right)\right\}} E\{|x_1|^2\}. \tag{10.61}$$

We now start to evaluate the expectation under exploitation of the zero-mean property of **e** and by making use of (10.43) to obtain

$$\text{SINR}(\hat{g}_{\text{IRC},a}) = \frac{\gamma^2 + E_{\mathbf{e}}\left\{\left|\mathbf{e}^H\Phi_{z_1 z_1}^{-1}\overline{\mathbf{h}}_1\right|^2\right\}}{\gamma + E_{\mathbf{e}}\left\{\mathbf{e}^H\Phi_{z_1 z_1}^{-1}\mathbf{e}\right\}} E\{|x_1|^2\}. \tag{10.62}$$

With the identity $\mathbf{a}^H\mathbf{e} = \text{tr}(\mathbf{e}\mathbf{a}^H)$ [HJ99] and the general definition for Φ_{ee} from (10.58), we obtain the final general SINR result

$$\text{SINR}(\hat{g}_{\text{IRC},a}) = \frac{\gamma^2 + \overline{\mathbf{h}}_1^H \Phi_{z_1 z_1}^{-1} \Phi_{ee} \Phi_{z_1 z_1}^{-1} \overline{\mathbf{h}}_1}{\gamma + \text{tr}\left(\Phi_{ee}\Phi_{z_1 z_1}^{-1}\right)} E\{|x_1|^2\}. \tag{10.63}$$

The intention now is to put this SINR into a relationship with $\text{SINR}(\mathbf{g}_{\text{IRC}})$ from (10.42), to obtain the general SINR ratio

$$\frac{\text{SINR}(\hat{\mathbf{g}}_{\text{IRC},a})}{\text{SINR}(\mathbf{g}_{\text{IRC}})} = \frac{1}{\gamma} \cdot \frac{\gamma^2 + \overline{\mathbf{h}}_1^H \boldsymbol{\Phi}_{z_1 z_1}^{-1} \boldsymbol{\Phi}_{ee} \boldsymbol{\Phi}_{z_1 z_1}^{-1} \overline{\mathbf{h}}_1}{\gamma + \text{tr}\left(\boldsymbol{\Phi}_{ee} \boldsymbol{\Phi}_{z_1 z_1}^{-1}\right)}. \tag{10.64}$$

In a final step, we make use of our channel estimation error model with specific definition for $\boldsymbol{\Phi}_{ee}$ from (10.58), which allows for substituting $\boldsymbol{\Phi}_{ee} \boldsymbol{\Phi}_{z_1 z_1}^{-1} = \mathbf{I}/G$ to obtain the SINR ratio

$$\frac{\text{SINR}(\hat{\mathbf{g}}_{\text{IRC},a})}{\text{SINR}(\mathbf{g}_{\text{IRC}})} = \frac{1}{\gamma} \cdot \frac{\gamma^2 + \gamma/G}{\gamma + N_{\text{ue}}/G} = \frac{\gamma + 1/G}{\gamma + N_{\text{ue}}/G}. \tag{10.65}$$

This is valid for the specific model that channel estimation error exhibits the same spatial correlation like data, but only scaled by a scalar processing gain. It needs to be understood that this is purely the ratio of post-combining SINR. The actual losses measurable in error rate or throughput will usually be higher and they will also depend on the specific type of signal constellation and the post-combining demodulation strategy. As such, the ratio in (10.65) manifests a fundamental lower bound for the non-recoverable losses caused during combining by weight mismatch due to imperfect channel estimation. We will treat the additional demodulation loss from phase- or amplitude mismatch in the next paragraph.

Further, we can read from (10.65) that the SINR ratio is equal to 1 for $N_{\text{ue}} = 1$ UE receive antenna, which is reasonable, since the post-combining SINR does not suffer from an amplitude or phase error. It is only the demodulation process itself, which later causes losses. Another sanity check is that the SINR ratio converges to 1 for $G \to \infty$ irrespective of a finite N_{ue}.

Demodulation Loss

In order to consider the impact of the demodulation loss, the channel estimation error is introduced into the system model established in (10.38):

$$\tilde{x}_1 = \mathbf{g}^H \left(\widehat{\mathbf{h}}_1 - \mathbf{e}\right) x_1 + \mathbf{g}^H \mathbf{z}_1$$
$$= \mathbf{g}^H \widehat{\mathbf{h}}_1 x_1 + \mathbf{g}^H \left(\mathbf{z}_1 - \mathbf{e} x_1\right). \tag{10.66}$$

The useful part for demodulation is only $\mathbf{g}^H \widehat{\mathbf{h}}_1 x_1$, while $\mathbf{g}^H \mathbf{e} x_1$ has to be treated as noise. Then, the SINR expression for arbitrary filter weights $\mathbf{g}$ based on an extension of (10.41) with the model from (10.66) reads

$$\text{SINR}(\mathbf{g}) = \frac{\left|\mathbf{g}^H \widehat{\mathbf{h}}_1\right|^2}{\mathbf{g}^H \left(\boldsymbol{\Phi}_{z_1 z_1} + \boldsymbol{\Phi}_{ee} E\{|x_1|^2\}\right) \mathbf{g}} E\left\{|x_1|^2\right\}. \tag{10.67}$$

With the specific channel estimation error model introduced in (10.58), this SINR expression can be further simplified to

$$\text{SINR}(\mathbf{g}) = \frac{\left|\mathbf{g}^H \widehat{\mathbf{h}}_1\right|^2}{\mathbf{g}^H \mathbf{\Phi}_{z_1 z_1} \mathbf{g}} \cdot \frac{1}{1/E\{|x_1|^2\} + 1/G}. \qquad (10.68)$$

With the chosen model being valid for not too low SNRs, the demodulation loss is independent of the SNR that data transmission is subject to (but of course dependent on the SNR channel estimation is subject to) and the actually chosen filter weights. So, when comparing the achievable SINR for different weight selection approaches, the demodulation loss has no impact since it effects each approach equivalently. It is a significant contribution to the overall loss, though, as we will see in Fig. 10.3 in the next paragraph.

Numerical Evaluation of SINR Loss with Weight Mismatch
The channel estimation interpolation gain achievable in a general OFDMA system depends on the topology of reference signals in the time-frequency grid and the two-dimensional correlations of channel coefficients parameterized by time dispersion properties (e.g., delay spread) and time variation (e.g., Doppler frequency) of the channel (see Section 9.1). It can be shown by analysis of the Wiener solution [HKR+97b] that for the specific reference signal situation and channel model parameters in LTE, the interpolation gain by means of UE-side implementable filtering with limited span in frequency direction ranges roughly between 3 dB in a fairly dispersive channel still compliant with a normal cyclic prefix and 12 dB in a non-dispersive channel. Further interpolation gain may be achieved by means of filtering in time direction.

Fig. 10.3 shows the SINR ratio for exemplary parameters. With a processing gain of 6 dB for $N_{\text{ue}} = 2$ UE receive antennas, we have an acceptable combining loss of 0.11 dB when targeting $10 \log_{10}(\gamma) = 10$ dB. For $N_{\text{ue}} = 4$, the combining loss increases to 0.31 dB. To guarantee the same post-combining SINR with $N_{\text{ue}} = 4$ instead of $N_{\text{ue}} = 2$, the processing gain needs to be improved by approximately 5 dB. This result is related to [RCP09], and it shows that the cost for channel estimation is rising with the number of antennas when the combining loss with respect to ideal performance shall be limited.

The combining loss to be expected with realistic channel estimation is small compared to the significant gains offered by ideal IRC, as shown in Fig. 10.2. It should also be noted that the requirements on processing gain are dominated by the demodulation loss, if we consider $N_{\text{ue}} \leq 4$ and the higher SINR range, e.g. above 8 dB. Hence, irrespective of combining loss due to channel estimation errors, IRC implementation still yields substantial improvements.

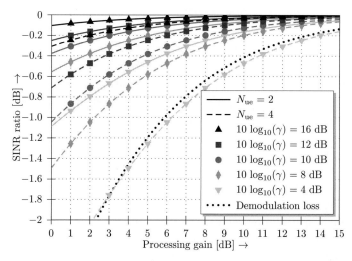

Figure 10.3 SINR ratio versus processing gain G. Combining losses from (10.65) for $N_{ue} \in \{2, 4\}$ and a selection of typical values for γ are shown together with the demodulation loss from (10.68) for $E\{|x_1|^2\} = 1$.

10.2.4 Implementation Losses from Spatial Interference-and-Noise Covariance Estimation

IRC employs a spatial interference-and-noise covariance corresponding to (10.39) and, hence, requires knowledge of either (i) channel state corresponding to all or at least dominant interfering signals or (ii) interference-and-noise spatial covariance matrix. Estimation of the channel state for a certain number of interfering signals was discussed in Sections 5.1 and 6.3, but may not be practical as it involves rather high computational complexity especially in the case of multiple interferers. Hence, this subsection focuses on the estimation of the spatial interference-and-noise covariance matrix.

The spatial interference-and-noise covariance matrix can be estimated by averaging the Hermitian outer product of interference-and-noise vector samples, which yields the so-called sample covariance matrix. The sample covariance matrix is the maximum likelihood estimate of the true interference-and-noise covariance matrix with the properties that the estimator is unbiased and the matrix is positive semi-definite. However, the interference as perceived by the UE depends on the channel, power and time-frequency resources allocated for co-channel interferers. The power and time-frequency resources used for co-channel interferers are changing with the resource allocation granularity to exploit the throughput gains from link adaptation and multi-user diversity. Hence, only samples within the transmission time interval share the same spatial interference-and-noise covariance matrix, which in turn also limits any sample averaging to the duration of the scheduling interval, e.g. 1 ms in LTE.

In [JH09, LZLG08, MMK07], some sample covariance matrix enhancement techniques are suggested for MIMO-OFDM systems which make use of the limited delay spread of the channel. However, for numerical robustness it needs to be guaranteed that any smoothed sample covariance matrix remains positive semi-definite. From the sample covariance matrix enhancement techniques, the approach in [JH09] satisfies this criterion and has significantly lower computational complexity when compared to the alternatives in [LZLG08, MMK07]. In particular, it is suggested in [JH09] to average the entries of the sample covariance matrix over a set of sub-carriers, whereby the averaging length is chosen as a trade-off between noise reduction and distortion of the auto/cross-correlation terms. In practice, however, different interferers may be present per allocated resource block. Hence, only samples within the resource allocation size share the same spatial interference-and-noise matrix, which in turn also limits any sample averaging to the resource block size in frequency direction. In LTE, the resource block size corresponds to 12 sub-carriers with 15 kHz sub-carrier spacing.

The ideal spatial interference-and-noise covariance can be written as

$$\Phi_{z_1 z_1}(q,o) = E_{\mathbf{x},\mathbf{n}}\left\{\left(\mathbf{y}_1(q,o) - x_1(q,o)\overline{\mathbf{h}}_1(q,o)\right) \cdot \left(\mathbf{y}_1(q,o) - x_1(q,o)\overline{\mathbf{h}}_1(q,o)\right)^H\right\}, \quad (10.69)$$

where q denotes the sub-carrier and o denotes the OFDM symbol, and $E_{\mathbf{x},\mathbf{n}}\{\cdot\}$ denotes expectation with respect to noise and transmit symbols. Equation (10.69) can be approximated with the sample covariance matrix as

$$\hat{\Phi}_{z_1 z_1}(q,o) = \frac{1}{N_{\text{scm}}} \sum_{\tilde{q}\in\mathcal{Q}, \tilde{o}\in\mathcal{O}} \left(\mathbf{y}_1(\tilde{q},\tilde{o}) - x_1(\tilde{q},\tilde{o})\widehat{\overline{\mathbf{h}}}_1(\tilde{q},\tilde{o})\right) \cdot \quad (10.70)$$

$$\cdot \left(\mathbf{y}_1(\tilde{q},\tilde{o}) - x_1(\tilde{q},\tilde{o})\widehat{\overline{\mathbf{h}}}_1(\tilde{q},\tilde{o})\right)^H$$

$$= \frac{1}{N_{\text{scm}}} \sum_{\tilde{q}\in\mathcal{Q},\tilde{o}\in\mathcal{O}} \left(\sum_{k=2}^{K} x_k(\tilde{q},\tilde{o})\overline{\mathbf{h}}_k(\tilde{q},\tilde{o}) + \mathbf{n}(\tilde{q},\tilde{o}) - x_1(\tilde{q},\tilde{o})\mathbf{e}(\tilde{q},\tilde{o})\right) \cdot$$

$$\cdot \left(\sum_{k=2}^{K} x_k(\tilde{q},\tilde{o})\overline{\mathbf{h}}_k(\tilde{q},\tilde{o}) + \mathbf{n}(\tilde{q},\tilde{o}) - x_1(\tilde{q},\tilde{o})\mathbf{e}(\tilde{q},\tilde{o})\right)^H,$$

where the summation is over a set of sub-carriers $\tilde{q} \in \mathcal{Q}$ and a set of OFDM symbols $\tilde{o} \in \mathcal{O}$ and $\widehat{(\cdot)}$ denotes estimates, respectively.

The cardinality of the sets $\mathcal{Q}, \mathcal{O}$ defined as $|\mathcal{Q}|$ and $|\mathcal{O}|$ corresponds to the number of samples $N_{\text{scm}} = |\mathcal{Q} \times \mathcal{O}|$ used to compute the sample covariance matrix. The number of used sub-carriers $|\mathcal{Q}|$ is limited in practice by a bandwidth smaller than the coherence bandwidth and by the smallest resource block allocation size, while the number of used OFDM symbols $|\mathcal{O}|$ is limited by the time duration smaller than the coherence time and by the scheduling interval. Note that the sample covariance is not stationary across the boundaries of the resource allocations and, hence, the sample covariance needs to be estimated within the

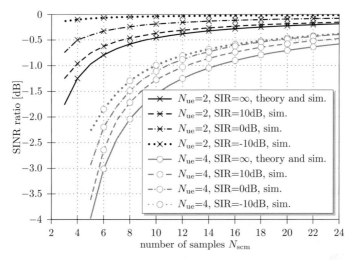

Figure 10.4 SINR ratio due to covariance matrix mismatch for varying number of samples at SNR = 0 dB, with one MPSK modulated interferer.

scheduling interval in time direction and within the resource block allocation size in frequency direction. Some latency constraints may further reduce the number of samples that can be used in estimating the spatial interference-and-noise covariance. Note that interference can also be vastly different between the time-frequency resources used for control signaling and those used for data transmission and, hence, estimation of the spatial interference-and-noise covariance also needs to be restricted to the appropriate time-frequency resources. Furthermore, the computation of the sample covariance matrix requires knowledge of some symbols x_1 either from transmitted reference symbols or from decision feedback of the desired received signal, which may also limit the number of samples that can be used for estimation. For IRC, the sample covariance matrix needs to be invertible and, hence, $N_{scm} \geq N_{ue}$. Note that there are typically only between 4 and 12 samples available for spatial interference-and-noise covariance estimation when using UE-specific RSs and without decision feedback in the LTE Release 8 downlink [3GP07b].

The set $\mathcal{Q}$ of sub-carriers is ideally centered at the sub-carrier q where the sample covariance matrix is computed. Similarly, the set $\mathcal{O}$ of OFDM symbols is ideally centered at the OFDM symbol o where the sample covariance matrix is computed. However, centering the sets $\mathcal{Q}, \mathcal{O}$ at q, o is not possible at the boundaries of the time-frequency resource allocations where the spatial interference-and-noise covariance may change. Moreover, it is desirable to use the same covariance matrix over an entire range of sub-carriers q and OFDM symbols o. The partitioning of the time-frequency resources into tiles with identical sample covariance matrix determines the main computational complexity as given by the number of sample covariance computations and matrix inversions of size $N_{ue} \times N_{ue}$ required for IRC.

Performance Loss from Covariance Estimation

For the evaluation of the performance impact of the limited number of samples available for spatial interference-and-noise covariance estimation, the channel of the desired signal is assumed to be perfectly known, i.e. $\widehat{\overline{\mathbf{h}}}_1(f,t) = \overline{\mathbf{h}}_1(f,t)$ for all f, t and, hence, $\mathbf{e}(f,t) = \mathbf{0}$. In particular, the SINR can then be evaluated as

$$\text{SINR}(\hat{\mathbf{g}}_{\text{IRC,b}}) = \frac{\overline{\mathbf{h}}_1^H E_{\mathbf{n}}\left\{\hat{\mathbf{\Phi}}_{z_1z_1}^{-1}\overline{\mathbf{h}}_1 \cdot \overline{\mathbf{h}}_1^H \hat{\mathbf{\Phi}}_{z_1z_1}^{-1}\right\}\overline{\mathbf{h}}_1}{\overline{\mathbf{h}}_1^H E_{\mathbf{n}}\left\{\hat{\mathbf{\Phi}}_{z_1z_1}^{-1}\mathbf{\Phi}_{z_1z_1}\hat{\mathbf{\Phi}}_{z_1z_1}^{-1}\right\}\overline{\mathbf{h}}_1} \tag{10.71}$$

where $E_{\mathbf{n}}\{\cdot\}$ denotes expectation with respect to noise as part of the sample covariance matrix. Note that we have dropped the explicit dependency of the sample covariance with respect to the sub-carriers q and OFDM symbols o because the sample covariance matrix can be assumed to be constant over an entire tile in time-frequency resources. For simplicity of the notation, we have also dropped the dependency of the channel for the desired UE with respect to the sub-carriers q and OFDM symbols o, which corresponds to a block-fading approach in frequency and time directions. Equation (10.71) cannot be simplified in general and the performance loss from imperfect covariance matrices needs to be evaluated by simulations. However, there exists a simple and elegant solution for the SINR loss from covariance estimation in case the sample covariance matrix is complex non-central Wishart distributed, which is the case when the interference-and-noise samples can be assumed to be Gaussian distributed with zero-mean [TC94]. This assumption holds (i) for the case of no interferers and (ii) for the case of interferers with complex Gaussian signal alphabet, i.e. when $K = 1$ or $x_k(\tilde{q}, \tilde{o})$ is Gaussian distributed in (10.70). Furthermore, the assumption of zero-mean Gaussian distributed interference-and-noise samples is good approximation for the case of multiple interferers and interferers with higher order modulation. For complex non-central Wishart distributed sample covariance matrices, the SINR ratio can then be derived [TC94] as

$$\frac{\text{SINR}(\hat{\mathbf{g}}_{\text{IRC,b}})}{\text{SINR}(\mathbf{g}_{\text{IRC}})} = \frac{N_{\text{scm}} - N_{\text{ue}} + 1}{N_{\text{scm}}} \quad \text{for} \quad N_{\text{scm}} > N_{\text{ue}} + 1. \tag{10.72}$$

Note that the SINR ratio from estimating the spatial interference-and-noise covariance matrix is independent of the SNR and SIR for zero-mean Gaussian distributed interference-and-noise samples. Equation (10.71) is evaluated by simulations for the case that the interference-and-noise samples are zero-mean Gaussian distributed without interferer, i.e. SIR $= \infty$, and for the case that there is one interferer with, e.g., MPSK modulation, such that the samples are no longer Gaussian distributed. Fig. 10.4 illustrates the theoretical and simulated SINR losses for the cases of $N_{\text{ue}} = 2$ and $N_{\text{ue}} = 4$ dependent on the number of samples N_{scm}, and for SNR $= 0$ dB. The theoretical results from (10.72) are shown as solid lines, and the simulated results are depicted with crosses for $N_{\text{ue}} = 2$ and circles for $N_{\text{ue}} = 4$, respectively. From the simulation results with one interferer, the loss observed from sample covariance estimation is upper bounded by the loss due to zero-mean Gaussian distributed interference-and-noise samples.

10.2.5 Implementation Losses from Channel and Interference Estimation Errors

In the previous two subsections, the impact of channel estimation errors and covariance estimation errors has been investigated separately. In this subsection, these estimation losses are jointly evaluated by simulations, which show the robustness of IRC with respect to receiver imperfections and provide some guidelines for the implementation losses to be expected. The instantaneous SINR resulting from combining-weight mismatch is given as

$$\text{SINR}(\hat{\mathbf{g}}_{\text{IRC},c}) = \frac{E_{n,e}\left\{\left|\hat{\mathbf{h}}_1^H \hat{\boldsymbol{\Phi}}_{z_1 z_1}^{-1} \overline{\mathbf{h}}_1\right|^2\right\}}{E_{n,e}\left\{\hat{\mathbf{h}}_1^H \hat{\boldsymbol{\Phi}}_{z_1 z_1}^{-1} \boldsymbol{\Phi}_{z_1 z_1} \hat{\boldsymbol{\Phi}}_{z_1 z_1}^{-1} \overline{\mathbf{h}}_1\right\}} \cdot E\{|x_1|^2\}. \tag{10.73}$$

and we evaluate the performance loss with the SINR ratio given by $\text{SINR}(\hat{\mathbf{g}}_{\text{IRC},c})/\text{SINR}(\hat{\mathbf{g}}_{\text{IRC}})$ by randomizing all relevant variables.

Fig. 10.5 shows the results for the SINR ratio versus SNR for one interferer with MPSK modulation and SIR $= 0$ dB, $N_{\text{ue}} = 2, 4$ and processing gain $G = 6$ dB and 15 dB for the channel scenario outlined in Section 10.2.2. The number of samples for covariance estimation is chosen as $N_{\text{scm}} = 6$ and $N_{\text{scm}} = 18$ to cover a range typical for LTE as well as a scenario with reduced losses from covariance mismatch. The combined losses from channel estimation and spatial interference-and-noise covariance estimation are smaller than the sum of the individual losses. From comparing Fig. 10.5 with Fig. 10.4, it can be seen that at low SNR the channel estimation errors even compensate for some covariance matrix mismatch such that the combined losses become smaller than the losses from covariance matrix mismatch alone. The losses are not negligible but rather small as compared to the gains from IRC versus MRC outlined in Section 10.2.2. In particular, in the low SIR regime, the benefits of IRC versus MRC outweigh the implementation losses by far. Note that the implementation losses are substantially increased with a larger number of antennas N_{ue} which may also need to be considered in overall performance and throughput evaluations for MIMO systems.

10.2.6 Summary

This section has focused on the link-level assessment of IRC in the presence of estimation errors due to low-complexity implementation. First, SINR expressions for IRC and MRC were derived, showing that the MMSE approach yields exactly the same SINR value as the studied IRC method. A basic performance analysis in terms of achievable SINR after combining was conducted comparing IRC and MRC when perfect knowledge on channel and interference characteristics is

Efficient and Robust Algorithm Implementation

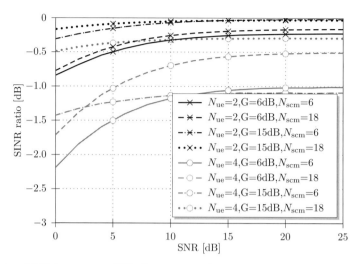

Figure 10.5 SINR loss versus SNR for one interferer with MPSK modulation and SIR $= 0$ dB, $N_{ue} = 2, 4$ and processing gain $G = 6$ dB and 15 dB.

assumed. In chosen interference-limited scenarios with one dominating interferer, substantial gains for IRC on link-level were observed.

In order to assess one effect of low-complexity implementation, the SINR loss due to channel estimation errors for the case of IRC was derived, indicating a simple dependency on channel estimation processing gain and the number of terminal receive antennas. A numerical study showed that, for $N_{ue} = 2$ receive antennas and processing gains of 6 dB and higher, the loss due to imperfect channel estimation is always below 0.5 dB even at low SINR operating points such as 4 dB. So, general statements about the superiority of IRC versus MRC still holds if imperfect channel estimation is taken into account.

An additional effect of low-complexity implementation is the erroneous estimation of the interference-and-noise covariance. Again, the SINR loss for the case of IRC was derived as a function of complexity, where the latter is expressed in terms of number of samples used for estimation. Depending on the operating SINR point, 6 to 10 covariance samples are required to keep the resulting SINR loss below 0.5 dB when $N_{ue} = 2$ receive antennas are used.

Finally, the section was wrapped up with a joint assessment of errors from channel and interference covariance estimation. The general finding is that the combined loss for the different estimation aspects is slightly smaller than the sum of the individual ones. An important overall conclusion is that the losses due to estimation errors become significantly larger when $N_{ue} = 4$ terminal antennas are considered. Hence, when designing MIMO systems, not only the cost of many antennas at the UE side matters, but also performance degradation due to estimation errors becomes more and more evident.

11 Scheduling, Signaling and Adaptive Usage of CoMP

In this chapter, we address how CoMP can be applied selectively and adaptively to well-chosen sets of terminals in a mobile communications system. While Section 11.1 focuses on scheduling approaches, where a central scheduling unit performs multi-cell resource allocation in the context of non-cooperative or joint transmission in a cellular downlink, Section 11.2 looks into radio link control and signalling aspects connected to establishing CoMP on-demand. Finally, Section 11.3 ventures into the field of ad-hoc CoMP, where cooperation is established flexibly after uplink transmission has already taken place.

11.1 Centralized Scheduling for CoMP

Tarcisio Maciel, Ricardo B. dos Santos and Anja Klein

In this section, we discuss centralized multi-cell scheduling for a system using either non-cooperative transmission or joint transmission in the downlink. After Subsection 11.1.1 motivates the topic, Subsection 11.1.2 presents the studied scenario and its main models. Subsection 11.1.3 introduces some relevant scheduling problems, where the aim is to maximize system throughput. Subsection 11.1.4 analyzes the problems introduced earlier through system level simulations. Finally, Subsection 11.1.5 adds some final remarks on the problem of centralized scheduling and its extension to uplink scenarios.

11.1.1 Introduction

In previous chapters, we have typically observed transmissions between multiple base stations (BSs) and user equipments (UEs) on a single orthogonal frequency division multiplex (OFDM) sub-carrier, assuming that the assignment of system resources to the communicating entities has already taken place. While the question of which BSs should in principal be clustered, and hence enabled to cooperate, was already addressed in Chapter 7, we now want to look into the question of how UEs can be assigned to system resources efficiently, such that the performance under a particular transmission scheme is maximized. More specifically, we will investigate

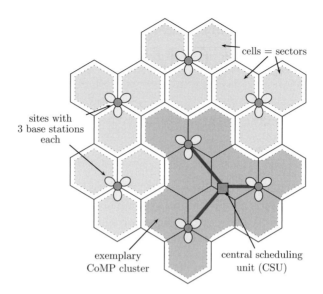

Figure 11.1 A CoMP system setup with an exemplary CoMP cluster.

- which UEs should simultaneously use the same physical resource block (PRB) in different cells in the case of *conventional, non-cooperative transmission*, or
- which UEs can be efficiently served on the same resources if *downlink joint transmission (JT)* (see Section 6.3) is used.

In this respect, scheduling may explore the degrees of freedom of choosing tuples of UEs whose links will be lightly affected by the mutual co-channel interference in the conventional, non-cooperative transmission case, or who can be served efficiently on the same resources in the JT case. In the following, some heuristic algorithms are described, and their results will show that, by making use of the information fed back by the UEs and made available to a central scheduling unit (CSU) through backhaul links, centralized scheduling provides considerable gains compared to conventional, individual scheduling by the BSs.

11.1.2 System Model

In this subsection, the system modeling considered for the study of scheduling algorithms is described. The downlink of a CoMP-enabled system with M BSs is considered. These BSs are grouped into C clusters, where the sets of BSs included are denoted as $\mathcal{M}_c$, $c = 1, \ldots, C$. All BSs within one cluster have a backhaul link to a dedicated CSU. A setup where one exemplary CoMP cluster is highlighted is shown in Fig. 11.1.

In the sequel, we shift our focus to this one cluster c, and assume it has a set of UEs $\mathcal{U}_c$ that are served by the BSs in $\mathcal{M}_c$ and may be assigned to R available PRBs, which are indicated by $r = 1, 2, \ldots, R$. For simplicity, in this section only single-antenna BSs and UEs are considered, hence $N_{\text{bs}} = N_{\text{ue}} = 1$ according to the notation used in previous chapters. Let us denote as $\mathcal{K}_r$ the set of all UEs in the system that are assigned to resource r, and $\mathcal{K}_{c,r} \subseteq \mathcal{K}_r$ as the subset of these UEs that are served by cluster c. Vector $\mathbf{h}_k^c(r) \in \mathbb{C}^{[|\mathcal{M}_c| \times 1]}$ denotes the channel coefficients connected to resource r, representing the links from UE k to each BS m belonging to cluster c. Similar as derived before in Section 5.1, the downlink signal-to-interference-and-noise ratio (SINR) experienced by a UE k belonging to cluster c and assigned to resource r can be stated as

$$\gamma_k(r) = \frac{\left|(\mathbf{h}_k^c(r))^H \mathbf{w}_k^c(r)\right|^2}{\underbrace{\sum_{j \in \{\mathcal{K}_{c,r} \setminus k\}} \left|(\mathbf{h}_k^c(r))^H \mathbf{w}_j^c(r)\right|^2}_{\text{Intra-cluster interference}} + \underbrace{\sum_{j \in \{\mathcal{K}_r \setminus \mathcal{K}_{c,r}\}} \left|(\mathbf{h}_k(r))^H \mathbf{w}_j(r)\right|^2}_{\text{Inter-cluster interference}} + \sigma^2}, \quad (11.1)$$

where $\mathbf{w}_k(r)$ is the precoding vector employed at the BS-side to serve UE k on resource r, of which $\mathbf{w}_k^c(r)$ is the sub-part connected to the transmission originating from the BSs in cluster c. Note that in the case of non-cooperative transmission, the transmission to each UE can only originate from one BS, i.e. each precoding vector $\mathbf{w}_k(r)$ is zero in all elements except one, whereas $\mathbf{w}_k(r)$ can be non-zero for all elements connected to one and the same cluster for JT.

Considering (11.1) and the setup in Fig. 11.1, one sees that co-channel interference comes from BSs belonging to the same cluster, denoted as *intra*-cluster interference, and from BSs belonging to other clusters, denoted as *inter*-cluster interference. Intra-cluster interference can be estimated well or even predicted by the CSU based on channel state information (CSI) fed back through the backhaul, thus enabling the use of intelligent resource reuse through centralized scheduling. On the other hand, inter-cluster interference might only be estimated and cannot be directly controlled.

In the following, models for adaptive modulation and block error rate (BLER) assessment are presented. Both are based on SINR values estimated using the CSI available at each CSU. Note that the modulation choice depends on the SINR, which is an "outcome" of the resource assignment. Consequently, adaptive modulation becomes strongly dependent on the resource assignment.

Due to the resource reuse across the system, link quality will vary according to resource assignments done by other clusters, thus making inter-cluster interference harder to estimate. As a consequence, packet losses might occur due to imperfect selection of modulation schemes caused by considering wrong inter-cluster interference values. In order to capture packet errors, the following model is employed. According to [Cal04], considering uncoded QAM with q bits/symbol and assuming that co-channel interference is Gaussian-distributed with its power added directly to the additive white Gaussian noise (AWGN) power, the bit error rate (BER) for a link with SINR γ can be approximated as

$$\text{BER}(\gamma) \approx 0.2 \exp\left(-\frac{1.6\gamma}{2^q - 1}\right). \quad (11.2)$$

Then, assuming packets of L symbols being transmitted on a single PRB during one transmit time interval (TTI), the BLER can be written as

$$\text{BLER}(\gamma) = 1 - (1 - \text{BER}(\gamma))^{L \cdot q}, \quad (11.3)$$

which is used to model whether a given transmission has been successful. Here, adaptive modulation takes into account 2-, 4-, 16- and 64-QAM as available modulation schemes. Since the focus here is on enhancing total throughput, adaptive modulation selects the QAM of order $Q = 2^q$ yielding the highest average throughput, i.e.,

$$Q^\star = \arg\max_{Q \in 2^{\{1,2,4,6\}}} (1 - \text{BLER}(\gamma)) \cdot L \cdot q. \quad (11.4)$$

11.1.3 Centralized Scheduling Problems

In this subsection, some relevant scheduling problems are described. In general, scheduling UEs in a CoMP system is a very complex optimization problem involving multiple dimensions – *BSs, UEs, PRBs, transmit/receive antennas, transmit powers, CoMP schemes, etc.* – and incorporates several sub-problems:

- the **PRB assignment problem**, which selects PRBs to be allocated to each UE and also defines the reuse in centralized scheduling.
- the **CoMP decision problem**, i.e. to determine whether or not and which CoMP scheme is to be applied.
- the **power allocation problem**, which corresponds to distributing the available power among UEs and PRBs efficiently.
- the **precoding problem**, which corresponds to determining precoding vectors as to spatially multiplex signals intended to different UEs.
- the **link adaptation problem**, e.g. the problem of determining a suitable modulation and coding scheme (MCS) for each transmission.

These subproblems, especially the first four, cannot be separated without incurring some loss of optimality. If the channels of the UEs sharing a PRB are highly spatially uncorrelated, these UEs can be efficiently separated in space using precoding. However, if their channels are highly correlated, these UEs cannot be spatially separated efficiently and strongly interfere with each other. Further, the interference between UEs sharing a given resource through spatial division multiple access (SDMA) is a function of the power distribution among the UEs, as well as the power distribution among PRBs. Considering a certain amount of power available for a PRB, allocating more power to one UE enhances the quality of its signal, e.g., in terms of SINR, but also reduces the SINR of the other UEs using the PRB. Analogously, allocating more power to a certain

PRB enhances the SINR of the UEs sharing this PRB, but reduces the SINR of the UEs allocated on the other PRBs. Finally, the spatial compatibility among UEs is PRB-dependent, i.e., UEs that are spatially compatible on a given PRB might be incompatible on another [LZ06, MK10].

These relationships illustrate the strong interdependency among the above subproblems, and the challenge of jointly solving them usually leads to computationally prohibitive solutions. All previously referred aspects affect signal and/or interference levels and consequently the system performance. Moreover, even for some subproblems, optimum solutions can already be very complex. Thus, sub-optimal scheduling solutions are often preferred [FDH07, MK10].

Different objectives may be pursued by the scheduler (spectral efficiency, fairness, quality of service (QoS) requirements, etc.) and each objective results in its own problem which may not have a known optimal solution. While conventional schedulers share the same scheduling objectives, the additional information available to a centralized scheduler allows this to consider the impact of the resource allocation at one BS on the other BSs within the same cluster. Moreover, the additional control of joint scheduling introduces other degrees of freedom which are exploited, e.g., by adapting the set of simultaneously transmitting BSs.

In this section, we focus on the *resource allocation* problem comprising PRB assignment and precoding sub-problems, and we target at *maximizing system throughput*. If we assume a fixed and equal power distribution among PRBs, the problem may be decoupled and solved separately for each PRB. This approach is, in fact, used in the presentation of all algorithms in this section. We further consider a fixed power control explained later and also fix the used CoMP scheme and choice of precoders to the following two options:

- **Conv. transmission (CT)**: PRBs are reused by multiple BS-UE links within a cluster, but each active UE is served exclusively by only one BS.
- **Joint transmission (JT)**: PRBs are reused by multiple BS-UE links within a cluster, with all BSs sending linearly and jointly precoded data to all UEs and, consequently, based on a zero-forcing (ZF) filter.

The remaining resource allocation problem is hence: For the case of conventional transmission (CT), the CSU needs to determine which BS-UE links can simultaneously use a same PRB. Assuming that the CSU has CSI on all links within a cluster, it is able to estimate the impact of the intra-cluster interference induced by the PRB reuse, and can dynamically determine which PRBs should be assigned to which UEs served by which BSs. For the JT problem, the CSU needs to find sets of UEs with good compound channel properties, which can be efficiently served with spatial multiplexing on the same PRB. Maximizing system throughput assuming CT or JT becomes a combinatorial problem whose optimal solution may be computationally complex to find. Therefore, only sub-optimal, but rather efficient solutions are considered herein.

Table 11.1. Greedy scheduling algorithm for conventional transmission (CT).

1. Find the BS-UE link $\{k^*, m^*\} = \arg \max_{k \in \mathcal{U}_c,\, m \in \mathcal{M}_c} |h_k^m(r)|^2$ with highest gain.
2. Assign PRB r to the link $\{m^*, k^*\}$ by
   - defining set of scheduled UEs of CoMP cluster c on PRB r as $\mathcal{K}_{c,r}^* = \{k^*\}$ and
   - defining set of scheduled BSs of CoMP cluster c on PRB r as $\mathcal{M}_{c,r}^* = \{m^*\}$.
3. Estimate the total rate $R(\mathcal{K}_{c,r}^*, \mathcal{M}_{c,r}^*)$ using (11.1) to (11.4).
4. Find $\{m^*, k^*\} = \arg \max_{k \in \mathcal{U}_c \setminus \mathcal{K}_{c,r}^*,\, m \in \mathcal{M}_c \setminus \mathcal{M}_{c,r}^*} R(\mathcal{K}_{c,r}^* \cup \{k\}, \mathcal{M}_{c,r}^* \cup \{m\})$.
5. If $R(\mathcal{K}_{c,r}^* \cup \{k^*\}, \mathcal{M}_{c,r}^* \cup \{m^*\}) \geq R(\mathcal{K}_{c,r}^*, \mathcal{M}_{c,r}^*)$, set $\mathcal{K}_{c,r}^* = \mathcal{K}_{c,r}^* \cup \{k^*\}$, $\mathcal{M}_{c,r}^* = \mathcal{M}_{c,r}^* \cup \{m^*\}$ and go to the previous step, otherwise finish.

Heuristic Scheduling Algorithm for Conventional Transmission

In this subsection, a heuristic scheduling algorithm for the CT case is described. This introduces flexibility to decide, for each PRB, whether all BSs associated to the CSU will be used or only some of them in order to reduce co-channel interference and transmission failure probabilities. As mentioned before, equal power assignment among different PRBs is commonly assumed for simplicity. Although not optimal, it leads to only marginal performance degradation compared to optimum power allocation if adaptive modulation is employed [JL03, ZL04], and allows to consider resource allocation individually for each PRB whenever throughput maximization is being pursued. Some particular cases of joint precoding, power allocation, and scheduling are addressed in [SB04, TC04, SBO06, CC07, MSLT07].

Greedy scheduling algorithms assigning resources to the BS-UE links with highest gain on each PRB offer a sub-optimal solution for throughput-oriented scheduling in CoMP systems. Indeed, in the absence of intra-cluster interference, a greedy algorithm would be even optimal for orthogonal frequency division multiple access (OFDMA)-based systems. In any case, it is assumed that the CSU has knowledge about the gains of all BS-UE links, so that it can efficiently estimate the impact of intra-cluster interference on achievable rates.

Greedy algorithms usually solve a problem in stages where at each stage a decision for the "best solution" is made, assuming that decisions taken at previous stages were "optimal". In other words, greedy algorithms make a locally optimal choice at each stage and hope that these will lead to the global optimum [CLRS01]. Clearly, this optimum is not necessarily reached in this way.

Nevertheless, for the problem of maximizing the system throughput, a greedy algorithm can be employed to schedule, within each CoMP cluster, a set of UEs with high channel gains by estimating their rates after each allocation. In other words, it starts by scheduling the BS-UE link with highest channel gain within the whole cluster. Then, based on the resulting SINRs, it calculates the potential cluster throughput of scheduling each available BS-UE link within the same cluster together with the previously scheduled link. Finally, the scheduled

Table 11.2. Greedy scheduling algorithm for joint transmission (JT).

1.	Find the UE $\{k^\star\} = \arg\max_{k \in \mathcal{U}_c} \|\mathbf{h}_k^c(r)\|$ with highest channel vector norm.
2.	Assign initially the PRB r to the UE $\{k^\star\}$ by defining the set $\mathcal{K}'_{c,b}$ of scheduled UEs of CoMP cluster c on PRB r as $\mathcal{K}'_{c,r} = \{k^\star\}$.
3.	While $K' \leq K^\star$ a. Find $k^\star = \arg\max_{k \in \mathcal{U}_c \setminus \mathcal{K}'_{c,r}} \phi(\mathcal{K}'_{c,r} \cup \{k\})$. b. Set $\mathcal{K}'_{c,r} = \mathcal{K}'_{c,r} \cup \{k^\star\}$.
4.	Set $\mathcal{K}^\star_{c,r} = \mathcal{K}'_{c,r}$.

link is the one which leads to the highest throughput and, in case of ties, the link with highest channel gain is chosen. This procedure continues adding new links as long as the cluster throughput increases. Otherwise, it finishes and goes to the next PRB. The greedy algorithm for each PRB r in the case of conventional, non-cooperative transmission can be stated as shown in Table 11.1.

Heuristic Scheduling Algorithm for Joint Transmission

In this subsection, a heuristic, sub-optimal scheduling algorithm for the case of JT is described, where the discussion is restricted to a single CoMP cluster c and PRB r, on which a group $\mathcal{K}^\star_{c,r}$ is built. For simplicity of notation, the indices c and r are omitted in the sequel. For JT, the data symbols x_k intended for all scheduled UEs k are made available to all BSs of the CoMP cluster and are precoded using their associated precoding vectors $\mathbf{w}_k$ before transmission. For the spatial signal separation, linear precoding can be employed [PNG03, JJT+09], while the efficiency of such separation strongly depends on the characteristics of the channel vector of the scheduled UEs. Therefore, a JT scheduler that only allows sharing PRBs among UEs with uncorrelated channels is usually employed [MK10]. Thus, the problem to be solved is choosing a group of $K^\star \leq M$ UEs that are spatially compatible, i.e., that can efficiently share the same PRB. JT schedulers for this problem are usually heuristic and composed by two elements: a spatial compatibility metric and a user selection algorithm [MK10].

In the following, the spatial compatibility metric is discussed. It is employed by the CSU to measure the spatial compatibility among UEs. In general, spatial compatibility metrics are functions of the CSI (available at the CSU through the backhaul) that try to map the characteristics of the spatial channels of the UEs to a scalar value quantifying how efficiently these UEs can be separated in space [MK10]. Such groups of UEs are termed a *compatibility group*.

When ZF precoding is employed, it has been shown that the sum of channel gains with null-space successive projections represents an effective measure of spatial compatibility, especially when aiming at maximizing the system throughput [MK10, TUBN06, YG05]. For a compatibility group $\mathcal{K}' = \{1, 2, \ldots, K'\}$, the use of successive null-space projections imposes that the channel vector

$\mathbf{h}_{k'}$ of UE $k' \in \mathcal{K}'$ be projected onto the null-space of the channels of all UEs $k \in \mathcal{K}', k = 1, 2, \ldots, k' - 1$ [TUBN06, YG05], i.e., a vector space orthogonal to channels of all UEs $k = 1, 2, \ldots, k' - 1$.

Since signals conveyed through orthogonal channels do not interfere with each other, the more orthogonal to $\mathbf{h}_k$ the channel vector $\mathbf{h}_{k'}$ of UE k' is, the less its squared norm (i.e. the channel gain) will be affected by the projection and the more spatially compatible to the UEs $k = 1, 2, \ldots, k' - 1$ the UE k' will be.

Denoting as $\mathbf{T}_{k'} \in \mathbb{C}^{[M \times M]}$ the matrix that projects the channel vector $\mathbf{h}_{k'}$ of UE k' onto the null-space of the channels of UEs $1, 2, \ldots, k' - 1$ [TUBN06], one has

$$\mathbf{T}_{k'} = \begin{cases} \mathbf{I}_{[M]}, & \text{for } k' = 1, \\ \mathbf{T}_{k'-1} - \dfrac{\mathbf{T}_{k'-1}^H \mathbf{h}_{k'-1}^H \mathbf{h}_{k'-1} \mathbf{T}_{k'-1}}{\|\mathbf{h}_{k'-1}\mathbf{T}_{k'-1}\|^2}, & \text{for } k' = 2, \ldots, K', \end{cases} \quad (11.5)$$

where for $k' = 1$ no projections are needed, i.e., $\mathbf{T}_1 = \mathbf{I}_{[M]}$. Then, using (11.5), the sum of channel gains with null-space successive projections $\phi(\mathcal{K}')$ considered in this subsection is written as

$$\phi(\mathcal{K}') = \sum_{k'=1}^{K'} \|\mathbf{h}_{k'}\mathbf{T}_{k'}\|^2. \quad (11.6)$$

Note that according to (11.6), the higher the gain of the channels of the UEs belonging to $\mathcal{K}'$ is, and the more orthogonal to each other they are, the larger $\phi(\mathcal{K}')$ becomes. Altogether, these high orthogonality degrees and high channel gains result in an increased system throughput, rendering $\phi(\mathcal{K}')$ a suitable spatial compatibility metric, especially for algorithms oriented towards throughput maximization, at it is the case herein.

In the following, the user selection algorithm considered in this subsection is discussed. Its task is to arrange the UEs of the CoMP cell in a compatibility group by using the spatial compatibility metric. Often, the optimum compatibility group can only be found through an exhaustive search over all possible groups, so that sub-optimal, but rather efficient user selection algorithms are desired. One such algorithm is the best fit algorithm [STKL01, DS04, Cal04], which is also a greedy algorithm.

Starting from a compatibility group containing only an initial UE, the best fit algorithm extends the group by sequentially admitting the most spatially compatible UE with respect to those UEs already admitted to the compatibility group. Let $\mathcal{K}' = \{k'\}$ be the initial compatibility group containing only the UE k', which is chosen as the UE with the highest channel norm because this leads to the highest throughput for single-user transmission. Let K' be the size of the compatibility group $\mathcal{K}'$. Then, the best fit algorithm computes the spatial compatibility metric $\phi(\mathcal{K}' \cup \{k\})$ for each UE $k \in \mathcal{K}_c \setminus \mathcal{K}'$. Then, the UE $k^\star$ which leads to the highest value for the spatial compatibility metric $\phi(\cdot)$ is admitted to the group $\mathcal{K}'$. After that, the same procedure is repeated with the remaining

UEs and an additional UE is admitted to the group, and so on until the group size K' reaches the target compatibility group size $K^\star$. When using ZF precoding, the choice of $K^\star$ is fundamentally limited by the number of transmitting antennas, which is the maximum number of UEs that can be multiplexed in this case. However, additional restrictions such as a maximum number of scheduled UEs per PRB may be applied to limit the amount of control information to be exchanged at each TTI.

Combining the spatial compatibility metric described by (11.6) and the best fit algorithm, an overall scheduling algorithm for JT can be derived, which is presented in Table 11.2. Similarly to the scheduling algorithm for the CT case, PRBs are processed sequentially. It should be noted that the scheduling algorithm does not compute precoding vectors $\mathbf{w}_k$ for any UE, thus avoiding a considerable amount of computations [MK10]. Additionally, it also does not involve power allocation. This allows the algorithm to be more easily combined with different precoding and power allocation schemes. However, precoding and power allocation should also be oriented to the same objective of the algorithm, namely throughput maximization. Beyond the single-antenna case, the algorithm can be straightforwardly adapted to cases considering multiple antennas at the communicating nodes by extending the channel vectors $\mathbf{h}_k(r)$ accordingly [TUBN06]. The effective channel (including the effect of transmit precoding) might be estimated at the UEs using pilot symbols, as discussed in Section 9.1. Alternatively, fixed receive filters at the UEs might be considered at the transmitter using, e.g., a receiver-oriented design of the BSs' precoding vectors [MBQ04].

11.1.4 Analyses and Results

The scheduling algorithms described in Section 11.1.3 are now analyzed using system level simulations. The performed simulations employ the models described in Section 11.1.2. $C = 7$ CoMP clusters composed of 21 BSs each and organized in 3-sectored sites with a inter-site distance of 1 km are considered as the system setup. A CSU controls the operation of each CoMP cluster. A wrap-around model is used to avoid border effects on interference among CoMP clusters, as described in detail in Section 14.1.

The CoMP system considers a carrier frequency f_c of 2 GHz and $B = 15$ PRBs, each composed of 12 sub-carriers spaced by $\Delta f = 15$ kHz [3GP07a]. Each sub-carrier transmits 14 symbols per TTI, which has a duration of 1 ms. There is no power control, i.e., we consider equal power allocation on all resources. Moreover, perfect CSI on the channels of all links within a CoMP cluster is assumed to be available at the CSU. It is worth mentioning that, if only compressed CSI is available at the CSI, e.g., due to backhaul capacity constraints or limited feedback through the air interface, the performance of centralized scheduling algorithms might be reduced considerably, as discussed in Section 5.2.

Table 11.3. Simulation parameters.

Parameter	Value
Number of CoMP clusters	7
Number of sites per CoMP cluster	7 (i.e. 21 BSs and cells per cluster)
Inter-site distance	1 km
Minimum BS-UE distance	50 m
Carrier frequency	2 GHz
Sub-carrier spacing	15 kHz
Number of PRBs	15 (with 12 sub-carriers each)
Number of symbols per TTI	14
Effective TTI duration	1 ms
Pathloss model	$35.3 + 37.6 \log_{10}(d)$ in dB
Shadowing standard deviation	8 dB
UE speed	3 km/h
Channel power-delay profile	TU [3GP08b]
Spatial precoding	ZF
Average signal-to-noise ratio (SNR) at cell-edge	6 dB
Snapshot duration	1 s
Number of UEs per BS	3 to 12
Target compatibility group size	21

The path loss and shadowing are modeled according to [PDF+08] alongside various other simulation parameters in Table 11.3, and BS antenna patterns are modeled according to [3GP07a], i.e.,

$$G^{(a)}(\theta_{k,m,c}) = -\min\left\{12\left(\frac{18}{7\pi}\theta_{k,m,c}\right)^2, 20\right\} \text{ [dB]}. \quad (11.7)$$

Fast fading considers an average UE speed of 3 km/h and employs the typical urban power-delay profile to model frequency selectivity [3GP08b]. When considering JT, linear ZF precoding is adopted as precoding technique due to its simplicity and its ability to suppress intra-cell interference [MK10, YG05]. The precoding vectors $\mathbf{w}_k^c(r)$ computed for the scheduled UEs according to the ZF criterion are first scaled to become unit-norm vectors. After that, because each BS has a limited transmit power available per PRB, all precoding vectors within a CoMP cluster are scaled so that no BS spends more than this total transmit power per PRB. This is easily accomplished as follows. First, the precoding vectors $\mathbf{w}_k^c(r)$ for all UEs $k \in \mathcal{K}_{c,r}$ scheduled to receive on PRB r within the CoMP cluster c are organized in a precoding matrix $\mathbf{W}^c(r)$. Then, the total power spent by a BS m corresponds to the sum of the squared absolute values of the weights it applies to each transmit signal, i.e., $\text{tr}\{\mathbf{W}^m(r)(\mathbf{W}^m(r))^H\}$. Since the power ratio among elements of each column may not be changed in order to preserve the properties of ZF, the precoding matrix $\mathbf{W}^c(r)$ is simply scaled down so that power constraints are fulfilled. Note that this means that the transmit power of some BSs might not be fully used, which is sub-optimal.

For each BS, a number of UEs is uniformly distributed over the coverage area of the cell. The transmit power of the BSs is set as to grant an SNR of at least 6 dB at the cell-edge considering the effects of pathloss, antenna pattern

11.1 Centralized Scheduling for CoMP

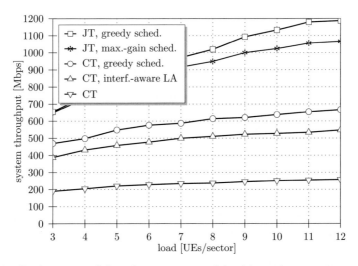

Figure 11.2 Performance of the schemes compared in this section.

and shadowing (95% of reliability). It is assumed that BSs always have data to transmit to the UEs, which make use of a best-effort service. Several snapshots are considered in each simulation. During each snapshot, large-scale fading is assumed constant while small-scale fading variation is modeled using Jakes' model [Jak74]. A sufficient number of snapshots is simulated in order to get reliable statistics about the system throughput.

Fig. 11.2 shows the total throughput of the system averaged over all simulated snapshots as a function of the system load in UEs per cell. The following transmission, scheduling and link adaptation schemes are compared:

- **Conventional transmission, scheduling and link adaptation**. This corresponds to a conventional cellular system in which there is no coordination or communication among sites. Each BS uses a local, greedy scheduler. For a given PRB, it schedules the UE with the highest channel gain at each BS. Thus, a full reuse of frequency resources is observed. In this scenario, link adaptation is based on the interference perceived during the last transmission to a UE on a PRB.
- **Conventional transmission and scheduling, but interference-aware link adaptation**. Here, the same local schedulers are used, but link adaptation is based on the assumption that each BS knows the scheduling decisions of the other BSs and can precisely predict intra-cluster interference, as proposed in Section 5.2.2 for the uplink.
- **Conventional transmission, centralized scheduling and interference-aware link adaptation**, using the greedy scheduler for CT proposed before.
- **Joint transmission, conventional scheduling and interference-aware link adaptation**, where the UEs with the highest channel gains in the cluster

are scheduled for JT, and two UEs may not be scheduled to the same BS on the same resource.
- **Joint transmission, centralized scheduling and interference-aware link adaptation**, using the greedy scheduler for JT proposed before.

Note that for all schemes, *inter*-cluster interference is estimated as interference perceived during the last transmission to the UE on the PRB.

Regarding non-cooperative transmission, we can see that a large performance gain of 100 % to 120 % can already be achieved if knowledge on intra-cluster interference is used for link adaptation, as also observed in Section 5.2 for the uplink. Performance can further be increased by around 20% if the proposed centralized scheduling algorithm for CT is used. Joint transmission in general performs significantly better than non-cooperative transmission, as also observed in Sections 6.3 and 6.4, but we can see that a throughput improvement of about 10% can additionally be obtained if the greedy, centralized scheduling scheme proposed in this section is used, as opposed to classical scheduling at each BS.

11.1.5 Summary

In this section, multi-cell centralized scheduling algorithms oriented towards the maximization of system throughput have been presented, for a downlink system based on non-cooperative or multi-cell joint transmission. While both algorithms are heuristic and have low complexity, the results presented in this section have illustrated the huge potential of intelligent scheduling to provide high data rates in CoMP systems.

This section has concentrated on the downlink. However, similar relative performances are expected in the uplink if a sufficient amount of CSI is available at each cluster. The studies considered here have assumed relatively idealized conditions. Real systems have to deal with further challenges such as backhaul constraints, signaling overhead, limited or outdated CSI and synchronization issues which degrade system performance, as discussed in various other parts of the book.

11.2 Decentralized Radio Link Control and Inter-BS Signaling

Christian Hoymann and Laetitia Falconetti

This section discusses radio link control and signaling aspects of practical implementations of CoMP schemes in cellular systems at the example of 3GPP LTE. The radio link towards a user equipment (UE) is controlled by its serving base station (BS), and this section discusses the potential modifications of existing control loops when applying CoMP.

The section focusses on decentralized radio link control[1], where each BS typically controls the UEs of its cell, though some aspects also apply to centralized control, where a central node controls all UEs of one or more CoMP clusters. In general, uplink (UL) and downlink (DL) transmissions need radio link control, but some control loops only refer to UL transmissions, e.g., UL power control, UL timing advance, etc. In 3GPP LTE, frequency division duplex (FDD) and time division duplex (TDD) use the same radio link control and they face basically the same problems; only the radio link measurements may differ if the channel is reciprocal. *Signaling* refers to the direct communication of cooperating BSs when using decentralized radio link control and to the indirect communication via a central node when centralized radio link control is applied.

11.2.1 Resource Allocation

Resource allocation, as considered from a physical layer point of view in Section 11.1, is the process where BSs allocate radio resources in time and frequency to certain UEs. The BSs need to answer the question which and how many resources to allocate. The decision on which resources to allocate is based on information about the (predicted) radio link quality, e.g., the current radio channel state including slow and fast fading as well as the interference situation. The decision on how many resources to allocate is based on information about the traffic demand, e.g., buffer fill levels and quality of service (QoS) requirements. All information needs to be available at the serving BS performing resource allocation.

In cooperative transmission and reception schemes as introduced in Chapters 5 and 6, the *resulting* radio link quality changes compared to a non-cooperative transmission, and the resource allocation should be based on the quality of the cooperative radio link between the UE and the BS antennas in multiple cells, see also Chapter 9. Inter-BS communication could be used to exchange channel information between cooperating BSs so that they know the quality of the cooperative radio link. However, such an information exchange between BSs might not be possible, e.g., due to the lack of an appropriate interface (limited capacity and/or long delays). In that case, the serving BS could try to estimate the improvement of the radio link quality due to cooperation, e.g., an offset could be added to the estimated signal-to-interference-and-noise ratio (SINR).

Having estimated the quality of the cooperative radio link, each BS can allocate resources independently of the other BSs, or a set of BSs can do some form of joint resource allocation in order to better take interference into account, as proposed in Section 5.2. In the former case, the process does not differ from regular

---

[1] Note that the distinction of centralized and decentralized radio link control does *not* correspond to the classification of centralized and decentralized CoMP schemes, which refers to the place where decoding (uplink) and encoding (downlink) are performed, see Section 4.

resource allocation, and strongly interfering transmissions in neighbor cells could be allocated to the same resource. In the latter case, a hierarchical scheduling can be applied, which relies on detailed knowledge of co-channel interference. For instance, a first cell (or group of cells) starts scheduling its own UEs. Then the schedule, i.e., the allocation of radio resources to UEs probably extended by information on transmit power, precoding vectors etc., is forwarded to a second cell, which in turn schedules its own UEs considering the known interference caused by the first cell. The schedules of the first and second cell are forwarded to a third cell and so on. By doing so, strongly interfering transmissions in close-by cells could be allocated onto different resources. As a drawback, hierarchical scheduling requires additional signaling between cooperating BSs and increases scheduling delay.

The amount of allocated resources depends on the traffic demand. If the demand exceeds available capacity, some form of prioritization needs to be performed. This is particularly important if QoS requirements have to be guaranteed. The traffic demand in DL can be deducted from the buffer fill levels at the BS. The traffic demand in UL can be acquired from buffer status reports or scheduling requests sent by the UEs. With traffic-dependent QoS, knowledge about the traffic demand is required per service-class. A common differentiation is to separate signalling-, real-time-, and best-effort traffic. Each serving BS is aware of the traffic demand for all attached UEs. Coordinated scheduling schemes, which would need to take the traffic demand into account, require access to the traffic demand per service-class per UE and per cooperating BS, which would require additional signalling.

Having acquired knowledge on the traffic demand, the same QoS-aware resource allocation schemes as in non-CoMP systems, such as proportional fair or maximum rate scheduling, can be applied in CoMP-enabled systems.

11.2.2 Link Adaptation

The selection of modulation and coding schemes (MCSs) is carried out by the serving BS. Based on the estimated SINR, the BS selects the MCS that maximizes the user throughput. Since CoMP can increase the SINR perceived at the receiver, link adaptation in CoMP-enabled networks should not be based on the SINR of the BS-UE link, but on the increased SINR after cooperation. Thereby, the BS can select a more aggressive MCS resulting in a higher throughput.

One way to estimate the channel quality after cooperation is to measure the radio links involved in the cooperation and to gather and combine the measurements at the serving BS, see also Section 9.1 and details provided later in Section 11.2.3. Besides channel quality, the SINR is also determined by interference, which can either be estimated from previous transmission attempts, or be more accurately predicted if cooperating BSs exchange their schedules prior to link adaptation (see Section 5.2). Exchanging schedules between BSs of course requires additional signaling and increases delay.

Alternatively, the serving BS could estimate the SINR increase due to cooperation without additional inter-BS or UE to BS signaling. For instance, mobility measurements, which are anyway reported by a UE, give an indication on the pathloss to candidate BSs. Such a technique was for instance described in Section 7.2. For CoMP transmissions, which last over several hybrid automatic repeat request (HARQ) round-trip times (RTTs), the number of HARQ re-transmissions that indicates the actual block error rate (BLER) can be considered when setting the MCS. During such a transmission period, the MCS could be adapted to better meet the BLER target, which maximizes throughput. If the MCS selection is not adapted to the increased SINR obtained with cooperation, CoMP only reduces the BLER leading to fewer re-transmission requests. Although this slightly reduces the packet delay, it is desirable to operate the HARQ at a more spectrally efficient BLER.

11.2.3 Radio Link Measurements

As described above, the CoMP-specific adaptation of control loops may be based on radio link measurements. In UL, BSs can measure the radio links and exchange the measurement reports between cooperating BSs. In DL, the UE can measure the links towards cooperating BSs and report the measurements to the serving BS. If supported by the radio interface, the UE might report the measurements to multiple BSs, a concept considered later in Section 13.4. If the channel is reciprocal, UL measurements could be leveraged for adapting DL transmissions. In that case, UEs would not need to transmit their reports over the air. However, since interference is not reciprocal, UE reporting and inter-BS signaling is desirable to obtain an estimate of the DL interference experienced by a UE.

The amount of information contained in the measurement data influences the accuracy of the radio link control: the more the better. However, the required signaling capacity and therewith the signaling delays increase as well. Especially for DL CoMP, where UEs report the measurements over the air, the signaling overhead is a very serious issue. In order to balance control loop accuracy against signaling overhead, several measurements can be defined. First of all, detailed channel state information (CSI), which includes frequency selective phase information for all antennas, can be measured and reported. From the CSI reports, BSs can extract all kinds of information required for various purposes: relative quality of resource blocks for the purpose of resource allocation, channel rank, corresponding optimal precoders, and SINR after cooperation for the purpose of link adaptation etc.

Since CSI reports are huge and generate lots of signaling traffic over the air (as well as on the backhaul), measurements can be tailored for specific purposes. Such measurements contain only the required information and hence they generate much less signaling traffic. Examples of such radio link measurements in LTE are rank indicator (RI), precoding matrix indicator (PMI), and channel quality indicator (CQI). One could even further reduce the signaling load by reporting only long-term measurements such as pathloss coefficients.

In general, the information exchange between the nodes can be on-demand or periodic. In the former case, the serving BS can request the required information for a specific cooperation attempt with specific BSs on-demand. BSs not involved in the cooperation and UEs attached to those BSs are not required to measure or report anything. With periodic signaling, all candidate links have to be measured and reported independently of the actual need for cooperation, and UEs and/or BSs continuously exchange CoMP-related information.

11.2.4 Uplink Power Control

UL power control is the process to adjust the UE transmit power so that signals are received at the BS with an appropriate power level. The power level should be selected to maximize spectral efficiency by balancing achieved link bitrate and generated interference to co-channel cells. In 3GPP LTE, the UE transmit power for the UL data channel consists of an open-loop and a closed-loop component, and the UE sets it according to [3GP09f]:

$$P_{\mathrm{TX}} = \min\left\{P_{\max},\ P_0 + \alpha\,\mathrm{PL}_{\mathrm{DL}} + 10\log_{10}(R) + \Delta_{\mathrm{MCS}} + \delta\right\}. \quad (11.8)$$

Here, $P_{\max}$ is the maximum UE transmit power, which is of course the upper limit of the actual UE transmit power. P_0 can be seen as the (cell-specific) desired receive power, which is transmitted by the BS as part of the LTE system information. The term $10\log_{10}(R)$ reflects the fact that for a larger number of assigned resource blocks R a higher received power and thus a correspondingly higher transmit power is needed. The parameter Δ_{MCS}, configured by the BS, adds an MCS-dependent power offset, which reflects the different SINR requirements per MCS. $\alpha\,\mathrm{PL}_{\mathrm{DL}}$ is part of the open-loop power control component, where each UE selects an appropriate transmit power to compensate a fraction α of the estimated DL pathloss to the serving cell, $\mathrm{PL}_{\mathrm{DL}}$. The DL pathloss is derived from the signal strength of the DL reference signals. Without cooperation, a UE only detects the DL reference signals of its serving BS to estimate the pathloss.

With cooperation, this component should also take the pathloss to supporting BSs into account. This is tricky because UEs would then need to know which cells actually cooperate, which would require additional BS-UE signaling to pre-configure supporting BSs. This signaling would reduce the serving BS's ability to react quickly to changing transmission conditions by selecting supporting cells on-demand on a subframe basis. Furthermore, UL CoMP can be designed to be transparent to the UE, which would allow the support of legacy UEs. Including the pathloss to supporting BSs in the open-loop power control component would not be transparent to UEs, and hence it would not be backward-compatible. See details on UE-aware clustering concepts in Section 7.2.

The term δ is part of the closed-loop power control component, where the UE transmit power is adjusted by the serving BS by means of power control commands. Without CoMP, the term δ is used, e.g., to compensate for UE errors or for UL multi-path fading, which is not reflected in the estimated DL pathloss [DPSB08]. With CoMP, the closed-loop component could be used to adjust the UE transmit power to the increased SINR resulting from UL CoMP: the UE transmit power and therewith the interference level could be reduced while keeping the same MCS and BLER.

11.2.5 Uplink Timing Advance

The BS advances the UE transmission timing so that UL signals can be received time-aligned at the BS. More specifically, in order to maintain UL orthogonality between UEs in a cell, any timing misalignment should fall in the cyclic prefix (CP) duration [DPSB08]. The timing advance depends on the signal propagation delay, which basically depends on the distance between the BS and the UE.

With UL CoMP, the UE signal is received by several BSs and, in general, the distance from the UE to the BSs is different. If the misalignment due to the different signal propagation delays is shorter than the CP duration, UL orthogonality between UEs can be maintained even when a BS receives signals from UEs of several cells. If the misalignment is larger than the CP duration, UL orthogonality will be degraded and UEs signals will interfere each other, as investigated in detail in Section 8.2. This constraint imposes an upper limit on the potential distance between cooperating BSs.

11.2.6 HARQ-related Timing Constraints for UL CoMP

Cooperation between distant sites introduces additional delays, which result mainly from the delay caused by the transmission of data and control information between cooperating sites. The actual delay depends on the inter-site distance, on the core network deployment topology to be discussed in Section 12.2, and on the backhaul technology to be discussed in Section 12.3. This extra CoMP delay prior to each transmission is a potential threat to the strict timing constraints required by HARQ protocols. In 3GPP LTE, DL HARQ is based on asynchronous re-transmissions, hence a re-transmission for an erroneous initial DL transmission can be triggered anytime. In contrast, UL HARQ is based on synchronous re-transmissions, where a re-transmission can only be triggered in the subframe associated with the same HARQ process as the initial transmission. The UL re-transmission is triggered by a negative HARQ acknowledgement (NACK) or an UL grant. In LTE FDD, these messages are transmitted 4 ms before the re-transmission or 4 ms after the initial transmission [3GP09f].

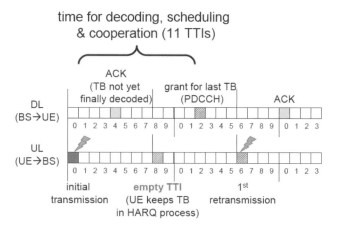

Figure 11.3 Suspension of HARQ process for uplink CoMP in 3GPP LTE.

With UL CoMP, the entire process of cooperative UL reception hence needs to be completed 4 ms after the UL transmission so that the serving BS can send a relevant HARQ feedback (or UL grant) to the UE. This timing requirement is very challenging, and it is most likely not possible to be met if distant sites cooperate. However, there are some possibilities to relax the transmission timing and hence give more time for the cooperative reception. Two alternatives are briefly described in the following.

In order to gain more time for cooperation, the serving BS could send a positive HARQ acknowledgement (ACK) to the UE, thereby *suspending the HARQ process*. Fig. 11.3 shows that the BS sends the ACK in subframe 4 (4 transmit time intervals (TTIs) after the initial transmission), although the transport block (TB) has not yet been decoded. Due to the reception of an HARQ ACK, the UE does not perform a re-transmission for the corresponding HARQ process. A new initial transmission for the corresponding HARQ process can also not be triggered by the BS because the UE has to keep the TB in the HARQ buffer for a potential re-transmission in case the cooperative reception fails. For the particular UE, the corresponding TTI has to remain empty. By suspending the HARQ process, the time for cooperative joint detection increases by 8 ms and the serving BS has more time to gather the required CoMP information from supporting sites and to perform the CoMP reception. One RTT after the HARQ feedback, the CoMP processing should be finished, and the serving BS should send relevant feedback, which either triggers a re-transmission or which acknowledges the successful data transmission. In Fig. 11.3, the initial transmission could not be decoded successfully, and the BS triggers the re-transmission by transmitting a scheduling grant for the last TB to the UEs.

With this alternative, the time for the entire process of UL CoMP can be extended from 3 to 11 ms without any changes in the LTE specifications. However, this alternative refrains a CoMP UE to utilize certain TTIs, while the

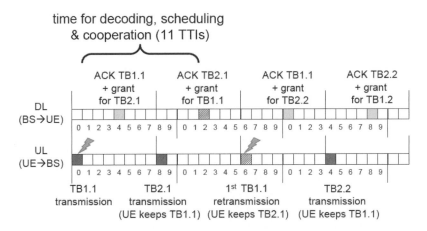

Figure 11.4 Usage of two transport blocks per HARQ process for uplink CoMP.

corresponding HARQ process is suspended. This is visible in Fig. 11.3, where every second TTI of the considered HARQ process is empty.

Therefore, from UE perspective, it is beneficial to apply UL CoMP with HARQ process suspension only if the achievable throughput with CoMP is more than twice as large as the throughput in a non-CoMP mode, where every TTI can be used. Nevertheless, from network perspective, UL CoMP with HARQ process suspension could improve cell spectral efficiency when the resource blocks (RBs) in empty TTIs can be assigned to other UEs.

Alternatively, a second transport block per HARQ process could be used, which allows interlacing the transmission of the other transport block while the cooperative reception is ongoing. Figure 11.4 shows that the UL transmission starts in subframe 0 with the transmission of the first TB of the HARQ process indexed with 1.1. Like in the previous scheme, the BS sends an ACK after 4 TTIs to avoid a non-adaptive retransmission. At the same time, the BS sends a scheduling grant for the second TB indexed 2.1. Eventually, 11 TTIs after the initial transmission, the BS has to send meaningful feedback for TB 1.1. This solution also doubles the HARQ RTT, but in contrast to the above alternative, the HARQ process can be used continuously by the UE. This solution requires that the corresponding control signaling is in place. The concept of having two TBs per HARQ process is known from the DL multiple-input multiple-output (MIMO) transmission mode in LTE [3GP07b]. There, either one or two transport blocks can be transmitted in a single subframe. UL MIMO transmission is currently being specified in LTE Release 10 [3GP10d], and the stated control signaling might be re-used for UL CoMP.

11.2.7 Handover

In cellular systems, such as LTE, a given terminal is associated to one serving cell. In general, the serving cell is chosen based on the BS-to-UE radio link quality. UEs regularly measure the signal strength of neighbor cells and report to their serving BS. As soon as the signal of the serving cell is received with lower signal strength compared to the signal of a neighbor cell, a handover is triggered by the serving BS. In general, thresholds and a hysteresis are used to avoid ping-pong effects. After the handover, the target cell with the best radio link is in charge of the UE and becomes its new serving cell.

By means of CoMP, the effective signal quality and thereby user and cell throughput increases. In order to cooperate, data and control has to be exchanged via the transport network connecting the sites. If the transport network, especially the serving BS's backhaul link, is highly loaded, the serving BS is not able to perform CoMP. Therefore, improved user and cell throughput due to CoMP may be limited by the backhaul link capacity. Different BSs may have different limitations on their backhaul link, e.g. some may be connected via fibre, other via leased telephone lines (E1/T1).

The handover algorithm could mitigate this limitation by considering the backhaul capacity and the current backhaul load. For instance, a UE that is not supported by CoMP due to the limited (copper) backhaul of its serving BS could be handed over to a different serving BS with free (fibre) backhaul resources. With CoMP support at the target BS, the performance could be enhanced. An adapted algorithm could trigger a handover for active UEs with a backhaul-limited serving BS such that the new (target) serving BS has free backhaul capacity to cooperate. Accordingly, the cell-selection criterion of UEs in idle mode could be modified. The thereby selected cell may not have the best (non-cooperative) radio link quality, which is typically used as a criterion to select a serving cell, but due to BS cooperation the resulting user and cell may be optimized.

11.2.8 Inter-BS Signaling

According to LTE operation, each UE is associated to one serving cell, and the corresponding BS controls the radio link, e.g., resource allocation, link adaptation, etc. With CoMP, transmissions in multiple cells are coordinated, and the radio link control can be done in a centralized or distributed manner.

With *centralized* radio link control, a master entity controls and coordinates the transmissions in multiple cells. In order to perform resource allocation, link adaptation, power control etc., the master requires the above mentioned radio link measures, such as CQI, RI, CSI, buffer fill levels, etc. Such information has to be centrally collected by the master, while control commands have to be distributed to the sites. In such a setup, where the radio link control is centralized, the baseband data processing could also be centralized. In that case, UL signals are forwarded from the antenna sites to the master, which jointly processes them;

11.2 Decentralized Radio Link Control and Inter-BS Signaling

DL signals are fed from the master to the sites. The sites are then equipped with simple nodes, such as remote radio heads (RRHs), and all complex processing is performed in powerful BSs, see also Section 11.1. With such a scheme, fixed CoMP clusters are determined by the cells coordinated by one master, also refer to Section 7.1. Within that cluster, all transmissions are coordinated, but coordination across cluster boundaries is not possible. Since there is a logical tree topology between the master and each site of the cluster, UL and DL signals are transmitted only once between a site and the master (in contrast to the following distributed radio link control). However, since the network control is located in the master, data (and control) needs to be transferred between the antenna sites and the master for every UL and DL transmission irrespective of the CoMP gain (in contrast to the following distributed radio link control).

With *distributed* radio link control, the communication with the UE is still controlled by the serving cell, although the UE signal can be jointly received or cooperatively transmitted by several cooperating cells. As a result, there is a logical mesh topology between cooperating cells, which is composed of multiple individual tree topologies between each serving cell and its supporting cells. Since a given cell can support multiple serving cells at the same time, UL and DL signals may be exchanged among multiple cells. However, network control remains in the BSs, and each BS is capable of handling UL and DL transmissions on its own, i.e., CoMP can be disabled. This is especially useful for UEs with very good channel conditions to the serving cell, which would not benefit much from CoMP. A distributed control scheme allows adapting (i.e. decreasing and increasing) the cluster size: the selection of supporting cells can be done in a UE- or cell-specific manner.

In a cell-specific selection, cells which benefit most from the CoMP scheme, e.g., cells with large overlapping areas, would be clustered. This could be done by the operation and maintenance system based on network planning and the result would be fixed clusters, which corresponds to the centralized scheme above. The cell-specific selection can be made more dynamic by re-configuring the clusters during operation based on measurements, such as UE location and signal quality. Here, cells would be clustered so that certain hotspots or certain areas with bad link quality benefit from cooperation, see also Section 7.2. However, with cell-specific clustering the supporting cells are not optimal for all UEs of a cell.

In a UE-specific cell selection, the serving BS could request cooperation from one or more supporting BSs for certain UEs. Here, each UE always has the optimal cluster of supporting cells for the given cooperation mode. Consequently, there are no cluster boundaries anymore where transmissions cannot be coordinated; each UE is always in the middle of its own cluster.

An example of UE-specific signaling for decentralized UL joint detection based on IQ sample exchange is shown in Fig. 11.5. Basics on UL joint detection were provided in Sections 6.1 and 6.2, and a simulative performance evaluation of the particular scheme considered here can be found in Section 14.3. First, the

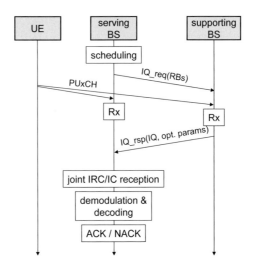

Figure 11.5 Message sequence chart of requesting IQ samples from a supporting BS [HFG09]. © 2009 IEEE.

serving BS does the scheduling. As described above, resource allocation, link adaptation and power control can be adapted to the mode of cooperation. Then, the serving BS requests support from one or more supporting cells for a particular UE transmitting on certain resources. The corresponding message to request IQ samples for certain RBs is named *IQ_req(RBs)* in Fig. 11.5. UEs requiring cooperation and the corresponding supporting cells are selected on-demand. As it will be described in the sequel, UE and BS selection can be based on different parameters, such as location, pathloss, actual channel realization, etc. Having received the UEs data (physical uplink shared channel (PUSCH)) or control channel (physical uplink control channel (PUCCH)) on the indicated resources, the supporting BS transfers the requested IQ samples to the serving BS. The corresponding response carrying IQ samples and other optional parameters is named *IQ_rsp(IQ, opt. params)* in Fig. 11.5. The serving BS performs joint reception (using maximum ratio combining (MRC) or interference rejection combining (IRC)) based on the IQ samples received from cooperating BSs in conjunction with its own received signal. Finally, it checks if the reception was correct and prepares the transmission of HARQ feedback.

A hybrid approach combining a centralized control scheme with a decentralized CoMP scheme has the advantage that the cluster size can be adapted in the sense that the actual CoMP cluster can be a subset of the control cluster. However, it still has the drawback of a new network entity requiring a continuous exchange of control information with the sites.

UE Selection
A UE-specific CoMP scheme with distributed control requires a proper selection of UEs. In CoMP schemes where the backhaul traffic and the computational

complexity scales with the number of supported UEs, it could be beneficial not to select all UEs of a cell for cooperation. If too many UEs are selected for CoMP, the resulting backhaul load might be overwhelming, or processing time may become a critical resource. If the wrong UEs or if too few UEs are selected, then the potential gain of CoMP cannot be fully exploited. Various different methods could be used aiming at optimizing different parameters:

Relative channel quality: This method aims at selecting UEs for which the quality of the channels, e.g., pathloss, towards cooperating cells are relatively close to the quality of the channel towards the serving cell. This method allows selecting UEs that are close to the cell-edge. These UEs are usually suffering from high co-channel interference.

Current radio link performance: This method aims at selecting UEs that have bad radio link performance, e.g., bad absolute channel quality to the serving cell or very active co-channel cells generating lots of interference.

Data rate improvement: This method aims at selecting UEs that would experience the largest throughput increase due to CoMP. The selection would be based on the difference of the (estimated) user throughput with and without CoMP support. Depending on the expression used to measure the throughput increase, this method can lead to maximum cell capacity.

Geographic location: This method uses the geographic location to select UEs for CoMP mode, which can be obtained by means of different techniques, e.g., cellular location methods (see Section 15.1), or global positioning system (GPS) measurements reported by outdoor UEs.

Type of application: This method uses service or application-specific parameters to select UEs to operate under CoMP mode. UEs in CoMP mode perceive increased data rates, but they might see slightly higher packet delays. Furthermore, the setup of the CoMP mode may take time. So CoMP perfectly suits for all kind of services requiring large bit rates over a certain period of time. Such services are, e.g., file download, file sharing, (high definition) video streaming, software updates, mailbox synchronization, harddisk backup, etc.

Supporting BS Selection

A UE-specific CoMP scheme requires the proper selection of supporting BSs for each selected UE. Even a cell-specific CoMP scheme requires the selection of supporting BSs for a given serving BS. Again, the careful selection of supporting BSs is important to not overload the backhaul network and the BS processors. This holds for CoMP schemes where the backhaul traffic and the computational complexity scales with the number of supported BSs. BSs of the same site are less critical to select, since the information can be exchanged at no backhaul expense, whereas information between BSs of different sites are exchanged via backhaul. As an example, two different approaches to select supporting BSs are detailed in the following:

Maximize received signal energy: This approach aims at selecting supporting BSs that can collect most of the carrier signal energy. This method uses the characteristics of the links between non-serving BSs and UEs of the serving BS as selection criterion. By cooperating with BSs having good link quality, the serving BS can increase the received carrier signal energy. This method can improve cell-edge throughput, especially in noise-limited scenarios.

Control severe interference: A second approach aims at selecting supporting BSs that create most of the interference. This approach is based on the characteristics of the link between the serving BS and UEs of non-serving BSs. By cooperating with BSs generating strong interference, the serving BS can control and mitigate the interfering signals. This method is especially useful in interference limited scenarios.

11.2.9 Summary

In this section, the potential modifications of existing radio link control loops and signalling aspects of practical implementations of CoMP schemes in cellular systems at the example of 3GPP LTE were discussed. Most radio link control loops are affected when introducing CoMP. The biggest challenge is to get the right radio link measures, i.e., the measures of the radio channel between antennas of multiple BSs and the UE, at the right place, i.e., the place where the radio link is controlled. Only the HARQ and UL timing advance procedures impose more strict constraints.

The signalling corresponding to the communication between cooperating BSs mainly depends on whether control is centralized or not. When using decentralized radio link control, the corresponding signalling can trade-off complexity, backhaul load and latency. An example signalling scheme for a backhaul-efficient, UE-specific, on-demand decentralized UL CoMP scheme has been introduced.

11.3 Ad-hoc CoMP

Michael Grieger, Patrick Marsch and Gerhard Fettweis

In this section, the concept of ad-hoc CoMP is introduced for the cellular uplink. In this concept, a certain cooperation strategy is decided upon after uplink transmission has already taken place. Furthermore, the extent of cooperation may be progressively adapted until successful decoding is possible. Both concepts make use of the fact that the channel knowledge during detection and decoding is more accurate than at the time of scheduling. The topic is motivated in Subsection 11.3.1, after which concrete schemes are proposed for two particular scenarios in Subsections 11.3.2 and 11.3.3, respectively. It is discussed in Subsection 11.3.4, to which extent the concept of ad-hoc CoMP may shed a different light on hybrid automatic repeat request (HARQ), followed by a summary in Subsection 11.3.5.

11.3.1 Introduction

Opportunistic Communication and Scheduling

The volatile nature of the wireless channel has long been seen as a burden, complicating the life of wireless system engineers. In recent years, however, the perspective has changed, bringing wireless channels into a more favorable light as summarized by David Tse and Pramod Viswanath in [TV05]. In a cellular system, available time and frequency resources are shared among multiple-users. Provided the channel state can be tracked with sufficient accuracy, channel fluctuations can be exploited by scheduling users on those time and frequency resources for which channel conditions are the best. A particular channel condition can then be exploited by matching modulation and coding schemes (MCSs) as well as signal parameters such as transmit power to the channel state, a concept referred to as link adaptation. Centralized scheduling schemes that exploit channel fluctuations for improved CoMP system performance were already presented in Section 11.1.

Imperfect CSI at the Scheduler

It is obvious that the performance of these scheduling concepts depends strongly on the amount and quality of channel information that is available at the scheduler, which should be as *up to date*, *accurate* and *extensive* as possible, i.e., ideally including information on interference, distortion, radio frequency (RF) imperfections, etc. In reality, however, the picture that is available at the scheduler is fairly noisy, and is akin to an image seen through a narrow lens. For CoMP systems in particular:

- Channel links to different base stations (BSs) have diverse gains and are, therefore, hard to estimate. Additionally, there might be interference between pilot signals (see Section 9.1) which further impedes accurate channel estimation.
- Joint scheduling, which promises huge gains in terms of total throughput and fairness by exploiting multi-user diversity, requires that channel state information (CSI) be forwarded to a central scheduling node and the scheduling decision be forwarded to the user equipments (UEs). Due to this scheduling delay, the scheduler bases its decision on outdated information.

Hence, while making its decision, the scheduler relies on imperfect CSI. Consequently, transmission errors are inevitable, because there is a probability that the scheduler assigns a transmission rate which is not supported by the channel. As described in Section 11.2, in contemporary standards of cellular systems like LTE, the impact of transmission errors is reduced by using HARQ techniques.

CoMP under Backhaul Constraints
In Chapter 2, it was already emphasized that a major economic hurdle challenging a substantial deployment of CoMP is the extensive backhaul infrastructure that is required for information exchange among BSs. Hence, the identification of backhaul efficient cooperation strategies is an important challenge, in order to keep additional costs for an upgrade of the existing backhaul infrastructure low. The results presented in Section 4.3.1 show that a flexible usage of different CoMP schemes is beneficial in the uplink of a backhaul constrained system, because their performance depends on current channel conditions. For example, it is known that the exchange of decoded information for the purpose of interference cancelation is very effective (in terms of CoMP gain vs. required backhaul) for asymmetric scenarios where the most severe interference links can be canceled. An exchange of compressed receive signals to a joint decoder, however, as observed in Section 6.1, allows achieving higher data rates when the available backhaul capacity is large. Hence, the scheduler should take all available CoMP options and backhaul constraints into consideration while assigning physical resources and MCSs.

Scheduling, Ad-hoc CoMP and HARQ
In Section 11.1, we have seen that the functionality of a scheduler may consist of a variety of tasks, namely resource allocation, decision on CoMP schemes, power control, choice of precoders, and link adaptation. However, as stated above, in a mobile environment the scheduler has imperfect CSI, which results in non-optimal scheduling. While most decisions made by a scheduler have to be kept fixed during transmission, we here want to point out that the choice of a particular CoMP scheme can indeed still be altered after uplink transmission has already taken place. This choice comprises two main aspects:

- The size of the cooperation cluster — Intuitively, it makes sense to flexibly choose more and more supporting BSs (see, e.g., Section 11.2.8), as opposed to a fixed cluster of CoMP cells, until a UE can be decoded successfully.
- The cooperation scheme that is used, and parameters involved — As shown in the previous chapters, many different uplink CoMP schemes exist with different performance / backhaul trade-offs, different properties regarding latency etc.. Furthermore, different uplink CoMP schemes are suitable for different channel conditions. Hence, it also makes sense to adapt the CoMP strategy according to the *better* CSI (and possibly first decoding feedback) available after transmission.

In this section, we show how an ad-hoc decision on the CoMP mode, which we refer to as *Ad-Hoc CoMP*, leads to a more efficient usage of backhaul infrastructure. Thus, system performance can be increased if the availability of backhaul capacity limits the number of terminals that can be served with CoMP. To this end, we investigate a model for the cellular uplink that includes three aspects: scheduling, Ad-Hoc CoMP, and HARQ, as shown in Fig. 11.6.

11.3 Ad-hoc CoMP

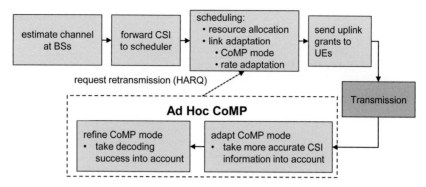

Figure 11.6 Scheduling, Ad-Hoc CoMP, and HARQ process.

CSI impairments can be divided into two different classes:

- Impairments that only affect the CSI available at the scheduler, thereby resulting in more accurate CSI being available for Ad-Hoc CoMP.
- Impairments that affect the CSI at the scheduler as well as the CSI available for Ad-Hoc CoMP.

In the remainder of this section, we will address both of these effects separately. In Subsection 11.3.2, impairments that affect the CSI at the scheduler will be considered. Examining a distributed antenna system (DAS), where quantized receive signals are exchanged between BSs, we analyze the effect that these kinds of impairments have on achievable throughput, and we propose adaptive compression as a means for more efficient usage of the available backhaul capacity. In Subsection 11.3.3, we address CSI impairments that effect the CSI available at the scheduler and for Ad-Hoc CoMP. Once again, we examine these impairments more closely for the example of a DAS, and observe how feedback on decoding success can be used for a more efficient backhaul usage. Obviously, in any real system, both kinds of CSI impairments will occur together. Ad-Hoc CoMP is thus divided into two successive processes as depicted in Fig. 11.6. First, the CoMP scheme is adapted by taking new channel knowledge into account. If this is not sufficient for successful decoding, the usage of CoMP is refined. In a conventional system, the only way to achieve reliable communication despite having imperfect CSI at the scheduler is to employ HARQ. However, in a CoMP system, there is an additional degree of freedom: the extent of cooperation. The link between Ad-Hoc CoMP and HARQ is discussed in Subsection 11.3.4.

11.3.2 Ad-Hoc CoMP With More Accurate CSI

Joint scheduling requires the distribution of CSI to a central scheduling node and forwarding of the scheduling decision to the UEs. Due to this delay and the time varying nature of the mobile channel, the scheduler bases its decision on

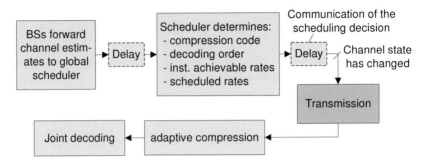

Figure 11.7 Scheduling and ad-hoc decoding process.

CSI that does not describe the channel that is used for the transmission of user data perfectly. In cellular systems, the channel is variant for two main reason:

- fast fading.
- time varying interference, particularly in systems with little interference averaging such as LTE.

Assuming that these are the only kinds of CSI impairments that could occur, the CoMP mode can be adapted such that the backhaul capacity available is used optimally. For all users that can be decoded locally, CoMP would not be used at all. The same is true for users that could not be decoded even with the maximum CoMP support available. At the same time, different uplink CoMP schemes such as distributed interference subtraction (DIS) or DAS as introduced in Section 4.3.1 would be used whenever they deemed most effective. In the sequel, we study the effect of outdated CSI in a distributed antenna system with centralized decoding (see Section 4.3.1).

Example: Adaptive Compression in a Distributed Antenna System
As introduced in Section 4.3.1, in a DAS with centralized decoding, one of the BSs functions as a joint decoder of codewords transmitted by all UEs in the CoMP cluster, and all other BSs function as remote radio heads (RRHs), forwarding their receive signal. We consider that the backhaul connecting the BSs is limited in its capacity. Thus, the signals received at all RRHs have to be compressed prior to their exchange over the backhaul.

The scheduling, transmission, and decoding process is depicted in Fig. 11.7. Since the problem of resource allocation was described in detail in Section 11.1, we here assume that resources have already been allocated to UEs by any arbitrary algorithm. For this reason, we consider simplified scenarios of few BSs and UEs as drafted in Fig. 4.4(c). The influence of inter-cluster interference is neglected in this model, and we only observe outdated CSI due to fast fading effects.

Since we consider the effect of outdated CSI at the scheduler only, we assume that the channel is perfectly estimated for every transmission block. As depicted in Fig. 11.7, in a mobile time-variant environment, the scheduler has access only to CSI that is outdated by n_d transmission blocks, because certain delays for the exchange of channel estimates and for the communication of uplink grants are inevitable. Based on this outdated channel information, the scheduler estimates achievable transmission rates and assigns appropriate MCSs. For simplicity, in the remainder of this section, we assume that the number of possible MCSs is unlimited, ignoring the fact that in real systems only a certain granularity of MCSs is available. Transmission errors occur if the rate of the assigned MCS is too high to be successfully decoded. Since achievable rates in a DAS depend on the compression accuracy, the scheduler has to find a trade-off between throughput and the required backhaul capacity. Note that in the multi-user case with the employment of successive interference cancelation (SIC) at the decoder, the rate of each UE also depends on the decoding order.

In the following paragraphs, we investigate the benefit of adaptive compression. In particular, two strategies are compared:

1. fixed compression: a fixed backhaul rate is used for the exchange of the compressed signals from the RRH to the decoder.
2. adaptive ad-hoc compression: the updated CSI after transmission is taken into account to decide which backhaul rate (and therefore compression accuracy) is sufficient for successful decoding.

If the adaptive scheme is employed, we exploit the fact that the RRHs have full knowledge of the current channel state after the transmission. They are therefore able to adapt the compression appropriately and to enable successful decoding with as little information exchange as possible. The gain of the adaptive scheme is indeed two-fold: besides achieving higher throughput due to the reduced probability of transmission errors; backhaul consumption is reduced because the adaptive scheme exploits all cases where decoding is possible with a backhaul rate that is lower than the fixed rate. Additionally, in the case where successful decoding could not be achieved even under full cooperation, the backhaul is not used at all, enabling its potential usage for other terminals.

A comparison of the maximum sum-rate that can be achieved for a certain average backhaul rate is shown in Fig. 11.8, where we consider a scenario with $M = 2$ BSs with $N_\mathrm{bs} = 1$ receive antennas each, and either $K = 1$ double-antenna UE (Fig. 11.8(a)) or $K = 2$ single-antenna UEs (Fig.11.8(b)). The UEs use fixed per antenna transmit power P, and the received signal is distorted by additive white Gaussian noise (AWGN) with variance σ_v^2. In general, a rich scattering environment leading to complex Gaussian channel realizations (Rayleigh channel) that are spatially independent is assumed. The UEs are assumed to be located at the cell-edge. In order to model the time-variance of the channel, it is assumed that the 1 or 2 UEs are moving at a constant speed v. We employ the widely used Jakes' spectrum to model the effects of the Doppler spread [JC94]. Furthermore, we assume coding over a complete transmission block of 1 ms and

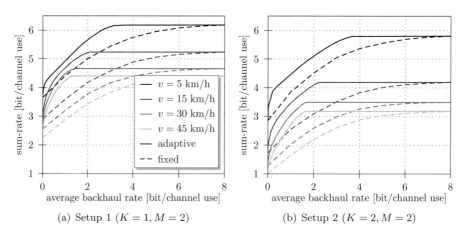

Figure 11.8 Comparison of sum-rate vs. backhaul for the adaptive and fixed schemes for different time varying Rayleigh channels ($f_c = 2.68$ GHz, $\sigma_v^2 = 0.1$, $P = 1$).

a scheduling delay of 3 ms. The results are based on information-theoretic models that include the use of best-known compression techniques that also utilize side-information by Wyner-Ziv coding [dCS09], and a rather simple scheduler that makes a decision based on the channel that was observed n_d codewords earlier and considers a backoff-factor which is chosen such that throughput is maximized. For further details, we refer to [GMF10a].

As expected, Fig. 11.8 shows a throughput loss that increases with the time-variance of the channel. However, when the ad-hoc cooperation scheme is employed, we see strong gains in terms of the throughput/backhaul trade-off. Indeed, ad-hoc cooperation allows us to achieve almost maximum throughput for much lower backhaul rates than with fixed cooperation, which in the low backhaul regime mitigates the negative impact of time varying channels on the achievable throughput. As expected, the backhaul savings of the ad-hoc scheme increase with increasing mobility.

When the achievable gains for the one UE case (Fig. 11.8(a)) and the two UE case (Fig. 11.8(b)) are compared, we see that the possible gains of Ad-Hoc CoMP are reduced. The reason for this observation is that, in the two user case, it is not possible to adapt the backhaul rate such that the rates of both users are equal to the scheduled rate separately because the backhaul rate is increased until both users can be decoded successfully. Hence, the gains from the proposed adaptive cooperation scheme decreases with an increase in the number of UEs that are decoded jointly. However, this occurs when only single antenna BSs are employed. By using multiple BS antennas, the backhaul rate can be distributed on the spatial dimensions of the receive signal in a way that less backhaul rate is utilized for the compression of user signals beyond the accuracy required for successful decoding as shown in [GMF10c].

11.3.3 Ad-Hoc CoMP with CSI Impairments

In the case of outdated CSI, the CSI available for Ad-Hoc CoMP after the transmission is more accurate than the CSI available at the scheduler. However, perfect CSI will never be achieved because of

- channel estimation errors,
- noise covariance estimation errors, and
- RF-impairment estimation errors.

Hence, in practice it is not possible to adapt the CoMP mode perfectly as done in the previous subsection. At the same time, we still strive for efficient backhaul use and a minimization of transmission errors. A possible solution to this problem relies on error detection coding schemes that indicate the decoding success. In practice, this decision is based either on the output of an outer error detection code (e.g. cyclic redundancy check (CRC)), or by observing reliability information delivered by soft output decoders. The idea is to increase the use of CoMP techniques progressively until successful decoding is achieved. In the following example, we use this concept in a DAS where compression accuracy is progressively increased by using successive refinements.

Example: Distributed Antenna System
In this example, we concentrate on the effects of channel estimation errors. In this case, the scheduled rates are determined based on the estimated channel state. The channel estimation error, which is constant during the reception of a codeword, leads to an additive transmission impairment with an unknown variance. A transmission error occurs if the scheduled data rate chosen is too high to be supported by the channel. In order to decide on the scheduled rate, the scheduler tries to predict the estimation distortion. In a practical system, a simple solution to this problem is to consider a signal-to-interference-and-noise ratio (SINR) margin. The CSI that is available for Ad-Hoc CoMP is impaired as well. Hence, optimal adaptation of the backhaul rate is not possible. However, the usage of a fixed backhaul rate is still inefficient because we can utilize the information on the decoding success in conjunction with successively refinable source coding schemes [EC91], as demonstrated in the following paragraph.

The algorithm used for the progressive refinement of the CoMP mode is depicted in Fig. 11.9. It is based on the employment of successively refinable source coding schemes [EC91]. The quantization accuracy is progressively refined as long as decoding is unsuccessful. It is known that successive refinements are possible without any rate loss [EC91] for Gaussian sources, however, other sources are also refinable. Prominent examples are all scalar quantizers. Here, some of the most significant bits can be exchanged initially. If the quantization distortion turns out to be too high for successful decoding, the representation can be refined by an exchange of the next most significant bits, and so forth. Outage occurs only if the finest possible quantization accuracy is insufficient for

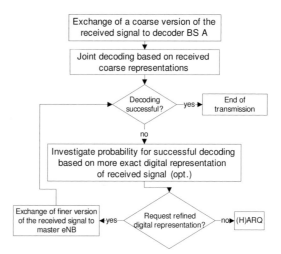

Figure 11.9 Flowchart of the progressive ad-hoc cooperation algorithm.

successful decoding. A downside of the proposed scheme is that it requires a feedback mechanism between the decoder and the forwarding BSs, which introduces an additional delay.

For the rest of this subsection, we will compare two approaches:

1. fixed compression: the same backhaul rate c_{fix} is always used for the exchange of compressed signals from BS 2 (the RRH) to BS 1 (the decoder). The throughput is maximized (in this case) by choosing an optimal signal-to-noise ratio (SNR) gap at the scheduler.
2. progressive cooperation: a progressive refinement of the exchanged signal is used to achieve the lowest backhaul rate c_{pro} that enables successful decoding.

When latency and complexity are not constrained, the most backhaul-efficient scheme would be to refine the accuracy of the forwarded information in very small successive steps. However, in real-world systems, a good trade-off between throughput, backhaul rate, and latency is desired. Therefore, we need to find other methods that limit the number of iterations. A straightforward approach is a simple three step scheme. The signal is first quantized with the rate $\frac{c_{\text{fix}}}{2}$. If decoding is unsuccessful, the exchanged signal is refined to a total rate of c_{fix}. If decoding is still not successful, in the last step, further refinement to a rate of $2c_{\text{fix}}$ is used. The transmission is in outage if even this rate is not sufficient for decoding. Further information is given in [GMF10b].

As mentioned earlier, we assume a block fading channel, such that the channel (as well as the channel estimation and the channel estimation error) are constant for the transmission of one codeword, and successive channel realizations are assumed to be uncorrelated. Fig. 11.10(a) shows the Monte-Carlo simulation results for the setup that was already observed in Section 11.3.2. In addition to the case of perfect channel estimation ($\sigma_{\text{est}}^2 = 0$), we consider channel estimation errors with variance $\sigma_{\text{est}}^2 = \{0.02, 0.05, 0.1\}$. The relatively large gap between

11.3 Ad-hoc CoMP

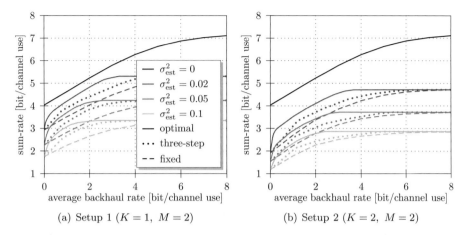

Figure 11.10 Comparison of sum-rate vs. backhaul for the optimal and the heuristic three-step progressive scheme as well as the fixed scheme ($\sigma_v^2 = 0.1$, $P = 1$).

the throughput for perfect channel estimation and for imperfect CSI is a consequence of the fact that the variance of the estimation distortion is unknown, resulting either in transmission errors or scheduled rates that are far from the achievable rates.

Fig. 11.10(b) shows that the gain of progressive ad-hoc cooperation scheme decreases with the number of UEs that are decoded jointly, because the backhaul rate is increased until both users can be decoded successfully. The reasons and potential countermeasures are the same as in the case of imperfect CSI due to a scheduling delay.

11.3.4 Ad-Hoc CoMP and HARQ

Even with the usage of ad-hoc CoMP, there is a certain chance that decoding is not possible. In these cases, a retransmission is required as indicated in Fig. 11.6. In current cellular standards, HARQ is used because the extra costs in terms of system complexity are justified by large throughput gains. As discussed in Section 11.2, when CoMP systems are considered, HARQ is more problematic because the information exchange required over the backhaul network adds additional delay to the total signal processing time until decoding is successful, potentially violating the demands of existing standards. Although these new requirements could principally be considered in new versions of cellular standards, this is undesirable, since the application of uplink CoMP does not necessarily require further changes to the standard. Thus, the use of Ad-Hoc CoMP is not only contingent on backhaul rate but also to backhaul latency constraints that need to be considered in the selection of appropriate CoMP modes. In particular, progressive schemes introduce additional delays and their practical feasibility needs to be proven in practice. The results presented in Sections 11.3.2

and 11.3.3 show that the number of retransmissions can be reduced by using Ad-Hoc CoMP. Future research will show if an ad-hoc use of CoMP along with coordinated scheduling might have the same potential as that of reliable communication without HARQ on the first two layers.

11.3.5 Summary

In this section, the concept of Ad-Hoc CoMP for the cellular uplink was introduced. The key concept is to adapt the CoMP strategy *after* transmission has taken place in order to exploit channel information that is more recent than the one available at the time of scheduling. In this way, a more efficient use of backhaul can be achieved. The potential gains of using Ad-Hoc CoMP were shown for the example of a distributed antenna system where base stations exchange quantized receive signals for centralized decoding, and where two particular scenarios were considered.

In the first scenario, assuming that *perfect* CSI is available to the BSs after transmission, while only inaccurate CSI was available at the time of scheduling, it was shown that the employment of an adaptive backhaul compression rate can greatly increase backhaul efficiency.

In the other scenario, now assuming that the CSI available at the time of decoding is also subject to estimation errors, it could be shown that a progressive ad-hoc cooperation scheme is highly beneficial in terms of backhaul savings. Here, successively refined information is passed over the backhaul in multiple iterations, until successful decoding of the terminal transmissions is possible. Clearly, this leads to a trade-off between latency (number of iterations) and sum backhaul rate. A simple three-step approach was introduced, and its performance relative to the optimal progressive scheme and a naive fixed cooperation scheme was shown.

For the scenarios observed, the results indicate that an adaptive and progressive use of CoMP promises to reduce the required backhaul rate by about 50 %.

12 Backhaul

In this chapter, we address a last, but absolutely not least important challenge connected to CoMP, namely the fact that most base station cooperation schemes require information exchange over a backhaul infrastructure. Depending on the existing infrastructure of a mobile operator, both backhaul capacity and latency requirements of some CoMP schemes may be the main cost drivers or potential show stoppers on the roadmap towards CoMP. The chapter starts with addressing fundamental aspects of backhaul-constrained cooperation in Section 12.1, after which concrete backhaul capacity and latency requirements of various uplink and downlink CoMP schemes and their scaling behavior are derived in Section 12.2. Finally, Section 12.3 gives an overview on existing and upcoming backhaul technology options, and hence gives the reader a feeling of whether particular CoMP schemes can be expected to be technically and commercially feasible in the near future or not.

12.1 Fundamental Limits of Interference Mitigation with Limited Backhaul Cooperation

I-Hsiang Wang and David Tse

As we have seen in previous parts of this book, cooperation among base stations (BSs) via infrastructure backhaul networks can help mitigate interference by forming distributed multiple-input multiple-output (MIMO) systems, while the rate at which BSs cooperate is limited in wide-band cellular systems. How much interference can one bit of backhaul cooperation mitigate? In this section, we study the two-user Gaussian interference channel with limited backhaul cooperation to answer this question in a simple setting. We identify two regions pertaining to the fundamental gain from backhaul cooperation: linear and saturation regions. In the linear region, cooperation is efficient and provides a *degrees-of-freedom* gain, which is either *one cooperation bit buys one more bit* or *two cooperation bits buy one more bit* until saturation. In the saturation region, cooperation is inefficient and provides a *power* gain, which is at most a constant regardless of the rate at which BSs cooperate. The conclusion is drawn based on the characterization of the capacity region to within a constant gap[1].

12.1.1 Introduction

Why is Backhaul Limited?

One of the common misconceptions about backhaul cooperation is that the backhaul provides near unlimited cooperation capability, so that base stations can cooperate in an unlimited manner. To refute this, we shall use a simple example to illustrate that in a wide-band cellular system, backhaul cooperation is usually limited.

Consider a wide-band orthogonal frequency division multiplex (OFDM)-based cellular system with a bandwidth of 20 MHz. To attain near unlimited cooperation, the received signal at a base station should be quantized finely enough so that it can be recovered with a negligible distortion at other base stations. Let us do a back-of-the-envelope calculation to get a sense of the rate that should be used to convey these quantization outputs. Suppose we use 8 bit/s/Hz to quantize the signal and use the backhaul to exchange them. The total throughput required in the backhaul is then $20 \times 8 = 160$ Mbits/s. Even for optical carriers in synchronous optical network (SONET)/synchronous digital hierarchy (SDH), such a high data rate is only supported beyond OC-12, and not to mention other technologies such as digital subscriber line (DSL) that cannot support it. For wireless technologies with growing bandwidth, since the backhaul link capacity does not increase with wireless spectra, from the above calculation we conclude that backhaul cooperation should be considered limited, and understanding how to make use of backhaul cooperation efficiently for interference mitigation becomes important.

Gaussian Interference Channel with Backhaul Cooperation

The simplest information-theoretic model for studying the fundamental limits of a communication system in the presence of interference is the *interference channel (IC)*. In its simplest form, an interference channel consists of two transmitter-receiver pairs, and each receiver is only interested in retrieving information from its own transmitter. Therefore, one user's information-carrying signal becomes interference for the other user. A *Gaussian* IC is one where the second user's signal x_2 interferes with the first user's signal x_1 in an additive fashion and vice versa, along with additive white Gaussian noises at both receivers. Mathematically, the Gaussian interference channel is defined as follows:

$$y_1 = h_{11}x_1 + h_{12}x_2 + z_1, \quad y_2 = h_{21}x_1 + h_{22}x_2 + z_2 \tag{12.1}$$

---

[1] The difference between inner and outer bounds are within a constant number of bits, which does not depend on channel parameters.

12.1 Fund. Limits of Interf. Mitigation with Limited Backhaul Coop.

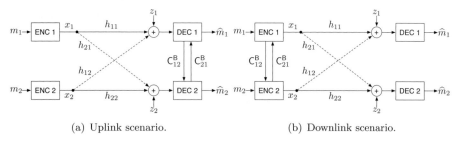

(a) Uplink scenario. (b) Downlink scenario.

Figure 12.1 Channel model considered. Dashed lines denote interfering links.

are the received signals, where two mutually independent additive noise processes $\{z_i[k]\}_{k=1}^{N}$ ($i = 1, 2$) are independently and identically distributed (i.i.d.) with $\mathcal{N}_{\mathbb{C}}(0, 1)$ over time. In this section, we use $[\cdot]$ to denote time indices. Transmitter i intends to convey message m_i to receiver i by encoding it into a block codeword $\{x_i[k]\}_{k=1}^{N}$, with transmit power constraints

$$\frac{1}{N} \sum_{k=1}^{N} |x_i[k]|^2 \leq 1, \ i = 1, 2, \tag{12.2}$$

for an arbitrary block length N. Messages m_1 and m_2 are independent. Define channel parameters

$$\mathsf{SNR}_i := |h_{ii}|^2, \ \mathsf{INR}_i := |h_{ij}|^2, \ i, j = 1, 2, \ i \neq j. \tag{12.3}$$

In the uplink scenario, since BSs serve as receivers, backhaul cooperation is modeled as *receiver cooperation*. On the other hand, in the downlink scenario, backhaul cooperation is modeled as *transmitter cooperation*. Backhaul cooperation links are modeled as noise-free links with finite *cooperation capacity* $\mathsf{C}_{ij}^{\mathsf{B}}$ from BS i to j, for $(i, j) = (1, 2)$ or $(2, 1)$. The models are depicted in Fig. 12.1.

Despite the simplicity of this model, even in the scenario *without cooperation*, exact characterization of the capacity region has remained open since its introduction in 60's. In this section, we will not pursue the *exact* characterization of the capacity region. Instead, we aim at a *uniformly approximate* characterization where the capacity region is determined to within a *constant gap*, meaning that the difference between inner and outer bounds is within a constant number of bits. The constant gap is not dependent on channel parameters, and hence the fundamental limit in the interference-limited regime (that is, at high signal-to-noise ratio (SNR)) is fully characterized. Etkin *et.al.* characterize the capacity region of the Gaussian IC to within 1 bit/s/Hz per complex dimension [ETW08]. Wang *et.al.* characterize the capacity region of the Gaussian IC with limited backhaul cooperation to within 2 bit/s/Hz and 6.5 bit/s/Hz, for the uplink receiver cooperation scenario [WT09a] and downlink transmitter cooperation scenario [WT10], respectively.

A Deterministic Approach to the Gaussian Interference Channel

Throughout the section, we shall employ the *linear deterministic model* [ADT07, BT08] for the Gaussian interference channel to study the problem and illustrate high-level intuitions. For our system, the corresponding linear deterministic model is parameterized by six integers $\{n_{11}, n_{12}, n_{21}, n_{22}, k_{12}, k_{21}\}$, where

$$n_{ij} := \left(\lfloor \log |h_{ij}|^2 \rfloor\right)^+, \quad i,j \in \{1,2\}, \quad k_{12} := \lfloor C_{12}^B \rfloor, \quad k_{21} := \lfloor C_{21}^B \rfloor. \qquad (12.4)$$

For the interference channel part, the transmit signal at transmitter i is $\mathbf{x}_i \in \mathbb{F}_2^q$, for $i = 1, 2$. Here $\mathbb{F}_2$ denotes the binary field $\{0,1\}$. The received signals are

$$\mathbf{y}_1 = S^{q-n_{11}} \mathbf{x}_1 + S^{q-n_{12}} \mathbf{x}_2, \quad \mathbf{y}_2 = S^{q-n_{21}} \mathbf{x}_1 + S^{q-n_{22}} \mathbf{x}_2, \qquad (12.5)$$

where additions are modulo-two component-wise, $q = \max\{n_{11}, n_{12}, n_{21}, n_{22}\}$, and $S \in \mathbb{F}_2^{q \times q}$ is the shift matrix

$$S = \begin{bmatrix} 0 & 0 & 0 & \cdots & 0 \\ 1 & 0 & 0 & \cdots & 0 \\ 0 & 1 & 0 & \cdots & 0 \\ \vdots & & \ddots & & \vdots \\ 0 & \cdots & 0 & 1 & 0 \end{bmatrix}. \qquad (12.6)$$

An interpretation of this model considers the *binary* expansion of signals. The effect of additive white Gaussian noise is modeled by *truncation* of the signal below the noise level. The effect of superposition with interference is modeled by the modulo-two component-wise addition of the bits, where the carry-over in real addition is not captured for simplicity.

Fundamental Gain from Limited Backhaul Cooperation

We identify two regions pertaining to the gain from limited backhaul cooperation: linear and saturation regions, as illustrated by a numerical example in Fig. 12.2. The example is symmetric with $\mathsf{SNR}_1 = \mathsf{SNR}_2 = \mathsf{SNR} = 20$ dB, $\mathsf{INR}_1 = \mathsf{INR}_2 = \mathsf{INR} = 15$ dB, and $C_{12}^B = C_{21}^B = C^B$. In the linear region, backhaul cooperation is *efficient*, in the sense that the growth of user data rate is roughly linear with respect to the capacity of the backhaul links. The gain in this region is the *degrees-of-freedom* gain that CoMP systems provide. On the other hand, in the saturation region, backhaul cooperation is *inefficient* in the sense that the growth of user data rate becomes saturated as one increases the rate in the backhaul links. The gain is the *power* gain of a constant number of bits at best, and the constant is independent of the channel strength and the backhaul cooperation rate. We will focus on system performance in the linear region, not only because the rate at which base stations can cooperate is limited in most scenarios, but also because the gain from cooperation is more significant.

With the constant-gap-to-optimality result, we find the fundamental gain from cooperation in the linear region as follows: either *one cooperation bit buys one more bit* or *two cooperation bits buy one more bit* until saturation, depending

12.1 Fund. Limits of Interf. Mitigation with Limited Backhaul Coop.

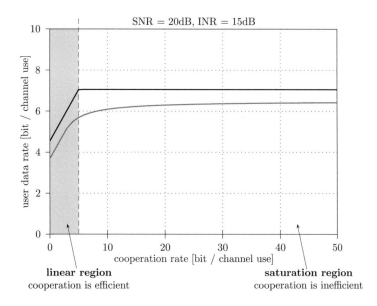

Figure 12.2 The gain from limited backhaul cooperation.

on channel parameters. This will be elaborated and explained in the last part of this section.

The rest of this section is organized as follows. First, we describe the cooperation strategies between base stations that achieve the capacity regions to within 2 bits and 6.5 bits in uplink and downlink scenarios respectively. Next, we show that there is an uplink-downlink reciprocity between the two scenarios, and hence there is no difference in the fundamental gains obtained from receiver or transmitter cooperation. We then quantify the degree-of-freedom gain by characterizing the number of *generalized degrees of freedom* in the system. Finally, we use a couple of linear deterministic examples to illustrate the high-level intuitive reasons why there are two different kinds of behaviors of the gain from backhaul cooperation in the linear region.

12.1.2 Uplink Scenario: Receiver Cooperation

We shall not give the full expression of the inner and outer bounds of the capacity region due to space constraints. We point interested readers to reference [WT09a] for more details. Instead, we first use a linear deterministic example to motivate the strategy and then describe the strategy that achieves capacity to within a constant gap in the Gaussian scenario.

Linear Deterministic Examples
Consider the following symmetric channel: $\mathsf{SNR}_1 = \mathsf{SNR}_2 = \mathsf{SNR}$, $\mathsf{INR}_1 = \mathsf{INR}_2 = \mathsf{INR}$, and $\mathsf{C}_{12}^B = \mathsf{C}_{21}^B = \mathsf{C}^B$. Set INR to be 2/3 of the SNR in dB scale, that is, $\log \mathsf{INR} = \frac{2}{3} \log \mathsf{SNR}$. Set $\mathsf{C}^B = \frac{1}{3} \log \mathsf{SNR}$. The corresponding linear deterministic channel (LDC) is depicted in Fig. 12.3. Bits at the levels of transmitters/receivers

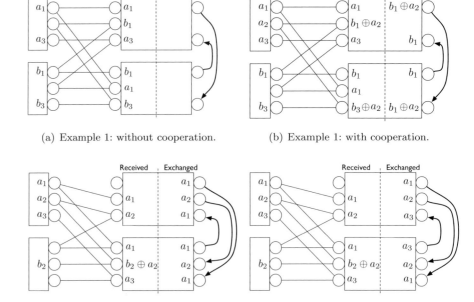

Figure 12.3 Example channels. $\{a_k\}$ denote user 1's bits, while $\{b_k\}$ denote user 2's. Index k denotes the k-th level at the corresp. transmitter. [WT09b] © 2009 IEEE.

can be thought of as chunks of binary expansions of the transmitted/received signals. Note that in this example, one bit in the LDC corresponds to $\frac{1}{3}\log$ SNR in the Gaussian channel.

We begin with the baseline scenario where two receivers are not allowed to cooperate. Transmit signals are naturally split into two parts: (1) the common levels, which appear at both receivers, and (2) the private levels, which appear only at its own receiver. Each transmitter splits its message into common and private parts, which are linearly modulated onto the common and private levels of the signal respectively. Each receiver then decodes both user's common messages and its own private message by solving the linear equations it received. This is shown to be optimal in the two-user interference channel [BT08]. In this example (Fig. 12.3(a)), bits a_1 and b_1 are common, while a_3 and b_3 are private. The sum-capacity without cooperation is 4 bits. Since all levels at both receivers are occupied, one cannot turn on bits a_2 or b_2 without causing collisions.

With receiver cooperation, the natural split of transmitted signals does not change. This suggests that the encoding procedure and the aim of each decoder remain the same. Each receiver with help from the other receiver, however, is able to decode additional information. Since each user's private message is of no interest to the other receiver, an obvious scheme for receiver cooperation is to exchange linear combinations formed by the signals *above* the private signal level so that the undesired signal does not pollute the cooperative information. In this example, as illustrated in Fig. 12.3(b), with one-bit cooperation in each direction in the LDC, the optimal sum-rate is 5 bits, achieved by turning on one more bit a_2. This causes collisions at the second level at receiver 1 and at the third level at receiver 2, which can be resolved with cooperation: receiver 1 sends $b_1 \oplus a_2$ to receiver 2, and receiver 2 sends b_1 to receiver 1. Now, receiver 1 can solve (a_1, a_2, a_3, b_1), and receiver 2 can solve (b_1, b_3, a_1, a_2). In fact, the exchanged linear combinations are not unique. For example, receiver 1 can send $(b_1 \oplus a_2) \oplus a_1$ and receiver 2 can send $b_1 \oplus a_1$, and this again achieves the same rates. As long as receiver 1 does not send a linear combination containing the private bit a_3 and the sent linear combination is linearly independent of the signals at receiver 2 (and vice versa for the linear combination sent from receiver 2 to receiver 1), the scheme is optimal for this example channel. The above discussion regarding the scheme in the LDC naturally leads to an implementable one-round scheme in the Gaussian channel, where both receivers quantize-and-bin their received signals at their own private signal level.

In the above example, it is optimal that each receiver sends to the other, linear combinations formed by its received signal above its private signal level. Is this optimal in general? The answer is no. Consider the following asymmetric example: $\mathsf{SNR}_2 = \mathsf{INR}_2$, SNR_1 is 2/3 of SNR_2 in dB, and INR_1 is 1/3 of SNR_2 in dB. $\mathsf{C}_{12}^\mathsf{B} = \frac{2}{3} \log \mathsf{SNR}_2$ and $\mathsf{C}_{21}^\mathsf{B} = \frac{1}{3} \log \mathsf{SNR}_2$. The corresponding LDC is depicted in Figs. 12.3(c) and 12.3(d), where one bit in the LDC corresponds to $\frac{1}{3} \log \mathsf{SNR}_2$ in the Gaussian channel. First consider the same scheme as in the previous example. Note that if receiver 2 just forwards signals above its private signal level, it can only forward a_1 to receiver 1 and achieves R_1 up to 2 bits. On the other hand, if receiver 2 forwards a_3 to receiver 1, which is below user 2's private signal level, it achieves $R_1 = 3$ bits. From this example, we see that whenever there is "useful" information (which should not be polluted by the receiver's own private bits) that lies *at or below* the private signal level (in this example, the bit a_3), the one-round scheme described in the previous example is sub-optimal. To extract the useful information at or below the private signal level, one of the receivers (in this example, receiver 2) can first decode and then form linear combinations using (decoded) common messages *only*.

Without loss of generality it turns out that, the above situation (there is useful information for the other receiver that lies at or below the private signal level) only occurs at one of the two receivers. In other words, there exists a receiver where no useful information (for the other receiver) lies at or below the private signal level. The reason is the following:

1. It is not difficult to see that the capacity region is convex, and hence if a scheme can achieve $\max_{(R_1,R_2) \in \mathcal{C}} \{\mu_1 R_1 + \mu_2 R_2\}$ for all $\mu_1, \mu_2 \geq 0$, it is optimal. Here $\mathcal{C}$ denotes the capacity region.
2. If $\mu_1 \geq \mu_2$, we weigh user 1's rate more. Since the private bits are cheaper to support in the sense that they do not cause interference at receiver 2, user 1

should be transmitting at its full private rate, which is equal to the number of levels at or below the private signal level at receiver 1. Therefore, all levels at or below the private signal level are occupied by user 1's private bits and there is no useful information for receiver 2 at receiver 1.

3. Similarly if $\mu_1 \leq \mu_2$, there is no useful information for receiver 1 at receiver 2, at or below the private signal level.

Hence, the following two-round strategy is optimal in the LDC: if $\mu_1 \geq \mu_2$, receiver 1 forms a certain number (no more than the cooperative link capacity) of linear combinations composed of the signals above its private signal level and sends them to receiver 2. After receiver 2 decodes, it forms a certain number of linear combinations composed of the decoded common bits and sends them to receiver 1. If $\mu_1 \leq \mu_2$, the roles of receiver 1 and 2 are interchanged. Depending on the operating point in the capacity region, we use different configurations, implying that time-sharing is needed to achieve the full capacity region.

From the above discussion, a natural and implementable two-round strategy for Gaussian channels emerges. For transmission, we use a superposition Gaussian random coding scheme with a simple power-split configuration, as described in [ETW08]. For cooperation, one of the receivers quantizes-and-bins its received signal at its private signal level and forwards the bin index; after the other receiver decodes with the side information that helps it, it bins-and-forwards the decoded common messages back to the first receiver and helps it decode.

Coding Strategy

The scenario is depicted in Fig. 12.1(a). The strategy consists of two parts: (1) the transmission scheme, describing how transmitters encode their messages, and (2) the cooperation scheme, describing how receivers exchange information and decode messages. We give an overview of the strategy below.

Transmission Scheme. We use a simple superposition coding scheme with Gaussian random codebooks. Each transmitter splits its own message into common and private (sub-)messages. Each common message is aimed at both receivers, while each private message is aimed at its own receiver. Each message is encoded into a Gaussian random codeword with certain power. For transmitter i, the power for its private and common codewords is Q_{ip} and $Q_{ic} = 1 - Q_{ip}$, respectively, for $i = 1, 2$. As [ETW08] points out, since the private signal is undesired at the unintended receiver, a reasonable configuration is to make the private interference at or below the noise level so that it does not cause much damage and can still convey additional information in the direct link if it is stronger than the cross link. When the interference is stronger than the desired signal, simply set the whole message to be common. In other words, for $(i, j) = (1, 2)$ or $(2, 1)$, $Q_{ip} = \min\left\{\frac{1}{\mathsf{INR}_j}, 1\right\}$ if $\mathsf{SNR}_i > \mathsf{INR}_j$, and $Q_{ip} = 0$ otherwise.

Cooperation Scheme. The cooperation scheme is two-round. We briefly describe it as follows: for $(i, j) = (1, 2)$ or $(2, 1)$, at the first round, receiver j

quantizes its received signal and sends out the bin index (described in detail below). At the second round, receiver i receives this side information, decodes its desired messages (both users' common messages and its own private message) with the decoder described in detail below, randomly bins the decoded common messages, and sends the bin indices to receiver j. Finally receiver j decodes with the help from the receiver-cooperative link. We call this a two-round strategy $\text{STG}_{j \to i \to j}$, meaning that the processing order is: receiver j quantizes-and-bins, receiver i decodes-and-bins, and receiver j decodes. Its achievable rate region is denoted by $\mathcal{R}_{j \to i \to j}$. By time-sharing, we can obtain an achievable rate region $\mathcal{R} := \text{conv}\{\mathcal{R}_{2 \to 1 \to 2} \cup \mathcal{R}_{1 \to 2 \to 1}\}$, the convex hull of the union of two rate regions.

There is a simple way to understand the strategy from an engineering perspective. To achieve $\max_{(R_1, R_2) \in \mathcal{R}} \{\mu_1 R_1 + \mu_2 R_2\}$ for some non-negative (μ_1, μ_2), the processing configuration can be easily determined: strategy $\text{STG}_{j \to i \to j}$ should be used, where $i = \arg\min_{l=1,2}\{\mu_l\}$ and $j = \arg\max_{l=1,2}\{\mu_l\}$. To summarize, the receiver which decodes last is the one we favor the most.

In the following paragraphs, we describe each component in detail, including quantize-binning, decode-binning, and their corresponding decoders. For simplicity, we consider strategy $\text{STG}_{2 \to 1 \to 2}$.

Quantize-binning: Upon receiving its signal from the transmitter-receiver link, receiver 2 does not decode messages immediately. Instead, serving as a relay, it first quantizes its signal by a pre-generated Gaussian quantization codebook with a certain distortion, and then sends out a bin index determined by a pre-generated binning function. How should we set the distortion? As discussed previously, note that both its own private signal and the noise it encounters are not of interest to receiver 1. Therefore, a natural configuration is to set the distortion level equal to *the aggregate noise plus the private signal power level*.

Decoder at receiver 1: After retrieving the receiver-cooperative side information, that is, the bin index, receiver 1 decodes the two common messages and its own private message, by searching the transmitters' codebooks for a codeword triple (indexed by the two common messages and the user's own private message) that is jointly typical [CT06] with its received signal and some quantization point (codeword) in the given bin. If there is no such unique codeword triple, it declares an error.

Decode-binning: After receiver 1 decodes, it uses two pre-generated binning functions to bin the two common messages and sends out these two bin indices to receiver 2.

Decoder at receiver 2: After receiving these two bin indices, receiver 2 decodes the two common messages and its own private message, by searching the transmitters' codebooks for a codeword triple such that it is jointly typical [CT06] with its received signal and the common messages that both lie in the given bins.

12.1.3 Downlink Scenario: Transmitter Cooperation

Once again, the complete expression of the inner and outer bounds of the capacity region will not be presented. Reference [WT10] contains more details. Instead, the strategy that achieves capacity to within a constant gap is described.

Coding Strategy

The scenario is depicted in Fig. 12.1(b). A natural cooperation strategy between transmitters is that, prior to each block of transmission, two transmitters hold a conference to tell each other a part of their messages. Hence the messages are classified into two kinds: (1) *cooperative* messages, which are known to both transmitters due to the information exchange, and (2) *non-cooperative* ones, which are unknown to the other transmitter since the cooperative link capacities are finite. On the other hand, messages can also be classified based on their target receivers: (1) *common* messages, which are aimed at both receivers, and (2) *private* ones, which are aimed at their own receiver. Hence there are, in total, four kinds of messages for each user, and seven codes for the whole system[2]. Now the question is, how do we encode these messages?

Our strategy turns out to be a simple superposition coding scheme, consisting of a pair of *non-cooperative* common and private codes and a pair of *cooperative* common and private codes (similar to the scheme proposed in Section 6.4.2). For the non-cooperative part, the Han-Kobayashi scheme [HK81] is employed, and the common-private split is such that the private interference is at or below the noise level at the unintended receiver [ETW08]. For the cooperative part, we use a simple linear beamforming strategy for encoding private messages, superimposed upon the common codewords. Below, we describe the strategy from a high-level perspective and leave the details to [WT10]. For the cooperative common message, we modulate it onto a two-dimensional vector code and use both transmitters to send it. Denote the cooperative common signal by $\underline{x}_o$. We choose $\underline{x}_o$ to be Gaussian with zero mean and a covariance matrix which has diagonal entries (values of transmit power) that are comparable with the total transmit power. For the cooperative private signal $\underline{x}_h$, we shall make it a superposition of *zero-forcing* vectors

$$\underline{v}_{1z} = \begin{bmatrix} h_{22} \\ -h_{21} \end{bmatrix}, \quad \underline{v}_{2z} = \begin{bmatrix} -h_{12} \\ h_{11} \end{bmatrix} \tag{12.7}$$

and *matched-filter* vectors

$$\underline{v}_{1m} = \begin{bmatrix} h_{11}^* \\ h_{12}^* \end{bmatrix}, \quad \underline{v}_{2m} = \begin{bmatrix} h_{21}^* \\ h_{22}^* \end{bmatrix}. \tag{12.8}$$

---

[2] There is only one cooperative common code carrying both cooperative common messages.

Hence, the overall cooperative signal transmitted by *combining both transmitters* is the following:

$$\underline{x}_{oh} = \underline{x}_o + \underbrace{w_{1z}\underline{v}_{1z} + w_{2z}\underline{v}_{2z} + w_{1m}\underline{v}_{1m} + w_{2m}\underline{v}_{2m}}_{\underline{x}_h}, \qquad (12.9)$$

where w_{iz} and w_{im} are independent Gaussian random codes carrying a part of the cooperative private message for user i. For the non-cooperative part, we simply transmit the superposition of two independent Gaussian random codes x_{ic} and x_{ip} for the non-cooperative common and non-cooperative private messages for user i, respectively. Hence, the overall non-cooperative signal transmitted by transmitter i is $x_{icp} = x_{ic} + x_{ip}$, for $i = 1, 2$. Overall, the transmit signal from transmitter i is the superposition of cooperative and non-cooperative signals, i.e.

$$x_i = \underline{x}_{oh}(i) + x_{icp}, \quad i = 1, 2. \qquad (12.10)$$

For the power allocation, note that the interference caused by the other user's *cooperative* private signal should be nulled out approximately, that is, its variance is at or below the noise level. Moreover, the interference caused by the other user's *non-cooperative* private signal should also be at or below the noise level. With this guideline, we can determine the power allocation policy. For more details we point the readers to [WT10].

The decoding procedure, compared to the uplink scenario, is much simpler. Each receiver decodes all common messages and its own private messages jointly.

12.1.4 UL-DL Reciprocity and Generalized Degrees of Freedom

It turns out that there is a nice reciprocity between uplink and downlink scenarios. First, we define the *reciprocal* downlink(uplink) system with respect to an uplink(downlink) system.

Definition (Reciprocal Systems). *Given an uplink(downlink) system with channel matrix $\mathbf{H}$ and cooperation capacities C_{12}^B from BS 1 to 2 and C_{21}^B from BS 2 to 1, its reciprocal downlink(uplink) system is specified by channel matrix $\mathbf{H}^*$ and cooperation capacities C_{21}^B from BS 1 to 2 and C_{12}^B from BS 2 to 1.*

The reciprocal property is summarized in the following theorem:

Theorem (Uplink-Downlink Reciprocity). *The capacity regions of the reciprocal uplink-downlink systems are within a constant gap to each other. Therefore, their system performances in the linear region are the same.*

Proof. Comparison of the outer bounds proves the result. See [WT10] for details. □

Based on reciprocity, we investigate performance in the linear region by characterizing the optimal *generalized degrees of freedom* available in the system, and demonstrate the fundamental gain from limited backhaul cooperation in the rest of this section. The notion of generalized degrees of freedom is originally proposed in [ETW08]. For simplicity, we consider a symmetric set-up, where

$$\mathsf{SNR} = \mathsf{SNR}_1 = \mathsf{SNR}_2, \ \mathsf{INR} = \mathsf{INR}_1 = \mathsf{INR}_2; \ C^\mathsf{B} = C^\mathsf{B}_{12} = C^\mathsf{B}_{21}, \tag{12.11}$$

and a standard performance measure is the symmetric capacity

$$C_{\text{sym}} := \sup \{R : (R, R) \in \text{capacity region}\}. \tag{12.12}$$

We begin with the definition.

Definition (Generalized Degrees of Freedom)**.** *Let*

$$\lim_{\mathsf{SNR} \to \infty} \frac{\log \mathsf{INR}}{\log \mathsf{SNR}} = \alpha; \quad \lim_{\mathsf{SNR} \to \infty} \frac{C^\mathsf{B}}{\log \mathsf{SNR}} = \kappa, \tag{12.13}$$

and define the number of generalized degrees of freedom per user as

$$d := \lim_{\substack{\text{fix } \alpha, \kappa \\ \mathsf{SNR} \to \infty}} \frac{C_{\text{sym}}}{\log \mathsf{SNR}}, \tag{12.14}$$

if the limit exists[3].

With the constant-gap-to-optimality result, the generalized degrees of freedom (g.d.o.f.) an be characterized via straightforward calculations:

Theorem (Number of Generalized Degrees of Freedom Per User)**.**

$$d = \begin{cases} \min\{1, \max(\alpha, 1-\alpha) + \kappa, 1 - \alpha/2 + \kappa/2\}, & 0 \leq \alpha < 1 \\ \min\{\alpha, 1 + \kappa, \alpha/2 + \kappa/2\}, & \alpha \geq 1 \end{cases} \tag{12.15}$$

Numerical plots for the g.d.o.f. are given in Fig. 12.4. We observe that the gain from cooperation varies at different values of α. By investigating the g.d.o.f., we conclude that at high SNR, when interference-to-noise ratio (INR) is below 50% of SNR (in dB), one-bit cooperation per direction buys roughly one-bit gain per user until full receiver cooperation performance is reached, while when INR is between 67% and 200% of SNR (in dB), one-bit cooperation per direction buys roughly half-bit gain per user until saturation.

---

[3] In fact, the limit does not exist when $\alpha = 1$, where the phases of the channel gains matter. In particular, its value can depend on whether the system MIMO matrix is well-conditioned or not. To overcome this issue, we pose a reasonable distribution, namely, i.i.d. uniform distribution, on the phases, show that the limit exists *almost surely*, and define the limit to be the number of *generalized degrees of freedom* per user. See [WT09a] for more details.

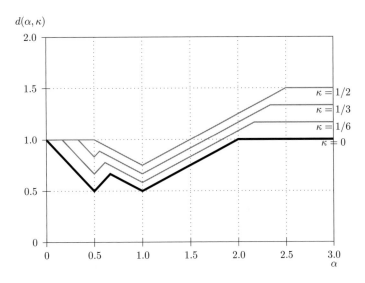

Figure 12.4 Generalized degrees of freedom. [WT09b] © 2009 IEEE.

Gain from Limited Cooperation

The fundamental behavior of the gain from limited backhaul cooperation is explained in the rest of this section, by looking at two particular points: $\alpha = \frac{1}{2}$ and $\alpha = \frac{2}{3}$ in the uplink scenario. We further use the LDC for illustration.

At $\alpha = \frac{1}{2}$, the plot of d versus κ is given in Fig. 12.5(a). The slope is 1 until full receiver cooperation performance is reached, implying that one-bit cooperation buys one more bit per user. We look at a particular point $\kappa = \frac{1}{4}$ and use its corresponding LDC (Fig. 12.5(b)) to provide insights. Note that 1 bit in the LDC corresponds to $\frac{1}{4} \log \mathsf{SNR}$ in the Gaussian channel, and since $\mathsf{C}^\mathsf{B} \approx \frac{1}{4} \log \mathsf{SNR}$, in the corresponding LDC each receiver is able to send one-bit information to the other. Without cooperation, the optimal way is to turn on bits not causing interference, that is, the *private* bits a_3, a_4, b_3, b_4. We cannot turn on more bits without cooperation since it causes collisions, for example, at the fourth level of receiver 2 if we turn on bit a_2. Now with receiver cooperation, we want to support two more bits a_2, b_2. Note that prior to turning on a_2, b_2, there are "holes" left in receiver signal spaces, and turning on each of these bits only causes one collision at each receiver. Therefore, we need 1 bit in each direction to resolve the collision at each receiver. We can achieve 3 bits per user in the corresponding LDC and $d = \frac{3}{4}$ in the Gaussian channel. We cannot turn on more bits in the LDC since it causes collisions, while no cooperation capability is left.

At $\alpha = \frac{2}{3}$, the plot of d versus κ is given in Fig. 12.5(c). The slope is $\frac{1}{2}$ until full receiver cooperation performance is reached, implying that two-bit cooperation buys one more bit per user. We look at a particular point $\kappa = \frac{1}{3}$ and use its corresponding LDC (Fig. 12.5(d)) to provide further insight. Now, note that 1 bit in the LDC corresponds to $\frac{1}{3} \log \mathsf{SNR}$ in the Gaussian channel, and since

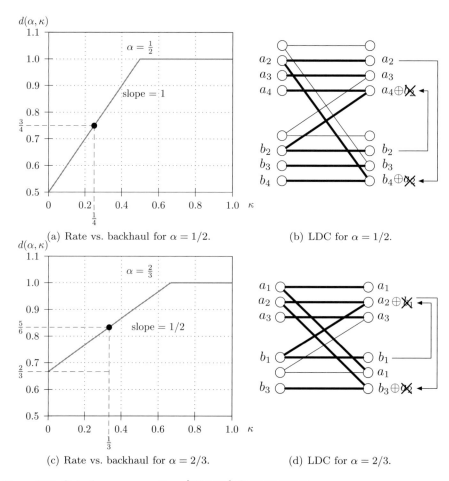

Figure 12.5 Gain from cooperation. [WT09b] © 2009 IEEE.

$C^B \approx \frac{1}{3} \log \text{SNR}$, each receiver is able to send one-bit information to the other in the corresponding LDC. Without cooperation, the optimal way is to turn on bits a_1, a_3, b_1, b_3. We cannot turn on more bits without cooperation since it causes collisions, for example, at the second level of receiver 2 if we turn on bit a_2. Now with receiver cooperation, we want to support one more bit a_2. Note that prior to turning on a_2, there are no "holes" left in receiver signal spaces, and turning on a_2 causes collisions at *both* receivers. Therefore, we need 2 bits in total to resolve collisions at both receivers. We can achieve 5 bits in total in the corresponding LDC and $d = \frac{5}{6}$ in the Gaussian channel. We cannot turn on more bits in the LDC since it causes collision while no cooperation capability is left.

From above examples and illustrations, we see that whether *one cooperation bit buys one more bit* or *two cooperation bits buy one more bit* depends on whether there are "holes" in receiver signal spaces before increasing data rates. This provides the high-level intuition for why there are two different behaviors of the interference-mitigation gain from limited backhaul cooperation.

12.1.5 Summary

In this section, we investigated fundamental limits of interference mitigation with limited backhaul cooperation by studying a canonical information-theoretic model, namely, the two-user Gaussian interference channel with limited backhaul cooperation. In this simple setting, we identify two regions pertaining to the fundamental gains from backhaul cooperation: linear and saturation regions. By characterizing the capacity region to within a constant gap, we establish uplink-downlink reciprocity and determine the optimal generalized degrees of freedom available in the system. We conclude that cooperation in the linear region is efficient and provides a *degrees-of-freedom* gain, which is either *one cooperation bit buys one more bit* or *two cooperation bits buy one more bit* until saturation, whereas cooperation in the saturation region is inefficient and provides a *power* gain, which is at best a constant regardless of the rate at which base stations cooperate.

12.2 Backhaul Requirements of Practical CoMP Schemes

Christian Hoymann, Laetitia Falconetti and Patrick Marsch

After having ventured into information-theoretic aspects about backhaul-constrained cooperation, this section now discusses backhaul requirements in terms of capacity and latency related to the usage of different CoMP schemes under more *practical* considerations. More precisely, we investigate We first distinguish between different types of data exchanged over the backhaul and look into general scaling laws in Subsection 12.2.1, and then determine specific backhaul requirements for exemplary CoMP schemes in Subsection 12.2.2. Backhaul topology issues are then discussed in Subsection 12.2.4, and latency issues in Subsection 12.2.3. The topic is concluded in Subsection 12.2.5.

12.2.1 Types of Backhaul Data and Scaling Laws

In both uplink and downlink, we can generally distinguish between backhaul requirements for the exchange of

- (partially) processed user data or receive signals,
- channel state information,
- scheduling information, or
- signaling information.

The extent of these different types of backhaul data depends on the particular CoMP scheme used, and will be elaborated in the sequel. Note that all backhaul quantities (and scaling terms) in this subsection always refer to *backhaul per cell*. As the backhaul requirements for signaling are in most cases negligible in terms of capacity, we focus on the other three types of backhaul data in the remainder of this section.

Let us first consider the CoMP schemes based on *interference coordination*, as introduced in Chapter 5. In the case of *uplink joint scheduling* (see Section 5.2) within a cluster of M cells, exactly $M-1$ base stations (BSs) have to forward quantized channel state information (CSI) as well as buffer status information to the central scheduling unit (CSU)[4], which sends back the performed scheduling decision. Clearly, the backhaul required for CSI exchange increases linearly in the number of users U per cell, the system bandwidth B, and in the number of antennas per BS, i.e. $O(UBN_{\text{bs}})$. The requirement for feeding back the scheduling decisions simply increases according to $O(B\log U)$, as $\log_2 U$ bits are needed to distinguish terminals. Note that from BS perspective, both backhaul quantities do not scale in the cooperation size. While this is not explicitly covered in this book, the backhaul required for *downlink* joint scheduling obviously has the same scaling behavior as in the uplink.

For *downlink coordinated beamforming* (see Section 5.3), cooperating BSs exchange indices of precoding vectors that should be avoided in order to alleviate inter-cell interference. Backhaul requirements hence scale with the system bandwidth B, the number of bits needed to represent the index of a precoding vector, which we denote as C, and the cooperation size M, as each BS needs to distribute the data to all other BSs.

Let us now observe CoMP schemes based on *joint signal processing*, as introduced in Chapter 6. For *uplink centralized joint detection* (see Section 6.1), $M-1$ BSs forward quantized receive signals to one master BS in the cluster. Here, the required backhaul scales with $O(B \cdot \min(N_{\text{bs}}, MN_{\text{ue}}))$. The $\min(\cdot)$ expression is due to the fact that it does not make sense to quantize more scalar values per orthogonal frequency division multiplex (OFDM) symbol than the number of Eigenmodes the channel offers. Hence, if the number of overall transmit antennas of the user equipments (UEs) jointly processed on the same physical resource is larger than the number of receive antennas per BS, the signals at each receive antenna should be quantized independently. Otherwise, however, it is more efficient to perform a Karhunen-Loève transform [GDV02] on the received signals to remove the signal correlation. In the joint detection case, it appears intuitive to quantize all signals received by the BSs in frequency domain, including the pilots contained therein, so that an explicit exchange of CSI is not required.

In the case of *uplink decentralized interference cancelation* (see Section 6.2), decoded data bits of the UEs (or coded or uncoded soft bits) are exchanged between BSs. In the extreme case that each BS iteratively forwards soft bits connected to its assigned UE to all other BSs in the cluster, the backhaul effort increases with $O(BML)$, if L is the number of iterations performed. Decentral-

---

[4] Joint scheduling could of course also be performed in a decentralized way, but we here take as an example the particular scheme introduced in Section 5.2.

12.2 Backhaul Requirements of Practical CoMP Schemes

Table 12.1. Scaling behavior of the backhaul requirements of different CoMP schemes.

Schemes based on interference coordination	User data / rx/tx signals	CSI / PMI Scheduling data
Uplink joint scheduling (see Section 5.2)	-	$O(UBN_{\text{bs}})$ $+O(B\log U)$
Downlink coordinated beamforming (see Section 5.3)	-	$O(BCM)$
Schemes based on joint signal processing	**User data / rx/tx signals**	**CSI / PMI Scheduling data**
Uplink centralized joint detection (see Section 6.1)	$O(B\min(N_{\text{bs}}, MN_{\text{ue}}))$	implicit in signals
Uplink decentralized joint detection (see Section 6.2)	$O(BML)$	-
Downlink centralized joint transmission (see Section 6.3)	$O(B\min(N_{\text{bs}}, MN_{\text{ue}}))$	-
Downlink distributed joint transmission (see Section 6.3)	$O(BM)$	$O(BN_{\text{bs}}N_{\text{ue}}M^2)$
Downlink decentralized multi-user transmission (see Section 6.4)	$O(BM)$	$O(BN_{\text{bs}}N_{\text{ue}}M^2)$

ized interference cancelation, as it has been introduced in Section 6.2, does not require the exchange of explicit channel information.

In *downlink centralized joint transmission*, as introduced in Section 6.3, precoding for multiple BSs is performed by a central network entity or by one of the involved BSs, and analog signals (known as radio over fibre (RoF)) or quantized transmit signals are forwarded to each of the other BSs. In the latter case, the backhaul scales with $O(B\min(N_{\text{bs}}, MN_{\text{ue}}))$, as in the centralized uplink case. The $\min(\cdot)$ operation is again based on the fact that the backhaul depends on the rank of the channel seen by each BS. In the case where, for example, a single UE is served by multiple BSs with multiple antennas each, it is best to forward the symbols to be transmitted to the terminal and the precoding vector separately, instead of quantizing the transmit signal of each antenna individually.

In the case of *downlink distributed joint transmission*, which was also discussed in Section 6.3 and is equivalent in terms of performance to centralized joint transmission, each BS performs precoding individually. Backhaul is required for the distribution of user data among the BSs, which scales with the system bandwidth

B and and cooperation size M, hence $O(BM)$. Furthermore, the BSs have to exchange channel information, i.e. each BS has to forward local channel matrices with $N_{bs}MN_{ue}$ coefficients each to all other cooperating BSs, and this of course scales with the system bandwidth B, leading to an overall $O(BN_{bs}N_{ue}M^2)$. While downlink joint transmission is typically not so critical from a backhaul point of view as joint detection in the uplink, this quadratical scaling in the cooperation size M can indeed become an issue, as pointed out in [MFF10].

The backhaul scaling behavior of the various CoMP schemes discussed before is summarized in Table 12.1. While this gives a first insight into general backhaul tendencies connected to the different schemes, we will focus on the exact extent of backhaul required in particular CoMP scenarios in the next subsection.

12.2.2 Specific Backhaul Requirements of Exemplary CoMP Schemes

We now want to derive concrete backhaul quantities for an exemplary subset of the CoMP schemes mentioned before. For this, we observe a particular CoMP cluster of $M = 3$ cooperating BSs, where each BS has $N_{bs} = 4$ transmit and receive antennas. Such a cluster can for example be created through static or dynamic clustering techniques, as introduced in Sections 7.1 and 7.2, respectively. As also stated in that context, current cellular systems typically employ sectorization, i.e. 3 BSs pointing towards different azimuths are usually grouped to one site, so that backhaul links between co-located BSs can be assumed to be a negligible issue. In the sequel, we assume that the observed cooperation cluster consists of two co-located BSs and one BS belonging to a different site, as shown in Fig. 12.6, so that only one logical inter-site backhaul link has to be considered. We furthermore assume that the cluster may use a system bandwidth of 20 MHz. If resource partitioning schemes are used as in Section 7.1, or a different bandwidth is otherwise available, all following results scale accordingly. Note that the backhaul quantities stated in the sequel refer to the overall bi-directional backhaul traffic on the logical inter-site backhaul link (i.e. we do *not* normalize this to a per-cell quantity as before, as we are now more interested in the overall capacity the link has to provide).

Downlink Coordinated Scheduling / Coordinated Beamforming

Let us first consider a downlink interference coordination scheme called coordinated scheduling / coordinated beamforming (CS/CB), which extends the coordinated beamforming concept mentioned before and introduced in Section 5.3 by coordinated scheduling. Here, all BSs use local precoding to transmit non-cooperatively to their assigned UEs, but exchange information on precoding vectors that may cause strong interference towards other intra-cluster cells. The BSs then perform coordinated scheduling based on the precoding constraints that have been exchanged. Applied to our exemplary CoMP cluster, let us assume that at each point in time exactly one of the 3 cluster BSs has the role of a

12.2 Backhaul Requirements of Practical CoMP Schemes

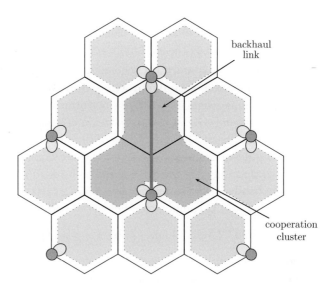

Figure 12.6 Example cluster with one inter-site link considered for backhaul capacity requirement calculations in this section.

master, and the other 2 are slaves. These roles then change cyclically. The UEs are assumed to continuously report the indices of precoding vectors (precoding matrix indicators (PMIs)) their assigned BS should use to maximize signal-to-interference-and-noise ratio (SINR), as well as the worst possible precoding vectors the other cells could use in terms of interference, referred to as worst companion indicator (WCI). In each transmit time interval (TTI), the *master* BS first assigns its UEs to physical resource blocks (PRBs) and applies the UE-chosen precoding vectors. It then forwards the indices of the precoding vectors (PMIs) and the WCIs to the 2 slave cells, along with information on how much SINR gain can be expected if the worst precoders are avoided. The slave BSs then perform their own scheduling and choice of precoding vectors (knowing through the PMI information obtained from the master BS and the WCI information from the own UEs which resource-UE combinations to avoid, and knowing from the WCIs from the master BS which precoders to avoid). The scheme will be elaborated in more detail in the context of system level simulations in Section 14.4.3. Assuming that all stated values are reported once for each PRB and require 4 bit each, the bi-directional backhaul capacity required in our cluster is given as

$$\frac{100 \text{ PRBs} \times (4 \text{ bit PMI } + 4 \text{ bit WCI } + 4 \text{ bit SINR impr.})}{1 \text{ ms}} = 1.2 \text{ Mbit/s} \quad (12.16)$$

Uplink Centralized Joint Detection

As stated before, the BSs here forward quantized receive signals to one central BS that performs multi-user and multiple-antenna receive signal processing, as if all signals would originate from its own antennas. We first investigate most straightforward quantization schemes, and then discuss how the required backhaul can be reduced through more sophisticated signal processing.

The most straightforward scheme is to **quantize and exchange samples of the received signals in time domain**, independently for each receive antenna. This would typically be performed after quadrature down-conversion in the baseband, i.e. we obtain a stream of samples for both the in-phase (I) and quadrature-phase (Q) components of the signal. In our example setup, it is obviously most backhaul-efficient to let the decoding BS be one of the 2 co-located BSs, so that only the third BS needs to forward received signals. For a 20 MHz system bandwidth, a sampling rate of 30.72 MHz is typically used, and assuming 8 bit per I and Q component, the overall backhaul traffic sums up to

$$30.72 \text{ MHz} \cdot 4 \text{ ant.} \cdot (8 \text{ bit I} + 8 \text{ bit Q}) = 1.97 \text{ Gbit/s}. \qquad (12.17)$$

Time domain IQ samples contain information about the whole channel bandwidth. In OFDM-based systems, where scheduling allows partial bandwidth allocation, part of the information contained in the time domain IQ samples may in fact not be relevant for joint detection. Furthermore, the time domain samples contain guard bands and guard intervals, which - after synchronization by the supporting BS - are also not relevant for the detection process. In OFDM, it hence may be more adequate to exchange **frequency domain IQ samples** that are obtained by the supporting BS by processing the receive signal up to the fast Fourier transform (FFT) (see Fig. 12.7). This also enables to limit the backhaul traffic by selecting and transferring only the IQ samples corresponding to the sub-carriers of interest. Assuming the received symbols in all sub-carriers are forwarded, the before stated backhaul requirement reduces to

$$1200 \text{ sub-carriers} \cdot 4 \text{ ant.} \cdot (8 \text{ bit I} + 8 \text{ bit Q}) \cdot 14 \text{ kHz} = 1.08 \text{ Gbit/s}. \qquad (12.18)$$

From information theory, it is known that the ratio of quantization noise vs. the number of invested quantization bits can be improved if multiple scalar values are quantized jointly, referred to as **vector quantization** [GG92]. The reason is simply that then all possible discretized signal values can be distributed over a signal space of higher dimension, such that the Euclidian distance of each possible receive signal to the nearest discrete representation is decreased. This is illustrated in [CT06]. Furthermore, vector quantization can exploit signal correlation, if for example applied across the signals received in the same OFDM symbol by multiple antennas. In this case, either a specific quantization codebook is needed, which has been designed for a particular correlation, or (equivalently) the receive signals are decorrelated through a Karhunen-Loève transform before quantization with a standard codebook is applied.

In principle, backhaul requirements can be further reduced if not only the correlation between receive signals at each quantizing BS is exploited individually, but also that between *all* supporting BSs *and* the signals received by the

12.2 Backhaul Requirements of Practical CoMP Schemes

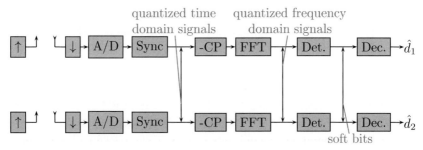

Figure 12.7 Physical layer of an LTE uplink receiver, with the different points at which inter-BS information change can take place.

decoding BS. This requires source coding by the quantizing BSs. A particular (though theoretical) scheme is to perform **distributed Wyner-Ziv compression** [Gas04], where it is known that (for Gaussian signals) the backhaul required from each quantizing BS to the decoding BS only corresponds to the entropy of the quantized signal at the quantizing BS, *conditioned* on the quantized signals at all other BSs. Though some practical algorithms have been proposed that can approach Wyner-Ziv performance [XLC04], it must be noted that such source coding schemes are highly questionable in practical systems, as the encoder requires knowledge of the extent of correlation of all signals, which may change entirely between successive TTIs due to scheduling. Furthermore, source coding further increases the backhaul signaling delay, which is already a critical factor (see Section 11.2).

Uplink Decentralized Joint Detection
Let us now consider the case of *decentralized* joint detection, where each BS detects and decodes its assigned UEs, but with the aid of other BSs, as introduced in Section 6.2. This can be based on an on-demand mechanism as described in Section 11.2, where for each UE to be decoded there is one *serving* and (possibly multiple) *supporting* BSs. Cooperation can in this case also be based on the exchange of quantized receive signals in time or frequency domain, as discussed before, and the same options of signal quantization and compression apply. Note that decentralized joint detection is rather inefficient from a backhaul perspective, as pointed out in Section 4.3.3, as then receive signals have to be simultaneously exchanged in different cooperation directions, instead of being gathered at one central point. However, a decentralized scheme is advantageous in the way that it is closer to an LTE Release 8 system in terms of core network interfaces. For our example scenario illustrated in Fig. 12.6, and in the extreme case that each BS requests receive signals from both other BSs for all PRBs, the overall bi-directional backhaul needed for the exchange of frequency domain signals would increase three-fold from (12.18) to

$$1200 \text{ sub-carriers} \cdot 12 \text{ ant.} \cdot (8 \text{ bit I} + 8 \text{ bit Q}) \cdot 14 \text{ kHz} = 3.24 \text{ Gbit/s}. \quad (12.19)$$

Especially in the context of decentralized joint detection, backhaul requirements can be strongly reduced if **soft coded bits** are exchanged between BSs instead of IQ samples. Any supporting BS processes the received signal from a supported user up to detection (more precisely, up to demodulation), as shown in Fig. 12.7. The obtained coded bits are quantized and become soft bits which are transferred to the serving BS. In OFDM transmission, each sub-carrier conveys one modulated symbol and is therefore associated with a certain number of coded bits. Consequently, and by contrast to the previous mode based on IQ sample exchange, the backhaul traffic generated by this joint detection mode is specific to a certain data stream sent by the considered UE. It depends on the modulation and the number of sub-carriers used to transmit this stream. Assuming a quantization rate of 5 soft bits per coded bit, that all 1200 sub-carriers of the system are assigned to UEs employing 16-QAM modulation for the transmission, and that all 3 BSs request soft coded bits from both other BSs and for all resources, the bi-directional backhaul traffic requirement can be calculated as

$$4 \text{ UEs} \cdot 1200 \text{ sub-carriers} \cdot 4 \text{ bits/symbol} \cdot 5 \text{ bits} \cdot 14 \text{ kHz} = 1.34 \text{ Gbit/s}. \quad (12.20)$$

The factor 4 at the beginning of (12.20) is due to the fact that 3×2 sets of UE-specific soft coded bits have to be exchanged, but 2 BSs are co-located, so that effectively, 4 sets of soft bit data have to be exchanged over the inter-site backhaul link. Both IQ sample and soft coded bit based decentralized joint detection are also covered in the context of system level simulations in Section 14.3.

Downlink Centralized Joint Transmission

In terms of backhaul usage, centralized joint transmission in the downlink is analog to uplink centralized joint detection treated before. Transmit signals are computed by one central point in the network (which could be a BS by itself), and then forwarded to the other cluster BSs. We again have the degrees of freedom of forwarding IQ samples in time domain, or, more efficiently, in frequency domain, where (12.17) or (12.18) apply. As the transmit signals in the downlink, however, are not obscured through noise, but are the result of precoding, exploiting co-antenna signal correlation becomes even more important. This is quite obvious in the exemplary case where one UE is served by multiple BSs with many antennas each. Here, it is clearly more efficient to forward the transmit signals connected to the UE and a precoding vector separately, rather than quantizing the transmit signals for each antenna separately.

Downlink Distributed Joint Transmission

In this case, all cluster BSs perform precoding redundantly, and backhaul is required for the exchange of channel knowledge and user data bits. Let us first focus on the backhaul effort required for CSI exchange. As discussed in Chapter 9, CSI for a single-input single-output (SISO) link is most efficiently captured by observing the most significant taps of the channel impulse response (CIR) in

12.2 Backhaul Requirements of Practical CoMP Schemes

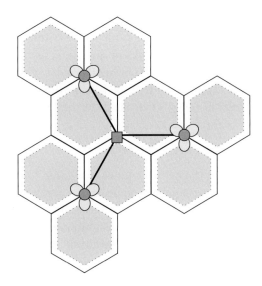

Figure 12.8 Star-like network topology.

time domain. For a wide range of urban scattering scenarios, it may be sufficient to capture the 4 dominant taps, with 4 bit per tap and I/Q dimension for the corresponding coefficient and 10 bit per tap for the delay. In our example cluster of 3 BSs of which 2 are co-located, each BS has to forward CSI *once* over the backhaul. Assuming that CSI needs to be updated each TTI, and $N_{\text{bs}} = 4$ and $N_{\text{ue}} = 1$ as before, we obtain a backhaul rate for CSI exchange of

$$\frac{4 \text{ taps} \cdot (4 \text{ bit I} + 4 \text{ bit Q} + 10 \text{ bit delay}) \cdot N_{\text{bs}} N_{\text{ue}}}{1 \text{ ms}} = 576 \text{ kbit/s}. \quad (12.21)$$

The backhaul requirement for distributing the user data to all involved BSs obviously depends on the achieved rates. We here just want to state some exemplary numbers to get an idea of the orders of magnitude. Assuming that a non-cooperative system with $N_{\text{bs}} = 4$ and $N_{\text{ue}} = 1$ and PMI-based precoding achieves roughly 2 bit/s/Hz/cell, and we can increase the average spectral efficiency through CoMP in our small cluster of 3 cells by 50%, then this leads with the system bandwidth of 20 MHz to an additional backhaul traffic of 180 Mbit/s. Clearly, this is significantly more than the backhaul required for CSI exchange.

12.2.3 Backhaul Latency Requirements

In mobile communication systems such as LTE, a short term hybrid automatic repeat request (HARQ) mechanism requires the timely transmission of feedback by the receiver to signal correctly or incorrectly decoded data. In the FDD LTE uplink, the BS must send its HARQ feedback exactly 4 ms after receiving the UE signal. To meet the defined fixed HARQ timing and considering the additional signal processing delay required at supporting BSs (in particular for the soft bit exchange mode), backhaul latency should not exceed 1 ms. This limits the choice of backhaul technologies suitable for BS cooperation to optical fiber or extremely low-latency Microwave links or air lasers, as will be discussed in 12.3. However, methods to relax the latency requirement while keeping the fixed HARQ timing can be applied as described in Section 11.2. This enables the usage of BS cooperation in today's systems that may not have an optical fiber backhaul.

12.2.4 Backhaul Topology Considerations

Depending on the network topology chosen to connect sites with each other, the overall backhaul capacity requirements become more or less stringent. A star-like topology as illustrated in Fig. 12.8 can be considered as the worst case scenario, since a single backhaul link must support all incoming and outgoing backhaul traffic generated by BS cooperation. If BS cooperation involves BSs located at the same site, the messages exchanged between cooperating BSs are not conveyed through the backhaul. Therefore, intra-site BS cooperation is applicable to any network regardless of the capabilities of their backhaul technology.

If BS cooperation is applied in an already deployed network, procedures needed for CoMP have to be adapted to the existing network topology, leading to suboptimal CoMP operation. An example of affected procedures is the CoMP cell clustering discussed in Chapter 7. If the backhaul infrastructure between two BSs cannot support CoMP operation because of its capabilities regarding capacity or latency, these two BSs can not belong to the same CoMP cell cluster, although CoMP between them could be beneficial for certain users.

An improved network dimensioning that better leverages the potential of CoMP is hence preferred for next generation networks. In areas where CoMP is expected to be beneficial, e.g. to enhance cell capacity in hotspots or improve performance in areas where users experience strong inter-cell interference, backhaul requirements should be considered when dimensioning the network. In a first step, a CoMP scheme, its parameters and potential CoMP cell clusters could be decided based on traffic statistics, geographical, and further considerations. In a second step, a network topology and infrastructure would then be chosen that reflects the backhaul requirements for the decided CoMP setup.

12.2.5 Summary

In this section, we initially investigated the scaling laws of the backhaul traffic generated by different CoMP schemes. From Table 12.1 it can be seen that the CoMP cell cluster size M is a main backhaul driver for most of the schemes, being another reason why CoMP should only be performed in reasonably dimensioned clusters. Based on exemplary CoMP schemes, the discussion also emphasized that the generated backhaul load strongly fluctuates from a few Mbit/s to a few Gbit/s depending on the considered CoMP scheme. Therefore, the backhaul infrastructure of already deployed networks may restrict the set of CoMP schemes that an operator can apply on a short term. When selecting a CoMP scheme, not only the related backhaul requirements must be considered, but also the achievable enhancement in spectral efficiency and cell-edge user throughput. The strongly differing backhaul requirements hence support the motivation of adaptive, or possible ad-hoc CoMP, as stated in Chapter 4 and Section 11.3, respectively. In general, it appears desirable to design the backhaul infrastructure such as to support CoMP between those cells where it is most beneficial.

Note that backhaul capacity requirements of a few CoMP schemes are also analyzed in conjunction with spectral efficiency gains in Section 14.3.

12.3 CoMP Backhaul Infrastructure Concepts

Torsten Fahldieck and Mark Doll

In the last section of this chapter, we finally discuss current and upcoming backhaul technologies that appear promising to satisfy the backhaul requirements in terms of capacity and latency stated in the last section.

Traditional backhaul equipment deployed for 3G cellular systems consists of time division multiplex (TDM) technology based on plesiochronous digital hierarchy (PDH) or synchronous digital hierarchy (SDH)/synchronous optical network (SONET). They provide multiple (up to about 100) E1/T1 (2 Mbit/s/1.5 Mbit/s) or a single STM-1/OC-3 (155 Mbit/s) up to STM-4/OC-12 (622 Mbit/s) respectively, over copper (E1/T1 only), fiber and wireless microwave links. For LTE-A and beyond, such access technologies merely provide sufficient bandwidth just for the user data rates possible with LTE-A, not to mention the additional backhaul bandwidth that is required for CoMP operation, as shown in the last section. Furthermore, inter-base station (BS) communication is currently not as sensitive to communication latency as CoMP, i.e. a latency in the order of 10 ms is deemed sufficient. Backhauling for a CoMP-enabled radio access network (RAN) therefore requires new backhaul technologies.

In the following four Subsections 12.3.1 to 12.3.4, selected high speed link technologies, summarized in Table 12.2, that could satisfy future requirements set out by CoMP are presented. Afterwards, the current inter-BS interface X2 is introduced and possible deficits with respect to CoMP are identified in Subsection 12.3.5. Finally, Subsection 12.3.6 proposes possible approaches to reduce inter-BS communication latency as well as increase the available bandwidth for CoMP in the backhaul, before this section concludes with with a short summary.

12.3.1 Ethernet

Ethernet [IEE08b] is not only the dominant technology for wireline based local area network (LAN) but is actually more and more deployed in metropolitan area networks (MANs) and wide area networks (WANs) replacing legacy MAN and WAN technologies like PDH and SDH/SONET. Alternatively to the 'dark fiber' approach, existing SDH/SONET can be reused, either as a transport network via the generic framing procedure (GFP) [ITU08b]/virtual concatenation (VCAT) [G.707]/link capacity adjustment scheme (LCAS) [ITU06a] approach or by utilizing the 10GBASE-W physical layer family [IEE08b] to connect to SDH/SONET line equipment directly on the physical layer using it as an exclusive point-to-point link. If the more stringent synchronous Ethernet timing characteristics [ITU10b] are fulfilled, a 10GBASE-W interface can even be client of a SDH/SONET and share lines with others.

Ethernet offers data rates from 10 Mbit/s up to 100 Gbit/s [IEE10b] and uses short copper lines as well as optical fibers of up to 40 km in length. The requirements of current and future RANs make 1 Gbit/s and higher speed Ethernet based on optical fibers convenient for RAN implementation. Copper based Ethernet does not provide the required link distances, but may be an option (as well as Ethernet in the backplane) for intra-site communication among those BSs co-located at the same site. Like other wireline technologies, Ethernet provides a very good error performance. The error rate of accurately installed optical fibers can be neglected. Ethernet switches may cause packet loss due to internal buffer overflows. These error cases can be avoided by appropriate provisioning to prevent such (short-term) overload. Latency is introduced by

i) propagation delay in the fibers,
ii) non-preemptive transmission of Ethernet frames,
iii) switching delay and
iv) queueing delay.

i) Propagation delay is about 5 μs/km, since fibers have a refraction index around 1.5. ii) Since an already ongoing transmission of a lower prioritiy packet is not preempted, a higher priority packet has to wait up to the duration of a maximum sized packet. Starting with 1 Gbit/s line speed, jumbo frames have been introduced. At 1 Gbit/s, a frame size of 9000 byte corresponds to a frame duration of 72 μs, that has to be added in the worst case for each hop. iii) Switching delay of high performance cut-through (or more precisely wormhole) switches is in the order of a few microseconds independent of frame sizes, while in store-and-forward switches switching delay can amount to multiple frame durations. Finally, iv) the queueing delay depends on the offered load, assigned service class, scheduling strategy and how traffic merges and diverges along the path through the backhaul, i. e. is influenced by the network topology. By appropriate provisioning, queueing delays can be kept small in the order a few to no frame durations, at least for highly prioritized service classes.

To conclude, many latencies connected to Ethernet depend on the frame duration and are therefore inversely proportionally to line speed. At 1 Gbit/s and below, these latencies can be larger than the propagation delay in the fibers, while

12.3 CoMP Backhaul Infrastructure Concepts

Table 12.2. Overview of different backhaul technologies.

Technology	Max. bandwidth	Min. node latency	Link latency	Max. len.
Ethernet	100 Gbit/s	a few[a] μs	5 μs/km (fiber)	40 km
XGPON	10 GBit/s shared	~100 μs (upstr.)	5 μs/km (fiber)	20 km
VDSL2	mult.[b] 100 Mbit/s	1.25...2 ms[c]	$\ll$ node latency	~0.4 km[d]
Microwave	1 Gbit/s	~100 μs	3.3 μs/km (air)	a few[e] km

[a] cut-through switch, [b] using bonding and phantom mode, [c] interleaving disabled, [d] for the given bandwidth, [e] depending on required resilience to heavy rain

at 10 Gbit/s and above, propagation delay dominates the minimum achievable latency of an Ethernet.

12.3.2 Passive Optical Network

A passive optical network (PON) is a point-to-multipoint, fiber based access network architecture. Passive optical splitters are used to enable a single optical fiber to serve a multitude of nodes. In the RAN context, a PON consists of a single central optical line termination (OLT), which connects the PON to other parts of the RAN, and a number of remote optical network units (ONUs), each of them connecting to and co-located with the BSs at one site. OLT and ONUs are interconnected by the optical distribution network (ODN), a tree of optical single mode fibers with an optical splitter at each branching point. A PON reduces the amount of fiber and equipment required compared to a point-to-point architecture like Ethernet. On the other hand, the available bandwidth per node is reduced with respect to the available line rate, since all nodes share one wavelength (per direction): downstream signals are broadcasted to all ONUs within one PON, upstream signals are combined using a multiple access protocol, usually time division multiple access (TDMA). Statistical multiplexing can to some extent regain parts of the bandwidth depending on the peak to average rate ratio of the traffic. The physical broadcast in downstream direction renders this access network technology particularly suitable for implementation of data and control distribution for downlink CoMP. Downlink data and control can be natively broadcasted to each of the ONUs respectively BSs, neutralizing the drawback of shared bandwidth at least in the downstream direction.

Currently 10-Gigabit-capable PON (XGPON) [ITU10c], more specifically XGPON1 [ITU10d], provides the highest transmission speeds of 10 Gbit/s downstream (OLT to ONU) and 2.5 Gbit/s upstream. XGPON2, which will provide 10 Gbit/s also in upstream direction is left as a future option, when current technology challenges to obtain low cost 10 Gbit/s upstream burst mode building blocks will have been overcome. The competing IEEE PON standard [IEE09] (called Ethernet PON (EPON) there), also provides 10 Gbit/s in downstream, while upstream can be either 1 Gbit/s or 10 Gbit/s. The error performance of a PON is similar to other fiber based wireline technologies like Ethernet.

Regarding latency, downstream is far less critical than upstream. While the OLT continuously transmits in downstream direction, an ONU has to wait until it receives an upstream allocation from its OLT before it may transmit. The OLT solicits buffer occupancy reports from its ONUs to decide about bandwidth assignments and/or monitors Gigabit-capable PON (GPON) encapsulation method idle frames to identify unused upstream allocations. Restoring the required upstream bandwidth for an ONU, e.g. after it has gone idle, introduces additional latency. The GPON dynamic bandwidth assignment (DBA) algorithm [ITU08c][5] distinguishes between fixed, assured, non-assured and best effort bandwidth. Only fixed bandwidth assignments are unconditionally available without requiring prior indication by the ONU for additional bandwidth. For assured bandwidth, a target, i.e. a *should*, of 2 ms (with a limit, i.e. a *must*, of "a few ms") is stated by the standard for the time until an ONU's assured bandwidth is restored after a period of lower bandwidth usage. For non-assured and best-effort bandwidth, the target of 6 ms (with a limit of 10 ms) is even less strict. Latency sensitive CoMP schemes may hence require fixed bandwidth assignments to meet the required latencies. The drawback is that the possible statistical multiplexing gain is reduced, especially if those CoMP schemes also require high data rates, i.e. a high fixed bandwidth assignment.

Additional latency is introduced by the capability of automated new ONU recognition and ranging by so-called *quiet windows*. These are inserted regularly at the discretion of the OLT, e.g. about once a second, with all scheduled upstream transmissions ceased. Their proposed length is 2 GPON transmission convergence (GTC) frames, i.e. 250 μs, but can be halved to 1 GTC frame if the OLT to ONU distance is limited to 7.5 km (or if the OLT a priori knows the distance to the nearest ONU and the difference in distance between that nearest and the farthest ONU is no more than 7.5 km). Quiet windows only make up a small fraction of all frames and therefore impact the latency only on a small fraction of the overall traffic (unless the load approaches 100 %).

Upstream transmissions cannot be arbitrarily short due to their physical layer overhead (Tx disable/enable, Rx settling, i.e. setting the decision threshold, clock recovery, frame delineation). Physical layer overhead is below one microsecond (possibly well below, depending on OLT's Rx capabilities). Assuming a high bandwidth demand per BS, low splitting factors of e.g. 8 may be reasonable. Since the GTC frame frequency is 8 kHz, each ONU may be served once within each frame respectively within each 125 μs period, resulting in about 15 μs per ONU and an acceptable physical layer overhead of about 5 %. This way, latency introduced by TDMA in upstream direction may be limited to 125 μs in the worst case (and half of that on average). A forward error correction (FEC) is optional and only required for larger distances to increase the link budget. Even

---

[5] The respective algorithm for XGPON [ITU10e] has not yet been approved, but supposedly the same or at least a similar algorithm will be adopted for XGPON.

if enabled, the block size is only 255 bytes, i. e. less than 1 μs. There is no cyclic redundancy check (CRC) for payload data (there is a bit-interleaved parity, but solely for the purpose of measuring the post FEC GTC frame error rate). Detection and discarding of erroneous frames is instead provided by e. g. the Ethernet multiple access channel (MAC) in case Ethernet frames are transmitted over the PON; accordingly, the latency of native Ethernet has to be added.

To sum up, the minimum latency of a PON in downstream is comparable to native Ethernet over fiber, but larger on average, because the medium is shared and transmission to a certain ONU thus may have to wait until transmissions to other ONUs have been finished. In upstream direction, latency is dominated by PON's TDMA nature. If carefully configured, especially by using fixed bandwidth allocations, a small splitting factor and short upstream bursts, latency for a large fraction of the overall traffic can be reduced to the order of 100 μs.

12.3.3 Digital Subscriber Line

So-called digital subscriber line (DSL) technologies reuse (the last part of) the plain old telephone service (POTS) twisted pair lines, which are installed between central exchange and subscribers, i. e. BS sites. DSL technologies are very restricted in terms of line length because of the attenuation of the POTS lines. Currently, very-high-speed digital subscriber line (VDSL)2 [ITU06b] is the latest step in the DSL evolution path. Depending on the actual POTS line, about 100 Mbit/s over a single unshielded twisted pair line of about 400 m length can be achieved. By applying vectoring [ITU10f, OSC+10], which is essentially CoMP for wireline, that rate can be achieved even under heavy far-end crosstalk found in large bundles. Vectoring requires essentially all lines within a bundle to be under control of the same DSL access multiplexer (DSLAM), i. e. all lines should be within the same vectoring group. Furthermore, all lines must by terminated by a vectoring-capable customer premises equipment, and especially must not be left open and unterminated. Otherwise, performance quickly drops to that without vectoring. Regulatory issues may be a hindrance for such a deployment, e. g. if a network operator is forced to allow competitors access to the physical copper lines instead of bitstream access.

Several DSL lines can be bonded to form a logical data pipe of $n \cdot 100$ Mbit/s, using one of the G.998.x bonding techniques like *time division inverse multiplexing* [ITU10g]. By applying an analogue technique called 'phantom mode', which distributes the differential DSL signal over two other lines, thereby forming a virtual third line, N physical lines can be complemented by $N - 1$ virtual lines, increasing the available bandwidth to $2N - 1$ times the rate of a single line. A phantom mode signal is equally added to both wires of a physical line (a twisted pair of copper wires), and is therefore eliminated by its receiver, which only considers the difference signal of both wires, while vice versa the phantom mode signal can be extracted by adding the signals of the two wires. The same is done on a second physical line to transmit both the positive and the negative signal of the differential VDSL2 signal in phantom mode. Since a phantom mode signal is differential itself, two of them (or one phantom mode and one physical line) can carry another phantom mode differential signal and so on.

To achieve the required short line lengths of a few 100 meters, a VDSL2 access network is realized in a hybrid way based on optical fibers and copper lines. Optical fibers are installed between the core network and street cabinets or underground chambers, which are located closer to the BS sites. A DSLAM terminates the VDSL2 line on the network side, while aforementioned fiber technologies are used for the fiber part up to the central office. The POTS copper lines are thereby only used for the remaining relatively short distance between the street cabinet and the BS sites.

The ITU-T requirement for the one way latency of a VDSL2 link in fast mode, i.e. with interleaving disabled, is 2 ms [ITU06b]. In a practical system, a lower bound can be assumed to be between 0.75 ms and 1.25 ms, respectively 3 to 5 VDSL2 symbols (VDSL2 symbol frequency is 4 kHz): 3 symbols are required for construction of a symbol at the transmitter, transmission of that symbol and processing of it at the receiver. One more is required from time to time when a synchronization symbol is inserted (once every 257 symbols) and another in case a FEC block spreads over 2 symbols, because the FEC block (32 to 255 byte, configurable) is normally smaller than a symbol (64 to 8192 QAM constellations partitioned to upstream and downstream and each encoding up to 15 bit) but not an integer fraction of it. Compared to that, propagation delay and delays on the fiber part can be neglected. With interleaving, the latency increases, depending on the interleaver block size and interleaver depth, up to multiple tens of milliseconds. With interleaving enabled, VDSL2 provides protection against impulse noise and achieves an error performance similar to that of Ethernet, without, it depends on the actual noise conditions. VDSL2 therefore allows for multiple latency paths sharing the same VDSL2 line, so that latency critical traffic can be handled with no or limited interleaving, while other traffic can be transmitted with a lower error rate on a second latency path with a larger interleaver block size and/or depth.

12.3.4 Microwave

Many universal terrestrial RAN (UTRAN) and enhanced UMTS terrestrial radio access network (E-UTRAN) installations in the field use microwave links for backhauling. This technology offers 40 to 160 Mbit/s with a typical latency of 100 μs. New evolved microwave solutions have been announced, which use the E-Band (71 – –76 GHz, 81 – –86 GHz) and provide up to 1 Gbit/s. Microwave links require line of sight and the achievable throughput depends on the actual weather conditions. To provide availability of the link even under heavy rain, distances that can be bridged with sufficient resilience are typically limited to a few kilometers.

12.3.5 The X2 Interface

After having looked into possible physical layer solutions to backhaul, let us now describe the logical interface LTE defines between two enhanced Node Bs (eNBs). The eNB is the entity within the E-UTRAN that comprises one or multiple BSs. The logical definition of the X2 interface is independent from the physical deployment of the E-UTRAN. Two different protocol stacks are specified according to different functions. One protocol stack, X2-U, is defined for user data transfer, the other, X2-C, for control data transfer. Both protocol stacks are identical up to the transport layer. X2-U specifies user datagram protocol (UDP) on the transport layer and GTP user plane (GTP-U) [3GP10g] above. The main functions of GTP-U are data packet transfer, encapsulation and tunneling, data packet sequencing and path alive check. X2-C specifies stream control transmission protocol (SCTP) [IET07] on the transport layer, which provides reliability and flow control. Current applications for the X2-C in LTE Release 8 are handover coordination, interference coordination, self organization and radio resource management (RRM). SCTP may also be applicable for control purposes of CoMP applications especially if reliability is required.

Neither SCTP nor GTP-U provide on time delivery checks, which may be required by CoMP. For that purpose, it may be necessary to implement this functionality in a future CoMP application layer protocol or define a further transport layer protocol for CoMP. To provide interoperability between eNBs of different vendors, protocols and stacks have to be standardized, as well as the definition of CoMP application layer protocols and new information elements.

12.3.6 Backhaul Topology Concepts

As pointed out in Section 12.3, a low inter-BS communication latency is essential for deploying CoMP. If BSs are located at different sites, the length of optical fibers and the number of switches/routers on the backhaul connection between them becomes critical and should be kept as small as possible.

For illustration, a backhaul based on a switched optical point-to-point Ethernet with a tree structure (cf. Fig. 12.9) is assumed. A number of sites are connected to a switch as a first level of aggregation. These switches are in turn are connected to a switch of the next higher aggregation level switch and so on until the core network is reached. A typical case is 3 aggregation layers, which corresponds to switches per neighborhood, city district and a central switch per city, respectively. With such a tree structure, a problem in regard to communication latency can arise. Two sites, although geographically close to each other, and therefore potential CoMP cooperation partners, might be connected to dif-

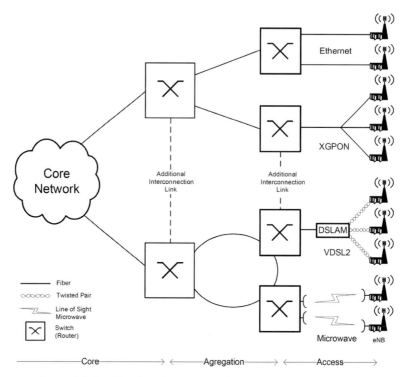

Figure 12.9 A possible backhaul architecture tailored to support CoMP could add shortcuts (dashed lines) on the first or second aggregation level.

ferent branches of the tree, e.g. because they are located on opposite sites of a district border. The communication path between the two then goes up multiple levels in the aggregation tree, e.g. all the way up to the city-wide switch, and from there down again into the other branch to the neighboring site. This results in a long path with a large number of switches passed through, and results in high communication latency. Furthermore, traffic concentrates near the root of the tree, which can be challenging especially for high bandwidth traffic of coherent CoMP schemes exchanging radio samples. For CoMP, such a detour can be avoided by providing a more direct communication path.

Direct inter-Site Connections
One possibility for a more direct path is the direct connection of coordinated sites with a dedicated link. This gives the smallest possible latency, but for a coordination area size of n_{coord} sites this requires a node degree, i.e. connections ending at a site, of $n_{\mathrm{coord}} - 1$ and in total $Nn_{\mathrm{coord}}/2$ interconnections within a RAN of N sites. Furthermore, $n_{\mathrm{coord}} - 1$ network ports per site are required, i.e. a switch needs to be co-located at each site, introducing additional switching delay. Another drawback is that it is necessary to have a priori knowledge about which sites should be coordinated.

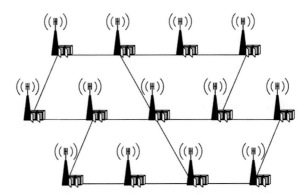

Figure 12.10 Directly interconnecting selected neighboring sites establishes a mesh on the site level, here with cycle length 5 and average node degree of 10/3.

To reduce the number of required links compared to such a full mesh with direct links for each pair of coordinated sites, links could be placed only between selected neighboring sites. In case there is a line of sight between clustered neighbors, such links could even be wireless microwave links instead of wireline links. The more links are used to mesh up sites, the shorter the maximum latency between two sites becomes. Obviously at least a node degree of 2, i.e. as many links as sites, is required to fully connect all sites among each other. This results in one large cycle of length N, i.e. a classical ring topology, and $N/2$ hops between two sites in the worst case. An obvious more delay-friendly approach is to connect just neighboring sites in the typical hexagonal grid of sites. A node degree of 6 respectively three times the number of sites is required (less for border sites). Communication between n-tier neighbors then always requires n hops.

A possible compromise between these two may be to introduce $5/3 \approx 1.7$ as many links as sites respectively an average node degree of $10/3 \approx 3.3$ (ignoring border sites). Again assuming site locations to follow the typical hexagonal grid, all sites are then connected by a mesh with cycles of length 5 as depicted in Fig. 12.10. The number of hops between 1-tier neighbors is 1 or 2 (on average $25/18 \approx 1.4$) and between 2-tier neighbors 2 to 4 (on average $62/27 \approx 2.3$). If cooperation is limited to 1-tier and 2-tier neighbors, this means that between 2 to 5 switches are passed.

Shortcuts between Aggregation Level Switches

A more conservative approach to provide a compromise between latency (number of switches along the path and optical fiber length) and cost (low number of required additional connections) while still being flexible enough not to require a priori knowledge about which BSs cooperate can be achieved by interconnecting switches at the first or second aggregation level as depicted in Fig. 12.9. By adding shortcuts on lower aggregation levels, traffic between sites is kept local, thereby reducing latency and increasing the available bandwidth between coordinated BSs. Only switches that aggregate geographically neighboring sites need to be connected to each other unless two sites need to cooperate that are farther apart than the diameter of the geographical area covered by one of the interconnected aggregation level switches. If shortcuts follow a hexagonal grid (the sites itself may be located arbitrarily) and are deployed throughout the complete RAN, 6 extra ports at each switch (less at the RAN border) and additional optical fibers no more than 3 times the number of switches on the respective aggregation level are required. Such shortcuts between neighboring switches limit the number of passed switches between communicating BSs to 2 (first) respectively 4 (second level). The total fiber length along the path between the BSs is reduced to a few diameters of the geographical area covered by an interconnected switch (taking into account a factor of about 2 for some detour due to indirect cable duct layout following the streets).

12.3.7 Summary

At least from a technical point of view, today's backhaul technologies are powerful enough to support CoMP. While fiber based Ethernet and passive passive optical network provide 10 Gbit/s and beyond as well as sub-millisecond per link latency (utilizing fixed bandwidth allocations in case of a PONs), both microwave and (bonded and vectored) VDSL2 may pose a noticeable limit on the bandwidth, as these technologies currently provide below 1 Gbit/s. Additionally, VDSL2 may have an impact on latency critical CoMP schemes, because its minimum per link latency lies above 1 ms, even in fast mode. The main challenge to backhauling in respect to CoMP, though, is that in the future queueing delays must (and can) be prevented through prioritization and load limitation of the latency critical CoMP traffic, possibly complemented by the introduction of additional backhaul links to limit the length of every physical path between any two cooperating base stations to a few hops.

Part IV

Performance Assessment

13 Field Trial Results

In this chapter, we finally provide field trial results for different CoMP schemes discussed in this book. While Section 13.1 observes the performance of successive interference cancelation (SIC) algorithms and uplink macro diversity in laboratory and outdoor drive tests, Section 13.2 provides results on a large-scale field trial in Dresden, where two terminals were moved within a setup of 16 base stations, and the gain of uplink joint detection is assessed. Sections 13.3 and 13.4 then present two slightly different implementations of downlink joint transmission CoMP, where the former puts a primary focus on various challenges encountered, while the latter section discusses an evaluation methodology that enables to predict downlink CoMP performance over larger areas. The chapter is concluded with a review on the lessons learnt through field trial implementation and test in Section 13.5.

13.1 Real-time Implementation and Trials of Advanced Receiver and Uplink CoMP Schemes

Uwe Dötsch and Johannes Koppenborg

In this section, field trial activities connected to uplink CoMP schemes and advanced receiver algorithms are described. These new features are promising methods to increase performance for LTE-A networks and have been implemented in a Bell Labs eNodeB prototype. In a first step, an uplink (UL) successive interference cancelation (SIC) algorithm is considered in a single-cell scenario. Simulations and lab tests are used to select the best algorithm approach and to optimize involved parameters. To show the feasibility and to assess the advantages of such a SIC receiver algorithm in a real deployment scenario, field trials in the EASY-C research test bed in Dresden [I+09] have been conducted. In a second step, the performance of distributed UL macro diversity is assessed through a field trial in the EASY-C test bed in Berlin. Here, a central processing unit (in this case a base station or *eNodeB* itself) was connected to distributed remote radio heads (RRHs) at two remote sites.

13.1.1 Real-time Implementation and Lab Tests

The used LTE-A platform is based on Alcatel-Lucent's base station hardware. Within the EASY-C project, duplicates of this platform have been deployed in the field in the test beds in Dresden and Berlin. With this platform, a variety of advanced beyond-LTE physical and MAC layer algorithms and concepts can be studied. All algorithms and concepts are evaluated in real-time, as they are implemented in field programmable gate arrays (FPGAs) and digital signal processors (DSPs) on the baseband board. For the uplink trials, advanced receiver algorithms and UL multi-user MIMO (MU-MIMO) schemes were implemented on the baseband board. Lab tests of these new algorithms were done to evaluate the implementation and technologies. The validation of the implementation and the communication with test mobiles were done with a test setup, which allowed high signal-to-noise ratio (SNR) at the receiver side. Quantitative measurements were performed with various fading channels for vehicular and pedestrian users in urban and suburban scenarios. The radio channels were emulated by one or several fading emulators. The outcome of these measurements are performance values like the physical layer throughput, with a special focus on the throughput performance at the cell-edge and the block error rate (BLER). The advantage of quantitative lab trials as opposed to field trials is that the former are more accurately controllable and reproducible. This allows the usage of identical fading scenarios for simulation and tests. On the other hand, lab measurements with their simplified radio environment are not as realistic as field trials. Therefore, field trials were used to test the new developed algorithms in the test beds in Dresden and Berlin.

13.1.2 Uplink Successive Interference Cancelation (SIC) Receiver

One challenging requirement for LTE-A is to improve the uplink throughput in the network and enhance the user experience. SIC receiver algorithms are promising in the uplink to improve the user experience at the cell-edge and decrease the residual interference for MU-MIMO. Therefore, we here show a proof of feasibility of a SIC receiver algorithm in a realistic LTE environment and the evaluation of gains from SIC receiver processing in the field with optimized parameter settings. SIC is a method where you in a first step decode one stream and then subtract this part from the original received signal, as stated in Section 3.4.1. This improves the signal-to-interference-and-noise ratio (SINR), because the signal from the first stream is interference for the second stream, unless it is subtracted. Practically, a lot of different implementation approaches are possible, depending on complexity and latency, as discussed in Section 6.1.2. We have chosen an approach where we initially aim at decoding the first stream and check whether this was successful via a cyclic redundancy check (CRC). In the case of success, we then encode the decoded bits and use the channel estimation weights to rebuild the signal and subtract the corresponding part from the

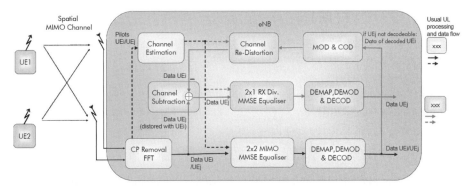

Figure 13.1 Basic SIC principle.

aggregated receive signal. One very intuitive strategy to improve the throughput performance is the following. After multiple-input multiple-output (MIMO) equalization, both streams are decoded independently, and in case that one is successfully and the other not successfully decoded, we use the SIC algorithm to improve the signal quality for the not successfully decoded stream. We modified this approach to get a simplified SIC receiver algorithm. Depending on the channel quality, the more reliable stream is decoded first and if this was successful, the encoding is started directly to reduce latency. In case of an unsuccessful decoding of stream 2, the SIC algorithm will improve the signal, and stream two will be decoded again. The SIC receiver algorithm handling is executed on a transmit time interval (TTI) basis, which means that each hybrid automatic repeat request (HARQ) retransmission is handled as a separate transmission.

Lab measurements with UL MU-MIMO using the advanced minimum mean square error (MMSE)-SIC receiver have been done with 2 test user equipments (UEs) and a channel emulator. The lab measurements showed an improvement of the uplink throughput of about 10% by the additional usage of the MMSE-SIC receiver. Another interpretation of the result is that MMSE and SIC needs up to 3 dB less SNR for the same throughput compared to the MU-MIMO case without SIC receiver. A comparison of the lab tests to simulations showed a very good compliance of the results.

To assess the advantages of a SIC receiver algorithm in a real deployment scenario, field trials in the EASY-C research test bed in Dresden were conducted. One base station (BS) was located at the site "Main station" and connected with cross-polarized (X-pole) antennas covering a cell in a dense urban surrounding. The X-pole antenna means that the antenna uses two polarization directions with 45 degrees shift compared to the vertical direction. The carrier frequency was 2.6 GHz, with a system bandwidth of 10 MHz. For the trials, one stationary test UE was placed at a distance of about 500 m from the site on a parking lot with an UL SINR varying between 8 and 15 dB, while a second UE was moved in a test van. The drive route was chosen in such a way that the UL SINR of the moving mobile varied between 0 and 30 dB. With this trial scenario, the

Field Trial Results

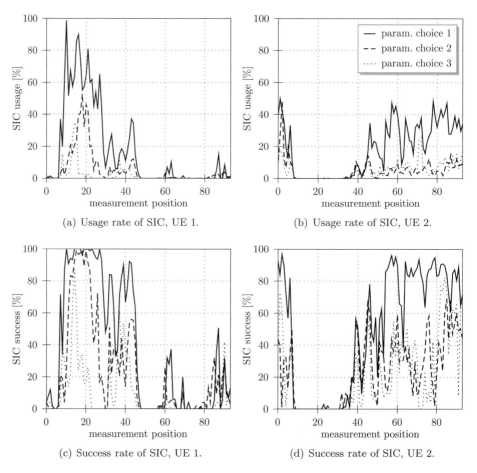

Figure 13.2 Usage and success rate for the SIC algorithm.

UL SINR of the moving terminal is better than the one of the static terminal for a part of the drive route, and worse for the other parts of the drive route. To analyze the impact of the SIC algorithm on the performance, two statistics are very useful. To improve the performance, the SIC algorithm should be used frequently, because only then the receiver can cancel some of the interference which results in a higher throughput. But also important for the performance improvement is the successful decoding after applying SIC. These statistics are shown in the following figures. In the upper part of Fig. 13.2, the percentage of SIC usage is shown for the static (left) and moving test terminal (right). For a good choice of link adaptation parameters, a SIC usage of around 60% for the static terminal and around 40% for the moving terminal can be seen. In the lower part of the figure, a SIC receiver success rate of up to 80% can be seen for both terminals.

13.1 Real-time Impl. and Trials of Adv. Receivers and UL CoMP

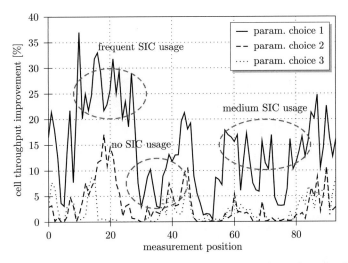

Figure 13.3 Used link adaptation parameters influence the throughput performance.

Test drives with MU-MIMO configuration with different link adaptation settings were conducted with and without using the advanced SIC receiver algorithm. In Fig. 13.3, you can see a cell throughput improvement by using the SIC receiver of about 25% for these specific field conditions. Clearly, there can be no disadvantage of using the SIC algorithm, as this is only applied in the case where conventional decoding was unsuccessful. The only price paid for the SIC receiver is some additional baseband processing power. Additionally, it is important to point out that the SIC receiver has no impact on the UE side, hence it can be used without any change of standardization.

13.1.3 Uplink Macro Diversity Trials with Distributed RRHs

As a different uplink CoMP technique, UL macro diversity was studied in a live field trial in the LTE test network in Berlin. This test was a first step to prove the feasibility of UL CoMP in a realistic environment and to quantify potential macro diversity gains in the field. The aim of the trials was to investigate especially the enhancement of the cell-edge spectral efficiency in the uplink, meaning a more consistent quality of service (QoS) throughout the network and for the end-user an enhanced user experience. CoMP is particularly promising in the uplink due to the UE transmit power limitation at the cell-edge. Coherent signal combining can be realized without changes of the LTE air interface and without increasing the total number of antennas in the network. The used field trial setup consisted of a central processing unit (a base station or *eNodeB*) located at the site "HHI", which was connected to distributed RRHs at two well separated sites "HHI" and "TUB". The inter-site distance of these 2 sites was about 570 m. The RRHs were connected via common public radio interface (CPRI) optical fibre links to the central processing unit. To avoid delay differences of the signals in the two

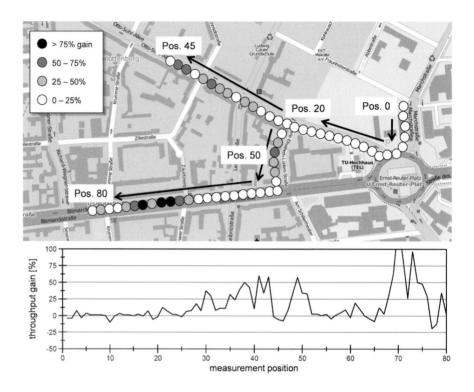

Figure 13.4 Overview over the test bed drive route and the relative throughput gain of macro diversity vs. rx diversity. © OpenStreetMap & contributors, CC-BY-SA.

fibres, which could potentially lead to inter-symbol interference (ISI) as discussed in Section 8.2, the fibre length balance was accurately matched. Each RRH was connected to one X-pole antenna. The carrier frequency was 2.6 GHz, and the system bandwidth 5 MHz.

For the macro diversity tests, the signals of the test terminal were received by the two RRHs with 2 receive antennas each (meaning 4 receive antennas in total) and coherently combined and processed in the same TTI by the central processing unit. The results were compared to reference measurements with 2 receive antenna reception at the two sites with RRHs and the assumption of an ideal handover (0 dB margin), which means that always the best results from both sites were taken into account.

In the upper part of Fig. 13.4, the drive route in the Berlin test bed is visualized. The route has been chosen such that the distance from both sites is in the same regime. In the lower part of Fig. 13.4, the uplink throughput gain for macro diversity is shown w.r.t. the 2 receive antenna handover scenario. Significant macro diversity gain can be seen for most parts of the drive route.

13.1.4 Summary

The usage of MU-MIMO and SIC with an optimum link adaptation setting of the eNodeB leads to significant cell throughput improvements. In the Dresden text bed, we observed an increase of the cell throughput by 25% for good field conditions. The discussed trials in Berlin have further shown that uplink CoMP in a realistic LTE environment is feasible in this case with a central processing unit and with distributed RRHs which are more than 500 m apart from each other. Furthermore, the baseband data transmission over optical fibers of about 3 km length has shown to work without losses, and there seem to be no issues with optical transmission delays when the system is designed properly. The uplink coherent combining for an MMSE receiver has also been validated in the field, and the receiver was able to handle the challenging delay spread and alignment of the timing advance. Especially for cell-edge users, significant uplink throughput enhancements could be shown. For users with good channel conditions, we see a potential for further throughput improvement by using 64-QAM modulation in the uplink. In further steps, the system performance shall be improved by combining both methods. We expect an additional gain by using also a SIC receiver in the case of multi-cell joint detection, and by pairing MU-MIMO users over different cells in a smart way in the context of coordinated scheduling, as for example addressed in Sections 5.2 and 11.1. All these options should be investigated in future trials.

13.2 Assessing the Gain of Uplink CoMP in a Large-Scale Field Trial

Patrick Marsch, Michael Grieger and Gerhard Fettweis

In this section, we present results from a large-scale field measurement campaign on different uplink (UL) CoMP schemes in the EASY-C test bed in Dresden. While the first proof-of-concept of schemes such as multi-cell joint detection (JD) was already delivered earlier and published in [GMRF10], the aim is now to determine

- the urban environmental scenarios, the frequency, and the extent of gains that can be expected from different uplink CoMP concepts, and
- how large the cooperation size should be chosen.

In the sequel, the measurement setup is described in Subsection 13.2.1, after which details on the signal processing architecture and evaluation concept are provided in Subsection 13.2.2. In Subsection 13.2.5, the actual field trial results are presented, after which a summary is provided in Subsection 13.2.6. Similar field trial results have also been published in [MGF11].

320 Field Trial Results

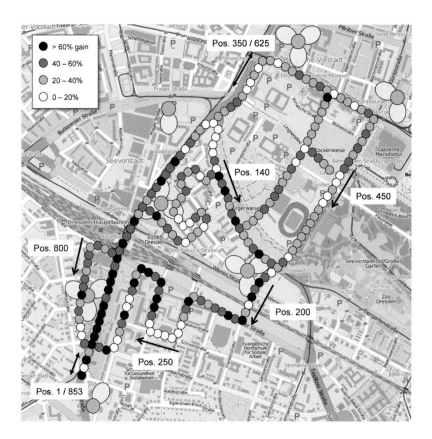

Figure 13.5 Field trial setup and measurement trajectory, indicating the spectral efficiency gain of a joint detection of 2 UEs by 3 BSs with $N_{\text{bs}} = 1$ using SIC vs. the LTE Release 8 baseline. Map data © OpenStreetMap & contributors, CC-BY-SA.

13.2.1 Measurement Setup

The measurement setup consists of 16 base stations (BSs) deployed at a total of 7 current universal mobile telecommunications standard (UMTS) sites in downtown Dresden, as shown in Fig. 13.5. The BS spacing varies in the range from 350 m to 600 m, and the antenna height from 15 m to 35 m. The BSs are synchronized through global positioning system (GPS) fed reference normals, and sites are connected through microwave links with a net bi-directional capacity of about 100 Mbit/s. Each BS is equipped with a cross-polarized directional antenna (70 degrees half-power beamwidth and 14 dBi gain), hence with $N_{\text{bs}} = 2$ individual antenna elements per BS. Key transmission parameters are listed in Table 13.1. The user equipments (UEs), employing one omnidirectional antenna each, transmit using orthogonal frequency division multiplex (OFDM) and a sequence of different modulation and coding schemes (MCSs), as listed later in Table 13.2. The UEs are carried on a measurement bus, and UE antennas are

13.2 Assessing the Gain of Uplink CoMP in a Large-Scale Field Trial

Table 13.1. Transmission parameters.

Parameter	Value
Carrier frequency	2.53 GHz
System bandwidth	20 MHz
Resource blocks (PRBs)	30
Sub-carriers per PRB	12
UE transmit power	12 dBm
Quantization resolution	12 bit per real dim.

assembled outside the bus in about 5 m distance, arranged such that these are maximally uncorrelated, and placed roughly 1.5 m above ground. Both UEs are scheduled to transmit on the same resources in time and frequency. The current implementation of the test bed does not have any handover functionality. Thus, in order to perform an uninterrupted trial, downlink control information such as uplink grants is sent from an additional BS, which is carried on the measurement bus as well. The receive signals of all other BSs are recorded for offline evaluation, which facilitates the investigation of different CoMP concepts and cooperation sizes for the same recorded signals. Clearly, the focus of this approach is on physical layer evaluation.

13.2.2 Signal Processing Architecture and Evaluation Concept

We will now briefly explain the general signal processing steps performed in the offline evaluation chain mentioned before.

Synchronization

As explained in Subsection 13.2.1, the carrier frequency of the BSs is synchronized by using GPS fed reference normals, which is accurate enough for remaining errors to be neglected [JWS+08]. The frequency offset of the UEs is pre-compensated based on reference signals that the UEs obtain over the downlink. Compared to the sub-carrier spacing, the remaining offset of less than 200 Hz is small enough to disregard inter-carrier interference (ICI). The remaining common phase error is taken into account by an appropriate interpolation of the channel estimates, as will be explained in the sequel. The frame and symbol synchronization is based on the auto-correlation properties of the OFDM time domain signal. For more details, we refer to [MKP07].

Channel Estimation

The pilot positions used for uplink channel estimation are the same as those used in LTE Release 8 [McC07], i.e. within each transmit time interval (TTI) consisting of 14 OFDM symbols, pilots are mapped on all sub-carriers of the 4th and 11th OFDM symbol. Provided that the channel coefficients of two neighboring sub-carriers are almost equal, interference between pilot symbols of different

UEs is avoided by a code-orthogonal design using Frank-Zadoff-Chu sequences. Accordingly, each user is identified by a user-specific phase rotation $\exp(j\pi qk)$ multiplied with the baseline pilot sequences, where q and k are the sub-carrier and user index, respectively. At the receiver side, multi-user channel estimation can be performed by a simple Hadamard approach, as is well-known from multiple-input multiple-output (MIMO) channel estimation theory [Kam08]. Due to the spreading factor of two, the channel of each UE is estimated for every other sub-carrier in frequency domain. To estimate the channel for all other sub-carriers, time and frequency interpolation and extrapolation are carried out separately. Due to synchronization and common phase errors between UEs and BSs, resulting in a linear phase drift over time and frequency, the channel coefficients are interpolated with respect to amplitude and phase separately. In our subsequent model, we assume unit transmit power per UE, i.e. the estimated channel links $\hat{\mathbf{h}}_k^m$ inherently contain the actual transmit power of the UEs.

13.2.3 Noise Estimation

The estimation of the noise variance, important for all detection schemes considered in this section, is based on the channel estimates $\hat{\mathbf{h}}_k^m$. Due to orthogonal pilot sequences, we are able to estimate the noise variance independent from interference of the other UE. The particular estimation procedure relies on the assumption that the channel of two adjacent sub-carriers is equal and the noise is uncorrelated. We can then exploit the auto-correlation properties of $\hat{\mathbf{h}}_k^m$ to separate noise and signal components, and compute their respective power. For a robust estimation, the values are averaged over both pilot sequences of each TTI and over the receive antennas. Using this approach, one noise variance $\hat{\sigma}_m^2$ is determined per BS. Note that noise *covariance* estimation would not make sense in out setup, as we have no spatially colored interference background.

13.2.4 Channel Equalization

If residual synchronization errors are neglected, and we assume a flat fading channel on each sub-carrier of bandwidth $\Delta F = 15$ kHz, the received signal of each symbol on a single OFDM sub-carrier at BS m can be stated as

$$\mathbf{y}_m = \mathbf{h}_1^m x_1 + \mathbf{h}_2^m x_2 + \mathbf{n}_m, \qquad (13.1)$$

where $\mathbf{y}_m \in \mathbb{C}^{[N_{\text{bs}} \times 1]}$ are the signals received by the N_{bs} antennas of BS m, $\mathbf{h}_k^m \in \mathbb{C}^{[N_{\text{bs}} \times 1]}$ denotes the channel gain vector from UE k to BS m, $x_k \in \mathbb{C}$ is a symbol transmitted by UE k, and $\mathbf{n}_m \in \mathbb{C}^{[N_{\text{bs}} \times 1]}$ denotes additive, uncorrelated noise of covariance $E\{\mathbf{n}_m \mathbf{n}_m^H\} = \sigma_m^2 \mathbf{I}$. The channel vectors include UE transmit power due to the assumption of $E\{x_k x_k^H\} = 1$. The set of BSs that form a cooperation cluster is denoted by $\mathcal{C}$ with elements $\{c_1, c_2\}$ or $\{c_1, c_2, c_3\}$, where the cluster size is chosen to be two or three, depending on the cooperation scheme observed.

13.2 Assessing the Gain of Uplink CoMP in a Large-Scale Field Trial

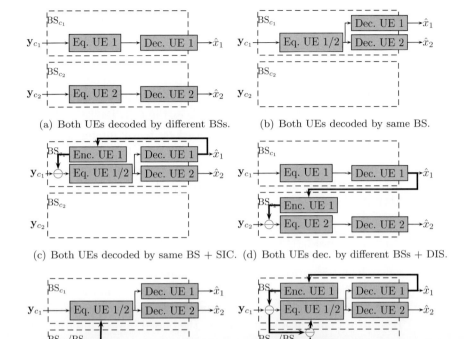

Figure 13.6 Signal processing setup for various detection and cooperation schemes, adapted from [MGF10]. © 2010 IEEE.

The corresponding transmission model for the cluster is given as

$$\begin{bmatrix} \mathbf{y}_{c_1} \\ \mathbf{y}_{c_2} \\ \mathbf{y}_{c_3} \end{bmatrix} = \begin{bmatrix} \mathbf{h}_1^{c_1} & \mathbf{h}_2^{c_1} \\ \mathbf{h}_1^{c_2} & \mathbf{h}_2^{c_2} \\ \mathbf{h}_1^{c_3} & \mathbf{h}_2^{c_3} \end{bmatrix} \begin{bmatrix} x_1 \\ x_2 \end{bmatrix} + \begin{bmatrix} \mathbf{n}_{c_1} \\ \mathbf{n}_{c_2} \\ \mathbf{n}_{c_3} \end{bmatrix}. \quad (13.2)$$

The signal processing architecture enables a variety of cooperation and equalization schemes, of which the following examples are illustrated in Fig. 13.6:

- Both UEs are decoded independently by different BSs, employing interference rejection combining (IRC) (see Fig. 13.6(a)).
- Both UEs are decoded by the same BS (see Fig. 13.6(b)), using a linear detector or successive interference cancelation (SIC), as was investigated in Section 13.1.2 (see Fig. 13.6(c)).
- Both UEs are decoded independently by different BSs, but one BS forwards decoded data bits to the other for distributed interference subtraction (DIS), a scheme introduced in Section 4.3.1 (see Fig. 13.6(d)).

- One or two BSs forward all received signals to another BS, where both UEs are decoded jointly. This can be done via linear equalization (see Fig. 13.6(e)), or using SIC (JD+SIC) (see Fig. 13.6(f)).

In the following, we will consider the union of all equalization options depicted in Figs. 13.6(a), 13.6(b) and 13.6(c), i.e. including the option of joint detection of both UEs at the same BS with SIC as the LTE Release 8 *baseline*.

Equalization itself is generally based on linear minimum mean square error (MMSE) filters, which depend on whether any kind of interference subtraction (i.e. through DIS or SIC) has been performed in advance. If UE k is locally detected at BS m, and still subject to the interference from UE $\bar{k} \neq k$, the biased MMSE filter for a particular sub-carrier is given as

$$\mathbf{G}^{[m,k]}_{\text{biased}} = \left(\hat{\mathbf{h}}^m_k\right)^H \left(\hat{\mathbf{h}}^m_k \left(\hat{\mathbf{h}}^m_k\right)^H + \hat{\mathbf{h}}^m_{\bar{k}} \left(\hat{\mathbf{h}}^m_{\bar{k}}\right)^H + \hat{\sigma}^2_m \mathbf{I}\right)^{-1}, \quad (13.3)$$

where $\hat{\mathbf{h}}$ and $\hat{\sigma}^2_m$ are estimates of the channel and noise, respectively, and $\bar{k}$ is the index of the interfering UE. Note that this implementation exploits the channel knowledge to each UE for the purpose of IRC. If the receive signals of multiple BSs are available at a joint receiver, the biased MMSE filter for UE k is given as

$$\mathbf{G}^{[k]}_{\text{biased}} = \hat{\mathbf{h}}^H_k \left(\hat{\mathbf{h}}_k \hat{\mathbf{h}}^H_k + \hat{\mathbf{h}}_{\bar{k}} \hat{\mathbf{h}}^H_{\bar{k}} + \begin{bmatrix} \hat{\sigma}^2_{c_1} \mathbf{I} & 0 \\ & \ddots & \\ 0 & & \hat{\sigma}^2_{c_3} \mathbf{I} \end{bmatrix}\right)^{-1}, \quad (13.4)$$

where $\hat{\mathbf{h}}_k$ denotes the channel from UE k to all BSs in the considered cluster. If interference of the other UE has already been canceled, filters in (13.3), (13.4) change to

$$\mathbf{G}^{[m,k]}_{\text{SIC/DIS,biased}} = \left(\hat{\mathbf{h}}^m_k\right)^H \left(\hat{\mathbf{h}}^m_k \left(\hat{\mathbf{h}}^m_k\right)^H + \hat{\sigma}^2_m \mathbf{I}\right)^{-1} \quad (13.5)$$

and $\mathbf{G}^{[k]}_{\text{SIC,biased}} = \hat{\mathbf{h}}^H_k \left(\hat{\mathbf{h}}_k \hat{\mathbf{h}}^H_k + \begin{bmatrix} \hat{\sigma}^2_{c_1} \mathbf{I} & 0 \\ & \ddots & \\ 0 & & \hat{\sigma}^2_{c_3} \mathbf{I} \end{bmatrix}\right)^{-1}, \quad (13.6)$

respectively. To avoid demapping errors for higher order modulation schemes, the bias has to be removed from all stated filters by applying $\mathbf{G} = (\Delta(\mathbf{G}_{\text{biased}}))^{-1} \mathbf{G}_{\text{biased}}$, where $\Delta(\mathbf{G})$ sets all off-diagonal elements of $\mathbf{G}$ to zero.

Soft Demodulation and Decoding

After equalization, signal-to-interference-and-noise ratios (SINRs) are estimated via an error vector magnitude approach [SRIA06], followed by soft demodulation. The demodulator output is fed into an LTE Release 8 compliant decoding chain using the codes listed in Table 13.2. Each codeword spans one TTI in time

Table 13.2. Modulation schemes and code rates used for transmission, assuming Turbo codes as used in LTE Release 8.

Num.	Mod. scheme	Code rate	Peak rate	bpcu
1	4QAM	3/16	1.3 Mbps	0.375
2	4QAM	1/2	3.46 Mbps	1.0
3	16QAM	2/5	5.62 Mbps	1.6
4	16QAM	4/7	7.99 Mbps	2.29
5	16QAM	3/4	10.6 Mbps	3.0
6	16QAM	6/7	12.3 Mbps	3.43
7	64QAM	3/4	16.3 Mbps	4.5
8	64QAM	7/8	18.72 Mbps	5.25

domain and 30 physical resource blocks (PRBs) in frequency domain. Decoding success is determined through an outer cyclic redundancy check (CRC)-code.

13.2.5 Field Trial Results

The route traversed by the measurement car is depicted in 13.5. It has a total length of 17 km and passes through surroundings of very different building morphology. Part of the trajectory is followed twice to see whether results are reproducible, which is indeed the fact when observing the results around measurement positions 350 and 625 in later figures. The car traveled at a speed of about 6 km/h during measurements. The UEs continuously transmitted codewords, each spanning 1 TTI (1 ms), switching cyclically between the MCSs given in Table 13.2. Given that the channel does not change significantly during the time it takes to loop through all MCSs, we are able to determine the instantaneous rate of the UEs that would be achieved under the assumption of optimal link adaptation. Due to limited memory capacity, we are not able to store all received signals at the BSs continuously. Instead, the BSs synchronously capture their received signal for a duration that is long enough for several iterations through all MCSs to obtain robust statistics of achievable rates for a small-scale area. Thus, we determine the rate (per position) that leads to the largest successful throughput on average, which is denoted as $r_{k,p}$ for UE k and position p. Even when SIC is used, we are able to determine the optimal MCS for each UE. This is possible even though both UEs actually transmit using the same MCS at the same time, because the transmitted codeword is known under field trial conditions. Thus, we can determine the highest rate MCS that is successfully decoded either with or without prior SIC (using a genie-SIC) and, assuming the channel did not change significantly during one loop through all MCSs, we can apply the decoding order that achieves the highest sum-rate. Note that we generally assume that SIC is applied only if one of the UEs is decoded successfully (i.e., *hard-SIC*), though in a commercial system one would probably consider *soft-SIC* as well. One receive signal dump was generated every 10 s for a total of $P = 853$ measurement positions, to observe large-scale effects.

In principle, we are able to investigate the performance if a UE were decoded by any arbitrary cluster of BSs because we use offline evaluation. However, to reduce evaluation complexity, for each transmitted codeword, the BSs that are considered for the decoding of the UEs and a particular codeword are determined by a minimum pathloss criterion. Regarding multi-cell JD, we consider cases where either two or three BSs, which are again heuristically chosen based on pathloss criteria, cooperate. In all the plots in this section, we show only sum-rate results, i.e. we compute the sum of the UE rates $(r_{\text{sum},p} = r_{1,p} + r_{2,p})$ at each position.

The achieved sum-rates for all investigated cooperation and equalization schemes are shown in Fig. 13.7. We here distinguish between a case where only the first antenna of each BS is used for signal processing (i.e. emulating a system with $N_{\text{bs}} = 1$, see Fig. 13.7(a)), or where both are used ($N_{\text{bs}} = 2$, see Fig. 13.7(b)). While the latter case is certainly the more commonly assumed setup (it is for example also considered for simulation in Sections 4.3.1 and 14.3), the former one is also interesting, as it resembles a case where the number of overall UE and BS antennas is equal if two BSs cooperate. The results obtained from this case can then be used to predict the gains in CoMP setups where multiple cooperating BSs serve two UEs *per cell* on the same resource, or where UEs with two transmit antennas each are employed. From information theory, this clearly leads to a larger sum-rate over all involved terminals or streams, but is a setup which could not yet be evaluated in the test bed. To focus on large-scale effects, we smoothed out small-scale variations of measured rates in Fig. 13.7 using a moving average filter with a length of 10 positions.

Clearly, the sum-rates of all compared schemes are significantly larger in the case of $N_{\text{bs}} = 2$ than for $N_{\text{bs}} = 1$, simply due to the additional degree of freedom at the BS side. In the case of $N_{\text{bs}} = 1$, non-cooperative, linear detection of course performs rather badly, as the BSs are not able to spatially separate the terminal transmissions. Here, one can see vast rate improvements by applying local SIC (on average 49.7%), as then one of the UEs can transmit at a low rate which is just decodeable despite interference, while the other can strongly benefit from interference subtraction and potentially use a high rate. Note that the impressive performance of SIC in this section is also due to the fact that we inherently assume perfect link adaptation as stated before, and operate in a fairly high signal-to-noise ratio (SNR) regime, as no background interference is generated[1]. The much smaller gains from non-cooperative, local SIC in the case of $N_{\text{bs}} = 2$ (on average 6.5%) show how well the users can already be separated in the linear detection case.

---

[1] Note that this is also the main reason for the different results compared to [MGF10], where UE transmit powers and hence also SNRs were significantly lower.

13.2 Assessing the Gain of Uplink CoMP in a Large-Scale Field Trial

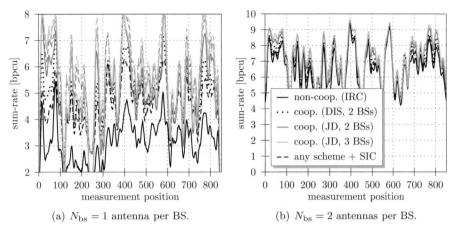

Figure 13.7 Achieved sum-rates for considered cooperation and detection schemes along the measurement trajectory.

By using DIS instead of SIC, i.e. by decoding the UEs at different BSs but performing interference subtraction via the backhaul, one can improve sum-rates by an additional +8% and +2.4% on average, for the single-antenna and double-antenna BS case, respectively. In future field trials, we will also consider UEs with a considerably larger spacing in order to obtain more asymmetrical interference scenarios, where it is known that DIS concepts become even more valuable. This was shown in simulations in, e.g., [MF09b, MF11], as well as in a previous small-scale field trial [MGF10].

An additional gain can then be obtained by using multi-cell joint detection, i.e. based on the exchange of quantized received signals, leading to an overall gain of 33.3% and 9.7% over the non-cooperative SIC baseline case for $N_{bs} = 1$ and $N_{bs} = 2$, respectively. In general, one can see that cooperation and SIC gains seem to complement each other, hence if an operator already invests into the infrastructure required for cooperative detection, then the system should definitely also make use of non-linear detection strategies. For $N_{bs} = 2$, it is clear that both cooperation and SIC yield significantly reduced gains, as the baseline already performs very well. For both antenna configurations, we can see that most cooperation gains are already obtained for clusters of 2 jointly detecting BSs, and using a cluster size of 3 only leads limited additional benefit, such as an extra array gain. However, we expect this result to be different if interferers are placed in each cell, as then a cooperation size of 2 will not be sufficient to capture the dominant interference each UE's transmission is subject to.

As expected, the achieved average rates vary as the UEs are moved through the test bed, attesting that fairness is a challenging issue in a cellular system. However, if the achieved rates are observed more closely, we see that low rate events are improved by using cooperation in most cases. A measure that is often

Table 13.3. Fairness (Jain's index), gain and backhaul of the compared schemes.

Decoder	Jain's index $N_{bs}=1/ N_{bs}=2$	Spectral eff. gain $N_{bs}=1/N_{bs}=2$	backhaul [bpcu] $N_{bs}=1$
non-coop. (linear)	0.61/0.89	−33%/−6.1%	0
non-coop. (SIC, **baseline**)	0.75/0.90	-	0
coop. (DIS, 2 BSs)	0.80/0.91	7.8%/2.4%	1.21
coop. (JD, 2 BSs)	0.78/0.91	8.0%/3.6%	24
coop. (JD+SIC, 2 BSs)	0.82/0.92	27.8%/7.8%	24
coop. (JD, 3 BSs)	0.79/0.91	17.1%/5.9%	48
coop. (JD+SIC, 3 BSs)	0.83/0.92	33.3%/9.7%	48

used to evaluate fairness in networks is the Jain's index, which is defined as

$$\text{fairness} = \left(\sum_P \sum_K r_{k,p}\right)^2 \bigg/ \left(PK \sum_P \sum_K r_{k,p}^2\right). \tag{13.7}$$

The Jain's index can take values between $\frac{1}{PK}$ (minimum fairness) and 1 (maximum fairness). It is usually used to evaluate fairness of different users over a period of time, but since we only observe two UEs and do not consider scheduling, we compute fairness over the measurement positions. Hence, the index reflects the achievable rate distribution over the measurement area and the two UEs. In the case of $N_{bs} = 1$, fairness is already increased strongly by using local SIC, as shown in Table 13.3, simply because the variance of (sum-)rates over the measurement positions is reduced. However, using SIC will always lead to the fact that the UE decoded last may obtain a much better rate than the first one, so that fairness among the two UEs is bad. This is the reason why JD in conjunction with SIC leads to a further fairness improvement. In this case, the UEs can already be spatially separated quite well due to the additional degree of freedom at the receiver side, and SIC can still improve rates, but does not require trading one UE's rate against the other to the extent that was the case for non-cooperative detection. As the rate improvements for the case of $N_{bs} = 2$ are quite small, anyway, we here also see only minimal improvements of the Jain's index.

Relative sum-rate gains that allow a comparison of different detection and cooperation schemes are depicted in Fig. 13.8. The plots once again emphasize the benefit of using non-linear detection, as also concluded in Section 13.1. When observing the benefit of using DIS over local SIC (i.e., the LTE Release 8 baseline), one has to take the additional backhaul traffic that is required for the exchange of decoded bits into consideration. In this field trial, the backhaul rate required for the use of DIS was 1.21 bpcu or 4.8 Mbit/s in the case of $N_{bs} = 1$, which is low when compared to 24 or 48 bpcu (12 bits per I/Q dimension and antenna), leading to 112 or 224 Mbit/s required for JD among 2 or

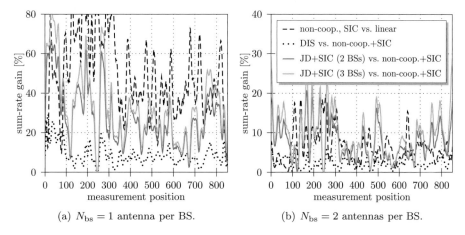

Figure 13.8 Relative gains between different pairs of compared schemes.

3 BSs, respectively. These numbers are based on the assumption of exchanging frequency domain signals, focusing on the 360 sub-carriers on which transmission has taken place. However, no form of compression is considered (i.e. the signals forwarded to the decoding BS have the same bit resolution the analog to digital conversion (ADC) uses), which would allow reducing backhaul under similar performance, as discussed in Section 12.2.

The relative gain of JD among three BSs with SIC compared to the LTE Release 8 baseline (detection of both UEs at different or the same BS with IRC or SIC) is indicated in Fig. 13.5 for $N_{bs} = 1$. Beside the fairly good average gain for this antenna configuration, we can see that the sum-rate gain rises up to 80% in some cell-edge areas. Having a commercial usage of uplink CoMP in mind, it is interesting to see that some cases of strong cooperation gain can be obtained with inter-sector cooperation at the same site, where proprietary and inexpensive solutions are thinkable (see, e.g., the measured gains around position 200 in Fig. 13.5).

Sum-rate cumulative distribution functions (CDFs) depicted in Fig. 13.9 support the conclusions that were made before. In the case of $N_{bs} = 1$, we can see an average sum-rate improvement by using BS cooperation of about 27.8% and 33.3% for a cooperation size of 2 and 3 BSs, respectively (using SIC in all compared cases). As discussed before, these numbers can give us an idea of the gains achievable in a system where two UEs are served on the same resource in each cell, or UEs with two transmit antennas are employed. The outages in Fig. 13.9(a) are cases where the single BS antennas were not even able to decode the lowest-rate MCS. For $N_{bs} = 2$, i.e. the antenna configuration typically assumed, cooperation gains strongly reduce to an average of 7.8% and 9.7%, respectively, due to the strong performance of the considered LTE Release 8 baseline in this case. In general, we have to point out that the field trial setup considered here only allows a limited generalization of conclusions because

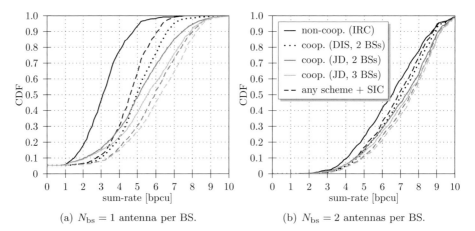

(a) $N_{\mathrm{bs}} = 1$ antenna per BS.   (b) $N_{\mathrm{bs}} = 2$ antennas per BS.

Figure 13.9 Cumulative distribution of sum-rates observed on the measurement route.

- UEs were located in close proximity and with fixed distance,
- no background interference has been considered so far, and
- no power control was employed.

We expect larger gains from cooperation if an interference background is considered, as then the additional spatial multiplexing and array gain provided by a larger number of virtual receive antennas comes into play. In a field trial with more terminals, we also expect a larger gain from cooperation size 3, as this will allow to capture one additional interferer in each detection process.

13.2.6 Summary

In this section, large-scale field trial results for different uplink CoMP schemes were presented, where two UEs were moved through an urban cellular test bed with a total of sixteen base stations (BSs). For the evaluation of results, either one or two antennas per BS were considered. While the latter corresponds to the setup also assumed in other sections of this book, observing the one-antenna case allows the prediction of gains in setups with two UEs per cell served on the same resource, or UEs with two transmit antennas each. Compared to non-cooperative linear detection, local SIC already increases average spectral efficiency by about 49% or 6.5%, for the different antenna setups, respectively. On top of this, multi-cell joint detection yields an average gain of 33.3% or 9.7%, for one or two BS antennas, respectively. As expected, particularly strong gains are visible at cell-edges, which can in some cases already be achieved through intra-site CoMP. Most of the cooperation gains are obtained with two cooperating cells, whereas a cooperation size of three yields a limited additional gain. We assume relative gains in general to be larger if background interference is added in future field trials, as then also the array gain from cooperation will become more visible than in the before observed scenarios of high SNR. We further expect the benefit from distributed interference subtraction (DIS) to be larger in cases where the terminals are spaced further apart than was possible in this measurement campaign. Future field trials shall focus on scheduling for uplink CoMP and compression schemes for backhaul-efficient joint detection.

The authors would like to thank the German Ministry for Education and Research (BMBF) for partial funding of the project EASY-C. Further, this work would not have been possible without Sven-Einar Breuer, Ainoa Navarro Caldevilla, Vincent Kotzsch, Eckhard Ohlmer, Matthias Pötschke and Joachim Heft.

13.3 Real-time Implementation and Field Trials for Downlink CoMP

Volker Jungnickel, Andreas Forck, Stephan Jäckel, Sander Wahls, Lars Thiele, Thomas Haustein, Wolfgang Zirwas, Heinz Droste and Gerhard Kadel

In this chapter, we share experience gained during real-time implementation and testing of distributed joint transmission (JT) CoMP in the downlink of an LTE-A trial network. We implemented and tested enabling features in real-time on top of an existing LTE trial system. These are distributed synchronization, cell- and stream-specific pilots and a fast backbone network for feedback provision as well as channel state information (CSI) and data exchange between the base stations (BSs). Interference-limited transmission experiments have been conducted using three multi-cell transmission modes. As a lower bound, we assumed no coordination between the cells, but apply per-antenna rate control with interference-aware equalization at the user equipments (UEs) as described in Section 5.1. Second, by using distributed JT CoMP as described in Section 6.3, the mutual interference between the cells is canceled. Third, we study isolated cells as an upper bound where the interference from other cells is switched off. We have investigated the benefits of JT CoMP in real-time transmission experiments over the air in indoor, outdoor-to-indoor and outdoor scenarios. Our experiments cover both intra- and inter-site setups. While outage is indeed a big problem at the cell-edge with full frequency reuse and interference-limited transmission, with JT CoMP it is not observed anymore. Between 27% and 78% of the isolated cell throughput is measured in both cells simultaneously. Average throughput gains by factors 4 to 22 are observed when using JT CoMP compared to interference-limited transmission. Of course, these gains are obtained only in the absence of interference from other cells (this case is considered further in system level simulations in Section 14.4). Nonetheless, our experiments demonstrate that the implementation challenges can be overcome and that similarly high gains as predicted by theory can be realized in practice.

13.3.1 Introduction

Inter-cell interference is the major bottleneck for the performance in mobile networks nowadays. If multiple terminals are served on the same radio resource in adjacent cells, they experience severe co-channel interference. At the cell-edge, the situation becomes critical. The signal of the serving BS is received with similar power compared to the sum of the signals received from other BSs.

Nonetheless, one may share resources even at the cell-edge [BMWT00, SZ01, WMSL02, HV04, ZD04, FHK+05, KFV06, JJT+09]. The cellular network can be considered as a distributed multiple-input multiple-output (MIMO) system where the BS antennas are the inputs and the UE antennas are the outputs. Provided that the channel is known at each BS and there is a high-speed backbone between the BSs over which channel information as well as data can be exchanged with minor delay, BSs may perform joint beamforming for all terminals served in all cells. Full frequency reuse is hence enabled since the mutual interference between the cells is canceled by coherent JT. These schemes are denoted as network MIMO [KFV06] and as JT CoMP in recent standardization documents [3GP10d]. For next generation mobile networkss (NGMNs), CoMP is considered as a tool to improve the coverage of high data rates and the throughput at the cell-edge.

Information theory considers a centralized network architecture, see Fig. 13.10. Over the backhaul, all BSs are connected to a central unit (CU). UEs estimate the multi-cell CSI in the downlink and deliver CSI feedback to the CU over the uplink. The CU pre-computes the transmitted waveforms for all coordinated BSs. BSs get their waveforms over a star-like network and act merely as remote radio heads (RRHs). Over the air, the desired signals add constructively while the mutual interference between the cells is canceled. By removing all interference, a performance can be achieved as in an isolated cell [HV04]. The centralized approach has a high backhaul effort since analog waveforms (radio over fiber) or IQ samples (baseband over fiber) are fed from the CU to all BSs. Moreover, waveforms need to be time-aligned on a few μs when being transmitted, as pointed out in Section 8.2.

There are several arguments towards limiting the coordination to a cluster of cells (see also Chapter 7). Realistic wave propagation implies that inter-cell interference is spatially limited to the nearest neighbor cells [JJT+09]. Beside the high pathloss exponents between 3 and 4 in urban areas, interference is further limited if we include 3D radiation characteristics of the antennas at real BSs. They form a narrow beam with a half-power beam-width of roughly 6 degrees vertically and 60 degrees horizontally (Kathrein 80010541). For LTE-A, it is assumed that the beam touches the ground at roughly 1/3 of the inter-site distance (ISD). Including these effects causes a smaller pathloss exponent in the cell, while the exponent gets as large as 6 or 7 outside the designated cell area due to the vertical beam shape [TWB+09]. Hence, the downtilt can be effectively

13.3 Real-time Implementation and Field Trials for Downlink CoMP

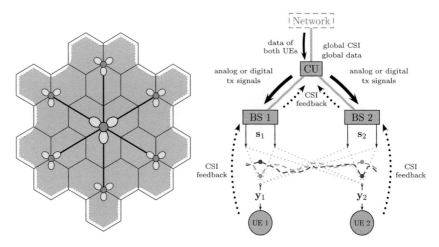

Figure 13.10 Centralized joint transmission (JT) CoMP setup.

used to control interference in adjacent cells. We assume that most of the CoMP gains can be realized with small or medium clusters of coordinated BSs.

Clustering is also a way to reduce the overhead. The number of orthogonal pilots for identifying the cells, the effort for the feedback over the uplink and the backhaul traffic between the BSs scale essentially with the cluster size. On the other hand, there is always residual interference from cells surrounding the cluster. Intuitively, we approach the information-theoretic limit if the cluster size tends to infinity.

There is a general trend of using distributed signal processing in mobile networks. Adaptation to the wireless channel can be much faster if it is performed at the serving BS. For downlink CoMP, we reduce the overall delay in the closed transmitter adaptation loop if the waveforms are generated at the serving BS. Ideas for distributed JT CoMP are developed in [ZKJ+07, HTJ08, JJT+09, JTS+08b, ZMS+09a, TWH+09], see Fig. 13.11. BSs use distributed synchronization derived from the global positioning system (GPS) or network protocols. UEs estimate the multi-cell CSI in the cluster and deliver CSI feedback to their serving BS. BSs exchange CSI as well as shared user data over a low-latency signaling network denoted as X2 [3GP10i]. The joint beamforming is computed independently at each BS, so that the transmitted waveforms are obtained locally from shared data. Same as in the centralized approach, the desired signals add constructively, while the mutual interference inside the cluster is canceled.

Shared data instead of IQ samples transfer is a lighter burden for the backhaul. In the distributed approach, latency is related to the ongoing aging process of the CSI while it is exchanged over the backhaul. A few ms may be tolerated for slowly moving UEs. Altogether, capacity and latency requirements for the backhaul are relaxed compared to the centralized approach. For these reasons, we believe that clustering and distributed implementation are the key features for introducing

334 Field Trial Results

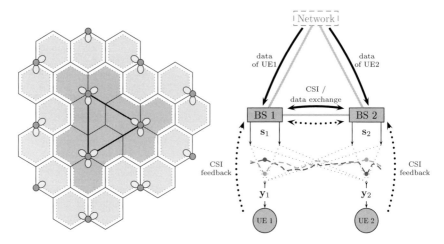

Figure 13.11 Distributed joint transmission (JT) CoMP.

JT CoMP in NGMNs. Our aim is to demonstrate that all enabling features in the closed control loop needed for distributed beamforming can be implemented with reasonable effort, and similarly high gains as predicted in theory can be realized in real propagation scenarios.

The section is organized as follows. At first, we explain the enabling features and how they can be implemented. Then we describe our field trials in indoor, outdoor-to-indoor and outdoor scenarios.

13.3.2 Enabling Features

Distributed Synchronization
A first requirement of JT CoMP is that all BSs are tightly synchronized. A precise time reference is available using an advanced GPS receiver provided that the line-of-sight (LOS) to the sky is free. The whole radio network can be synchronized with respect to the 1 pulses per second (PPS) signal. In advanced receivers, a phase-locked frequency reference (e.g. 10 MHz) is also provided.

Indoor BSs can be synchronized over the backhaul using a precision time protocol such as IEEE1588-2008, as discussed in Section 8.1. Based on the global reference at the grand master BS, the network propagation time with respect to the local clock at the slave BS is measured and corrected locally. In addition, the 10 MHz reference is distributed over the network. It is recoverable from the rising and falling edges of data bits continuously transmitted over the network using a clock-data recovery circuit at each network element. The recovered clock from the last hop is used for transmitting data bits over the next hop to the slave BS in the network. Network-wide synchronization is common in synchronous optical networks (SONETs) and synchronous Ethernet (SynchE) [Mil09].

13.3 Real-time Implementation and Field Trials for Downlink CoMP

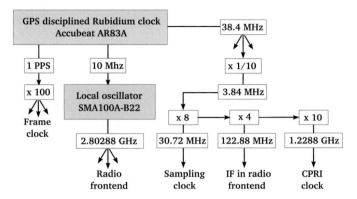

Figure 13.12 Synchronization and clock distribution at each site in the test bed [JFJ+10]. © 2010 IEEE.

Usually, the local timing reference (e.g. 1 PPS) is fed into a D-flip-flop jointly with the slope of the frequency reference. The stability of the frequency reference is therefore passed on to the timing reference as well. Hence, short-term stabilities of both references are related to each other. Rarely, the gating catches the next period of the frequency reference, and stability is often specified including these long-term effects while short-term stability is much better. For example, over a period of about 1000 s, we have measured a slow drift of 20 ns between two GPS-disciplined Rubidium clocks operated independently next to each other [JJT+09]. This translates into a short-term timing stability of $2 \cdot 10^{-12}$.

For JT CoMP, phase-coherent carriers are needed. It is intuitive that the allowable carrier frequency offset (CFO) between two BSs depends on the update interval of the CSI feedback. As it is 20 ms in our experiments, we set $N = 20$ in Section II.B in [JJT+09]. The tolerable short-term stability is 10^{-10} using 2 cooperative BSs targeting a signal-to-interference-and-noise ratio (SINR) of 30 dB. This is met using Rubidium or oven-controlled crystal oscillators (OCXOs) for stabilizing the reference [Mei]. Requirements depend on the desired SINR and become tighter with an increasing number of BSs in the cluster [MOJ10].

Independent phase noise is critical for JT CoMP. There is always a delay between channel estimation and application of the precoder. Assume that the precoder $\mathbf{W}$ is the pseudo-inverse of the channel matrix $\mathbf{H}$. Phase noise can be described by a matrix $\mathbf{\Phi}$ placed between precoder and channel, i.e.

$$\begin{pmatrix} h_{11} & h_{12} \\ h_{21} & h_{22} \end{pmatrix}^H \cdot \begin{pmatrix} e^{j\varphi_1(\tau)} & 0 \\ 0 & e^{j\varphi_2(\tau)} \end{pmatrix} \cdot \begin{pmatrix} w_{11} & w_{12} \\ w_{21} & w_{22} \end{pmatrix}.$$

If we set the random phase variations during the precoder update interval independently as $\varphi_1(\tau) \neq \varphi_2(\tau)$, the residual interference is hardly predictable. Only if $\varphi_1(\tau) = \varphi_2(\tau)$, the common phase variation can be tracked at the UEs.

In the trial system, all clock signals are created once at each site and shared between the sectors. As a local reference, we use a GPS disciplined Rubidium, see Fig. 13.12. It has a short-term stability of $5 \cdot 10^{-11}$. Following output signals are locked to GPS: 1 PPS clock (3x), 10 MHz reference (1x), 38.4 MHz reference (3x) and a serial interface providing location, time and other information in the common NMEA format (3x). The 10 MHz frequency reference feeds a low-phase noise local oscillator (LO) set to 2.80288 GHz (Rhode&Schwarz SMA100A), see Fig. 13.12. The 38.4 MHz frequency reference fulfills many purposes in the LTE signal processing unit (LSU). Dividing it by 10 results in the 3.84 MHz chip clock of the universal mobile telecommunications standard (UMTS). Multiplying this clock by 8, the 30.72 MHz sampling clock for the LTE signal processing chain is derived. Multiplication of 30.72 MHz by 4 results in the 122.88 MHz clock which is further multiplied by 10 to obtain the 1.2288 GHz clock for the common public radio interface (CPRI) interface between LSU and each radio frontend. The CPRI line receiver recovers the 122.88 MHz clock used also as an intermediate frequency (IF). The difference of LO and IF frequency provides our 2.68 GHz downlink carrier.

Cell-Specific Reference Signals

For JT CoMP, we need to know the downlink channel prior to the transmission. In time division duplex (TDD) mode, up- and downlink propagation is reciprocal. By means of calibration [LBC+04] or in specific transceiver configurations [JKI+04], reciprocity is realized in the baseband as well. In frequency division duplex (FDD) mode, the channel is no longer reciprocal. All BSs in the cluster provide CSI reference signals (CSI RSs) from which terminals estimate the multi-cell downlink channel. Terminals provide CSI feedback to their serving BS over the uplink.

In an orthogonal frequency division multiplex (OFDM) based air interface, the CSI RSs are easily designed in the frequency domain. We exploit the sampling theorem for the channel and propose an efficient use of CSI RSs in the multi-cell network, as discussed in Section 9.1. As a general rule, the channel shall be tested at least as often within the system bandwidth as there are independent fading taps resolved in the given bandwidth. The length of the cyclic prefix (CP) provides a lower bound for the density of pilots needed. The short CP in a 20 MHz LTE system has 144 samples, hence we have used a pilot spacing of 8 over 1200 sub-carriers. Note that advanced channel interpolation is able to reconstruct the channel between the pilots [SJ06], see also Section 9.1.

Our CSI RS design is shown in Fig. 13.13 [JMT+09]. With a spacing of 8, up to 8 cells can be identified without co-channel interference. We apply a cyclic shift of the regular pilot comb along the frequency axis by an integer number of sub-carriers, where the shift identifies the cell. Next consider the interference scenario in Fig. 13.13 (left), and the sequence assignment (center). If we identify each of the 7 cells by a shift from 0...6, shift 7 can be used for noise estimation.

1/7 is often used as a frequency reuse factor in mobile networks, see Fig. 13.13 (right). We propose to reuse the pilots in the second one after the next cell. For

13.3 Real-time Implementation and Field Trials for Downlink CoMP

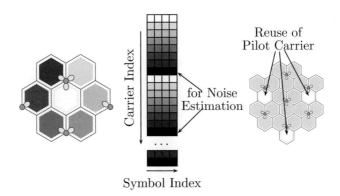

Figure 13.13 Design example for CSI reference signals (CSI RS) [JTW+09]. © 2009 IEEE.

this reason, interference from pilots in other cells is reduced by spatial distance. With frequency reuse 1/7, the CSI estimation error is improved by 11 dB at the cell-edge and 18 dB on average compared to full frequency reuse [JMT+09].

For identifying multiple antennas in each cell, we have used four consecutive OFDM symbols. On the same pilot sub-carrier and over consecutive OFDM symbols, another orthogonal sequence is transmitted for each antenna. We introduced this idea already in our early work [JFH05], and it has been adopted in the IEEE 802.11n standard. E.g. with four OFDM symbols, up to four antennas per cell can be identified.

There are limitations regarding the density of CSI RSs in time as well. Note that CSI RSs create a considerable amount of overhead and generate CSI feedback, accordingly. In LTE, there is a feedback delay of at least 7 ms. More than one sample of the channel in this time may increase our principal knowledge about the past channel. But predicting the future channel is imprecise beyond the channel coherence time. As a compromise, we use a feedback interval of 10 ms resulting in more feasible amounts of pilots and feedback overhead. This means that there is a maximum supportable UE velocity of 20 km/h sufficient for pedestrian and slowly mobile users.

CSI Feedback

Next we need to feed back the multi-cell CSI from each terminal to the serving BS. The provision of CSI feedback causes a significant overhead and some compression may be helpful. First, we shall provide feedback only for resource blocks (RBs) assigned to a user. This divides the amount of feedback by the number of terminals in a cell [JTS+08b]. Second, we may transform the channel from frequency to time domain. Less multi-path components are resolved if a fraction of the bandwidth is assigned to a user, i.e. both steps can be combined. Third, we can exploit correlation in time, frequency and spatial domains being likely at low mobility, in limited multi-path and for closely spaced antennas, as discussed in Sections 9.1 and 9.2.

To overcome the feedback delay, prediction shall be used. Normally we assume that the channel is continuous. Sometimes, this is not true for the CSI estimates. UEs irregularly apply individual automatic gain control (AGC) at each antenna and correct their timing and carrier frequency offsets. AGC corrections cause amplitude jumps, while timing corrections cause a frequency-dependent phase ramp over the whole bandwidth. CFO corrections are minor if they are phase-continuous. They can be regarded as a virtual Doppler spread. For minimal signal processing delay, these UE settings shall be either corrected out of the CSI feedback at the UE or reported in parallel. Based on corrected feedback, the BS can predict the channel evolution continuously after the last sample point of the channel.

In our trials, the channel is fed back once in each 10 ms radio frame. In the two-cell CoMP scenario and for 20 MHz bandwidth, feedback of the uncompressed complex-valued CSI for the 2x4 MIMO downlink channel to both BSs requires (4 Tx) x (2 Rx) x (144 pilots) x (2 x 16 bit) / 10 ms = 3.68 Mbit/s and effectively 4.6 Mbit/s including overhead. Neither feedback compression nor further quantization are used. Feedback is formatted sub-carrier-wise. For each pilot in the comb, the CSI for the entire 2x4 channel matrix is fed back.

Robust transmission is mandatory for avoiding feedback errors. We have used QPSK modulation and code rate 1/2 in 10 MHz bandwidth while splitting the 20 MHz uplink between both terminals. Inter-cell interference is hence avoided for CSI feedback transmission. Feedback may be delayed in case of decoding errors and the subsequent hybrid automatic repeat request (HARQ) process. We have introduced time stamps in our feedback packets. They are derived from the GPS time and the internal frame number and transmitted over the physical downlink control channel (PDCCH) to the UEs in each radio frame. Stamps are considered as follows: A new precoder is computed only if all CSI packets do have the same stamp and if the delay is 2 radio frames with respect to the local time at the BS. Otherwise, the precoder is considered corrupt and we stick to the most recent one.

Neither have we considered AGC, timing and frequency settings, nor implemented channel prediction. Instead, we have computed the precoder instantly from the feedback, so that zeros in space are set at previous locations of the UE antennas in mobile scenarios. We would need to consider them, of course, for optimizing JT CoMP for higher mobility.

Data Sharing
The data flows shall be strictly synchronized for all BSs in the cluster at the inputs of the local precoding units. In the centralized approach, data flows are terminated at the CU and synchronization is easier. In the distributed approach, however, synchronization of the flows is needed such that all BSs in a cluster

13.3 Real-time Implementation and Field Trials for Downlink CoMP

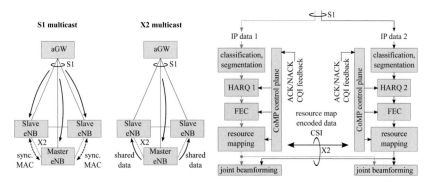

Figure 13.14 Organization of data flows for JT CoMP. Left: Multicast over S1 using synchronized MAC layers. Center: Unicast over S1 and exchange of shared data over X2. Right: The second option allows decentralized MAC layer processing and avoids unmanageable complexity for synchronizing the MAC layers inside the cluster.

transmit the same data symbol on the same resource element in the time-, frequency- and spatial domains.

There are two ways: First, one could multi-cast the data at the advanced gateway (aGW) to all BSs in the cluster, see Fig. 13.14, similar as in [IHJ08]. However, data has to pass several processing steps in the MAC layer before reaching the joint precoding operation located in the physical layer: Classification, segmentation, queue processing, HARQ, channel coding and mapping to resource elements. All these steps would need duplicate work at each BS in the cluster being fully synchronized. This can in principle be realized by introducing time stamps in the data packets at the aGW. However, all BSs would need to agree how, where and when a given data bit is transmitted over the air.

We have avoided this complexity by transmitting data in a first hop only to the serving BS where the multiple access channel (MAC) layer processing is terminated. We obtain encoded data bits for which it is known on which resource element they shall be transmitted jointly from all BSs. In a second hop, we forward a copy of this preprocessed data over X2 to the other BSs in the cluster, as indicated in Fig. 13.14 (center) [JJT+09]. In this way, even retransmissions in the HARQ process can be controlled locally and need no coordination between the BSs. A similar approach is recently investigated in information theory where it is denoted as conferencing encoder [WBBJ10]. The delay due to the transport over X2 is compensated by inserting a buffer at the precoder input for own data so that this is aligned with data obtained from other BSs. By using this approach, we were able to leave our MAC layer unchanged. Also, we managed to measure real throughputs using frequency-selective link adaptation and to support real-time applications over the air.

Flexible Backhauling

Exchanging both shared data and CSI turned out as an essential enabler for applying CoMP in the field with distributed BSs and UEs. In preliminary laboratory trials [JTW+09], the CSI feedback from both UEs had been decoded at each BS. When testing this configuration in the field, frequent feedback decoding errors were observed. These errors are related to the variable channel gains and delays between the distributed BSs and UEs. Note that optimized power control and timing advance is always used in mobile networks with respect to the serving BS, yet not for the other BSs in the cluster. At many locations in the field, consequently, the crossed links are not optimal and the feedback cannot be decoded correctly. Therefore, we have changed the CSI flow. Now the feedback of a UE is decoded only at the serving BS. Next, BSs exchange the CSI over the backhaul, as shown in Fig. 13.11. In this way, the feedback is much more robust for distributed setups in the field.

Therefore, we had to organize rather complex data flows. UEs are served with Internet protocol (IP) data which come over the S1 interface from the aGW. BSs exchange shared data over the X2 interface. Terminals transmit CSI feedback to their serving BS. CSI is multiplexed with shared data and transmitted over X2.

We have organized all flows consequently using the IEEE 802.3 Ethernet protocol enabling low-cost equipment in the backhaul. We take profit of an existing standard extension IEEE 802.1q denoted as virtual local area network (VLAN) to multiplex and demultiplex the different types of traffic. VLANs are based on a packet switching technique. While carrying a so-called VLAN identifier (VID), a packet is switched selectively from one input port of a switch to one or more output ports. The path of a packet in the network can be pre-configured depending on the VID. In this way, we can establish virtual sub-networks in which packets are multicast between selected BS. Therefore, we simply add a VID to the packets identifying the cluster as well as the type of data. If a given UE is served by a cluster of BSs, it is assigned a VID over the PDCCH. This VID is inserted into any CSI feedback packet. After uplink transmission and based on the VID, the switches in the network know how this information is conveyed to other BSs in the cluster. A second VID is used at the serving BS for the shared data multicast to other BSs in the cluster. Our proposal is based on standard protocols and allows pre-definition of potential cooperation clusters. Clustering can be changed on demand in a flexible and highly dynamic manner, as considered in Section 7.2. This is needed if UEs are mobile and their neighborhood relationships change.

In our trials, once in each 10 ms radio frame, a particular digital signal processor (DSP) at the UE takes over raw estimates of the multi-cell channel from the physical layer implemented in a field programmable gate array (FPGA). The DSP packs the CSI into four Ethernet packets, one for each 5 MHz bandwidth, by using a multi-cast Ethernet address. Packets denoted as C1 in Fig. 13.15 are further processed by the FPGA where they are sent to port 2 of the UE using 100 Mbit/s Ethernet. Next the VLAN switch at the UE adds a VID to

13.3 Real-time Implementation and Field Trials for Downlink CoMP

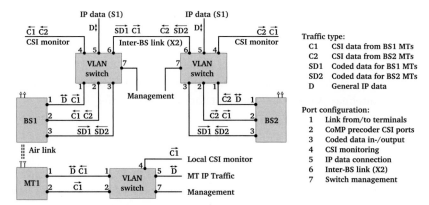

Figure 13.15 Setup of the virtual local area network (VLAN) for the CSI feedback and the data exchange between the BSs [JFJ+10]. © 2010 IEEE.

all C1 packets identifying them as CSI. By using another VID, application data (denoted as D) is identified at input 5. On output 1 of the switch, CSI is multiplexed with application data D. Port 1 at the UE can be regarded as a transparent Ethernet tunnel to the serving BS over the wireless link.

At the BS, the multiplexed streams are fed into another VLAN switch. Based on the VIDs, the switch splits CSI and application data packets and forwards them into different ports. The CSI stream C1 is copied into several output ports: The input of the precoder at BS 1 (2), a mirror of port 2 used for monitoring and measurements (4), and the X2 interface (6). In addition to the C1 stream, a shared data stream SD1 for UE 1 is also forwarded to port 6 of the switch. This data is taken out of the signal processing immediately after the MAC layer, see Fig. 13.14, right. The same packet switching is done simultaneously at BS 2. As a result, the data streams C2 and SD2 are arriving on port 6 of the switch and forwarded to port 2 and 3 of BS 1, respectively.

Compared to the delays introduced over the air, the use of VLANs in the backhaul has a negligible impact. Queuing delays in a 1 Gbit/s network can be estimated from the maximum packet length of 1500 bytes (12 μs) and the propagation time e.g. over 5 km fiber (25 μs). Cumulative delays over the air with respect to the beginning of a radio frame are: 0.3 ms for coarse multi-cell channel estimation, 1.7 ms until results are transferred to the DSP, 5 ms until CSI packet formation is completed, 12 ms until they are transmitted over the uplink and 19 ms until channel interpolation and computation of the precoder for 1200 sub-carriers is completed. There is severe potential for reducing the delay by using faster interfaces between signal processing and Ethernet, efficient feedback compression and faster processing.

Precoder

The precoder enables different multi-cell transmission schemes. It is realized as a linear matrix-vector multiplication originally described in [JFH05]. The incoming IQ signal constellations from all BSs are multiplied with a weight matrix before the waveforms are passed into the inverse fast Fourier transform (IFFT) at each antenna. Weights are different for each sub-carrier.

There is an algorithmic part executed in a companion DSP for computing the weights. The algorithm checks the consistency of the feedback and arranges the CSI in a channel matrix including the links between all BSs and all UEs in the cluster. Next, the channel is interpolated in frequency domain, see [SJ06]. Finally, the precoder is evaluated per each sub-carrier. Only rows relevant for local antennas are used. In the following, we describe the received signal vector $\mathbf{y}$ on each resource element as

$$\mathbf{y} = \mathbf{H}^H \mathbf{W} \mathbf{x} + \mathbf{n}, \tag{13.8}$$

where $\mathbf{H}$ is the multi-cell multi-user channel matrix as introduced in Section 3.5, $\mathbf{W}$ the precoder and $\mathbf{n}$ contains the noise at all antennas.

- **Full Interference**
  This setup realizes a 2-cell system without any coordination considered here as a lower bound for the performance in a system with unitary frequency reuse. The precoder is given as

$$\mathbf{W} = \begin{bmatrix} 1 & 0 & 0 & 0 \\ 0 & 1 & 0 & 0 \\ 0 & 0 & 1 & 0 \\ 0 & 0 & 0 & 1 \end{bmatrix} \tag{13.9}$$

- **Coordinated Multi-Point**
  By means of cooperation, we perform mutual interference cancelation between both cells. We consider spatial zero-forcing (ZF) [SSH04] by choosing $\mathbf{W}$ as the right-handed Moore-Penrose pseudo-inverse of the channel, i.e.

$$\mathbf{W} = \left(\mathbf{H}^H\right)^\dagger = \mathbf{H} \left(\mathbf{H}^H \mathbf{H}\right)^{-1}. \tag{13.10}$$

Depending on the channel, the ZF precoder causes signal fluctuations not present after the constellation mapper, see [JHJvH01, HvHJ+02, JHJvH02]. In order to avoid soft clipping in fixed-point arithmetics, the output is normalized. All columns of the precoder matrix have the same norm, i.e. all beams jointly provided by all BSs get the same power. Nonetheless, there is fluctuation in the frequency domain. As a numerical back-off, we have applied a downscale factor of 1/4 before passing the signal into the IFFT. After the IFFT, the signal is upscaled by factor 4 again.

- **Isolated Cells**
  Isolated cells are regarded as an upper bound for the performance. Precoders

for BS 1 and BS 2 are given as $\mathbf{W}_1$ and $\mathbf{W}_2$, respectively.

$$\mathbf{W}_1 = \begin{bmatrix} 1 & 0 & 0 & 0 \\ 0 & 1 & 0 & 0 \\ 0 & 0 & 0 & 0 \\ 0 & 0 & 0 & 0 \end{bmatrix} \quad \mathbf{W}_2 = \begin{bmatrix} 0 & 0 & 0 & 0 \\ 0 & 0 & 0 & 0 \\ 0 & 0 & 1 & 0 \\ 0 & 0 & 0 & 1 \end{bmatrix} \tag{13.11}$$

Demodulation Reference Signals

The precoding matrix $\mathbf{W}$ depends on the feedback in all cells in the cluster. Hence, it cannot be predicted by a terminal based on information available only in the own cell, how such CSI from other cells influences both signal and interference. Consequently, we need a second set of pilots denoted as demodulation reference signals (DRSs). DRSs enable the terminal to estimate the *effective*, i.e. the precoded multi-cell channel $\bar{\mathbf{H}} = \mathbf{H}^H \mathbf{W}$. Hence, DRSs identify the streams provided to the users in each cell.

Recall that CSI feedback is provided sparsely in time and frequency. Due to the time variant channel and the precoder delay, precoding is imprecise. There is residual cross-talk between co-channel streams served in the cluster and from other cells outside. The idea is to estimate the channels of all streams simultaneously on the air and to exploit multiple antenna techniques at the UE to reduce such interference. Moreover, amplitude and phase variations can be corrected in this way, so that the symbol constellations are properly reconstructed.

We have identified each stream according to the serving cell using another orthogonal sequence. The idea of our DRS design is described in [TSS+08]. There is a severe overhead problem when defining DRSs in a multi-cell scenario. Not only the BSs in the cluster shall be identified, as in case of the CSI RS, but also the streams transmitted in other cells, even outside the cluster. Moreover, the repetition rate of the DRSs shall be higher than for the CSI RSs. Residual interference is tracked on a ms time scale typical for channel equalization.

Our virtual pilot proposal is based on the same pilots used without JT CoMP for demodulation in one cell. In order to identify the streams, we have used a two-step approach based on code division multiplex (CDM). The first step is specific for our trial system and used to identify the streams at one site. In a slot duration of 0.5 ms, we transmit a stream-specific sequence on given pilot sub-carriers over multiple OFDM symbols. For example, in deployments with three sectors or cells per site and two transmit antennas per cells, this sequence would last over 6 OFDM symbols. Orthogonal sequences can be obtained from the rows of the 6x6 discrete Fourier transform (DFT) matrix. Each DRS is fed through the same spatial precoder as the corresponding data. An inverse discrete Fourier transform (IDFT) of the samples over the 6 consecutive OFDM symbols on a given pilot sub-carrier separates the precoded channels.

The second step can be used also with any other pilot pattern for identifying multiple sites. Therefore, we multiply all pilots in a slot by the same chip of a binary hyper-sequence taken from a Hadamard matrix. As an example, consider two sites. In the first slot, we multiply all pilots with +1 for both sites, while in the second slot, we multiply all pilots with +1 and −1 for sites 1 and 2, respectively. The correlation over two slots allows to distinguish from which site the stream is delivered. The value of virtual pilots is that no additional overhead is needed for the DRSs, rather the number of pilots is multiplied virtually owing to the hyper-sequence. However, the channel must not change during the period of the sequence. For this reason, virtual pilots are limited to low mobility. One particular aspect is that the estimation error can be reduced by optimizing the sequence assignment in the network. The particular assignment proposed in [TSS+08] is optimized in the way that Hadamard sequences being orthogonal already over a shorter correlation window are assigned nearby, while sequences with a longer correlation window are assigned more distantly.

Interference-aware Equalizer

We have used a simple interference-aware equalizer in our trials and determined the SINRs of the received data for the provision of channel quality indicator (CQI) feedback and link adaptation. For single stream (SS) transmission, the SINR is equal to that of the well-known optimum combiner [Win84]. In fact, we have implemented a linear minimum mean square error (MMSE) receiver useful for both SS and multiple stream (MS) transmission in a cell. Let

$$\mathbf{y} = \bar{\mathbf{H}} \cdot \mathbf{x} + \bar{\mathbf{H}}_I \cdot \mathbf{x}_I + \mathbf{n} \qquad (13.12)$$

be the received signal vector at a given UE, where $\bar{\mathbf{H}}$ and $\mathbf{x}$ are the 2x2 effective channel matrix and the transmitted 2x1 signal vector in the own cell, $\bar{\mathbf{H}}_I$ and $\mathbf{x}_I$ are the same in the interfering cell, respectively, and $\mathbf{n}$ is the 2x1 noise vector. The entries in $\bar{\mathbf{H}}$ and $\bar{\mathbf{H}}_I$ are estimated at the UE using the DRSs. Let us further decompose these matrices as

$$\bar{\mathbf{H}} = \begin{bmatrix} \bar{\mathbf{h}}_1 & \bar{\mathbf{h}}_2 \end{bmatrix} \qquad \bar{\mathbf{H}}_I = \begin{bmatrix} \bar{\mathbf{h}}_{I,1} & \bar{\mathbf{h}}_{I,2} \end{bmatrix}, \qquad (13.13)$$

where $\bar{\mathbf{h}}_i$ and $\bar{\mathbf{h}}_{I,i}$ with $i \in \{1, 2\}$ are the channel vectors corresponding to the first and second stream in the own and interfering cell, respectively. Further, we define the noise and interference covariance matrix

$$\mathbf{Z} = \begin{bmatrix} \sigma_1^2 & 0 \\ 0 & \sigma_2^2 \end{bmatrix} + \bar{\mathbf{H}}_I \left(\bar{\mathbf{H}}_I \right)^H, \qquad (13.14)$$

where σ_1^2 and σ_2^2 are the noise variances at antenna 1 and 2, which may differ due to independent AGC. Using $\mathbf{Z}$, we can compute the terminal-side MMSE receive filter as

$$\mathbf{G} = \bar{\mathbf{H}}^H \cdot \left(\bar{\mathbf{H}} \cdot \bar{\mathbf{H}}^H + \mathbf{Z} \right)^{-1} = \begin{bmatrix} g_{11} & g_{12} \\ g_{21} & g_{22} \end{bmatrix}, \qquad (13.15)$$

reconstructing our signal constellations as

$$\hat{\mathbf{x}} = \mathbf{G} \cdot \mathbf{y}. \qquad (13.16)$$

13.3 Real-time Implementation and Field Trials for Downlink CoMP

Using particular elements from (13.15), we can derive the estimated SINR on the spatially multiplexed stream k as

$$\boldsymbol{g}_k = \begin{bmatrix} g^*_{k1} \\ g^*_{k2} \end{bmatrix} \quad \text{SINR}_k = \frac{|\boldsymbol{g}^H_k \cdot \bar{\boldsymbol{h}}_k|^2}{\boldsymbol{g}^H_k \cdot \mathbf{Z} \cdot \boldsymbol{g}_k}, \quad (13.17)$$

where $k \in \{1, 2\}$. These equations correspond to the case of MS transmission in the desired cell. In case of SS transmission, the channel vector $\hat{\boldsymbol{h}}_i$ is set to zero if stream i is not transmitted in the serving cell. We always assume the conservative case that MS transmission is used in the interfering cell. The SINR obtained using (13.17) is used for estimating the achievable rates in the presence of the interference from the other cell.

Adaptive Rate Control

Next, we explain how the measured throughput is realized. On top of the physical layer, frequency-selective link adaptation and MIMO mode switching is implemented in real-time, based on CQI reports received each 10 ms from the terminals. UEs estimate the effective channels $\bar{\mathbf{H}}$ and $\bar{\mathbf{H}}_I$ using the DRSs. We compute the frequency-selective SINR for 16 groups where each group covers 75 sub-carriers. SINRs values are computed for a comb of three sub-carriers in each group for several transmission modes: SS transmission on either stream 1 or 2 and MS transmission in case of spatial multiplexing. The SINR values for all three options are mapped to achievable rates using two bits corresponding to on-off-keying (OFF), QPSK, 16-QAM or 64-QAM modulation. These bits are fed back to the serving BS as a compound frequency-selective CQI vector where the lowest rate among the three sub-carriers is selected. For SINR-to-rate mapping we have used a target of 10^{-1} for the uncoded bit error rate (BER).

The BS compares the achievable rates for all spatial modes. The rate is maximized selectively for each sub-carrier group by choosing the best mode and assigning the modulation format recommended for each stream by the terminal. In the sum over the whole bandwidth, a variable data rate is realized. All data bits transmitted in a transmit time interval (TTI) are considered as a transport block. The block is encoded using a fixed code rate of 1/2 in our setup and then passed through an interleaver spanning all resources assigned to a user in the TTI, hence supporting a variable codeword length. Actually measured error rates at the decoder input are in the order of 10^{-2} during our field trials, due to the conservative sub-carrier selection described above. This yields almost error-free transmission after decoding, i.e. retransmissions are rarely needed. The gross data rate is measured at the physical layer as the sum of the successfully transmitted data and parity bits per second excluding overhead for synchronization, multi-cell channel estimation and control signaling. The peak rate of the trial system is 141 Mbit/s in each cell if both streams are served with 64-QAM on all RBs. Throughputs, multi-cell channel coefficients and bit error rates are simultaneously recorded during field trials.

Table 13.4. Real-time system parameters.

Parameter	Value
Up-/downlink frequency	2.53/2.68 GHz
Bandwidth	20 MHz each
Number of cells	2
Number of BS antennas	2
Number of terminals	one in each cell
Number of terminal antennas	2
Feedback rate	5 Mbit/s in each cell
X2 data rate (two ways)	$\approx$ 300 Mbit/s

13.3.3 Real-time Implementation

The numerical complexity of JT CoMP is not as much increased compared to LTE Release 8 and it can be handled by state-of-the-art multi-core DSPs [KÖ7, MS09]. On the other hand, the scheme has a high functional complexity. It is working properly only if all enabling features mentioned in the previous subsection work hand in hand and in real-time. We consider JT CoMP as the most sophisticated closed-loop wireless transmission technique ever realized. Our implementation focuses on the physical layer, on top of which the other layers are operated almost unchanged. We assume that the set of BSs and UEs in the cluster is known and held fixed and that all feedback and X2 links have been pre-configured.

Managing the functional complexity has been a continuous learning process. Note early ideas concerning JT [JHJvH01, HvHJ+02, JHJvH02], initial JT experiments [HFG+04], the first 1 Gbit/s trials using MIMO OFDM [JFH05, JFH+06], the adaptation of OFDM parameters to LTE [JSF+07], the introduction of frequency-selective CQI feedback and MIMO mode switching [WJA08] and the first LTE field trials [VHW+08, WJF+08]. The innovation here is that we have realized multi-cell CSI feedback over the air and performed distributed coherent MIMO processing at spatially separated BSs.

Hardware Platform

We have implemented JT CoMP by extending our LTE trial system [JST+09] to a setup with two BSs and two UEs having 2 antennas each. Elementary system parameters are given in Table 13.4. The downlink is switchable between SS and MS transmission. Transmission mode and modulation can be set adaptively according to the frequency-selective CQI feedback provided by the terminals. The schematics of our JT CoMP implementation are available in [JTW+09].

Implemented CoMP Features at the BS

Both terminals are served in their cells via co-channel transmission. Data is sent to the serving BS over S1, realized as 1 Gbit/s Ethernet. Transmission over the

13.3 Real-time Implementation and Field Trials for Downlink CoMP

wireless link is transparent, i.e. functionally it replaces an Ethernet cable, but offers a variable data rate. After passing the data through the corresponding queue and HARQ processes, variable-length transport blocks are formed and fed through forward error correction (FEC) and interleaving, as described above. Data are mapped adaptively onto the RBs. Next we have inserted the DRSs, as described in detail in [TSS+08]. Pilot spacing is 4 sub-carriers. Antenna- and cell-specific Hadamard sequences of length 4 are used to identify the four streams in both cells.

After this stage, we have implemented synchronous data exchange. Exchange is realized over a physically separate 1 Gbit/s Ethernet link between both BSs. Shared data is packed as complete OFDM symbols into Ethernet packets. We have realized JT CoMP over the full 20 MHz bandwidth. Symbols on a given resource element are always exchanged using 6 bit, ignoring the modulation format actually used. This yields a constant data load of 300 Mbit/s on X2. Data in the own and from the other cell are synchronized at the input of the precoder using buffers. The added latency due to synchronous data exchange is 0.5 ms using an Ethernet cable between both BSs. Next the CSI RS are inserted. The design is taken from Fig. 13.13 (center). We have again used Hadamard sequences of length 4 over the first 4 OFDM symbols in the first slot of a radio frame. The same correlation as for the DRS has been reused. Just another sub-carrier mask is applied at the raw estimator output. CSI RS, precoded data and DRS are jointly fed into the MIMO OFDM transmit chain and sent over the air.

Implemented CoMP Features at the Terminal
For feedback generation, we have integrated a separate DSP in the UE. It has access to the raw channel estimates as well as to 100 Mbit/s Ethernet. The CSI packets are passed in a sequence of four Ethernet packets from the downlink receiver into a VLAN switch, where a specific VID is added, the CSI is multiplexed with uplink data using another VID, and then passed into the uplink transmitter. The CSI feedback is separated from uplink data in another VLAN switch at the BS and terminated twice: at the precoder DSP in the serving BS and, after transmission over X2, at the precoder DSP of the interfering BS. For the equalization at the receiver side, the precoded channel is estimated for all streams in the own and interfering cells using DRS. Interference-aware equalization and link adaptation is implemented as described above.

13.3.4 Field Trials

Setup and Scenarios
Scenarios in the field comprise specific indoor, outdoor-to-indoor, and outdoor configurations, as illustrated in Fig. 13.16. Indoors (scenario 1), both BSs are located in the same lab. For outdoor-to-indoor and in the field scenarios, we select two sectors either at one site or at two sites to realize intra- or inter-site

348 Field Trial Results

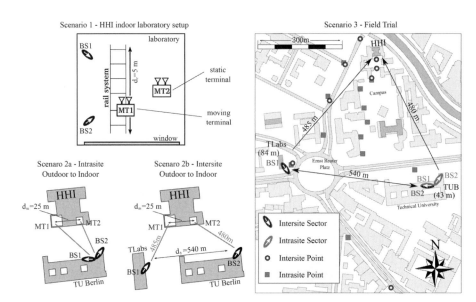

Figure 13.16 Measurement scenarios: Indoor laboratory scenario (top left), outdoor to indoor lab trials (bottom left), location of BSs and UE 1 in the field trials (right). Map data © OpenStreetMap & contributors, CC-BY-SA. From [JFJ+10]. © 2010 IEEE.

cooperation (scenarios 2a-b, 3a-b), respectively. Sites are located on the T-Labs building at Ernst-Reuter-Platz (84 m antenna height) and on the TU Berlin main building, Straße des 17. Juni (43 m, see Fig. 13.16). The average building height in the area is between 25 and 35 m. Sites are interconnected by an optical fiber with 4.5 km length.

For scenarios 1 and 2, both UEs are located on the 11th floor at the Heinrich-Hertz Institute (HHI) building. We have placed both UEs at the south front of the building with the window facing both BSs either in the same lab (scenario 1) or in two different labs 25 m separated (scenario 2). UE 2 is always at a fixed location. For capturing the local fading statistics, UE 1 moves at low speed of 3 cm/s on 5 m long rails in the lab (see Fig. 13.16, top left). In scenario 3, UE 2 is at a fixed position either at the ground in a van in front of the south side of the HHI building or in the 11th floor lab inside the HHI building. UE 1 is in a second van at variable but fixed locations in the field indicated in Fig. 13.16. The assignment of BS 1/2 and UE 1/2 is also available in Fig. 13.16. UE 1 is always connected to BS 1 and UE 2 to BS 2. Handover is regarded when selecting valid terminal positions for the statistical evaluation.

Results

Indoor and outdoor-to-indoor results are plotted in Fig. 13.17. Left in each row of figures we have plotted the SINR per data stream in the interference-limited case, i.e. without JT CoMP and after applying interference-aware equalization and link adaptation at the UE. Right, the throughput at each UE is plotted for

13.3 Real-time Implementation and Field Trials for Downlink CoMP

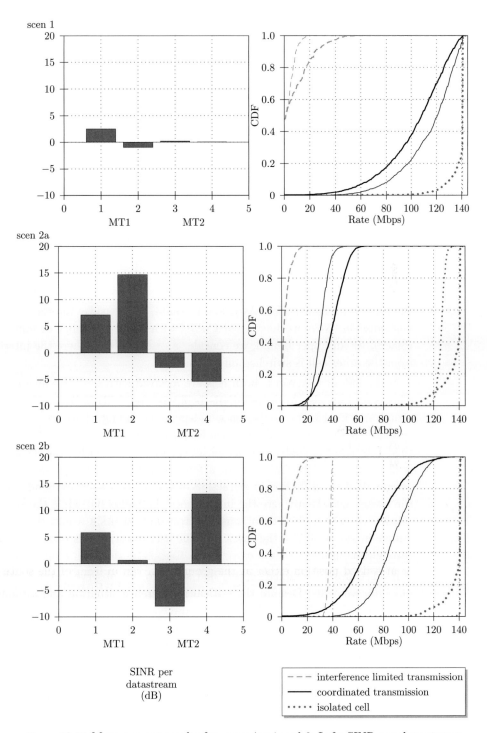

Figure 13.17 Measurement results for scenarios 1 and 2. Left: SINR per data stream. Right: Measured throughput for interference-limited transmission (dashed), JT CoMP (solid) and isolated cells (dotted). Thick lines denote UE 1, thin lines UE 2. From [JFJ+10]. © 2010 IEEE.

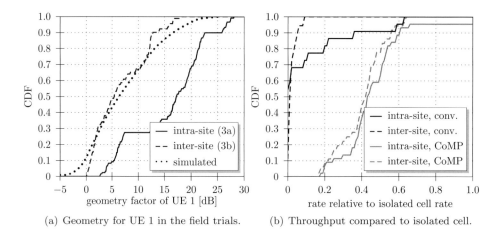

(a) Geometry for UE 1 in the field trials.   (b) Throughput compared to isolated cell.

Figure 13.18 Geometry and CoMP gain as observed through field trials [JFJ+10]. © 2010 IEEE.

three transmission modes. As a lower bound, we consider interference-limited transmission. Next we have used CoMP with a fixed number of four streams (black). As an upper bound, we consider an isolated cell where the interference from the other cell is switched off (dotted).

In the indoor scenario 1, both BSs are received with the same average power. This is obvious also from the SINRs being around 0 dB in both cells. However, signal and interference experience independent fading. This creates a critical throughput situation for a UE when frequency reuse 1 is applied: When moving the UE by a few cm only, we can realize situations where either the signal channel is flat while the interference is in a fade, and correspondingly the serving BS assigns data transmission in a fraction of the whole frequency band, as well as the reverse situation where the interference channel is flat while the signal is in a fade, so that no more data is usually transmitted. As a result, the UE suffers outage in about 50% of the cases, and the user experience is poor when moving through the lab.

If JT CoMP is enabled in this bad scenario, we observe dramatic throughput improvements. Despite the critical interference situation and although the data rate still varies, CoMP removes outage at the cell-edge completely. Using CoMP in scenario 1, both terminals have on average 18.7 times higher throughput and realize 78% of the rate they could achieve in their isolated cell (see Table 13.5). While doing these trials in the lab, we observed that CoMP establishes a new link quality at the cell-edge. The unpredictable on-off characteristics of the interference-limited channel is turned into a stable continuous data link with some residual capacity variation. Potentially, CoMP can be used for advanced real-time multimedia applications such as mobile video conferencing.

Next, consider the intra-site scenario 2a. It is typical in the distributed multi-cell network that the pathlosses are not equal for different pairs of BSs and

UEs. Nonetheless, the principal observations remain similar. In the interference-limited case, there is significant outage, and irregularly a few percent of the peak rate can be realized. Using JT CoMP, in contrast, both UEs can realize 27 and 21% of the peak rate on average. The performance of the inter-site scenario 2b is superior, i.e. 63 and 50% of the peak rate, despite distributed synchronization. This can be related to the higher correlation for co-polarized signals coming from the same site but different sectors, see [JFJ+10].

Let us finally consider the field results. We have selected only locations where both UEs are served by their strongest BS to consider the handover performed in a real network. The geometry factor (GF) is the ratio of the mean signal power to the mean interference power. It is shown for both the intra- and inter-site scenarios in Fig. 13.18(a). The fixed UE 2 position at the ground of the HHI building has a GF of -1.2 dB for the intra-site and -4.5 dB for the inter-site scenario. Likewise, the other UE 2 position at the 11th floor achieved -1.7 dB and $+7$ dB. Obviously, the GF statistics of the intra-site scenario are too optimistic, while the inter-site scenario contains a fairly realistic interference distribution similar to multi-cell simulations [TWS+09].

According to the higher GF, there are several locations in the intra-site scenario where a relatively high data rate is assigned despite the interference limitation. The scheduler is aware of both, the channels to the own and to the interfering BSs (see Fig. 13.18(b)). In the inter-site trials, the GF is more realistic and lower data rates are assigned in the interference-limited mode. In both cases, there is a certain probability of outage also in the field. If we use JT CoMP, instead, the differences in the geometries do no longer reflect the data rates. The observed rates have a similar distribution in both scenarios, possibly due to impairments such as residual time variance. On average, 46 and 40% of the isolated cell capacity are measured in both cells simultaneously, owing to the mutual interference suppression between both cells (see Tab. 13.5).

If one would use frequency reuse 1/2 in both cells, half of the isolated cell rate would be achieved. Strictly speaking, only in those cases where the CoMP rate is larger than 50% of the isolated cell rate, a true gain is observed. This is achieved for 20% of the intra-site and 30% of the inter-site scenarios in the field. On the other hand, the rates are much larger in the indoor-scenario where the path gains for signal and interference are similar. In a real system, we assume that further advanced precoding algorithms will be available, and that we may choose from multiple terminals in each cell having similar SINR. Well-conditioned multi-cell multi-user channel matrices with close-to-orthogonal channel vectors are then found. Including these additional algorithmic and multi-user gains, we expect that most UEs will benefit from CoMP as compared to frequency reuse 1/2.

Table 13.5. Mean rates for all scenarios [Mbit/s].

Scen.	Int. limited	CoMP	Isolated cell
1	5.8	109.3	139.6
2a	1.7	35.3	131.2
2b	21.1	80.1	139.1
3a	11.0	48.1	104.2
3b	2.1	39.1	98.9

13.3.5 Summary

We have implemented joint transmission CoMP in the downlink of an LTE trial system operated in FDD mode and measured the performance in real-world signal propagation scenarios. Our implementation follows a distributed concept reducing the backhaul significantly, as compared to a centralized concept. We have used distributed synchronization and linked base stations via Ethernet. Information exchange between base stations is realized using commercially available network equipment supporting IEEE 802.1q. Coherent interference nulling has been demonstrated over the air using distributed beamforming at 500 m inter-site distance and for hundreds of meters between BSs and UEs. This is the proof of concept that joint transmission CoMP can be integrated into the distributed mobile network architecture of LTE.

We have performed over-the-air indoor, outdoor-to-indoor and field transmission experiments and demonstrated that the advantages of CoMP are remarkable compared to interference-limited transmission. Indoors, we have observed that the unpredictable on-off characteristics of the interference channel next to the cell-edge is turned into a stable continuous data link with some residual rate variation if CoMP is enabled. Thus, CoMP enables a new link quality in mobile networks suitable e.g. for mobile video conferencing. We have realized on average between 27% and 78% of the isolated cell throughput in the investigated scenarios while running the system at unitary frequency reuse in both cells. The gains for the average data rate vary between factors 4 and 22 depending on the scenario. Our results show that the implementation challenges for CoMP can be overcome.

Note that the gains reported here are not yet realistic for large-scale mobile networks where cooperative BSs are surrounded by other cells. External interference has not been present in our trials, and it should be considered when assessing the performance of CoMP in real networks. First performance predictions are provided in Section 14.4, where the techniques implemented here have been investigated in large-scale system simulations.

Table 13.6. Physical layer parameters of the discussed downlink (DL) CoMP system.

Parameter	Value
Frequency band	2680 MHz (DL), 2530 MHz (UL)
DL bandwidth / PRBs	20 MHz / 10 PRBs shared by all UEs
UL bandwidth / PRBs	20 MHz / 30 PRBs for CSI per UE
SC per PRB / SC distance Δf	12 / 15 kHz
BSs (M) / BS tx antennas (N_{bs})	2 / 2 per BS
UEs (K) / UE rx antennas (N_{ue})	2 / 2 per UE
CSI feedback interval	1 TTI, i.e. 1 ms
CSI granularity	6 sub-carriers
OFDM symbols per TTI,	14
used for RS / PDCCH / PDSCH	4 / 1 / 9

13.4 Predicting Practically Achievable Downlink CoMP Gains over Larger Areas

Jörg Holfeld, Erik Fischer, Vincent Kotzsch, Patrick Marsch and Gerhard Fettweis

In this section, we share some complementary experiences that could be gained from a joint transmission (JT) CoMP implementation and subsequent measurement campaigns performed in the EASY-C test bed in Dresden, Germany. The approach pursued here also corresponds to distributed JT, as introduced in Section 6.3 and applied in Section 13.3. The main difference to the previous section, however, is that the closed-loop system considered now follows some different design paradigms, and a different evaluation methodology is used that enables an efficient prediction of downlink CoMP gains over larger areas. Details of the system setup, channel state information (CSI) feedback and precoding loop are discussed in Subsection 13.4.1, in particular those aspects differing from the implementation in Section 13.3, followed by the used measurement and evaluation methodology in Subsection 13.4.2 and finally the actual results in Subsection 13.4.3. Details on the employed linear precoder implementation at the base station side were already stated in Section 10.1.

13.4.1 Setup and Closed-Loop System Design

Overall Setup

In this section, we consider the same setup as in Section 13.3, i.e. a JT from $M = 2$ base stations (BSs) with $N_{\text{bs}} = 2$ transmit antennas each to $K = 2$ user equipments (UEs) with $N_{\text{ue}} = 2$ receive antennas each, enabling a transmission of up to $N_{\text{BS}} = N_{\text{UE}} = 4$ spatial streams over the same resources in time and frequency. Key parameters of the setup are stated in Table 13.6. Also as in Section 13.3, the BSs are synchronized in time and frequency via global positioning system (GPS)-fed reference normals.

354 Field Trial Results

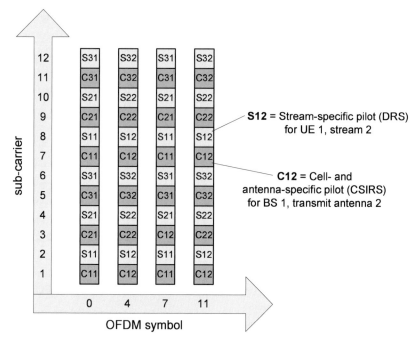

Figure 13.19 Pilot concept designed for downlink CoMP, for a maximum of 3 BSs and UEs with 2 antennas each.

Pilot Concept

As emphasized in Section 13.3, the JT CoMP concept envisioned in this book requires the usage of both *antenna- and cell-specific* pilots, i.e. CSI reference signals (CSI RSs) (such that the actual channel between BS side and each UE can be estimated and fed back to the BS side for precoder calculation) and *stream-specific* pilots, i.e., demodulation reference signals (DRSs) (needed by the UEs to estimate the *effective* channel after precoding, so that the actual data transmission can be equalized and decoded). Note that the latter kind of pilots are specifically required because the UEs are not aware of the precoding matrix chosen by the BSs.

The specific pilot concept employed in the Dresden test bed implementation is based on pilots that are orthogonal in time or frequency, and is illustrated in Fig. 13.19. The maximum setup possible here includes 3 BSs with 2 transmit antennas each, transmitting to 3 UEs with 2 receive antennas each. Hence, one can use up to 6 antenna- and cell-specific and another 6 stream-specific pilots. In the case of this maximum setup, one can see from the figure that the orthogonal frequency division multiplex (OFDM) symbols 4, 7, 11 and 14 are completely occupied with pilots, and each pilot appears four times, so that a reasonable number of supporting points is available for a channel inter- and extrapolation over time and frequency (see Section 9.1). As annotated in the figure, pilots denoted as Cma refer to the antenna- and cell-specific pilot connected to BS m

and antenna a, and Sks refers to the stream-specific pilot connected to UE k and stream s. For the reduced setup with $M = K = 2$ evaluated in this section, all pilot positions referring to BS 3 or UE 3 in Fig. 13.19 were left empty.

CSI Feedback Concept
Regarding the feedback of CSI from the terminals to the BS side, which is required for joint transmission, the approach pursued here follows a different paradigm than that in Section 13.3. The CSI feedback of the different UEs again takes place on orthogonal resources, so that this is not subject to inter-cell interference, but now all involved BSs attempt to decode the feedback from all UEs. In symmetric cell-edge scenarios, i.e. where the UEs are located at the cell-edge between the cooperating cells, it is rather likely that the involved BSs should be able to decode most of the feedback, if this is transmitted at a robust code rate. In this case, an additional exchange of CSI between BSs is not required, which means that not only one major complication connected to the implementation from Section 13.3 can be avoided, but also that the CSI which is finally used for precoding is less outdated. Clearly, the downside of this CSI broadcasting concept is the larger number of required uplink resources, and measurements have shown that BSs can only reliably decode feedback from UEs to which fairly strong links exist. However, we have seen in previous chapters that these are exactly the links that are most interesting in the context of joint signal processing. In future systems, one could potentially employ some kind of hierarchical CSI feedback concept, as suggested in Section 6.4, where the accuracy of CSI each BS can obtain on a particular UE increases with increasing link strength.

The actual CSI feedback structure employed in the Dresden test bed features a very flexible format of variable length and adaptable to different DL CoMP scenarios with different numbers of BS and transmit / receive antennas. An overview on the structure is provided in the sequel:

1. Sync sequence: Identification of message start (2 bytes)
2. SNR header: Noise power measurements
3. CSI header: Antenna configuration (2 bytes)
4. CSI payload size: Length of CSI payload (2 bytes)
5. CSI payload: CSI (up to 576 bytes)

The maximum length of the feedback message adds up to 584 bytes or 4672 bits. The feedback is sent on transmit time interval (TTI) basis, i.e. once per millisecond. Due to the large amount of data, the LTE physical uplink control channel (PUCCH) is not sufficient to store the complete feedback information, so that the LTE physical uplink shared channel (PUSCH) has to be used to transmit the feedback message. A fixed uplink transport format has been chosen with a transport block size of 5040 bits, QPSK modulation and code rate of 2/3, so that the feedback can be considered fairly robust. Note that this leads to a vast number of physical resource blocks (PRBs) used for CSI feedback in the uplink, and hence the concept cannot be considered as a concrete proposal for a commercial system. For our particular implementation, however, the main focus was to deliver the proof-of-concept for downlink CoMP in general, while the reduction of CSI feedback was left to future research. A simple single-input single-output (SISO) scheme is used for CSI transmission and detection, though this could of course be extended to multiple-input multiple-output (MIMO) and (possibly) uplink CoMP concepts in the future.

Signal Processing Approach
At the BS side, all decoded CSI feedback messages are passed to the order-recursive precoding matrix computation as described in Section 10.1. This minimum mean square error (MMSE) based transmit filter mitigates the spatial inter-stream interference as seen at the UE side. The resulting precoding matrices are passed to the linear precoder of the downlink transmitter that is responsible for the stream-to-antenna mapping, and applied to both data transmission (physical downlink shared channel (PDSCH)) and stream-specific pilots (DRS).

At the terminal side, initial synchronization, comprising the timing estimation for frame detection and carrier frequency offset (CFO) compensation, is applied. Then, the received antenna streams are OFDM demodulated and resources de-multiplexed to extract the physical channels from the corresponding time- and frequency resources. Before the data streams in the PDSCH can be decoded, the IQ symbols of the physical downlink control channel (PDCCH) must be processed. The control channel is transmitted by the serving BS and contains all relevant information about the modulation and coding scheme (MCS) within the current TTI. The PDCCH is equalized and decoded based on the CSI RSs from the first BS, as this has not been passed through the precoder. Before decoding the PDSCH, the channel on the corresponding resources has to be estimated based on the DRSs. Subsequently, an interference rejection combining (IRC) equalization is performed for the received PDSCH streams, as discussed in Section 10.2. It should be noted that the shared channel is decoded offline in our prototype CoMP system. For this purpose, the received baseband (I/Q) signals are dumped and decoded on a host PC using an LTE MATLAB chain, similar to the procedure used in Section 13.2. This approach is convenient for rapid prototyping of different receiver strategies, and sufficient to obtain relevant performance measures such as block error rate (BLER) or bit error rate (BER).

13.4.2 Measurement and Evaluation Methodology

The major drawback of extensive measurement campaigns is the large amount of time which is consumed for preparation, accomplishment and data evaluation. Especially in CoMP systems, one has to manage and control several distributed BSs and UEs. For scientific experiments, the reproducibility must furthermore be assured. But outdoor radio channel sounding campaigns, as described in Section 14.2, demonstrate that once a specific terminal constellation has been

13.4 Predicting Pract. Achievable DL CoMP Gains over Larger Areas

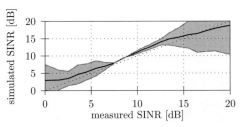

(a) Mapping of measured to simulated SINRs. The gray background area bounds the curve by its standard deviation.

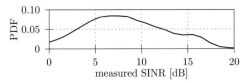

(b) Probability distribution of the measured SINR values.

Figure 13.20 The SINR mapping results and PDF for spatial interference situations.

altered, it is hardly possible to retrieve the original positions with equal CSI. For these and other reasons, all measurement results provided in this section have been obtained based on the following two-step measurement approach:

First, the complete closed-loop real-time transmission concept as explained before was carried out for each possible spatial configuration (i.e. different numbers of transmitted streams, one or two served UEs, different MCSs etc.) and for selected positions within the test bed, which is a very time-consuming process. Second, a channel measurement campaign followed to characterize the physical propagation environment for a substantially larger number of test bed positions. As this requires only one terminal, a fixed spatial configuration and fixed MCS, the measurement is significantly more efficient. Based on these data records, various performance indicators connected to downlink JT CoMP can then be accurately extrapolated to a larger area.

Mapping of Measured SINRs to Simulated SINRs

The entire closed-loop system described in Section 13.4.1 exists as a simulation environment with bit-true signal representations. This kind of signal processing environment is typically used for hardware verification, but can also be used to evaluate the measured CoMP performance. In Fig. 13.20(a), the measured resulting stream SINRs of the real-time closed-loop operation for a subset of positions are compared to simulated SINR values. For the measured values, the UEs use the DRSs to estimate not only the effective channels connected to desired streams, but also to the streams targeted towards the other UE, and calculate corresponding power ratios. A fairly good match between simulation and practice was achieved in a range from 0 to 20 dB SINR, which covers the majority of

358   Field Trial Results

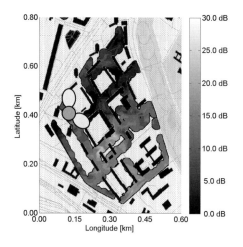
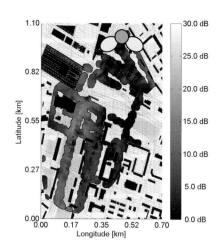

(a) Northern test bed area.   (b) Southern test bed area.

Figure 13.21 The measured geometry factor [dB] quantifies the cell-edges based on power ratios between the first and the second BS sector. All map data used in this section © OpenStreetMap & contributors, CC-BY-SA.

instantaneous SINR values observed in our measurements, as indicated in the probability distribution in Fig. 13.20(b). Generally, low SINRs are overestimated while large ones tend to become underestimated, but the average SINR prediction obtained through the simulation chain appears very accurate.

With this described evaluation methodology that consists of a hybrid of actual downlink CoMP measurements and channel measurement based performance prediction, the overall measurement time for the results presented in the sequel could be reduced from several months to about one week.

13.4.3   Measurement Campaign

In this subsection, we initially perform a general analysis of the channels observed during our measurement campaign, before the actual downlink JT CoMP performance is discussed. In the sequel, we generally distinguish between two measurement areas with different signal propagation properties and hence also different CoMP performance, which we refer to as *northern* and *southern* test bed area.

Radio Channel Characterizations

The geometry factor, as introduced in Chapter 7, is used to quantify the mean downlink receive power ratio between the serving BS and the interference level from the other BS over N observations, i.e.

$$\mathbf{G} = \frac{\sum_{n=1}^{N} |\mathbf{H}_1^1[n]|^2}{\sum_{n=1}^{N} |\mathbf{H}_1^2[n]|^2}. \tag{13.18}$$

13.4 Predicting Pract. Achievable DL CoMP Gains over Larger Areas

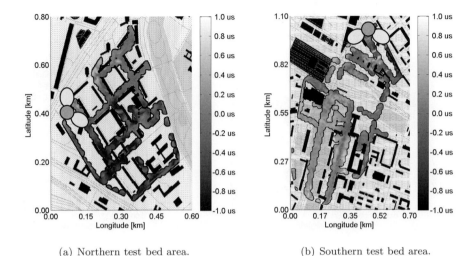

(a) Northern test bed area. (b) Southern test bed area.

Figure 13.22 The measured TDOAs can also be used to determine the cell-edges.

The geometry factor points out those places where the cell-edge with symmetric and cell-center with asymmetric receive power levels can be found. Fig. 13.21 emphasizes that only very few positions have almost equal power between serving BS and interference, i.e., a geometry of 0 dB, where CoMP gains can be expected to be largest. In both figures, the measured cell-edge fits quite well to the geometric cell-edge of the spanned sectors, though shadowing due to prominent buildings leads to more diverse instantaneous geometry factor realizations.

The cell geometry can additionally be described from a *timing* point of view, considering that radio waves propagate at the speed of light. Based on the pilot patterns, we estimated the power delay profile (PDP) of each transmission link and computed the *time delay of arrival (TDOA)* as difference of the mean excess delays between the two BSs, as observed at different potential UE locations. The excess delay is the first moment of the PDP [Rap02] defined as

$$\bar{\tau} = \frac{\sum_k P(\tau_k)\, \tau_k}{\sum_k P(\tau_k)}, \tag{13.19}$$

with $P(\tau_k)$ being the power of the k-th channel impulse response (CIR) tap and τ_k the corresponding delay. It can be seen in Fig. 13.22 that the cell-edges derived from the TDOA match quite well with those based on the geometry factor.

A second timing based quantity that we present here is the *root mean square delay spread* $\bar{\tau^2}$, which gives an impression of the scattering environment within the measurement areas. This is defined as the second central moment of the PDP [Rap02], i.e.

$$\sigma_\tau = \sqrt{\bar{\tau^2} - (\bar{\tau})^2} \quad \text{with} \quad \bar{\tau^2} = \frac{\sum_k P(\tau_k)\, \tau_k^2}{\sum_k P(\tau_k)}. \tag{13.20}$$

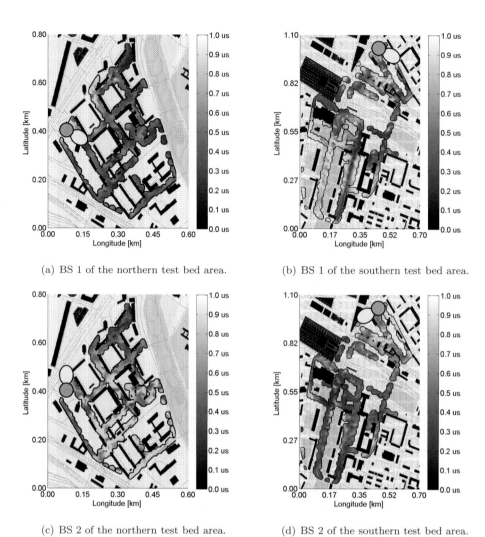

Figure 13.23 The measured delay spreads show the scattering environment as the influence of dominant CIR taps for the first and the second BS.

We can see in Fig. 13.23 the different characteristics of the northern and southern areas. The high density of buildings and the 120° direction of the main antenna beam leads for the northern BS 1 to a rich scattering with a low delay spread. BS 2 is mainly observed under non line-of-sight (NLOS) conditions. The southern part differs from the north in larger propagation delays such that $BS\ 1$ is subject to a line-of-sight (LOS) domination mainly in the cell-edge region and BS 2 can be observed with LOS on almost every position.

Throughput Comparison for Different CoMP Configurations

We finally provide extrapolated measurement results for different downlink CoMP configurations, based on the measurement and evaluation methodology explained in Section 13.4.2. Using the simulation chain described before, we obtain realistic stream SINR values for a wide set of measured channels, which are well calibrated with concrete DL CoMP performance values obtained for a subset of channels. The following results hence incorporate various potential performance degradations such as imperfect and outdated CSI at the BS side, precoding inaccuracies due to limited precision computation of precoding matrices etc. The only optimistic simplification made in the evaluation of the results is that Shannon capacity is now used to map the stream SINR values to stream *spectral efficiency* values. More precisely, we compute for each potential stream

$$C(\text{SINR}) = \frac{N_{\text{RE}} \cdot \log_2(1 + \text{SINR}) \,/\, T_{\text{TTI}}}{N_{\text{SC}} \cdot \Delta f}, \qquad (13.21)$$

based on the fact that coded transmission takes place over $N_{\text{SC}} = 120$ subcarriers of 10 PRBs that result in a block size of $N_{\text{RE}} = 1080$ resource elements. The results thus inherently assume perfect link adaptation, an infinite granularity of MCSs and capacity-achieving channel coding, but previous work in [MKF06] has shown that such a simplification is still valid to predict *relative* performance gains between different transmission concepts. Note that the calculation of stream spectral efficiencies based on Shannon capacity is also the reason why in the sequel it is in most cases best to employ a maximum number of streams (if the channel rank supports this), while this would not be the case in a practical system with a limited number of MCSs.

We compare the following stream configurations, which can easily be obtained in our concrete closed-loop setup by setting certain parts of the compound channel matrix fed back by the UEs to zero:

- **Non-CoMP**, where 4 streams in total are used, but each UE is only served by its assigned BS. Hence, the compound precoding matrix $\mathbf{W}$ must be block-diagonal, which we achieve by constraining the CSI feedback to

$$\mathbf{H}_{4\times 4,\text{non-CoMP}} = \begin{bmatrix} \mathbf{H}_1^1 & \mathbf{0} \\ \mathbf{0} & \mathbf{H}_2^2 \end{bmatrix}. \qquad (13.22)$$

- **SU-CoMP**, where only one UE is served by two streams from both BSs.
- Different **MU-CoMP** schemes with 2 to 4 transmitted streams, where the CSI feedback is correspondingly constrained to

$$\mathbf{H}_{4\times 2} = \begin{bmatrix} [\,\mathbf{h}_{1,1} \quad \mathbf{0}\,] \, [\,\mathbf{h}_{2,1} \quad \mathbf{0}\,] \end{bmatrix} \qquad (13.23)$$

$$\mathbf{H}_{4\times 3} = \begin{bmatrix} [\,\mathbf{h}_{1,1} \quad \mathbf{h}_{1,2}\,] \, [\,\mathbf{h}_{2,1} \quad \mathbf{0}\,] \end{bmatrix} \qquad (13.24)$$

$$\mathbf{H}_{4\times 4} = \begin{bmatrix} [\,\mathbf{h}_{1,1} \quad \mathbf{h}_{1,2}\,] \, [\,\mathbf{h}_{2,1} \quad \mathbf{h}_{2,2}\,] \end{bmatrix}. \qquad (13.25)$$

In general, if additional spatial streams and consequently additional receivers are served in a MIMO-CoMP system, then the transmit diversity is reduced, and the spatial stream interference is increased. Hence, the spectral efficiency per stream decreases, but in most cases, the spatial multiplexing gain compensates for the stated rate losses.

The UE positions used for the evaluations in Fig. 13.24 were restricted to those which have a geometry factor larger than -20 dB (as otherwise no gain from CoMP would be expected, anyway) and a mean signal-to-noise ratio (SNR) per transmit/receive antenna link larger than 0 dB (i.e. only areas with a reasonable coverage w.r.t. both involved BSs). Performance metrics are obtained by averaging random UE positions from the measured CSI. In case of JT, users are selected within a cooperation radius, i.e. the distances are set to 12.5% of the largest possible distances which results in the northern area in 80 m and in the southern in 120 m maximum cooperation distance. In case of a conventional non-cooperative transmission, the cooperation radius is defined to select users that are *outside* of this radius, as a reasonable selection criterion to avoid a substantial loss in data rates due to inter-cell interference.

Clearly, this is a form of evaluation which would be tedious to obtain through real-time DL CoMP measurements, even though it delivers a very important insight into the value of DL CoMP as a function of user density.

In Figs. 13.24(a) and 13.24(c), the spectral efficiency per stream is plotted as a complementary cumulative distribution function (CDF), for the northern and southern test bed area, respectively. For the CoMP schemes, the SINRs are reduced with an increasing number of spatially multiplexed streams, as already anticipated in Section 10.1 and in [HKF10], because the inter-stream interference mitigation gets worse as the condition number of the physical channel matrix increases. One can see that non-CoMP transmission performs significantly worse than all other modes, as both BSs only have local channel knowledge with regard to their UE, and hence inter-cell interference can neither be avoided nor exploited.

In terms of the *sum* spectral efficiencies over all streams, in Fig. 13.24(b) the northern scenario 4×4 CoMP transmission gains 42% over the non-CoMP case. In the southern scenario in Fig. 13.24(d), a huge improvement of almost 300% could be achieved, because we here have a larger percentage of cell-edge areas with a low geometry factor. However, one could argue that one could also serve both UEs on orthogonal resources (in Section 13.3 referred to as the *isolated cell* case), leading to a performance slightly worse than the SU-CoMP case. With respect to this alternative non-CoMP option, the CoMP gain is on the order of 48%. In general, the absolute spectral efficiencies in the LOS dominated scenario are lower, and in both test bed areas a mean performance increase of about 20% could be achieved with each additional stream in the CoMP case.

13.4 Predicting Pract. Achievable DL CoMP Gains over Larger Areas

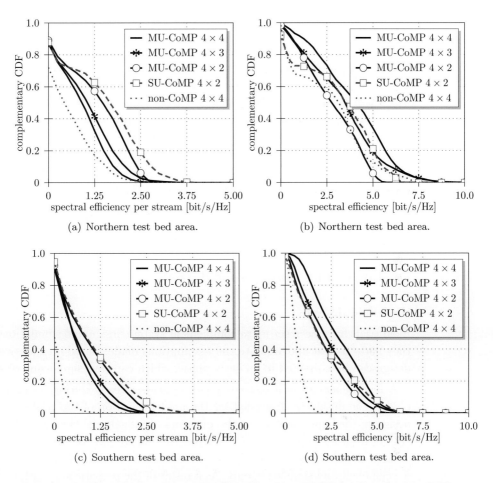

Figure 13.24 Comparison of spectral efficiencies for a NLOS (northern) and a LOS (southern) scenario and various CoMP setups.

13.4.4 Summary

In this section, a field trial implementation and results on downlink joint transmission CoMP were discussed, providing a complementary insight into these schemes in addition to Section 13.3. On one hand, the specific implementation discussed here follows a different and more simple design paradigm in terms of channel feedback, and on the other hand, a different evaluation methodology was used, which enables an accurate prediction of downlink CoMP gains over larger test bed areas at reduced measurement effort.

The fact that two individual and in many details slightly different implementations of downlink CoMP have lead to positive results further emphasizes the proof-of-concept of this promising CoMP scheme. In general, the CoMP gains revealed in this section are more moderate than in Section 13.3, due to the larger test bed areas over which the results were averaged. For downlink joint transmission from two cooperating base stations to two terminals, where each entity has 2 antennas, average spectral efficiency gains on the order of 40-50% could be observed. In must be noted, however, that in a commercial system such gains will only be observed for fairly static users, due to CSI feedback delay. The reason why these gains are significantly larger than those reported for uplink CoMP in Section 13.2 is mainly due to different antenna configurations and a much stronger baseline scheme considered in the uplink.

The authors would like to thank the German Ministry for Education and Research (BMBF) for the partial funding of the project EASY-C. Further, this work would not have been possible without the support from Sven-Einar Breuer, Ainoa Navarro Caldevilla, Martin Danneberg, Benedikt Nöthen, Julius Hoffmann, Martin Oemus, Matthias Pötschke and Joachim Heft.

13.5 Lessons Learnt Through Field Trials

Hans-Peter Mayer

Field trials have been set up to complement system design and simulations, since they allow to test and validate the assumptions made. Based on their results, a design can be corrected in an early phase, which can save significant effort.

A first benefit from a test campaign appears already during its preparation phase. The design work needed for the implementation of the trial system, the design of the experiments and the planning of a field trial challenges to see the system from an implementation perspective and to anticipate system design in a more detailed way than required for high-level system design or algorithm development and simulation only. As a result of the preparation work, a series of properties, limitations and issues of the system have been detected in an early project phase - and solved for most of the cases.

However, it is difficult to emulate an extended network with many cells in a trial set-up. As a result, the quantitative and statistical significance of trial results will be limited to some degree. By principle, trials will give information related to the particular scenario that has been set-up and tested. This holds in particular in cases where other cell interference has a major influence, the results will then depend on the size of the set-up. From system simulations it is known that results depend on the size of the playground chosen. Typically, different throughput values are obtained comparing 3GPP-type simulations done with 21 and 57 cells - even if a *wrap around* technique is used to virtually extend the playground. In a trial setup, similar numbers of cells are difficult to reach, especially within a research activity. The large-scale setup in Dresden described in Section 13.2 (currently supporting up to 17 base stations) was a challenging effort, but field trials with a large extent of background interference have yet to be performed, as it is complicated to generate a reasonably reproducible interference pattern in a field trial.

As a result, the size of the system being analyzed in a trial - in terms of network elements - is small compared to the one treated in system simulations, and the statistics provided by system simulation can be much richer.

In turn, the experiment gives direct access to the physics of radio equipment, antennas and radio channels. Within the scope of its design, it also includes the interactions of the different parts of the system. As an example, the real-time field tests described in Section 13.3 have well covered the interaction between the different parts of the system including radio channel, transceivers, PHY processing up to the MAC layer. In the functions from physical to MAC layers, the test set-up generally can represent a "real" system, without model assumptions, simplifications and the imperfections of models which impact the validity of simulations. Functions of OSI Layer 3, such as handover functions, were not part of the real-time test setup. Their properties and influence on the overall system behavior had to be taken into account again using model assumptions.

Another difference between simulation and field trial is given by the repeatability. In a drive test, one can run through the same course several times and never will measure exactly the same values on radio channel coefficients, signals or measured throughputs. We even found that the repeatability of measurements in subsequent drive test cycles in an urban environment as in Section 13.1 is higher for a moving mobile than for a static device in the same experiment, which may be explained by the fact that the fading environment itself changes - due to moving vehicles for instance. A moving terminal will average over the fading scenario and thus be less exposed to its variations. An example of a good reproducibility of field trial results was demonstrated in Section 13.2.

In some cases, early simulations have been clearly confirmed by the experiment. This was the case when studying the performance impact of user equipment (UE) velocity and time variant radio channels in systems using closed-loop channel feedback.

Laboratory tests take an intermediate position between drive tests and simulations. They are needed as a first step to validate the functions of the setup before going into a field deployment. Quantitative measurements can be done using emulation of the fading channels. This technique allows combining the exact modeling of the system that comes with a real-time hardware implementation with a repeatability of the experiments that comes very close to simulation.

In summary, we have found that field trials form an extremely valuable complement to system simulations. We have been able to use the leanings from the test campaigns to refine algorithms and the simulation models. The gains resulting from the use of a successive interference cancelation (SIC) receiver in single-cell multi-user MIMO (MU-MIMO) operation were characterized in a test based on a detailed real-time implementation. It was also shown that a SIC receiver can be within a base station from a complexity point of view. For uplink CoMP, the issues associated with its implementation have been largely solved. For example, methods for the time alignment of signals from different sites and the handling of delay spread were implemented and optimized within the real-time system. Alternatives for uplink MU-MIMO and CoMP methods (MU-MIMO with SIC receiver, distributed interference subtraction (DIS) and joint detection (JD)) have been compared with respect to their performance, resulting in a performance ranking which is in good agreement with the view obtained from system simulations.

As expected, differences are observed concerning the absolute throughput, which can be explained by simplifications of the modeling assumptions and by imperfections or limitations of the implementation on the trial side. If neighbor cell interference is properly taken into account, experiments typically yield lower throughput numbers compared to the simulation of a comparable scenario.

Downlink CoMP in the coherent case requires the mastering of a number of technical issues, the most important ones described in Part III of this book: clustering of the cells, synchronization of the antennas, channel estimation including the design of reference schemes that support multi-cell operation and the efficient feedback of the channel information from the mobile towards the base station. The tests undertaken on coherent downlink CoMP have shown that many of the enablers mentioned above are mastered, at least as the technical principles are concerned.

From the fact that there was successful CoMP transmission in the downlink, it can be seen that phase synchronization of transmit antenna signals is solved in principle. The fact that rubidium-stabilized clocks were used points to a cost issue that can be solved in a development phase. It was equally demonstrated that multi-cell channel estimation can be provided, as well as an efficient feedback of this channel information. At the same time, the tight limits experienced concerning the terminal mobility show the strong impact of channel ageing effects on coherent techniques. One consequence will be that this technique will be applicable to low-mobility users only; at the same time there are more efforts needed to improve the efficiency of channel feedback and to limit latencies. Finally, clustering could not be assessed within the tests due to the limited deployments, but this shall be addressed in the future in the Dresden test bed.

In total, the field tests have proven the principal feasibility of all considered CoMP techniques. The uplink techniques single-site MU-MIMO using SIC receivers, multi-site DIS and multi-site joint detection have been investigated and are sufficiently mature to be introduced in the coming years, depending on an economical assessment.

For coherent joint transmission in downlink direction, the situation is somewhat different: While its principal feasibility has been proven by the experiments as well, the complexity of this technique is significantly higher. Further research efforts will be required until a development phase can be started.

14 Performance Prediction of CoMP in Large Cellular Systems

While Chapter 13 has shown that various CoMP concepts discussed in this book do indeed work in practice and yield gains that match fairly well to theoretical predictions, any field trial result is of course always limited to a particular, not necessarily representative, scenario, and, more importantly, to a very limited number of terminals. Before an operator invests into the technology and infrastructure required for certain CoMP schemes, however, he will want to assess their performance in large-scale systems with a large number of mobile terminals and potentially complex traffic models.

In this chapter, we hence want to discuss how system level simulations can be conducted in order to assess CoMP performance in large system contexts at reasonable complexity. First, Section 14.1 introduces the standard assumptions and simulation methodology used by 3GPP for the simulation of LTE and LTE-A schemes. Section 14.2 then shows how channel sounding measurements or ray-tracing in a 3D city model can be used to parameterize channel models, especially as the simulation of CoMP systems shifts the focus to different large-scale parameters than that of non-cooperative systems. The chapter is concluded with system level results on a subset of the uplink and downlink CoMP concepts covered in this book in Sections 14.3 and 14.4, respectively.

14.1 Simulation and Link-2-System Mapping Methodology

Thorsten Wild and Andreas Weber

Before any mobile communication system is deployed in the real world, a lot of design decisions have to be taken, and the cost of features have to be balanced with the gain that they promise. As these decisions have to be taken before any measurements can be performed on a large-scale deployment with a substantial number of users, mathematical analysis and system simulations have to yield the required performance data.

The performance evaluation of mobile communication systems has become more and more complex in the past years. For example, GSM systems require network planing that reduces co-channel interference from neighbor cells significantly, where it is sufficient to analyze system performance based on mean signal qualities. Later, UMTS introduced code division multiple access (CDMA) and, consequently, the performance evaluation had to consider intra-cell and self interference from parallel transmissions using different spreading codes. GPRS, EGPRS, and high-speed packet access (HSPA) introduced a fast link adaptation that adapts the transmitted data rate on the instantaneous channel quality, which required an adequate fast fading model. Finally, LTE and LTE-A are based on orthogonal frequency division multiple access (OFDMA), and use the same frequencies in all cells. Features like frequency-selective scheduling, adaptive hybrid automatic repeat request (HARQ), multiple-input multiple-output (MIMO), precoding, CoMP, make a tremendous effort for the performance analysis of these 4G systems necessary.

For system performance evaluation, a system model is required that considers all relevant aspects of the real system that have a significant impact on performance. In case of LTE and LTE-A, for example, the model has to consider the spatial behavior of the channel. Furthermore, as we want to evaluate frequency-selective scheduling, the model has to consider channel dynamics in frequency and time.

It is very difficult to find a closed-form mathematical expression for the analysis of such a 4G system, especially if we want to evaluate the performance gain of enhanced features and the combination of those. From system simulation, we can expect a detailed insight into the behavior of the system, that supports system optimization.

14.1.1 General Simulation Assumptions and Modeling

System simulation assumptions for LTE and LTE-A are published in [3GP06c] and [3GP10d], respectively. System simulation is typically based on so-called Monte Carlo drops. A number of user equipments (UEs) are randomly dropped on a given area, in the following called *playground*. During a drop, the position of the UEs and, consequently, the pathloss between UEs and base stations (BSs) is kept constant. Furthermore, their so-called large-scale parameterss (LSPs) and smale-scale parameterss (SSPs) are also kept constant (see Section 14.1.2). LSPs and corresponding SSPs provide a two-stage randomization for channel delays, angles of arrival and departure, and Ricean K-factor for line-of-sight (LOS) channels. During a drop, fast fading is time variant, corresponding to a random velocity vector that is assigned to every mobile.

Different network layouts are proposed in [3GP06c] that differ in the inter-site distance (ISD) and in the pathloss model. In [3GP10d], new antenna models and new network layouts, mainly in the field of heterogeneous networks, are introduced. In the latter case, the position of micro BSs also differs between different drops. In the following, we will restrict ourselves to the macro network layout. In this case, the BS positions correspond to a hexagonal grid with a

14.1 Simulation and Link-2-System Mapping Methodology

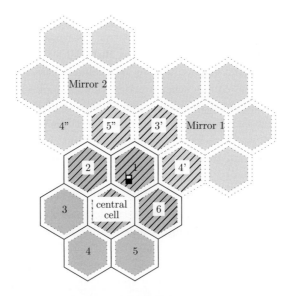

Figure 14.1 Network layout with wrap-around.

constant inter-site distance. It is proposed to consider one central site surrounded by two rings of other sites.

In order to avoid border effects, so-called *wrap-around* is applied. In this case, every mobile believes to be allocated to a center cell and, hence, every mobile, even if located at the playground border, delivers measurement values with the same level of interference as a mobile in the actual center of the playground. Wrap-around is realized by six mirror positions of every cell. Every mobile is surrounded by the closest cells or mirrors, which, e.g., results in the shaded area for the playground border UE shown in Fig. 14.1. Consequently, measurement values of UEs in all cells can be considered, instead of only those measurement values of the center cell UEs. This way, the number of usable measurement samples is increased, which reduces computing time drastically for the same level of accuracy.

Every mobile is associated to the cell delivering the strongest received signal level. Alternatively, when a handover margin is defined, a mobile is randomly assigned to one cell out of a set of cells which yield receive signal levels corresponding to a window defined by the receive signal level of the best cell and the predefined handover margin.

During a drop, an event driven simulation is performed. The granularity for the simulation is one sub-carrier in frequency and one orthogonal frequency division multiplex (OFDM) symbol in time. Lower granularities are possible, if the UE speed is slow and if the frequency selectivity of the channel is moderate. For downlink (DL) transmissions, the UE performs DL channel measurements in order to determine the number of preferred streams, optimal precoding, and corresponding channel quality. The according feedback message is periodically transmitted to the BS and delayed by a certain amount of time. The scheduler considers this feedback in order to determine the UEs that are scheduled during a frame and the mapping of these UEs to the corresponding resources. Furthermore, the link adaptation determines, based on the UE's feedback, the modulation and coding scheme (MCS) that is applied to a transmission. Using a link to system interface (c.f. Section 14.1.4), the UE determines if a transmission is successful or not. In the latter case, the BS is informed with a certain predefined delay about the failed transmission and the scheduler performs a HARQ retransmission in the following frame, at the earliest. The receiver improves the effective codeword signal-to-interference-and-noise ratio (SINR) by combining transmission and corresponding retransmissions.

The mostly used and simplest traffic model is the full buffer model which assumes that a file of infinite size is waiting for transmission for each UE. Other traffic models like, e.g. for voice over IP (VoIP), file transfer protocol (FTP), and web traffic have been defined. Depending on the traffic model, the key performance parameters have to be adapted (c.f. Section 14.1.5).

14.1.2 Channel Models and Antenna Models

2D- and 3D-antenna models

The single element gain $A(\phi, \Theta)$ as a function of elevation Θ and azimuth ϕ angles on system level classically was modeled using a certain cut in elevation $\Theta = \text{const}$, resulting in a 2-D antenna model. [3GP07a] is based on an inverted parabola in order to approximate the azimuthal (horizontal) antenna pattern, using a horizontal half-power beamwidth (HPBW) of $\phi_{3\text{dB}}$ (e.g. $70°$ for classical tri-sectorized sites) with a backward attenuation A_m in the order of 25 dB. The attenuation in dB as a function of azimuth ϕ can be modeled as:

$$A_\text{H}(\phi) = -\min\left[12\left(\frac{\phi}{\phi_{3\text{dB}}}\right)^2, A_\text{m}\right] \qquad (14.1)$$

The antenna gain has to be added, which is around 14 dBi for the $70°$ HPBW (including cable losses). As the downtilt plays an important role for cell separation, recent modeling [3GP10d] also takes into account the vertical pattern $A_\text{V}(\Theta)$ with a similar parabolic approximation, using a vertical HPBW of $10°$. The overall 3-D antenna model is approximated by adding up the gains from horizontal and vertical cut:

$$A(\phi, \Theta) = -\min\left[(A_\text{H}(\phi) + A_\text{V}(\Theta)), A_m\right] \qquad (14.2)$$

As this 3-D model results in realistic stronger cell isolation than a 2-D one, it is important to take it into account when judging CoMP gains.

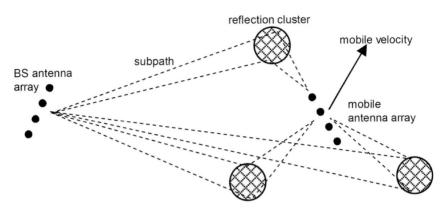

Figure 14.2 Illustration of a spatial channel model.

Spatial Channel Models

Fig. 14.2 shows a schematic top level view of a spatial channel model, convenient for MIMO system simulations. Every reflection cluster contains a number of sub-paths that differ in the angles of arrival and departure. Static phase shifts between the antenna elements of the multiple antenna array of transmitter and receiver are induced due to the angles of departure (AoD) and angles of arrival (AoA), respectively. At the UE side, dynamic phase shifts are induced due to the UE moving with a certain speed through the incoming waves.

In principle, every reflection cluster corresponds to a certain transmission delay. Consequently, in total we would have as many channel echos as reflection clusters. However, most of the models require the splitting of all or some clusters into sub-clusters, whereby the sub-clusters have their own delay, which is close to the original delay of the cluster. Sub-clustering extends the supported bandwidth of a channel model by improving the frequency correlation function within the simulated bandwidth [NKM+07].

At drop initialization, first the LSPs (delay spread, angular spreads, shadow fading, and Ricean K-factor) have to be determined. Large scale parameters are autocorrelated, i.e. UEs that are located closely have a higher probability to have similar LSPs compared to those with a larger distance between them. The autocorrelation can be exactly considered based on the correlation matrix and Cholesky factorization, or it can be approximately considered using an approach shown in [WIN03]. The latter approach consumes less computing time which is an important issue in case of a larger number of UEs and cells.

Based on angular spreads and delay spread, the realizations of angles at transmitter and receiver and the realization of the delays is determined. The resulting angles and delays are part of the so-called SSPs. Additionally, the cluster powers (or losses) are determined. Furthermore, a random coupling of angles of departure and arrival within one cluster is performed. The random generation of an initial phase for every sub-ray closes the drop initialization for a spatial channel. [3GP06c] and [3GP10d] present exhaustive parameter tables for mean values, standard deviations, and distribution functions for all LSPs, as well as distribution functions for all SSPs. During a drop, the channel impulse response (CIR) for every reflection cluster (or subcluster) n is calculated according to

$$\tilde{h}_{u,s,n}(t) = \sqrt{P_n} \sum_{m=1}^{M} F_{\text{rx},u}(\theta_{n,m,AoA}) e^{j\Phi_{n,m}} F_{\text{tx},s}(\theta_{n,m,AoD}) \cdots$$
$$e^{j d_s 2\pi \lambda_0^{-1} \sin(\theta_{n,m,AoD})} e^{j d_u 2\pi \lambda_0^{-1} \sin(\theta_{n,m,AoA})} e^{j 2\pi v_{n,m} t}. \quad (14.3)$$

Here, P_n corresponds to the (sub-)cluster power. M is the number of sub-rays of the (sub-)cluster. The antenna element at transmitter side is called s, and at the receiver side is called u. $F_{\text{rx},u}(\theta_{n,m,AoA})$ is the receive antenna gain for sub-ray m of (sub-)cluster n according to the angle at which the sub-ray is received. Similarly, $F_{\text{tx},s}(\theta_{n,m,AoD})$ corresponds to the antenna gain at the transmitter. $e^{j d_s 2\pi \lambda_0^{-1} \sin(\theta_{n,m,AoD})}$ is the phase shift between the considered transmitter antenna element s measured against the first antenna element of the transmitter, which depends on the angle of departure. Similarly, $e^{j d_u 2\pi \lambda_0^{-1} \sin(\theta_{n,m,AoA})}$ corresponds to the phase shift of the receive signal induced by the angle of arrival of the antenna element u of the receive antenna measured against the first receive antenna element. The Doppler part of the phase shifts is considered by $e^{j 2\pi v_{n,m} t}$, whereby $v_{n,m} t$ corresponds to the phase speed in which the UE runs through the incoming wave of sub-ray m of (sub-)cluster n.

Different types of spatial channel models are in use. All these models have been developed on the basis of channel measurements [WIN03, MRB06], such as those discussed in Section 14.2. The classical model is also called *spatial channel model* SCM and described in [3GP07a]. It offers models for suburban and urban macro, as well as for urban micro. It consists of 6 reflection clusters (paths) with 20 sub-paths each. No sub-clustering is applied, so that SCM is limited to system bandwidths up to 5 MHz due to its correlation properties in frequency domain. [3GP07a] includes a description of LOS channels, although this option is not used in the general simulation assumptions in [3GP06c]. An extention of SCM, called spatial channel model extended (SCME), is described in [BHD+05]. The main extension is the division of all of the six reflection clusters of the SCM into 3 sub-clusters, thus supporting higher bandwidth compared to the SCM. Both models, SCM and SCME, include a site-to-site correlation of shadowing which considers that a bad shadowing value to one cell might be due to a generally bad environment which results in rather bad shadowing values to other cells as well. Shadowing site-to-site correlation induces additional losses in signal quality, because the variance between signal and interference levels is reduced. The WINNER project published a number of enhanced channel models [NKM+07], which are the basis for the widely used ITU-R models published in [3GP10d]. These models consist of 15...20 reflection clusters, whereby the strongest two of them are split into 3 sub-clusters each, yielding 19...24 channel echos. Shadowing site-to-site correlation is assumed to be zero for ITU-R models.

14.1.3 Transceiver Techniques

As we have seen in various parts of this book, the spatial coloring of interference and useful signal plays an important role in the performance of MIMO and CoMP algorithms. Thus, we have to model spatial processing of both transmit and receive side as accurate as possible already on system level. Linear precoding introduces a transmit weight vector $\mathbf{w}$ per spatial layer. Channel models provide the realization of the MIMO/CoMP channel matrices $\mathbf{H}$ per user and frequency/time interval. Receive weights $\mathbf{g}$ (e.g. using minimum mean square error (MMSE) algorithms) can be computed as a function of the precoded channels. So for a given resource element and spatial layer l of a useful signal, the SINR can be computed by calculating the output of useful signal power at the receive combiner output over the sum of all interfering layers/users (including intra-cell interference by spatial division multiple access (SDMA) and self-interference by spatial multiplexing (SMUX)), indexed with l', plus the thermal noise power σ^2 after filtering:

$$\gamma_l = \frac{p_l \left|\mathbf{g}_l^H \mathbf{H}_l^H \mathbf{w}_l\right|^2}{\|\mathbf{g}_l\|^2 \sigma^2 + \sum_{\forall l' \neq l} p_{l'} \left|\mathbf{g}_l^H \mathbf{H}_l^H \mathbf{w}_{l'}\right|^2}. \tag{14.4}$$

This SINR calculation principle can also be extended to algorithms like successive interference cancelation (SIC), by removing all or a part of the interference of one or more successfully decoded sub-streams for the remaining sub-streams. In case of CoMP, typically assuming synchronized cells, the overall multi-cell precoders and channel matrices can be constructed by stacking the single cell components, as done in various parts of this book.

14.1.4 Link-to-System Interface

In system level simulations, modeling each link on sample level is not reasonably feasible due to computational complexity and thus simulation time. A certain abstraction of the link layer is required. However, we have also seen that the spatial coloring of interference and useful signal is especially important in the context of advanced receiver strategies. Additionally, the modeling of time- and frequency-adaptive scheduling demands to take into account the fast fading already on system level. So we have to be careful that abstraction still covers all of those aspects.

The remaining parts which can be abstracted are thus mainly the functionalities of the forward error correction (FEC). As channel decoding consumes a lot of computation, the system simulation time can be strongly reduced by abstracting the FEC. For this purpose, a natural point in a physical layer processing chain for placing the up-to-date link-to-system interface is between the output of the receive combiner and equalizer and the input of the FEC decoder. At this point, we can compute a set of SINRs $[\gamma_1, \ldots, \gamma_K]^T$ for the resource elements within a codeword, based on (14.4). The task is now to find an appropriate compression function which allows to map the vector of SINRs within a codeword to a scalar effective SINR value corresponding to a single-input single-output (SISO) additive white Gaussian noise (AWGN) link level performance curve. In the literature [BAS+05, TW05], a whole set of useful compression functions exists: E.g., EESM (exponential-), LESM (logarithmic-), and mutual information equivalent SINR mapping (MIESM). Those examples have in common that the compression itself is implemented via a non-linear averaging including a non-linear mapping function $I(\cdot)$ [WIN05]. Then, the effective SINR γ_{eff} is:

$$\gamma_{\text{eff}} = \frac{1}{\beta} I^{-1} \left(\frac{1}{K} \sum_{k=1}^{K} I\left(\frac{\gamma_k}{\beta}\right) \right). \tag{14.5}$$

Within the European WINNER project, such compression functions were compared to each other [WIN05]. The MIESM was seen as the most attractive compression function, as β is almost independent from the modulation order and it shows a very insensitive behavior towards the variation of β. Additionally, it has the advantage that the interface can be motivated from an information theory perspective. High soft bit reliability corresponds to a high value in mutual information. But for a certain modulation constellation (e.g. QPSK, 16-QAM etc.), the mutual information as a function of SINR saturates. It is e.g. not possible to carry more than two (or less than zero) bits of mutual information for QPSK. This sigmoid-shaped mutual information mapping curve I_{m_p} reflects these expected decoder properties. For a certain modulation constellation with m_p bits per symbol, the mutual information as a function of the SINR γ can be computed as [CTB96]

$$I_{m_p}(\gamma) = m_p - E_Y \left\{ \frac{1}{2^{m_p}} \sum_{i=1}^{m_p} \sum_{b=0}^{1} \sum_{z \in \mathcal{X}_b^i} \log \frac{\sum_{\hat{x} \in \mathcal{X}} \exp\left(-|Y - \sqrt{\gamma}(\hat{x} - z)|^2\right)}{\sum_{\tilde{x} \in \mathcal{X}_b^i} \exp\left(-|Y - \sqrt{\gamma}(\tilde{x} - z)|^2\right)} \right\}. \tag{14.6}$$

Hereby, $\mathcal{X}$ denotes the set of 2^{m_p} data symbols, $\mathcal{X}_b^i$ the set of symbols for which bit i equals b and $Y \sim \mathcal{N}_{\mathbb{C}}(0,1)$.

Fig. 14.3 shows a block diagram of the processing steps required for training and validating the model. The overall link layer processing chain is displayed, where the equalizer output is the point where the interface picks up the effective SINR calculation which can be mapped to a block error rate (BLER).

In summary, the presented link-to-system interface provides a decoder model. The compression function generates a scalar effective SINR which can be used in a one-dimensional lookup-table of a SISO AWGN performance curve in order to provide a corresponding BLER.

The described link-to system interface can be used beyond MIMO also for CoMP. The SINR computation of (14.4) can directly be generalized to CoMP. Potential open issues for future work here are e.g. models for phase shifts between

14.1 Simulation and Link-2-System Mapping Methodology

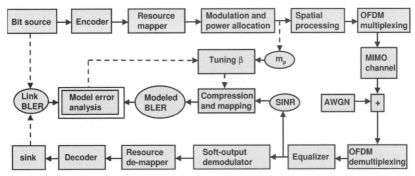

Figure 14.3 Block diagram for training and validating the MIESM mapping model.

BSs, as discussed in detail in Chapter 8.3. Additionally, for large cells and large cluster sizes, the inter-symbol interference (due to exceeding the OFDM cyclic prefix) has to be modeled properly. Furthermore, an appropriate model for multi-cell channel estimation is required.

14.1.5 Key Performance Indicators

The results of system level simulations serve for the comparison of different design options and features like, e.g. CoMP, multi-user MIMO (MU-MIMO), frequency-selective scheduling etc. It is important not to rely on one single performance value, but to co-optimize a set of possibly antagonistic performance values. As an example, it might not be advantageous to optimize system throughput, if, at the same time, the throughput fairness between users becomes worse.

The **user geometry** is the cumulative distribution function (CDF) of the DL long-term wideband SINR. It assumes a 1x1 antenna system and a full power allocation in every cell. The geometry mainly serves for the evaluation of different simulated environments and BS deployments, and was also considered, e.g., as a clustering criterion in Chapter 7. It does not consider, beneath others, any MIMO antenna setups or scheduling algorithms. **Cell throughput** is measured as the sum of all **user throughputs** of user equipments attached to a cell. The according CDF gives insight into the requirements to backhaul bandwidth. While average user and cell throughput are redundant, as long as the average number of users per cell is known, the according **user throughput CDF** is an important simulation result, because cell border throughput and user fairness can be deduced from it. The 5%-ile throughput, i.e. the user throughput that is not exceeded by 5% of the users is, by definition, the so called cell-border throughput. Please note that the location of these users is not necessarily close to the *geometric* cell-border. This is especially true in the case of CoMP, where the (compound) channel quality for UEs at the geometric cell-border is often better than for UEs in between BS and cell border. The **fairness requirement** used by 3GPP states that for all percentages x at maximum $x\%$ of the users may have a throughput of less then $x\%$ of the average throughput.

14.1.6 Summary

In this section, an overview on system simulation methodology for MIMO and CoMP systems was presented. Careful modeling of spatial channels and 3-D antenna characteristics is important in order to obtain meaningful performance results. Further, it is proposed to model transceiver and receiver processing as detailed as possible. E.g. in case of linear processing actual computation of transmit and receive weights should be performed in the simulator.

The suggested way to use the link-to-system interface is to provide an abstraction of the encoder and decoder of the system. It is possible to use several nonlinear compression functions in order to obtain an effective SINR, which can be mapped to a single block error probability. An attractive parametrizable mutual-information-based interface is proposed with high robustness.

All those described steps enable us to evaluate and optimize the performance of future systems which yet have to be built.

14.2 Obtaining Channel Model Parameters via Channel Sounding or Ray-Tracing

Richard Fritzsche, Jens Voigt, Christian Schneider, Stephan Jäckel and Carsten Jandura

As mentioned in the last section, geometry-based stochastic channel models from the SCM/WINNER family are widely used in system level simulations to investigate multiple-input multiple-output (MIMO) and CoMP technologies, see [NST+07] and references therein. These models are able to reproduce the statistical behavior of the radio channel for several environments. To parameterize these models, large-scale parameterss (LSPs) are used as input. These are:

- transmission loss (TL),
- shadow fading (SF),
- delay spread (DS),
- several angular spreads (AS),
- several cross-polarization ratios (XPR), and
- Ricean K-Factor as well as
- decorrelation distances of the LSPs, and
- several cross-correlation terms between the LSPs.

These LSPs and their corresponding distribution functions play a fundamental role, because as global scenario-dependent parameters they control the behavior of the modeled channel. This section introduces two approaches to extract LSPs from a certain environment: Channel sounding measurements and deterministic ray-tracing channel simulations based on a high-resolution 3D environment model. The motivation of using ray-tracing simulations instead of measurements

14.2 Obtaining Chn. Model Params. via Chn. Sounding or Ray-Tracing

is an extensive reduction in effort and cost. A special emphasis is put on those LSPs that are particularly interesting in the context of CoMP.

Specifics of both approaches are shown in the following. First, in Subsection 14.2.1, the calculation of the LSPs is shown. In Subsection 14.2.2, difficulties appearing by building up a CoMP measurement setup and processing the collected channel data are highlighted. Afterwards, the methodology of ray-tracing channel simulation using the ray-launching approach is described in Subsection 14.2.3. Finally, in Subsection 14.2.4, results from measurements and simulations in a real CoMP environment in central Europe are compared to each other. It is shown how LSPs extracted either from channel sounding measurements or from ray-tracing simulations fit to each other, after which this section is concluded in Subsection 14.2.5.

14.2.1 Large-Scale-Parameters

The classical calculation of LSPs within the SCM/WINNER framework is based on analysis in delay and power domain as well as the spatial domain. Whereby the results for the spatial analysis are derived after an intermediate processing: the high resolution multi-path parameter estimation [Ric05, LKT07]. Furthermore, the considered parameters are derived based on snapshot intervals covering $10 - 20\ \lambda$, so-called stationarity intervals. As this is typical for SCM/WINNER channel analysis and parametrization, the LSPs are based only on the estimated radio propagation paths between a transmitter and receiver point. In the sequel, we will base our nomenclature on downlink transmission, and hence denote the transmitter side as base station (BS), and the receiver side as user equipment (UE). The LSPs can also be calculated taken multiple BSs in a CoMP scenario into account. This is of particular interest for delay spreads in terms of synchronization aspects (see Section 8.2), and for the angular spread to analyze the spatial diversity, which will be the focus in the following, and is of key importance for CoMP. In the remainder of this section, a radio propagation path $l \in \{1, ..., L\}$ is characterized by the following properties:

a_l^{pq} — attenuation
p — polarization component at the base station
q — polarization component at the user equipment
φ_l^{bs} — azimuth angle at the base station
φ_l^{ue} — azimuth angle at the user equipment
θ_l^{bs} — elevation angle at the base station
θ_l^{ue} — elevation angle at the user equipment
τ_l — path duration

The polarization components p and q can, e.g., be captured by vertical (v), horizontal polarization (h), or any other polarization. From the entirety of all propagation paths between a transmit and a receive antenna, a channel impulse response (CIR) is given by

$$\tilde{h}^{pq}(\tau) = \sum_{l=1}^{L} a_l^{pq} \exp\left(-j2\pi f_c \tau_l\right) \delta(\tau - \tau_l), \qquad (14.7)$$

where f_c is the carrier frequency of the system. This notation does not consider any transmission bandwidth or any sampling frequency of the analog to digital conversion (ADC) at transmitter or receiver.

Transmission Loss

The transmission loss describes the attenuation in power between transmitter and receiver due to free space loss and several radio propagation phenomena. Considering polarization, the (wide-band) transmission loss can be stated as

$$\text{TL}^{pq} = 10 \cdot \log_{10} \left(\sum_{l=1}^{L} |a_l^{pq}|^2 \right). \qquad (14.8)$$

Shadow Fading

Fading of the receive signal around a local mean value of the pathloss due to shadowing from obstacles (e.g., buildings, vehicles, huge trees) is referred to as shadow fading. See, e.g., [APM02] on how to extract shadow fading for measurements and ray-tracing alike. For large to moderate cell sizes, shadow fading can in general be approximated by a log-normal, zero-mean distribution function [Gud91], of which the standard deviation is a further LSP.

Delay Spread

The root mean square (RMS) delay spread is the second central moment of the channel's power delay profile. It can be calculated by

$$\text{DS}_{\text{RMS}} = \sqrt{\frac{\sum_{l}^{L} \left(\tau_l - \left(\frac{\sum_{l}^{L} \tau_l |a_l|^2}{\sum_{l}^{L} |a_l|^2} \right) \right)^2 |a_l|^2}{\sum_{l}^{L} |a_l|^2}}. \qquad (14.9)$$

According to (14.8), the delay spread can be calculated polarization-wise, where we here omit the polarization indices for reasons of notational convenience.

Angular Spread

In analogy to (14.9), the RMS angular spread is the second central moment of the channel's power angular profile. It can be calculated in azimuth and elevation and also for BS and UE side, respectively. As an example, we give the formula

14.2 Obtaining Chn. Model Params. via Chn. Sounding or Ray-Tracing

for the elevation angular spread at the BS as

$$\mathrm{AS}_{\mathrm{RMS}}^{\mathrm{bs},\theta} = \sqrt{\frac{\sum\limits_{l}^{L}\left(\theta_{l}^{\mathrm{bs}} - \left(\frac{\sum\limits_{l}^{L}\theta_{l}^{\mathrm{bs}}|a_{l}|^{2}}{\sum\limits_{l}^{L}|a_{l}|^{2}}\right)\right)^{2}|a_{l}|^{2}}{\sum\limits_{l}^{L}|a_{l}|^{2}}}. \tag{14.10}$$

Through an exchange of θ_l^{bs} with θ_l^{ue}, ϕ_l^{bs}, and ϕ_l^{ue}, i.e. the elevation angular spread at the user equipment $\mathrm{AS}_{\mathrm{RMS}}^{\mathrm{ue},\theta}$, the azimuth angular spread at the transmitter $\mathrm{AS}_{\mathrm{RMS}}^{\mathrm{bs},\varphi}$, and the azimuth angular spread at the receiver, $\mathrm{AS}_{\mathrm{RMS}}^{\mathrm{ue},\varphi}$ can be calculated accordingly.

Cross-Polarization Ratio

Cross-polarized type antennas consist of multiple (usually two) equal antenna elements at the same position that use an orthogonal set of polarization directions for transmission or reception. Considering an orthogonal frequency division multiplex (OFDM) system where $h^{\mathrm{pq}}(f)$ is the Fourier transform of $\tilde{h}^{\mathrm{pq}}(\tau)$ in (14.7), polarization MIMO (transmission using multiple polarization directions) can be applied sub-carrier-wise if the coupling matrix

$$\mathbf{\Psi}_{\mathrm{H}}(f) = \begin{bmatrix} h^{\mathrm{vv}}(f) & h^{\mathrm{hv}}(f) \\ h^{\mathrm{vh}}(f) & h^{\mathrm{hh}}(f) \end{bmatrix} \tag{14.11}$$

has full rank. In (14.11), v and h denote the vertical and the horizontal polarization component, respectively. In the coupling matrix in (14.11), the main diagonal consists of links having the same polarization at transmitter and receiver. The performance of such a system depends on the correlation of the polarization modes. This correlation can be expressed by the cross-polarization ratio (XPR), where two different definitions exist: $\mathrm{XPR}_{\mathrm{out}}$, a measure of the power that goes to the cross-polarized receive antenna element, and $\mathrm{XPR}_{\mathrm{in}}$, a measure of the power that comes from the cross-polarized transmission antenna element. Both XPR values can be calculated as [LST+07]

$$\mathrm{XPR}_{\mathrm{out}} = 10\log_{10}\left(\frac{\sum\limits_{l}^{L}|a_l^{\mathrm{vv}}|^2}{\sum\limits_{l}^{L}|a_l^{\mathrm{vh}}|^2}\right) \quad \text{and} \quad \mathrm{XPR}_{\mathrm{in}} = 10\log_{10}\left(\frac{\sum\limits_{l}^{L}|a_l^{\mathrm{vv}}|^2}{\sum\limits_{l}^{L}|a_l^{\mathrm{hv}}|^2}\right). \tag{14.12}$$

Note that (14.12) shows exemplary expressions for vertical polarization. Switching h and v in both terms results in expressions for the horizontal component.

14.2.2 Measurement-based Parameter Estimation

Channel sounding data can be used for various research applications, e.g. channel analysis and modeling, verification of ray-tracing tools, link and system level evaluations [TST05]. For CoMP system layouts, it is desirable to measure multi-cell links simultaneously. This can be relaxed if no correlation between the CoMP links exist or if the subsequently following data processing is targeting an LSP analysis [NKK+10, NKJ+11].

For measurement and estimation of LSPs such as the angular spreads via SAGE or RIMAX [Ric05, LKT07], dedicated antenna arrays are necessary [Lan08]. Furthermore, a sophisticated antenna calibration and post-processing is necessary to perform the high resolution parameter estimation with highest accuracy [LKT07]. Parameter estimation allows to resolve several propagation paths from measured CIRs of every transmitter-receiver link considering the whole transmission bandwidth. The obtained paths are described in angle, delay, power, and polarization. Every path l is associated with a complex channel attenuation coefficient a_l for all polarization modes, the delay τ_l, the azimuth angle of departure θ_l^{bs} and arrival θ_l^{ue}, as well as the elevation angle of departure φ_l^{bs} and arrival φ_l^{ue}.

In general, post-processing of the data is necessary to extract noise from the measurement data correctly. Here, high-accuracy post-processing methods such as SAGE, ESPRIT, or RIMAX have a high computational complexity. Another method to handle the influence or beampattern of dedicated antenna arrays, which are included in the measured CIRs, is a simple post-processing method introduced by [JJ10, JTJ10], which is based on adaptive thresholding and weighted averaging over the measurement antennas. The resolution and the accuracy of this method is comparable and limited to the beamforming problem. Furthermore, it does not provide any multi-path parameter information needed for the LSP analysis in spatial domain, hence is only applicable to the delay and power domain LSP analysis discussed in Section 14.2.1.

14.2.3 Ray-Tracing based Parameter Simulation

Ray-tracing is a deterministic channel modeling method to simulate CIRs in time, angular and polarization domain. A precondition for any ray-tracing is a three-dimensional environment model with obstacles that are sufficiently larger than the wave length of the carrier frequency to be simulated. Having typical dimensions of obstacles in mind, this condition in fact rather limits the frequency range from some hundred MHz in outdoor environments to some hundred GHz in indoor environments. In case this condition is met, electromagnetic waves can be modeled as rays.

For the purpose of path finding through the environment model, we describe the *ray-launching* approach, which is better suited when the electromagnetic

14.2 Obtaining Chn. Model Params. via Chn. Sounding or Ray-Tracing

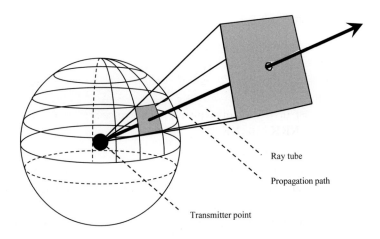

Figure 14.4 Ray tube principle and propagation of the electromagnetic field in ray-launching.

wave propagation shall be simulated to a multiplicity of receivers in a given area than point-to-point paths finding, see, e.g.,[GWBL95].

In ray-launching, a bundle of rays emanate from a transmitter that is modeled as a single point in the three-dimensional environment. Around such points, ray tubes are generated with a predefined angle of aperture, e.g., 1°, see Fig. 14.4. The rays are weighted with the complex value from the three-dimensional antenna pattern of the transmitter. The ray-launching algorithm uses a three-dimensional (vector data base) model of the environment (e.g., buildings and ground in an outdoor environment or walls and interior in an indoor environment) to determine the nearest obstacle in the current propagation direction of the ray. Once a ray hits an obstacle, the algorithm includes radio wave propagation effects such as *specular reflection, refraction, diffraction, penetration,* and *diffuse scattering*, taking the relative static permittivity $\epsilon_r(f)$ of the obstacle into account. Thus, in the ongoing calculations, the algorithms of *geometrical optics* and the *uniform theory of diffraction* [KP74] are applied. For diffuse scattering, the *effective roughness* approach can be used [DEGd+04] and [DEFVF07]. These rays are all traced along until some simulation stop conditions are met (e.g., the field strength falls below a defined noise threshold, or a defined number of reflections/diffractions is made, or alike) or a receiver is hit. See, e.g., [FVJF10b], for a discussion of the importance of certain ray-tracing model features in a CoMP environment. Receivers are modeled by horizontal square planes with a lateral size of several meters having complex antenna patterns. If a ray hits a receiver plane, it is registered at the receiver. When the simulation is finished, all collected rays are concluded to create a CIR for each transmitter-receiver pair. See Fig. 14.5 for an illustration of the simulated propagation paths in a virtual environment.

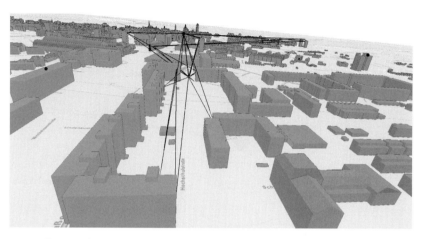

Figure 14.5 Ray paths through a 3D-environment building model.

Summarizing, the direct result of the ray-launching algorithm is a set of radio propagation paths for every transmitter-receiver combination, where each path is described by its properties (four complex polarimetric amplitudes, direction of departure (DOD), direction of arrival (DOA), and time delay of arrival (TDOA)).

14.2.4 Comparison between Measurements and Ray-Tracing

We now compare the results of LSP estimation based on channel measurements and simulations, in order to assess the accuracy of ray-tracing based parameter simulation in comparison to measurement based data. The analysis is based on a real-word scenario in downtown Dresden, Germany, as shown in Fig. 14.6.

Channel-Sounding Measurements

For channel sounding, the RUSK MIMO channel sounder by Medav, Germany [Med], was used. RUSK is a real-time radio CIR measurement system that supports multiple transmit and receive antenna element configurations. The RUSK MIMO channel sounder measures the CIR matrix between all transmitting and receiving antenna elements sequentially by switching between different transmit and receive antenna element pairs [THR+01]. With orthogonal sounding signals, the sounder is designed to handle distributed multiple transmitter or receiver units.

A first channel sounding campaign was conducted to survey, e.g., the interference geometry distribution and identify areas interesting for CoMP, and is here denoted as *overview* campaign. Thereupon, these areas were measured in a second *detailed* campaign [NKJ+11] with high-resolution antenna arrays, which enabled a double-directional channel characterization via RIMAX [Ric05, LKT07] parameter estimation.

14.2 Obtaining Chn. Model Params. via Chn. Sounding or Ray-Tracing

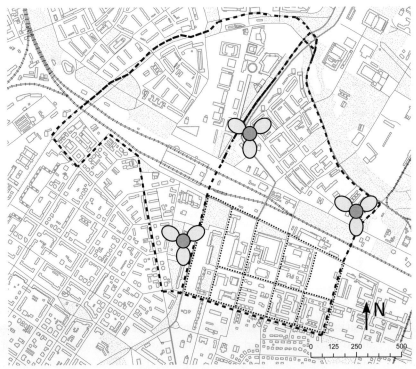

Figure 14.6 Measurement scenario in downtown Dresden, Germany. *Overview* campaign: dashed line (- - -), *detailed* campaign: dotted line (· · ·). The positions of the three sites are clearly marked. © OpenStreetMap & contributors, CC-BY-SA.

For the *overview* campaign, a cellular network with three sites separated in three sectors or cells each was used (see Fig. 14.6). The antenna heights were 55 m, 34 m, and 52 m, i.e. well above the average building height in this area. See, e.g., [FVJF10b] on a more detailed description of the measurement setup. The setup allowed the simultaneous and coherent measurement of the channels from the UE to all three sectors of one site.

For the *detailed* campaign, dedicated antenna arrays were used. At the BS side, a uniform linear array was used with eight dual polarized elements. Each polarized element consisted of a stack of four patches in order to form a narrow transmit beam in elevation. A resolution of the angle-of-departure in elevation and, thus, following the $AS_{RMS}^{bs,\theta}$ is not possible with this antenna, however. At the UE used for the measurements, a circular array with two rings of twelve patches (Stacked Polarized Uniform Circular Array - SPUCA) each having two polarization directions, was mounted on top of a car. Altogether, a 16 x 48 channel was measured. These dedicated measurement antenna types are necessary to be able to extract direction information in both, elevation (receiver side only) and azimuth (receiver and transmitter), and for angular spread estimation afterwards.

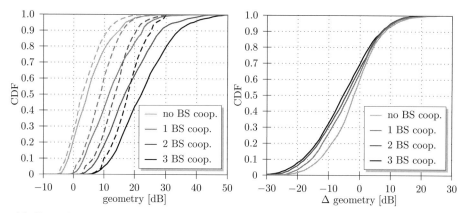

(a) Channel sounding measurements (dashed) vs. ray-tracing simulations (solid).

(b) Receiver-plane-wise differences between sounding and ray-tracing.

Figure 14.7 Evaluation of the interference geometry in a real-world scenario in Dresden.

Position data have been recorded for each track using two independent global positioning system (GPS) receivers. The GPS data sets were averaged afterwards, fitted to the map and interpolated to get a precise estimate of the terminal position. Both campaigns were operated in the 2.53 GHz range.

Ray-Tracing based Channel Simulation
A three-dimensional environment model of the same area in which the measurements took place was used in a ray-tracing simulator. Based on GPS information from the measurement campaigns, BSs and UEs were placed exactly into the environment, including the according antenna patterns.

In the measurement campaign, successive snapshots were taken in a time-constant distance of several ms, due to the fairly short channel coherence time. According to the velocity of the measurement car, successive snapshots were hence spaced several centimeters apart from each other. Instead, in the ray-tracing simulation, the area for collecting impinging rays is in the range of several meters and called a *receiver plane*. Consequently, for the following evaluation of the deviations between measurement and simulation multiple measurement, snapshots have to be mapped to receiver planes, see [FVJF10b] for more details and [FVJF10a] for extensions to the multi-user MIMO (MU-MIMO) case.

Results
In a first step, the distribution of the *interference geometry* is compared between channel measurements and channel simulations for the overview scenario, as this was introduced in Chapter 7. Hereby, the absolute power values (in mW) of the measurement points mapped to a ray-tracing receiver plane are averaged. The resulting cumulative distribution functions (CDFs) for a non-cooperative system

14.2 Obtaining Chn. Model Params. via Chn. Sounding or Ray-Tracing

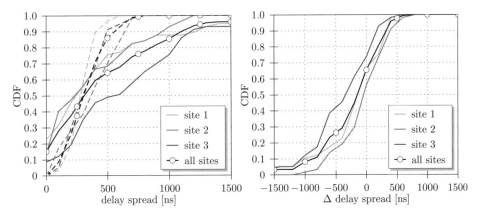

(a) Channel sounding measurements (dashed) vs. ray-tracing simulations (solid).

(b) Receiver-plane-wise differences between sounding and ray-tracing.

Figure 14.8 Evaluation of the delay spread based on a real world scenario in Dresden.

and for the assumption of (idealistic) joint signal processing between two, three, and four BSs are plotted in Fig. 14.7(a) for both, channel measurements and channel simulations.

The comparison shows an overestimation of the interference geometry by the ray-tracing simulation, which can be observed by the CDF (over all receiver planes) of the differences between the average of the measurement snapshots within one receiver plane and the simulation results of the same receiver plane in Fig. 14.7(b). The larger the CoMP cooperation size, the higher are the inaccuracies of the geometry simulation. This effect is based on an increase in inaccuracy with decreasing receive power, which is due to a higher average number of modeled propagation effects in the ray-tracing simulations, see [FVJF10b] for further discussions.

The difference between no cooperation and a CoMP cooperation size of three BSs is 11 dB at the median for our channel measurements. This result needs to be compared with the simulations presented in Chapter 7, where the difference is only 6 dB for the optimal clustering of three BSs in Fig. 7.2(a), and about 8 dB for the real-world scenario in Fig. 7.2(b). Furthermore, our measurements result in a further signal-to-interference ratio (SIR) gain of 5 dB for a cooperation size of four cells instead of three.

Further, important LSPs for the parametrization of spatial channel model extended (SCME)/WINNER channel models are the delay spread (14.9) and the azimuth angular spread (14.10). For LSP analysis, the statistics of a certain parameter are gathered by averaging over several snapshots assigned to one stationarity interval, as explained in Section 14.2.1. To compare the analysis based on measurement data and channel simulation, the selected snapshots for averag-

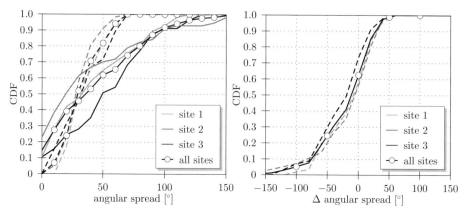

(a) Channel sounding measurements (dashed) vs. ray-tracing simulations (solid).
(b) Receiver-plane-wise differences between sounding and ray-tracing.

Figure 14.9 Evaluation of the angular spread at the receiver based on a real world scenario in Dresden.

ing have to be aligned with the receiver plan by the ray-tracing. By doing so, a fair comparison between the statistics from both data sources is ensured.

In Fig. 14.8(a), the CDF of the *delay spread* estimated from measurement and simulation data is shown. The analysis of the channel sounding measurement data results in a mean delay spread of about 300 ns. Variations between the different sites are in the range of less then 50 ns. Furthermore, the delay spread within a CoMP cooperation cluster considering all three sites is in the range of the single site delay spread.

Instead, the ray-tracing simulations again result in an overestimation of the delay spread also with much higher variations (20% of the simulated delay spreads are longer than 800 ns, which is not obtained from measurements at all). Between 10% and 23% of the receiver planes in the ray-tracing simulation possess a delay spread close to zero. For these points, only a single path or multiple paths from the same cluster during the channel simulation are received. The same effect was observed in the evaluation of the angular spread.

The analysis of the plane-wise differences between

- the delay spread averaged over all measurements snapshots within one simulation receiver plane and
- the delay spread of the simulated CIR obtained for the receiver plane

in Fig. 14.8(b) shows a bias at the median of about 150 ns between delay spreads analyzed from measurement and simulation data. The differences in angular spread from (14.10), as shown in Fig. 14.9, show a similar behavior.

14.2.5 Summary

As a pre-requisite for CoMP system simulation handled in this chapter, this section dealt with the parametrization of a widely used spatial channel model. The large-scale-parameters required for parametrization were introduced. Two alternatives for an extraction of these parameters from real environments were discussed: Channel measurements and channel simulation. Channel measurements directly reflect real world channels, but require a huge effort to be created. On the other hand, channel simulations include inherent modeling assumptions, but can be created quite easily. In the section, the main properties of both methods were highlighted, and extensions and specialities of measurements and simulations in CoMP environments were explained. Finally, the large-scale-parameters extracted from measurements and ray-tracing channel simulations in a sample CoMP environment in a European city were compared. This comparison has shown the applicability of well-designed deterministic channel simulations for large-scale-parameter extraction. The observed differences between the LSPs extracted from channel sounding measurements and ray-tracing simulations need to be traded-off with the effort of both methods. Also, the influence and sensitivity of the large-scale-parameters on system level simulations needs to be discussed in the following sections.

14.3 Uplink Simulation Results

Laetitia Falconetti, Christian Hoymann, Andreas Müller and Philipp Frank

After system level simulation methodology and the parametrization of channel properties have been explained in the last two sections, we now present specific simulation results for a subset of the *uplink* CoMP schemes introduced in Chapters 5 and 6. In this regard, we do not only show the achievable performance in terms of the average spectral efficiency and the cell-edge user throughput for different scenarios, but we also evaluate the associated backhaul load that is required with either approach due to the cooperation-related inter-base-station communication.

14.3.1 Compared Schemes

In the following, the compared uplink CoMP schemes are re-called from previous sections and described in more detail. We particularly focus on cooperation schemes which are suitable for implementation in real-world networks, and we take several practical constraints into account, such as limitations of the underlying backhaul network. Furthermore, in an attempt to reduce the backhaul load needed for cooperation, we do not only restrict the number of cooperating cells, but also the set of user equipments (UEs) which may participate and hence benefit from base station (BS) cooperation.

Cooperative Interference Prediction

Cooperative interference prediction has already been introduced in detail in Section 5.2.2. The basic idea of this approach is to predict the interference situation that will occur during a future data transmission already in advance and then to use this predicted interference rather than the currently measured one as the basis for link adaptation. This way, it is possible to better cope with the highly volatile nature of the inter-cell interference in cellular networks (especially in the uplink), which mainly arises from the fact that in different transmit time intervals (TTIs) completely different users may be scheduled in nearby cells, thus causing completely different levels of interference. Hence, with interference prediction the selection of appropriate modulation and coding schemes (MCSs) generally should become more accurate, as illustrated in Figure 5.12 in Section 5.2, and therefore the overall system performance should be improved.

Interference-Aware Joint Scheduling

The basic principle of interference-aware joint scheduling was presented in Section 5.2.1. In this regard, the idea is that by exchanging the channel state information (CSI) of all associated UEs within a certain cooperation cluster, a central scheduling unit (CSU) may be able to predict the inter-cell interference that would be caused by the various UEs to other cells within the same cooperation cluster. Based on this information, a global scheduling algorithm may then be applied in order to find the globally optimal or at least a close-to-optimal allocation of radio resources for the considered cluster. In general, UEs located near the cell-edge within such a cluster should benefit most from such an approach, since those UEs usually suffer most from inter-cell interference.

Joint Detection

We further assess the performance of CoMP schemes where multiple BSs perform joint signal processing to assist each other in detecting and decoding UEs. More precisely, we consider practical examples of *decentralized* joint detection (JD), related to the scheme introduced in Section 6.2, but significantly more simple. We assume that for each UE suffering from high co-channel interference and/or low received carrier signal strength, its serving BS sends a cooperation request to neighboring BSs specifying the desired cooperation mode. Three different cooperation modes are considered based on exchanging IQ samples, coded bits and decoded bits, respectively. UEs which are not selected for cooperation are served in a conventional, non-cooperative manner by their respective serving BS.

- **Joint detection based on IQ samples transfer**. A frequency-domain IQ sample is the complex representation of a constellation point of a given sub-carrier received at a given antenna. It is the output of the fast Fourier transform (FFT) of the orthogonal frequency division multiplex (OFDM) receiver and contains the amplitude and phase with which a particular sub-carrier has been modulated including the wireless channel impairment. In this cooperation mode, the serving BS must mention in which physical resource block (PRB) the signal of the supported UE was sent. The cooperating BS extracts the IQ samples of the indicated PRBs from its FFT module and transfers them to the serving BS. The serving BS processes the IQ samples of the supporting cell as if they were received by its own antennas. In a linear multi-antenna receiver, the signal arriving at each antenna is weighted with a complex factor before being combined with signals of other antennas. Different strategies are possible to determine the weights. In cellular systems where the signal impairment at the receiver contains correlated co-channel interference, interference rejection combining (IRC) outperforms maximum ratio combining (MRC) [Win84], as shown in Section 10.2.

- **Joint detection based on soft coded bits transfer**. In this cooperation mode, the cooperation request of the serving BS must contain not only the PRBs of the transmitted signal, but also its modulation and reference signals. Indeed, after equalizing and demodulating the received signal, the cooperating BS transfers the quantized soft values of the coded bits back to the serving BS. The serving BS combines the soft values of the supporting cell with its own, e.g. by applying Chase combining [Cha85], already introduced in cellular systems as one possible strategy to combine hybrid automatic repeat request (HARQ) retransmissions [PDF+01]. After combining, decoding takes place at the serving BS.

- **Joint detection based on decoded bits transfer**. In order to reduce the backhaul requirement, the serving BS can request decoded (hard) bits from a cooperating BS. As additional parameter compared to the case of soft bits exchange, the request message contains the coding scheme used by the UE. The cooperating BS detects, demodulates and decodes the signal. If the signal is correctly decoded, i.e. when the cyclic redundancy check (CRC) is successful, the cooperating BS transfers the decoded data back to the serving BS. After receiving the response message, the serving BS performs selection combining, meaning that the serving BS uses the decoded bits sent by the cooperating BSs only when it was not able to decode the data itself. Note that the exchanged data is used solely for selection diversity, and not for interference subtraction, as in the DIS scheme outlined in Section 4.3.1.

14.3.2 Simulation Assumptions and Parameters

The considered CoMP schemes have been investigated using a system level simulator as described in Section 14.1, observing a 3GPP LTE system operating in frequency division duplex (FDD) mode according to [3GP09e, HT09]. In particular, we consider a fully-loaded system with a carrier frequency of 2.0 GHz and a system bandwidth of 10 MHz. Both the Urban Macro 1 and Macro 3 cases as specified in [3GP06c] are studied, with inter-site distances (ISDs) of 500 m and 1732 m, respectively. The deployment scenario corresponds to a standard hexagonal grid with 19 sites, and each site is made up of three different cells

Table 14.1. Uplink system level simulation parameters.

Parameter	Value / Setting
Deployment scenario	19 sites with 3 cells per site
inter-site distance (ISD)	500 m (Macro 1) or 1732 m (Macro 3)
Carrier freq. / bandwidth	2.0 GHz / 10 MHz
Channel model	3GPP SCM
Distance-dependent pathloss	According to 3GPP TR 25.814
Shadowing standard deviation	8 dB
UE speed	3 km/h (quasi-static)
Default number of UEs/cell	10
Target BLER	10 %
UE / BS antennas	1 / 2 per cell
Traffic model	Infinite full buffer
HARQ	Synchronous, non-adaptive
Scheduling algorithm	Proportional-fair
Link-to-system interface	MIESM [BAS$^+$05]
BS antenna spacing	10 λ, λ: wavelength
BS noise figure	5 dB
UE category	5 (incl. 64-QAM support)
Power control	$P_0 = -58$ dBm, $\alpha = 0.6$ (Macro 1)
	$P_0 = -60$ dBm, $\alpha = 0.6$ (Macro 3)
	$P_{\max} = 24$ dBm
BS receiver types	LMMSE
Channel estimation	Ideal
Control channel overhead	Upper and lower 4 PRBs
Reference signal overhead	According to 3GPP TS 36.211

with $N_{\text{bs}} = 2$ receive antennas per cell, whereas all UEs are equipped with a single transmit antenna only. Multi-path fading is modeled by means of the 3GPP spatial channel model (SCM) [3GP07a] and we make use of the wrap-around technique to avoid any border effects, both of which were also described in Section 14.1. Further important simulation parameters are listed in Table 14.1.

The mutual information equivalent SINR mapping (MIESM), also explained in Section 14.1, is used for realizing the link-to-system interface [BAS$^+$05]. The link adaptation is modeled with a realistic delay of 2 ms, and the BSs may choose between several modulation schemes, namely QPSK, 16-QAM and 64-QAM, as well as a variety of different channel coding rates in order to accurately adapt the transmission format to the current uplink channel conditions. In this regard, the improved signal-to-interference-and-noise ratio (SINR) due to BS cooperation is already taken into account during the link adaptation stage and thus serves as the basis for the selection of appropriate MCSs. Besides, retransmissions are processed by means of a synchronous, non-adaptive HARQ protocol with incremental redundancy. In this regard, we assume that both the information exchange between the BSs and the joint signal processing are completed in time for all the considered cooperation schemes, so that the specified round-trip time (RTT) for a HARQ protocol process is not exceeded. Otherwise, the HARQ timing would have to be adjusted as discussed in Section 11.2, but this is not considered here. The transmit power of each UE in dBm is set according to a simple open-loop power control scheme as

$$P_{\text{TX}} = \min\{P_{\max}, P_0 + 10\log_{10} R + \alpha\,\text{PL}\},\qquad(14.13)$$

with $P_{\max}$ as the maximum transmit power, P_0 as a reference power level, R as the number of PRBs assigned to the UE, PL as the long-term attenuation of the channel between the UE and its serving BS, and finally α as a constant pathloss compensation factor. The values used for these parameters in our simulations are also given in Table 14.1. Please note that (14.13) can actually be obtained from the power control formula explicitly specified in [3GP09f] by neglecting all (optional) short-term components.

In order to reduce the amount of traffic generated on the backhaul, the number of cooperating cells per supported UE is limited to six for all considered cooperation schemes. In the case of *joint scheduling*, this means that even though a central scheduler is assumed for the entire system, this only takes a limited extent of interference information per UE into account. In the case of joint detection, the number of UEs benefiting from BS cooperation is further restricted through the introduction of a certain *cooperation range*. It is assumed that UEs regularly report to their serving BS a list of co-channel BSs whose received signal strength is within a certain cooperation range below the signal strength of the serving BS. This is similar to the UE-generated information considered for adaptive clustering in Section 7.2. UEs reporting a non-empty list are selected for cooperation. Depending on the given cooperation range value, the reported list, and therefore the set of cooperating cells, may contain less than six elements. If the limit of six cells is exceeded, only the six cells with the strongest links to the selected UE will be considered for cooperation. In the following, the cooperation range is set to 15 dB.

14.3.3 Backhaul Traffic

In general, cooperation between BSs requires an information exchange between the different sites by means of the underlying backhaul network, which was investigated in detail in Section 12.2. For the CoMP schemes considered in this section, the most important data that has to be exchanged between the different BS sites is listed in Table 14.2.

In case of *interference prediction*, every BS exchanges the resource allocation tables that have been fixed during the scheduling process with the set of cooperating BSs via the underlying backhaul network. We assume that the resource allocation information for each UE is indicated by a bitmap of 1 bit per PRB. Furthermore, the radio network temporary identifier (RNTI) of each UE is additionally signaled between the cooperating BSs in order to distinguish between all UEs located within the respective cooperation cluster. The transmit power of these UEs must also be made available among BSs of a cooperation clus-

Table 14.2. Required information exchange for considered uplink CoMP schemes.

CoMP scheme	Exchanged data	Assumptions
Joint detection	• IQ samples, or • coded bits, or • decoded bits	• 16 bit/IQ sample, or • 5 bit/soft coded bit
Interference prediction	• Resource allocation	• 1 bit/PRB, 16 bit/RNTI
Joint scheduling	• Multi-cell CSI • Resource allocation	• 16 bit/channel element • 1 bit/PRB, 16 bit/RNTI

ter on a more long-term basis as it is needed to estimate the expected received interference level.

In case of *joint detection*, the predominant part of the traffic generated on the backhaul comes from the cooperation response, where a cooperating BS sends IQ samples, soft coded bits or decoded bits to the requesting serving BS. If BSs exchange IQ samples, a certain cooperating BS transfers one IQ sample per receive antenna and per sub-carrier of the requested PRBs, where we assume a quantization granularity of 16 bit per IQ sample. Various ways to reduce the backhaul traffic generated by the transfer of IQ samples are discussed in Section 12.2. In case of soft coded bits exchange, the amount of backhaul traffic depends on the number of data streams sent by the supported UE, the number of requested PRBs and the modulation scheme from which the number of coded bits is derived. In the simulation, one soft coded bit is quantized with 5 bit, and we assume that only one data stream is sent by the UEs.

The data to be exchanged within a cooperation cluster in case of joint scheduling generally consists of two fundamental parts. On one hand, the CSU has to signal the scheduling decisions to the corresponding cooperating cells after completing the resource allocation and on the other hand the cooperating cells have to frequently report the current multi-cell CSI for each sub-carrier and each receive antenna to the CSU, e.g. with a 2 ms period as assumed in the following results. Clearly, the latter constitutes the most significant fraction of the overall backhaul traffic since the CSU has to be aware of the multi-cell CSI of all UEs located within the respective cooperation cluster. In this regard, we assume that each estimated frequency-domain channel coefficient of the multi-cell CSI is quantized with a resolution of 16 bit.

14.3.4 Simulation Results

Fig. 14.10(a) shows the gain in average and cell-edge user throughput of the considered CoMP schemes compared to the case of no BS cooperation for the Macro 1 scenario with an ISD of 500 m. Note that the cell-edge user throughput is defined as the 5-th percentile of the UE throughput distribution.

14.3 Uplink Simulation Results

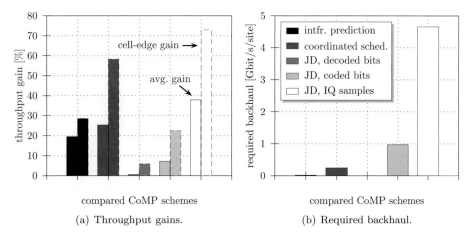

Figure 14.10 Relative performance gains and required backhaul for various uplink CoMP schemes and an ISD of 500 m (Macro 1).

Let us first observe the performance gains for the cooperative interference prediction and joint scheduling schemes. As already mentioned before, the cooperation cluster consists of 6 cooperating cells, which corresponds to the situation where each cell receives resource allocation tables from all surrounding cells in case of interference prediction, whereas each cell has only multi-cell CSI of the UEs associated with its six surrounding cells in case of joint scheduling, thus only the inter-cell interference caused by these UEs can be taken into account for the resource allocation by the central scheduling unit. Furthermore, we assume that for both schemes the number of UEs is not constrained, since the backhaul capacity requirements are much smaller compared to the joint detection scheme with coded bits or IQ samples exchange, which is also confirmed by Fig. 14.10(b). It can be seen from Fig. 14.10(a) that significant performance gains may be obtained with both schemes, where the joint scheduling always outperforms the interference prediction scheme. This is because the joint scheduling scheme combines both the cooperative resource allocation as well as the prediction of the inter-cell interference that is expected to occur during a future data transmission. Similar to the joint detection cases, we note that the relative gains in case of joint scheduling and interference prediction are also always higher in terms of the cell-edge throughput than in terms of the average throughput. This is because the users located at the cell-edge generally suffer more from the volatile nature of the inter-cell interference and therefore those users also benefit more from the joint resource allocation and interference prediction, respectively.

Considering the joint detection schemes, the gain obtained with the IQ sample exchange approach exceeds the one obtained with coded bits, which in turn outperforms the decoded bits exchange. With a 15 dB cooperation range and a maximum of 6 supporting cells, the cell-edge user throughput increases by 6%, 22.5% and 73% for decoded bits, coded bits and IQ samples exchange,

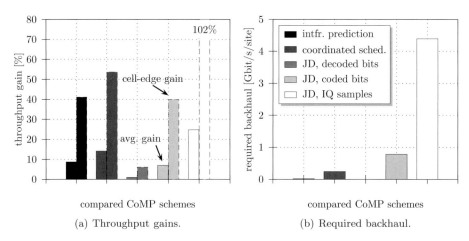

Figure 14.11 Relative performance gains and required backhaul for various uplink CoMP schemes and an ISD of 1732 m (Macro 3).

respectively, compared to a conventional LTE system without BS cooperation. The moderate gains of the decoded bit exchange mode can be explained by the following reasons. On one hand, this approach is based on the assumption that a BS may be able to decode data from a user in the neighboring cell. However, simulations showed that the decoding at at least one supporting BS is successful in 17% of all cases only. On the other hand, when a supporting BS is able to decode the user signal, the serving BS can not combine the decoded bits received from the supporting BS with the bits decoded by itself. The decoded bits received from the supporting BSs are effectively exploited only in the case where the serving BS is not able to decode the user signal alone. By contrast, the IQ sample approach outperforms the other joint detection schemes, since in this case the serving BS can virtually increase its number of receive antennas, enabling array and spatial multiplexing gain, as stated in Section 4. Moreover, it can be seen from Fig. 14.10(a) that for all joint detection schemes the gains in terms of average throughput are smaller than the gains in terms of cell-edge user throughput, as expected from Chapter 3. But here, the user selection based on a 15 dB cooperation range also contributes to this performance gap, as it tends to favorize selection of cell-edge users for CoMP. Non-selected users are detected by their serving BS conventionally without any BS cooperation.

The backhaul capacity requirement *per site* for the different CoMP schemes is shown in Fig. 14.10(b). While the backhaul traffic caused by joint detection based on decoded bits exchange is only minor, it is tremendously increased in case of coded bits and IQ samples exchange up to 0.975 Gbit/s and 4.7 Gbit/s, respectively. However, the latter ones therefore provide a higher gain in user data rate. Hence, a trade-off between (cell-edge) user throughput improvement and backhaul capacity requirement has to be found. When comparing the ratio of the required backhaul capacity per site to the gain in the cell-edge user throughput, exchanging decoded bits seems to be more attractive. However, the IQ samples exchange approach is still more promising since it yields higher user throughput gains and its backhaul traffic can be further reduced using methods described in Section 12.2. In the general, the results confirm very well the trade-offs between different uplink joint processing schemes pointed out in Section 4.3.3.

Finally, Fig. 14.11(a) illustrates the performance gains for the different BS cooperation schemes for the Macro 3 case with an inter-site distance of 1732 m. Since in the Macro 3 case the experienced inter-cell interference level usually is less dominant than for the Macro 1 case, the relative performance gains in terms of the average throughput are notably smaller than for Macro 1. Furthermore, if the distance between the BSs is increased in case of joint detection, the 15 dB cooperation range criterion is fulfilled by fewer users compared to the Macro 1 case. Consequently, only few users located at the cell-edge benefit from joint detection, resulting in a slight decrease of the joint detection performance in terms of average throughput, but also in a decrease of required backhaul. However, this does not hold for the gains in terms of cell-edge throughput, which even can be improved in case of joint detection compared to the Macro 1 case, reaching 6.2 %, 40 % and 102 % for decoded bits, coded bits and IQ samples exchange, respectively. This is because on the one hand the performance of cell-edge users in the reference case without cooperation decreases when the distance between BS and users increases. On the other hand, in the Macro 3 case the probability that cell-edge UEs become power-limited is higher than for the Macro 1 scheme, so that the limiting factor is not necessarily the inter-cell interference but rather the extremely low power with which signals from cell-edge users are received at the serving BS. With joint detection, the transmit signal is collected at all antennas of the cooperating BSs, thus increasing the effective receive signal power. This in particular is even more beneficial in case of large ISDs. Note that imperfect channel estimation is not considered for these results and may particularly impact the performance in a large ISD scenario, as suggested by the results in Section 4.2. Further, in scenarios of larger BS distance signal propagation delay based timing problems can arise, as discussed in Section 8.2, but these are not considered here.

14.3.5 Summary

In this section, potential spectral efficiency gains and backhaul requirements of different uplink CoMP schemes have been assessed through system level simulations. The results suggest that a good trade-off between system performance and backhaul capacity requirements is provided by the two interference coordination schemes, interference prediction and joint scheduling. Since these schemes require only a moderate backhaul load while still promising good performance gains, both represent very attractive options for future LTE-A systems. Nevertheless, they may not be easily introduced in cellular systems in a short-term.

By contrast to joint detection, which is based on existing signal combining techniques, interference prediction and joint scheduling require new and more complex link adaptation and scheduling algorithms. Moreover, interference prediction has a more stringent requirement on the backhaul latency than joint detection, (see Section 11.2), as the information exchange between BSs and the additional processing have to take place between the scheduling decision process and the transmission of grants to UEs. This time interval is usually in the range of 1 ms. Finally, the here considered joint scheduling scheme that assumes a CSU needs to be converted to a distributed approach to comply with the decentralized architecture of current cellular networks. This, however, would increase the amount of generated backhaul traffic.

14.4 Downlink Simulation Results

Jochen Giese, M. Awais Amin, Stefan Brück, Mark Doll, Thorsten Wild, Lars Thiele and Michael Olbrich

In this section, we finally provide system level simulation results for selected downlink (DL) CoMP schemes. As in Section 14.3, we assess both performance gains over non-cooperative schemes, as well as the amount of backhaul capacity required to achieve these.

14.4.1 Compared Schemes

In this section, we compare potential performance gains from the following downlink CoMP schemes:

Coordinated Beamforming / Coordinated Scheduling
The main idea of coordinated scheduling / coordinated beamforming (CS/CB), where theoretical principles were already introduced in Section 5.3, is that user equipments (UEs) not only feedback information about suitable precoding vectors with which their respective base stations (BSs) can serve them, but they also determine potential precoding vectors which would cause strong interference if used by BSs serving adjacent cells. If the BSs exchange this information and such precoding vectors are avoided, inter-cell interference can be reduced, which is particularly beneficial for cell-edge users. Before providing overall system level results for CS/CB at the end of this section, we perform a detailed analysis of potential gains and trade-offs connected to the scheme in Section 14.4.3.

Joint Transmission
In this section, we further observe potential performance gains from a particular multi-cell joint transmission scheme, as it was introduced in Section 6.3. We assume that multiple, i.e. M_c, BSs are grouped into a cluster c of cells, where all

BSs in this cluster cooperate in jointly transmitting precoded data symbols to multiple UEs, such that desired signals overlap coherently and the interference is partially canceled out. Joint processing is only allowed between BSs belonging to the same cluster, whereas BSs belonging to different clusters are not coordinated and thus produce residual inter-cluster interference. In principle, coordinated beamforming (CB) techniques may be used to deal with the interference between clusters, but this is not considered here.

We assume a dynamic and user-driven clustering method, as proposed in Section 7.2. An active set of users is selected according to the following metric: A set $\mathcal{U}_c$ of active multi-antenna terminals is uniformly distributed in the c-th cluster of the cellular environment. This set contains multiple disjoint user sets $\mathcal{K}_m \subset \mathcal{U}_c$, where the user in $\mathcal{K}_m$ experience highest channel gain to the m-th BS. Thus, each UE is connected to a master BS. Further, we emulate a cluster selection which is user-centric and dynamic over frequency: the M/C strongest channel gains of the users in $\mathcal{K}_m$ are the ones of the M_c BSs within the cluster, i.e. each UE is placed in a perfect cluster in the way that its strongest BS links are all covered by this cluster. A round-robin scheduling policy ensures that only N_{bs} UEs per BS $m \in \mathcal{M}$ are selected for joint transmission. This enables us to obtain gains from joint transmission that are separated from gains due to multi-user diversity.

14.4.2 Simulation Assumptions and Parameters

For performance evaluation, the same system level simulation methodology has been used as explained in Section 14.1, for example using a wrap-around technique, considering spatial channel model extended (SCME) radio channels for urban macro scenarios, and employing a mutual information equivalent SINR mapping (MIESM). 19 sites consisting of 3 cells are placed in a regular hexagonal layout with an inter-site distance of 500 m. The users are uniformly distributed over the whole area of the network layout with an average density of 10 UEs per cell, unless stated otherwise, and a full buffer assumption is used, as introduced in Section 14.1. As in the uplink in Section 14.3, a carrier frequency of 2 GHz and a system bandwidth of 10 MHz is assumed. Similar as in the uplink, the modulation and coding schemes (MCSs) applied by the BSs are adapted per transmission interval of 1 ms, based on information of the radio channel state provided by the UE, but larger delays occur, as stated later. In case of transmission errors, no retransmissions of the erroneous packets are modeled, but the packets are regarded as being lost.

While all schemes compared in this section assume $N_{\text{ue}} = 2$ omnidirectional antennas at the terminal side, the number and layout of BS antennas may differ, and will be stated explicitly for every simulation result. The (implicit or explicit) channel feedback provided by the terminals and the precoding applied at the BS side depends on the observed CoMP scheme, and is stated in the sequel:

In the case of *coordinated beamforming*, the UEs are assumed to feedback the most desirable precoding matrix from a predefined codebook, which follows the specification of the 3GPP LTE Release 8 system for $N_{\rm bs} = 2$ and $N_{\rm bs} = 4$ antennas. At the terminal side, minimum mean square error (MMSE) receive filters are assumed. The UEs further feedback so-called worst companion indicators (WCIs), which will be explained in detail in Section 14.4.3.

In the case of *joint transmission*, the UEs are assumed to apply the strongest eigenvector of the observed channel as a receive filter, a concept known as maximum Eigenvalue transmission (MET), and feedback the channel state information (CSI) connected to the resulting *effective* multiple-input single-output (MISO) channel, as described in detail in Section 6.3. In this context, the CSI is assumed to be of infinite granularity.

For both schemes, perfect CSI at the UE side is assumed, hence in particular the results for joint transmission can be assumed to be rather optimistic.

14.4.3 Detailed Analysis of Coordinated Scheduling/Beamforming

In this subsection, we initially state the WCI reporting scheme in more detail, and then CS/CB is analyzed in two steps: First, the benefit of avoiding certain precoding vectors in adjacent cells from the point of view of a terminal is observed in an isolated way. More precisely, this part focuses solely on the CB component, without looking into coordinated scheduling (CS), and ignoring the scheduling restrictions that the avoidance of precoding vectors poses towards other cells. The results obtained here can be seen as upper bounds, and they are observed for both homogeneous and heterogenous deployment scenarios. Then, scheduling restrictions are taken into account, and a particular joint CS/CB implementation based on cyclically prioritized scheduling is observed.

Worst Companion PMI Reporting

Multi-cell coordination of beamforming requires channel knowledge at the transmitter, including knowledge about interfering neighbor cells. For the frequency division duplex (FDD) scenario this knowledge has to be provided via uplink feedback signaling. The WCI feedback signaling [3GP08a], [3GP09b] efficiently provides this information using precoding codebook indices. The UE has to measure the channel of neighboring cells and reports cell and index of the strongest interfering precoder, as shown in Fig.14.12. A WCI is therefore the tuple of a cell identifier and a precoding matrix indicator (PMI). Additionally it provides the classical channel quality indicator (CQI) connected to the serving link plus a Delta-CQI, which indicates the estimated gain, in case the reported WCI is not used by the interfering neighbor cell, e. g. reflecting the difference in mean signal-to-interference-and-noise ratio (SINR) with and without that the WCI. In order to increase the interference avoidance gain for the coordinated cells, the UEs may report a set of WCIs and a corresponding Delta-CQI giv-

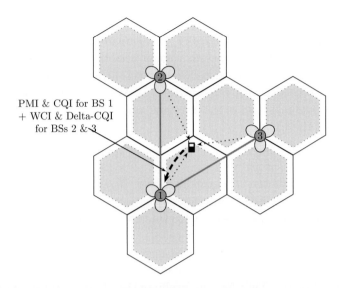

Figure 14.12 The 'worst companion' reporting principle.

ing the total gain, at the cost of an increased feedback rate [3GP09c, 3GP10b]. The UE can be configured by the BS to send the desired kind of reports e.g. semi-statically. Based on this additional information, the BSs then schedule UEs in the different cells (serving and interfering) on the same time and frequency resources using appropriate precoding/beamforming in such a way that the UEs observe lower inter-cell interference than in the uncoordinated case.

Performance Bounds for Homog. and Heterogeneous Networks

In this paragraph, we present the calculation of upper performance bounds for worst companion reporting schemes (see also [GA10]). A common measure to assess the receiver performance is the SINR at the output of an equalizer, which depends on the equalizer implementation as well as the noise and interference covariances, channel matrices, signal powers and precoding matrices. We consider one exemplary UE 1 which is served by BS 1, and potentially sees interference from various other BSs serving other UEs $k \in \{2 \ldots K\}$. We restrict the dependence of the SINR in our notation on the precoding matrices which are under the control of the transmitters, i.e. stating the SINR obtainable by UE 1 as

$$\text{SINR}_1 = \text{SINR}_1\left(\mathbf{W}_1, \mathbf{W}_2, \ldots, \mathbf{W}_K\right). \tag{14.14}$$

where $\mathbf{W}_1$ is the precoding matrix used by BS 1 to serve UE 1, and $\mathbf{W}_2, \ldots, \mathbf{W}_K$ are the precoding matrices employed by the other BSs to serve their users. Note that all $\mathbf{W}_k \in \mathcal{C}$ are taken from a codebook $\mathcal{C} = \{\mathbf{C}_1, \ldots, \mathbf{C}_N\}$ of N precoding matrices $\mathbf{C}_i$. The SINR determines the supportable data rate for the transmission and, therefore, the achievable throughput for this user assuming idealistic link adaptation. In order to investigate the gains achievable with beam coordination, we evaluate the gains in SINR at the UE under the assumption that interferers do not transmit at all or avoid specific precoding matrices that cause the largest SINR degradation. In particular, we compare the regular SINR value obtained when taking all interferers into account with the resulting SINR values when one interferer had been prevented from transmitting (expressed in our notation as using a precoder with zero entries, i.e., $\mathbf{W}_k = \mathbf{0}$), while all others transmit as before. The cell index of the interferer whose removal results in the largest SINR value at the UE, i.e.

$$d_1 = \arg \max_{k=2,\ldots,K} \text{SINR}_1 \big|_{\mathbf{W}_k=\mathbf{0}}, \tag{14.15}$$

is denoted as the *strongest* interferer. Based on the intuition that some precoding matrices used by interferer d_1 have a worse impact on the evaluated SINR than others, it is not necessary to switch off interferer d_1 completely, which would imply a loss in throughput for cell d_1. Instead, we only need to prevent interferer d_1 from using the most SINR-degrading codebook entries. This is beneficial because cell d_1 can then serve users where the remaining codebook is applicable with full power. Therefore, we evaluate the SINR for the entire codebook $\mathcal{C}$ and compute the set of SINR values $\left\{ \text{SINR}_1 \big|_{\mathbf{W}_{d_1}=\mathbf{C}_n} \right\}$ for $n = 1, \ldots, N$ together with the index function $i(k), k = 1, \ldots, N$ to order the SINR values

$$\text{SINR} \big|_{\mathbf{W}_{d_1}=\mathbf{C}_{i(1)}} \leq \text{SINR} \big|_{\mathbf{W}_{d_1}=\mathbf{C}_{i(2)}} \leq \cdots \leq \text{SINR} \big|_{\mathbf{W}_{d_1}=\mathbf{C}_{i(N)}} \leq \text{SINR} \big|_{\mathbf{W}_{d_1}=\mathbf{0}}.$$

Note that $N+1$ calculations of SINRs are needed. Preventing interferer d_1 from using the p most SINR-degrading codebook entries will result in an SINR which is at least as good as if $\mathbf{C}_{i(p+1)}$ had been used, i.e., $\text{SINR}_1 \geq \text{SINR}_1 \big|_{\mathbf{W}_{d_1}=\mathbf{C}_{i(p+1)}}$. The obtained SINR in comparison with the SINR value based on no interference coordination reflects the potential gain that is achievable at the UE with coordination of a single interferer. The approach can be generalized to multiple interferers when all possible combinations of precoding matrices used by the interferers are evaluated and a specific subset is excluded. We consider it more intuitive to assume that the coordination including an additional interferer is only reasonable if all other stronger interferers have been prevented from transmitting at all. Then, we can determine the second strongest interferer with index d_2 using the steps above assuming zero transmission power for cell d_1, i.e. $\mathbf{W}_{d_1} = \mathbf{0}$ and in that way successively compute the order of strongest interferers.

It should be emphasized that the evaluation of the SINR gain is based on the instantaneous SINR, which implies the optimum selection of interferer codebook entries to be removed on the shortest possible timescale. The obtained gains are therefore optimistic in the sense that for longer delays, misadjustment of the restriction of codebook entries will cause gain degradation. Moreover, the obtained throughput gains based on improved SINR are upper bounds for real system performance because the restriction of codebooks applies only to interfering cells in the calculation of the SINR at the UE but not to the scheduling in the interfering cells. Thus, the upper bounds neglect a loss in throughput due to the restricted codebook. They are therefore useful to assess the maximum potential of coordinating beam selection, but do not predict accurate system gains.

Homogeneous and Heterogeneous Network Models
Conventional wireless cellular networks consist of a number of macro cells providing coverage within a geographical area. The deployment of these cells follows careful and extensive radio planning to satisfy capacity and coverage criteria. In order to improve data throughput for cell-edge and indoor users by means of efficient spectral reuse and higher received power, network operators have been focusing on the introduction of low-power radio nodes (e.g., micro, pico and femto cells) to improve coverage. We call a network that contains cells with significantly different maximum amplifier powers a heterogeneous network in contrast to a homogeneous network containing only high-power macro BSs.

A standard model for simulating a homogeneous network places 3-sectorized sites on a hexagonal grid, i.e., each site uses sectorized antennas with possibly a three-dimensional antenna pattern. The users are dispersed uniformly within the network for a given average user density. We model a heterogeneous network by deploying low power nodes in addition to the macro network by dispersing these nodes randomly within the coverage area of each site. In particular, we model heterogeneous networks containing macro cells and femto cells. The femtos are placed in circular clusters. Within the area covered by a cluster, one or more femtos are placed with uniform distribution. Such an unplanned deployment of femtos (modeling installation by the user) can pose significant challenges in terms of interference caused to out-of-cell users. This can become particularly problematic if the access to a femto is restricted to a closed subscriber group (CSG), i.e., a small group of users that, e.g., live in the vicinity of the femto or own it. A femto may serve only those users that belong to its CSG, if defined. This restricted access can pose a serious problem to a user located close to a femto who cannot connect to it, because the user is not included in the corresponding CSG. Such a user may experience strong interference from the nearby femto resulting in poor downlink reception from the user's serving macro BS.

Users in the heterogeneous network can be differentiated based on the type of radio access point they are allowed to connect to. Macro-only (femto-unaware) UEs are allowed to connect to a macro BSs only whereas CSG (femto-aware) UEs are allowed to connect to either a specific femto or any of the macro BSs. Femto-aware UEs are also placed within the area covered by the corresponding cluster containing the femto node that the UE has special access to.

System level simulations were performed to find upper bounds on gains for CB using throughput as a performance metric. The SINR used in the simulation of the UE data detection was a hypothetical SINR that can be guaranteed if a defined number of strongest interferers is silent and the codebook of the next strongest interferer is restricted. The simulations were carried out for antenna

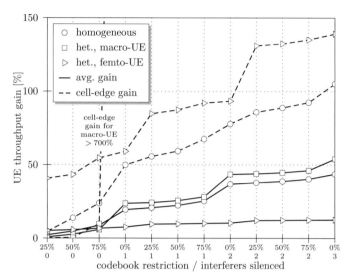

Figure 14.13 Upper bounds on UE throughput gains through CB concepts.

configurations with 4 vertical antenna poles at the BS and 2 vertical antenna poles at the UE side, all with $\lambda/2$ spacing. For a user scheduled on multiple physical resource blocks (PRBs), the hypothetical SINR was calculated assuming the same interfering precoding matrix was used over all scheduled PRBs.

Both an example homogeneous scenario as well as a heterogeneous scenario were investigated. The example heterogeneous layout contains 5 clusters per macro cell with one CSG user each and 5 macro-only UEs per cell on average. Note that the network load is the same as in the homogeneous network with 10 users per cell on average, albeit split up into an equal number of macro-only and CSG users. In each case, users are differentiated based on the type of cell they are connected to and termed as macro-UEs (M-UEs) and femto-UEs (F-UEs), respectively. Note that a femto-aware UE might still be connected to a macro BS and would be classified here as an M-UE.

The results in terms of gain in mean UE throughput as well as cell-edge UE throughput are illustrated in Fig. 14.13 both for the homogeneous and heterogeneous layouts. The gain in throughput is computed against the baseline case which does not involve any coordination between BSs. We consider the case that interferers are silenced and the codebook of the next strongest interferer is partially restricted, leaving out 25%, 50% and 75% of the worst codebook entries, respectively.

Comparing the gains using partial restriction of the codebook with the gains obtained by silencing an interferer in the homogeneous network, it turns out that the gains using partial restriction are relatively small, especially considering the mean throughput. It is now of interest to investigate how far higher gains can be achieved in a heterogeneous network due to the specific interference condition. Since the cell-edge macro users in a heterogeneous network suffer heavily from the interference caused by the femtos, the silencing of interferers turns out to be extremely beneficial with gains of more than 700% when silencing the strongest interferer. Essentially macro users in the vicinity of a CSG femto that suffer from strong interference can take advantage of coordinating their transmission with that strong interferer. Similar to the results for the homogeneous layout, partial codebook restriction does not provide significant gains for cell-edge M-UEs compared to interferer silencing. The mean throughput of F-UEs is limited in this simulation by a large number of equally strong interferers from the macro network which therefore does not allow significant gains using beam coordination. At the cell-edge, performance for the F-UEs does improve significantly with codebook restriction of the strongest interferer, showing throughput improvements of more than 30%.

In summary, the analysis illustrated the potential of beam coordination techniques which can lead to significant throughput improvements for cell-edge users. Due to the specific interference scenario in a heterogeneous network with femtos using CSG functionality, beam coordination can be extremely beneficial in this case. Note, however, that the results in Fig. 14.13 did not take scheduling restrictions due to avoided precoders or silenced interferers into account, which will be dealt with in the next paragraph.

Performance Gains for Cyclically Prioritized Scheduling
We now want to look into a particular CS/CB implementation based on cyclically prioritized scheduling. For determining the cooperation set, the neighborhood between cells is first determined without shadowing and precoding, i. e. based solely on pathloss (and antenna gain). That means we are only considering the 6 geometrically nearest neighbor cells, cf. Fig. 14.14. The shadowing and precoding is later included in the SINR calculation. In this case, the geometrically nearest neighbors are not always the strongest interfering cells. Because of this effect, the full performance improvement potential is not exploited. The advantage is that already 3 priorities are sufficient for the typical hexagonal cell layout with 3 cells per site, cf. the right side of Fig. 14.14. A lower number of priorities is beneficial, as this reduces the overall duration of a coordination cycle, as explained below.

In cyclically prioritized scheduling, scheduling among cooperating cells is serialized as far as it is necessary to prevent conflicting scheduling decisions, i. e. scheduling decisions that violate WCI constraints posed by other cells from a cell's cooperation set. Serialization assures that when scheduling is computed for a specific cell, no other cell within that cell's cooperation set computes its scheduling at the same time, but instead either earlier or later. In the simulated hexagonal layout, there are 3 priorities A, B and C. All cells of priority A schedule first and in parallel for a certain future sub-frame, then B priority cells and finally C priority cells, all for the same future sub-frame. The scheduling sequence is changed over time in a round robin fashion, i. e. cells with highest scheduling priority A will have lowest priority C in the next coordination cycle,

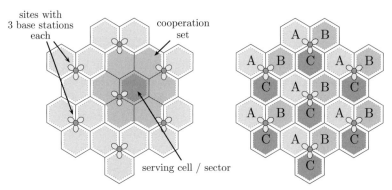

Figure 14.14 A cell's cooperation set consists of the serving cell itself and its geometrically nearest 6 neighbors (left). For such 7 cell sets, the cyclic prioritized scheduling scheme requires 3 scheduling priorities A, B and C (right).

B cells will become A cells and C cells will become B cells. Thereby each cell once has the opportunity to schedule first (priority A) without any constraints.

For determining the scheduling priority, the expected instantaneous throughput of each UE is calculated for each resource group by mapping its reported CQI and Delta-CQI values to a throughput using an approximated efficiency curve. The expected instantaneous throughput r_t of each UE is then α-proportionally weighted against its average throughput $\bar{r}_t$, which is calculated as the exponentially weighted moving average with 'forgetting' factor β according to $\bar{r}_t = \beta \bar{r}_{t-1} + (1-\beta)r_{t-1}$, giving a per UE $score = r_t/\bar{r}_t^\alpha$ per resource group. For each resource group, the UE with the highest $score$ and with a reported PMI that does not violate any PMI constraints given by other higher priority cells of the cooperation set is selected for transmission. The PMI used to transmit to the selected UE as well as the WCIs of the selected UE are then reported to lower priority cells of the cooperation set for interference estimation and consideration as constraints in their own scheduling, respectively. Besides this, no other information exchange is assumed between the cooperating BSs. The required bandwidth on the backhaul is therefore small.

In contrast, backhaul latency requirements for the presented scheme are more stringent. A coordination cycle, i. e. scheduling first by A, then B and finally C priority cells, needs to be completed before the transmission over the air can take place. Each BS within each coordination cycle therefore computes scheduling for a subframe as much further in the future as that coordination cycle lasts. In consequence, an increased backhaul latency and thereby coordination cycle duration has the same negative impact on performance as an increased uplink reporting delay on closed-loop precoding.

For performance assessment, again the simulation setup according to Section 14.4.2 was used. Reporting follows Release 8 granularity, i. e. CQI and Delta-CQI are reported frequency selectively per resource group of 6 PRBs, while PMI

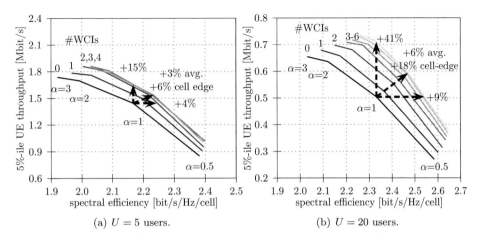

Figure 14.15 Cell-edge throughput vs. spectral efficiency for CS/CB in the 3GPP SCME case 1 scenario (500 m ISD, 3 km/h) using 4x2 V-pol. $\lambda/2$-spaced antennas.

and WCIs are reported frequency-diverse for all PRBs. Reporting delay and interval are 6 ms and 5 ms, respectively. Within each cell, per resource, a single UE was scheduled, i.e. the system was operated in single-user MIMO (SU-MIMO) mode. The resulting downlink throughput improvement is shown in Fig. 14.15 for 10 MHz system bandwidth and on average 5 or 20 UEs per cell at 3 km/h. We again observe a 4x2 antenna setup, where all BS and UE antennas are vertically polarized with $\lambda/2$ spacing. The number of PMI constraints per resource group is varied between 0, i.e. normal scheduling without any PMI constraints (lower left curve of each plot), and up to 6 PMI constraints (upper right curve of each plot). In any case, the number of constraints posed on a certain interferer, i.e. a certain neighbor cell, is limited to 3. This mitigates the effect that no PMI is left that fulfills the constraints of all higher priority neighbors. Furthermore, for each given number of constraints, the exponential alphas in the proportional fair scheduler were tuned through $\alpha \in \{0.5, 1, 2, 3\}$. This results in up to 7 curves with 4 data points each.

As can be seen, performance without CS/CB as well as the additional coordination gain increases with the number of UEs per cell. For 5 UEs, a gain of 3% in spectral efficiency and 6% at the cell-edge, i.e. for the 5%-ile of the thoughput cumulative distribution function (CDF), is achieved. For 20 UEs, the gain more than doubles to 6% and 18%, respectively. By increasing α, the gain in spectral efficiency can be converted into an additional cell-edge gain yielding e.g. in the 20 UEs per cell case 41% gain at the cell-edge while retaining spectral efficiency as without coordination. Alternatively, α can be reduced to improve overall spectral efficiency, e.g. in the 20 UEs case 9% are gained in spectral efficiency while cell-edge throughput is retained as is.

In multi-user MIMO (MU-MIMO) mode, the situation completely changes, as no considerable gain can be achieved by CS/CB in the presented simulation scenario. The reason is that the increasing number of WCI constraints lead to more and more resources operated in SU-MIMO mode. Since MU-MIMO achieves about one third higher spectral efficiency than SU-MIMO in the considered scenario, this impact cannot be compensated by the SINR improvement through coordination. For a 2x2 V-pol. antenna setup, the gains are considerably lower, about one half of the gains of 4x2 antennas, because the resulting beams are much broader than those of 4 transmit antennas. The same can be observed when rearranging 4 vertical antennas into two $\lambda/2$-spaced pairs of cross-polarized antennas.

14.4.4 Backhaul Traffic

In this subsection, we want to briefly assess the amount of backhaul required for the CS/CB and joint transmission techniques compared in this section.

Regarding *CS/CB*, one can see from Fig. 14.15 that most of the CS/CB gain is already achieved with 4 WCIs. Therefore, reporting per PRB pair 1 PMI and 4 WCIs, i.e. on average 4/6 WCIs per each of the 6 neighboring cells, is assumed. Both PMI and WCI typically have 4 bit resolution in the case of 4 BS antennas. The cell index of each WCI need not be transmitted, as it is the receiving cell itself. PMI and WCIs only need to be sent to lower priority cells at other sites. Ignoring any X2 protocol overhead, this results in (1 PMI $\cdot(3+2+0)/3$ other sites with at least one cell of lower priority *plus* 4 WCIs/6 cells $\cdot(4+2+0)/3$ cells of lower priority at other sites) $\cdot$4 bit per PMI respectively WCI $\cdot$50 PRB pairs of 10 MHz system bandwidth / 1 ms sub-frame duration $\cdot$3 cells per site = 1.8 Mbit/s net signaling rate per site, in as well as out.

For *joint transmission*, let us here focus on the backhaul required for data exchange, as that connected to CSI exchange is typically orders of magnitude less, as pointed out in Section 12.2. Assuming that BSs in two arbitrary adjacent cells perform joint transmission, this happens on-site with a probability of 1/3, requiring no backhaul at all. In the remaining cases, each involved BS has to forward the data bits connected to its UE to the other BS. With the spectral efficiencies obtained in Section 6.3, and assuming a bandwidth used for data transmission of 9 MHz, this leads to an overall backhaul rate of 26.7 Mbit/s/cell, or 80 Mbit/s/site. If three arbitrary adjacent BSs cooperate, we have a chance of 1/6 that all three are on-site, 1/2 that two sites are involved, and 1/3 that three sites are involved. Averaging over these probabilities, the average backhaul requirement is then 56 Mbit/s/cell or 168 Mbit/s/site.

14.4.5 Simulation Results

Figure 14.16 finally shows throughput gains and required backhaul quantities for four different CoMP schemes considered in this section:

- CS/CB, employing adaptive SU-MIMO,

14.4 Downlink Simulation Results

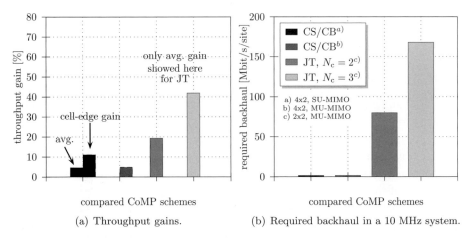

(a) Throughput gains. (b) Required backhaul in a 10 MHz system.

Figure 14.16 Relative performance gains and required backhaul for the compared downlink CoMP schemes and an ISD of 500 m (Macro 1).

- CS/CB, employing adaptive MU-MIMO,
- joint transmission, with a cooperation size of $N_c = 2$, and
- joint transmission, with a cooperation size of $N_c = 3$.

As in Section 14.3, Fig. 14.16(a) distinguishes between avg. and cell-edge throughput gains. While the CS/CB results are based on a 4x2 antenna setup, with 4 vertical antenna poles with $\lambda/2$-spacing at the BS side and cross-polarized antennas at the UE side, the joint transmission results are based on a 2x2 antenna setup with two vertical antenna poles with 4λ spacing at the BS side, and 2 parallel antenna poles with 0.5λ spacing at the UE side. Note, however, that all throughput gains are stated w.r.t. the performance of non-cooperative systems employing the same antennas, so that the relative gains should be comparable.

As stated before in Section 14.4.3, CS/CB appears to be only attractive in SU-MIMO mode (or in the context of heterogeneous systems, as suggested before), delivering average and cell-edge throughput gains of 5% and 11%, respectively. In general, coordination gain increases with the number of co-polarized BS antennas. In comparison to this, joint transmission among multiple cells can provide significantly larger gains w.r.t. the considered baseline of MET based $2x2$ non-cooperative MU-MIMO (see Section 6.3), as inter-cell interference is actually exploited as useful signal power. However, this of course comes at a certain price not only in terms of the backhaul capacity requirements visible in Fig. 14.16(b), but also in terms of various other issues identified in this book, namely

- the need for precise synchronization of the local oscillators (LOs) of cooperating BSs (see Chapter 8),
- a downlink pilot overhead increasing linearly in the cooperation size (see Section 9.1), and
- a potentially large extent of CSI-feedback (see Section 9.2).

14.4.6 Summary

In this section, two promising downlink CoMP concepts for LTE-A were assessed through system level simulation, namely coordinated beamforming / coordinated scheduling and joint transmission. The results revealed that CS/CB only delivers moderate average and cell-edge throughput improvements for particular antenna configurations and a fairly large number of scheduled users, as the avoidance of precoding vectors for the purpose of interference reduction always leads to reduced degrees of freedom in scheduling. However, the results in this section also suggest that CS/CB may be very valuable in the context of heterogenous networks. Joint transmission in clusters of 2 or 3 cooperating cells promises significantly larger spectral efficiency gains, but at the price of a strongly increased complexity, and many inherent challenges such as BS synchronization, pilot overhead, CSI feedback and CSI exchange between cooperating BSs.

Acknowledgements

Mark Doll and Thorsten Wild would like to thank Ralph Ballentin for providing simulation results as well as Christian Gerlach and Andreas Weber for many fruitful discussions on downlink CS/CB.

Part V

Outlook and Conclusions

15 Outlook

In this chapter, we give an outlook onto some further issues connected to CoMP, where major research is still ongoing. In Section 15.1, it is shown that CoMP can also be used for purposes other than spectral efficiency and fairness increase, namely for the localization of terminals in cases where GPS is not available. Section 15.2 then discusses the usage of CoMP in conjunction with relaying, showing that both technologies can complement each other well in providing efficient and ubiquitous mobile broadband access. Section 15.3 discusses how the planning and optimization of future networks will be affected due to the introduction of CoMP, while Section 15.4 addresses one essential yet unanswered aspect of CoMP: How does CoMP perform in terms of *energy* efficiency?

15.1 Using CoMP for Terminal Localization

Armin Dammann, Christian Mensing and Stephan Sand

Location information about user equipments (UEs) is a valuable characteristic and can be exploited in various layers of mobile communications systems. At the application layer, the availability of location information provides new market opportunities by enabling location based services. In lower system layers, performance improvements can be achieved using location information for optimizing handover or radio resource management for example. Nowadays, an increasing number of UEs are equipped with global positioning system (GPS) chipsets, which allows location determination at the terminals themselves. Satellite based localization reaches its limits for instance in urban canyon environments or even indoors. However, these environments typically show high user densities. To overcome these limitations, mobile communications systems themselves can be used for location determination. Similar principles like those used for satellite navigation can be applied, where the basic idea is to measure propagation delays of signals traveling from satellites to the receiver. Like for satellite based localization, the application of this principle requires a timely coordinated transmission of signal parts, which are capable for timing estimation. Compared to that, terrestrial communications systems provide higher receive power levels, which increases localization service coverage. However, mobile radio communication signals show more extensive multi-path and non line-of-sight (NLOS) propagation compared to satellite navigation. These propagation conditions typically decrease the accuracy of location determination.

15.1.1 Localization based on the Signal Propagation Delay

A straightforward method for UE localization is to use distance measures from the UE to several base stations (BSs), whose positions are well known. Considering M BSs, there is a system of equations whose solution yields the UE's position. In principal, distance measurements can be obtained from signal traveling time measurements if the signal traveling speed is known. For electromagnetic signals this is the speed of light, which is $c = 299\,792\,458$ m/s $\approx 3 \times 10^{-8}$ m/s in vacuum. In the atmosphere close to the Earth's surface, the speed of light is about 0.028% lower. Assuming a direct signal propagation path and a constant signal traveling speed c along that path, the distance between a transmitter and a receiver is

$$d = c(T_1 - T_0), \qquad (15.1)$$

where T_0 and T_1 are the moments of signal transmission and reception.

Propagation delay based localization methods require accurate signal timing estimation. Therefore, principles and algorithms are similar to time synchronization in communications receivers. The Cramér-Rao lower bound (CRLB) provides a fundamental performance limit of any estimation algorithm (e.g. [SG01]). The variance of any estimate $\hat{\tau}(\boldsymbol{y})$, based on N noisy measurement samples $\boldsymbol{y} = [y_0, \ldots, y_{N-1}]$ is lower-bounded by

$$\mathrm{VAR}\{\hat{\tau}(\boldsymbol{y})\} \geq \frac{1}{E\left\{\left(\frac{\mathrm{d}}{\mathrm{d}\tau} \ln \mathrm{p}(\boldsymbol{y}|\tau)\right)^2\right\}}, \qquad (15.2)$$

where $E\{f(\boldsymbol{y})\} = \int f(\boldsymbol{y})\mathrm{p}(\boldsymbol{y}|\tau)\,\mathrm{d}\boldsymbol{y}$ is the expectation operator over the noisy measurement vector $\boldsymbol{r}$ conditioned on a certain true delay τ. Let the noisy measurement samples (that could also be used, e.g., for synchronization or channel estimation) be $y_i = s(iT_\mathrm{S} - \tau) + n_i$, $i = 0, \ldots, N-1$, where $s(t)$ is a complex valued baseband transmit signal which appears at the receiver at a delay of τ, and T_S is the sampling time. Assuming complex additive white Gaussian noise $n_i \sim \mathcal{N}_\mathbb{C}(0, \sigma^2)$, the conditioned probability density function for $\boldsymbol{y}$ is

$$\mathrm{p}(\boldsymbol{y}|\tau) = (\pi\sigma^2)^{-N} \exp\left(-\frac{1}{\sigma^2}\sum_{i=0}^{N-1} |y_i - s(iT_\mathrm{S} - \tau)|^2\right). \qquad (15.3)$$

Using (15.2) and (15.3), we obtain the CRLB in time domain

$$\mathrm{VAR}\{\hat{\tau}(\boldsymbol{y})\} \geq \frac{\sigma^2}{2\sum_{i=0}^{N-1}\left|\frac{\mathrm{d}}{\mathrm{d}\tau}s(iT_\mathrm{S} - \tau)\right|^2}. \qquad (15.4)$$

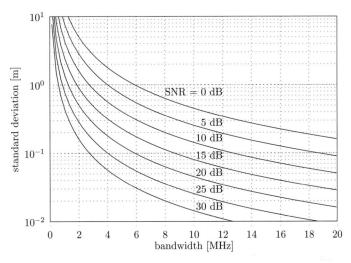

Figure 15.1 Cramér-Rao lower bound for TOA estimation.

In orthogonal frequency division multiplex (OFDM) systems, the complex baseband transmit signal is usually designed in frequency domain by complex valued symbols S_ℓ, modulated on sub-carriers with a sub-carrier spacing of $f_{\rm SC} = \frac{1}{NT_{\rm S}}$. The delayed and sampled version of this signal is obtained by an inverse discrete Fourier transform

$$s(iT_{\rm S} - \tau) = \frac{1}{\sqrt{N}} \sum_{\ell=-\lfloor \frac{N}{2} \rfloor+1}^{\lfloor \frac{N}{2} \rfloor} S_\ell \exp\left(j\, 2\pi\, \ell f_{\rm SC}(iT_{\rm S} - \tau)\right). \tag{15.5}$$

Using (15.5) in (15.4), we obtain the CRLB in frequency domain

$$\text{VAR}\{\hat{\tau}(\boldsymbol{y})\} \geq \frac{\sigma^2}{8\pi^2 f_{\rm SC}^2 \sum_{\ell=-\lfloor \frac{N}{2} \rfloor+1}^{\lfloor \frac{N}{2} \rfloor} \ell^2 |S_\ell|^2}. \tag{15.6}$$

Fig. 15.1 shows the CRLB for distance estimation in form of the standard deviation stddev $= c\sqrt{\text{VAR}\{\hat{\tau}(\boldsymbol{y})\}} = 3 \times 10^8\,\text{m/s} \times \sqrt{\text{VAR}\{\hat{\tau}(\boldsymbol{y})\}}$ versus the signal bandwidth $B = N f_{\rm SC} = N \times 15$ kHz, where N is the number of used sub-carriers for different receiver sub-carrier signal-to-noise ratios (SNRs) SNR $= \frac{E\{|S|^2\}}{\sigma^2}$.

Increasing the SNR by 20 dB results in a ranging performance improvement of a factor of 10. It is also worth to notice that a sufficient signal bandwidth is essential for an adequate distance estimation performance. Note that in LTE the synchronization channels occupy a bandwidth of 945 kHz.

Principles

If distance measurements to several BSs are available in form of signal arriving time estimates, we can apply the two principles time of arrival (TOA) and time delay of arrival (TDOA) in order to provide the UE's position. These principles are applicable in both the downlink and the uplink. In the downlink, the signal arrival times are measured at the UE, whereas in the uplink, signal arrival times are measured at different BSs. Subsequently, we will introduce the downlink variants of TOA and TDOA.

Time of Arrival (TOA)
Under ideal conditions, it is straightforward to directly use the TOA of signals received from different BSs. For each involved BS, we obtain an equation

$$\sqrt{(x-x_i)^2 + (y-y_i)^2} = c\,(T_i - T_0) = d_i, \quad i = 1, \ldots, M, \qquad (15.7)$$

where (x_i, y_i) is the position of BS i out of M BSs. T_i is the time of arrival of the signal coming from BS i and T_0 is the time of signal transmission at the BS. Receiving signals from M BSs, we obtain a system of M non-linear equations. Each of them is describing a circle of possible UE locations (x, y) with radius d_i. In real systems, the time of signal transmission is not known and must be estimated as well. Therefore, we need at least three equations or BSs in order to solve for the unknowns x, y and T_0. The generalization to three dimensions x, y and z is straightforward. The system of equation is usually solved by iterative algorithms and using the minimum mean squared error criterion. Since the equation system describes a set of circles (2D) or spheres (3D), this method is also called *circular localization*.

Time Difference of Arrival (TDOA)
We may eliminate the unknown time of signal transmission by subtraction of one of the M equations from the remaining ones. Without loss of generality we subtract equation $i = 1$ from the other ones and get

$$\sqrt{(x-x_i)^2 + (y-y_i)^2} - \sqrt{(x-x_1)^2 + (y-y_1)^2} = c\,(T_i - T_1) \qquad (15.8)$$

for $i = 2, \ldots, M$. For simplification, we consider one of the remaining $M - 1$ equations, say $i = 2$. We choose the coordinate system such that the two BSs are located symmetrically on the x-axis at positions $(\pm a, 0)$. From (15.8) we get

$$\sqrt{(x-a)^2 + y^2} - \sqrt{(x+a)^2 + y^2} = c\,(T_2 - T_1). \qquad (15.9)$$

By squaring this equation twice we arrive at

$$\frac{x^2}{\left(\frac{c\,(T_2-T_1)}{2}\right)^2} - \frac{y^2}{a^2 - \left(\frac{c\,(T_2-T_1)}{2}\right)^2} = 1. \qquad (15.10)$$

Based on the measurement of the TDOA $T_2 - T_1$, Equation (15.10) describes hyperbolas with foci in the locations $(\pm a, 0)$ of the involved BSs. Therefore, this principle is also called *hyperbolic localization*. In order to obtain the UE's position at least two (2D) respectively three (3D) independent hyperbolic equations or time difference measurements are required.

Challenges

In the sequel, the main challenges for localization with terrestrial communications systems will be discussed. As 2D terminal localization based on timing information requires interaction with at least three sites, there is a strong relation to CoMP aspects.

Non-Line-of-Sight and Multi-path Propagation

When the direct path is blocked or reflections due to multi-path propagation are present, the timing measurements are affected by a bias. In worst-case urban scenarios, this bias can be in the order of several hundred meters. Therefore, several approaches to mitigate these effects have been developed and analyzed for terminal localization systems (e.g., [STK05] and references therein). The multi-path effects are usually detected and compensated on the physical layer level, i.e., during the timing estimation by using, e.g., advanced channel estimation techniques. Hence, the individual links to the different sites are handled separately. The additional path length introduced by NLOS signal propagation is usually compensated on the localization level, i.e., when the timing information from several sites is used to determine the location.

Interference

Due to spectral efficiency reasons, it is of high interest to operate future communication systems like LTE with a frequency re-use of one. From a communications point of view this results in a problematic inter-cell interference situation at the cell-edge, where several sites are received with a similar power level. Here, CoMP concepts can be chosen to improve the performance. From a localization point of view, at every location in the network, three sites have to be detected. This is a problem in the cell-center, where a strong signal from the serving BS makes a reception of other sites difficult. The pilot sequences used for synchronization and channel estimation usually do not have beneficial auto- and cross-correlation properties for terminal localization as sequences for dedicated systems like GPS have. Therefore, techniques like interference cancelation [MSDU09] or interference avoidance are necessary to allow terminal localization in the whole network.

Network Synchronization

For terminal localization with timing measurements it is required that the network is synchronized at least in time. Compared to classical communications systems with a synchronization on frame or symbol level, the requirements for terminal localization are much higher. For instance, in a 20 MHz LTE system a deviation of one OFDM sample results in a systematic error of 9.76 m for the timing measurements. If the network is not synchronized (e.g., by design), so-called location measurement units located at known positions in the network can estimate the offsets and provide this information to the UE (cf. [3GP09d]). Another option for future networks might be the inclusion of GPS-equipped UEs to perform this task.

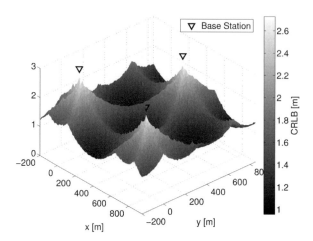

Figure 15.2 CRLB for TDOA localization, WINNER typical urban channel model.

Network Topology

Two factors account for the overall accuracy of a classical terminal localization system. The first is the accuracy of the single-link measurements (here: timing measurements). The second is the geometrical configuration of the problem. From a geometrical point of view, the UE lies on intersection of circles (TOA) or hyperbolas (TDOA) spanned by the locations of the involved BSs. For most accurate localization, the circles or hyperbolas have to intersect perpendicularly. If they intersect under a small angle, noisy timing measurements have a much higher impact on the overall localization accuracy. Hence, assuming a constant timing measurement accuracy, the performance for terminal localization is better at the cell-edge compared to the situation with a terminal close to a BS. The effect is visualized in terms of the CRLB for the localization accuracy in Fig. 15.2 for the TDOA approach employing the WINNER C2 typical urban model [IST07a]. As expected, the achievable accuracy increases when we move from the cell-edge to a BS. The average CRLB on the standard deviation is 2.17 m.

15.1.2 Further Localization Methods

Round-trip Time of Arrival

Previously, we have introduced the circular localization method *Time of Arrival*, where the distance to several BSs is measured. Solving a system of non-linear equations, which describe circles around the involved BSs, yields the position of the UE. A further circular localization method is the *round-trip time of arrival (RTOA)*. The distance between two transceivers is measured using two signal transmissions. Transceiver A transmits a signal at time T_0. This signal is received at transceiver B. After a defined processing time T_{proc}, B transmits another signal back to A, where the signal is received at time instance T_1. Assuming that the signal traveling time from A to B is equal to that from B to A, the overall delay, i.e., the round-trip delay $T_1 - T_0$ is two times the signal propagation delay from A to B plus the processing time. The distance between A and B is

$$d_{\mathrm{AB}} = \frac{T_1 - T_0 - T_{\mathrm{proc}}}{2c}. \tag{15.11}$$

Equivalently to TOA, these distances are used to setup and solve a non-linear equation system. Synchronization between the time bases of the two devices A and B is not required. Only the processing time has to be known to the device which is initiating the RTOA measurement.

Angle of Arrival

If devices are equipped with directive antennas, we are able to measure the direction of incoming signals. This leads to the *angle of arrival (AoA)* principle. In uplink direction, several BSs estimate the angle of signal arrival from a UE. Therefore, we get one equation for each BS, describing a straight line of possible UE positions. Solving the resulting linear equation system then yields the UE's position. In the uplink, it is reasonable to assume fixed BSs, where the orientation of the directive antennas are known. So two BSs are sufficient in order to determine the position of the UE. For the application of AoA in the downlink, the UE needs to resolve the angle of signal arrivals from different BSs. Hence, downlink AoA also requires an estimation of the UE's orientation. This makes AoA measurements to further BSs necessary. Major problems of AoA are multipath, non line-of-sight propagation, and — especially in the downlink — a huge angular spread. Signal reflections can completely change the signal propagation direction and, therefore, the AoA. This can lead to significant mis-estimations of the UE's position.

Fingerprinting

For localization we need to measure location dependent signal parameters. Such signal parameters characterize the UE's position. These characteristic parameters are called the fingerprint of a certain position and the method of mapping these fingerprint to a location *fingerprinting*. It is obvious that the localization methods which we have introduced previously also use location dependent signal parameters in order to solve the localization problem. For those methods the relation of the fingerprints to a position can be described mathematically with manageable complexity, e.g., by the circle equations in case of TOA. Therefore, fingerprinting is a more general description of all those methods. If the relationship of the fingerprint parameters and the position is too complex for a mathematical description, data bases can be used to provide these relation. An example for a highly complex fingerprint is the channel power delay profile. Some well-known methods, closely related to power delay profile based fingerprinting methods are:

- The **received signal strength (RSS)**. The received signal power, i.e., the integral over the power delay profile, is another characteristic for a certain position. In free space, the received signal power decreases at least with the inverse square of the distance between transmitter and receiver. The circular relation for this simple case can be described mathematically similar to TOA. However, in mobile radio systems, multipath propagation, shadowing, etc. leads to a rather complex map, which typically requires a data base.
- **Cell ID**. A further simplification of the RSS principle is to define whether a signal from a specific BS is received or not. This is similar to a coverage map for cellular wireless communications systems. If a mobile receives signals from different BSs, the intersection of the different coverage areas provide possible locations of the UE. The UE simply has to detect the BS, i.e., the cell ID, from which the signal has been transmitted.

15.1.3 Localization in B3G Standards

As of today, there exist several 3G or IMT-2000 standards [ITU08a], such as the universal mobile telecommunications standard (UMTS) of the 3GPP using code division multiple access (CDMA), CDMA2000 as direct evolution of Interim Standard 95 (IS-95), Enhanced Data rates for global system for mobile communications (GSM) Evolution (EDGE), and Digital Enhanced Cordless Telecommunications (DECT). Apart from DECT, all other standards support localization of the UE, e.g., through TDOA measurements in the downlink or uplink [3GP10f, 3GP04]. Further, [ITU08a] also considers Worldwide Interoperability for Microwave Access (WiMAX) as a 3G standard, which supports UE localization through downlink or uplink TDOA measurements [IEE10a]. Currently, UMTS is evolved in the 3GPP LTE and LTE-A standards, and WiMAX in the IEEE 802.16m standard. Both LTE and WiMAX use OFDM as basic transmission scheme. According to [ITU10a], LTE-A and IEEE 802.16m are currently the only candidates for IMT-Advanced, i.e., 4G. Both LTE and WiMAX consider CoMP [ITU10a]. Further, LTE has standardized a specific positioning reference signal (PRS) [3GP07b] to achieve the FCC E911 requirements [Fed99]. In the sequel, we will focus on UE-centric positioning with LTE.

LTE has been standardized in December 2008 [3GP10a]. The major change between LTE and UMTS is that orthogonal frequency division multiple access (OFDMA) is used instead of CDMA as basic modulation and multiple-access scheme. In the beginning of 2010, 3GPP has updated LTE with a dedicated PRS for downlink transmission from the BS to the UE [3GP07b] for improved localization of the UE. In combination with an idle period downlink (IPDL) [3GP10c], this enhanced IPDL (E-IPDL) enables a successful position fix in at least 86% of the cases even for a bad urban channel [3GP09a]. Further, the FCC E-911 requirements for handset-based positioning can be achieved with E-IPDL in an asynchronous network for a circular error probability (CEP) of 67%, i.e.,

$CEP67 \leq 50$ m, and in a synchronous network both for $CEP67 \leq 50$ m and $CEP95 \leq 150$ m [3GP09a].

LTE Frame Structure

The LTE downlink frame in Fig. 15.3 has a duration of 10 ms and is divided in 10 sub-frames with 1 ms duration. Fig. 15.3(b) displays an example LTE frame for the central 73 sub-carriers, which fill the minimum supported bandwidth of 1.4 MHz. Within this bandwidth, the following signals are transmitted:

- Primary synchronization signal (PSS, light gray) and secondary synchronization signal (SSS, very dark gray) for rough synchronization and cell detection,
- PRS (grey) for fine synchronization and localization,
- Cell-specific reference signals (common reference signals (CRSs), black) for channel estimation and fine synchronization,
- Broadcast and control channels (dark grey) for configuring and controlling data communication,
- Data channel (very light grey) for voice and data communication.

In the following, we present the PSS, SSS, PRS, and CRSs in more detail, because these are the synchronization signals known to the receiver, which can be exploited for localization.

LTE Synchronization Signals

Some of the LTE downlink reference signals are different for the individual BSs and used at the UE for cell search, time- and frequency synchronization, or channel estimation. For LTE, there exist three different kinds of reference signals [3GP07b], which are relevant for localization:

1. Synchronization signals are transmitted within the central 63 sub-carriers of an LTE frame (cf. 15.3(b)), thus occupying only a bandwidth of less than 945 kHz. Hence, these signals have a low complexity in initial detection, but offer only limited positioning accuracy due to the low bandwidth.
    a. The PSSs are generated from three different frequency-domain Zadoff-Chu sequences and mapped to the seventh OFDM symbol in the first and sixth sub-frame of an LTE frame as in 15.3(a) (see [WHE08]). Each of the three PSSs corresponds to one physical-layer cell-identity group. The PSSs can be used to estimate the integer and fractional frequency offset in the receiver ([BP07]). Besides that, as there exist only three sequences, a bank of matched filters can be used for precise timing estimation [TZGO07]. Note that this approach for timing estimation is not robust against frequency offsets.
    b. The SSSs are generated from 2×504 different sequences and mapped to the sixth OFDM symbol in the first and sixth subframe of an LTE frame as in 15.3(a) (see [WHE08]). The SSSs are centrally symmetric signals in the time domain. Thus it is possible to detect the start of the SSS with a reverse

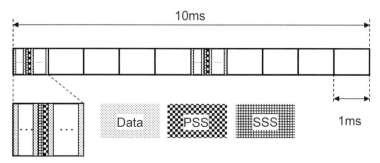

(a) Structure: Frame (10ms), subframes (10 with 1ms each), data (information symbols) and synchronization signals (PSS, SSS).

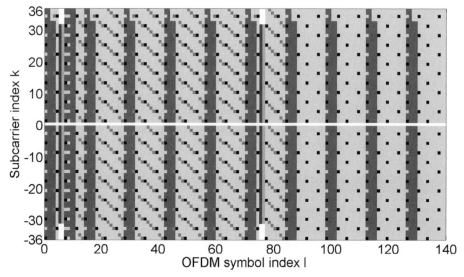

(b) Detail of 72 central sub-carriers: PSS (light gray), SSS (very dark gray), PRS (grey), CRS (black), broadcast and control channels (dark grey), data (very light gray).

Figure 15.3 LTE downlink frame.

differential correlation [BP07]. Although this approach may be less accurate than the matched filter correlator bank for the PSS, it has the benefit of being robust against frequency offsets. If we have corrected the frequency offset with the PSS and detected with the PSS the physical-layer cell identity group, we can search for the physical-layer cell identity by correlating the received SSS with the 168 physical-layer cell-identities belonging to one physical-layer cell-identity group. As the sequences between the first and sixth subframe differ, the SSS can be used to detect the frame start.

2. The CRSs are pilots scattered in time- and frequency over the complete LTE frame (cf. 15.3(b)). Thus, the CRSs occupy a bandwidth of up to 20 MHz with a frequency reuse of 1/6. In time direction, CRSs are contained in every third or fourth OFDM symbol. The CRSs can be used for synchronization or

channel estimation, and have a cell-specific frequency shift modulo 6. As there are 504 cell identities, 84 CRSs use the same time- and frequency resources. Hence, the CRSs should have good cross-correlation properties to separate the CRSs from different BSs (see the discussion on CSI reference signals (CSI RSs) in Section 9.1.2). Numerical evaluations show that the cross-correlation gain ranges between 5.9 dB and 17.15 dB. Hence, the correlation properties of the CRSs are not sufficient to overcome the near-far effect in the vicinity of the serving BS, but only at the cell-edge. Therefore, techniques like IPDL [3GP09a] or interference cancelation [MSDU09] are needed.

3. A PRS has been newly introduced for the purpose of positioning [3GP07b]. The PRS is transmitted in the downlink to measure observed TDOAs (OTDOAs). Positioning performance improves with an increasing number of reliable TOA measurements. In dense urban environments, typically, mobiles are able to hear only one or two BSs because of full reuse. To avoid the hearability problem (arising from the near-far effect) of a reuse-1 system, the UTRA standard provides IPDL [3GP10c], during which transmission of all channels from a BS is ceased. In these periods, the UE is able to receive the pilot signal of the neighbor BSs even if the serving BS signal on the same frequency is very strong. However, in LTE there is no continuous transmission of reference signals or control channels. Hence, the PRS has been defined to be transmitted in the idle periods. To enable simple receiver processing and to further enhance hearability, a frequency reuse of 1/6 is defined for the PRS similar to the CRSs. The PRS is designed such that it can be used jointly with the CRSs for an acquisition of $N_{\rm PRS}$ consecutive downlink subframes with $N_{\rm PRS} = 1, 2, 4, 6$. The PRS is punctured by PSS, SSS, broadcast and control channels (cf. 15.3(b)). The periodicity of the PRS can be configured to 160 ms, 320 ms, 640 ms, and 1.28 s. Thus, in combination with durations between 1 ms and 6 ms, the PRS has a duty cycle between 0.08% and 3.75%. Note that the PRS has the same correlation properties as the CRSs, as it is generated in the same way.

Example

Finally, we assess the terminal localization accuracy in an LTE system with a 20 MHz setup and frequency reuse of one. We employ the SSS for TDOA estimation with the three strongest BSs in a cellular network with an inter-site distance of 750 m. For multi-path propagation, shadow fading and long-term fading, we used the WINNER C2 typical urban model [IST07a]. The NLOS error was extracted from ray-tracing simulations for an urban scenario located in Munich (cf. [SMM+08]). Fig. 15.4 shows the localization accuracy in terms of cumulative distribution functions (CDFs). The CDF is defined as the probability that the absolute 2D position error is below a certain value, where it was averaged over several noise realizations and UE locations in the network. We observe that for the conventional (state-of-the-art) approach the localization error is higher than 300 m in 90% of the cases, as strong inter-cell interference limits

Outlook

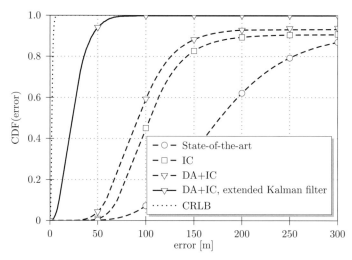

Figure 15.4 Localization accuracy for an LTE system, WINNER typical urban channel model.

the performance especially in the cell core. If we apply interference cancelation as discussed in [MSDU09], the interference from the serving BS is mitigated before timing estimation with the other BSs is performed. With that approach, the 90%-error can be reduced to 200 m. When we further apply data-aided positioning [MSDU09], where estimated data symbols are used as pilot symbols in another iteration, we can achieve around 160 m. If we include the mobility of the UE, we can additionally adapt tracking algorithms. Here, an extended Kalman filter is exploited assuming pedestrian users. The resulting 90%-error is 45 m. There still is a gap to the CRLB, which could be closed by further application of multi-path and NLOS mitigation algorithms. Note that with the proposed scheme the US FCC E-911 requirements as discussed before can be met.

15.1.4 Summary

In this section, we have briefly introduced terminal localization, in particular time of arrival (TOA) and time delay of arrival (TDOA) methods which are based on signal propagation delay estimation. For a first assessment of the achievable terminal localization accuracy the Cramér-Rao lower bound for OFDM systems has been presented. We have discussed challenges of terminal localization which are closely related to CoMP. Exemplarily, we have introduced the 3GPP-LTE frame structure and synchronization signals which can be easily used for TOA respectively TDOA based terminal localization. Simulation results using the WINNER C2 typical urban channel model have shown that the US FCC E-911 requirements for terminal localization accuracy can be met with interference cancelation and Kalman filtering methods.

15.2 Relay-Assisted Mobile Communication using CoMP

Peter Rost

In this section[1], relaying is discussed as a candidate technology to efficiently exploit the available resources, to flexibly organize wireless networks, and to increase the system capacity of a mobile communication network, which applies CoMP transmission and detection strategies. A joint approach is discussed, which efficiently integrates additional relay nodes and exploits the benefits of both cooperative multi-point and relaying. In order to provide a comprehensive analysis and discussion, a relay-assisted system is compared to a conventional system and CoMP in terms of its cost efficiency and energy efficiency. A main result is that relaying significantly improves the system's efficiency in the uplink while CoMP is the preferable choice in the downlink if the system's initial and operational expenditures or the system's energy consumption are normalized.

15.2.1 Introduction

The previous chapters have shed some light on the benefits and drawbacks of CoMP in the context of cellular networks. An alternative or rather complementary technique is relaying, which will be part of future mobile communication networks such as IEEE 802.16-2009 and 3GPP LTE Release 10.

As previously emphasized, CoMP offers a flexible way to exploit or cancel interference in both uplink and downlink. This implies a more homogeneous spatial signal-to-interference-and-noise ratio (SINR) distribution as a major part of the interference is avoided or subtracted. However, CoMP is not able to counteract the non-linearity of pathloss and the resulting decline of receive power with increasing distance. Relaying alleviates this effect by deploying additional intermediate nodes in order to improve the spatial power distribution within the network. These relay nodes receive the source signal, process it (analog and digital), and then forward the source information to the destination [CG79]. In this chapter, we focus on digital and single-path relaying. The former refers to decode-and-forward (DF), where the relay node decodes the complete source message, re-encodes it, and then retransmits it. Single-path relaying implies that the destination node does not exploit the direct connection between source and destination but only the link between relay node and destination.

The introduction of additional relay nodes in a mobile communication system implies the necessity for more advanced interference mitigation schemes and resource coordination algorithms. In addition, as the relay nodes considered in

---

[1] Part of this work has been performed in the framework of the IST project IST-4-027756 WINNER II, which has been partly funded by the European Union, and the Celtic project CP5-026 WINNER+. The author would like to acknowledge the contributions of his colleagues in WINNER II and WINNER+, although the views expressed are those of the author and do not necessarily represent the project.

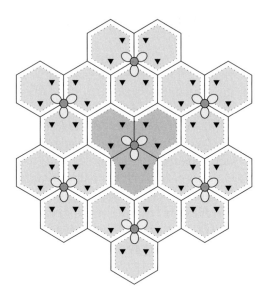

Figure 15.5 Macro-cellular deployment scenario used for numerical evaluation in this section. Ellipses indicate the main lobe direction of the individual base stations, and triangles indicate the relay position within each cell.

this chapter are only operating in half-duplex mode, the system must outweigh the loss in spectral efficiency due to the requirement that every bit must be transmitted twice – once to the relay node and once to the destination. This chapter discusses a system approach, which is partly able to counteract these effects. Through numerical analysis, we further elaborate on the question whether relaying is a beneficial extension for CoMP-based mobile communication networks.

This section analyzes the cost/benefit, energy/benefit, and computation/transmission power trade-off in the context of a macro-cellular environment using the assumptions defined by the European research project WINNER [WIN09]. In the first subsection, we discuss the individual system approaches, which are compared in this section. These approaches are then applied to the scenario defined in Subsection 15.2.2 and evaluated in Subsection 15.2.5, regarding the three different trade-offs mentioned before. The section is concluded in Subsection 15.2.9.

15.2.2 Reference Scenario

In this section, we present results based on the wide-area reference scenario defined by the European research project WINNER [WIN09, IST06][2], which corresponds to a hexagonal macro-cellular deployment with three base stations

---

[2] All referred WINNER documents are publicly available on http://www.ist-winner.org.

(BSs) serving three adjacent sectors at each site (as illustrated in Fig. 15.5). In addition, each sector is complemented by two fixed relay nodes, which are placed such that an almost uniform radio access point (RAP) distribution is achieved, i. e. if $d_{\text{isd}} = 1000$ m is the inter-site distance (ISD), the relay nodes are placed at distance $\frac{1}{3} d_{\text{isd}}$.

More specifically, we apply orthogonal frequency division multiplex (OFDM) [WE71, Cim85] at a carrier frequency $f_c = 3.95$ GHz with bandwidth $B_w = 100$ MHz and $N_c = 2048$ sub-carriers, which is the baseline assumption of the WINNER project. All further air parameters and channel models are described in detail in [IST06] and [RFL10]. Our numerical results are based on the assumption of Gaussian alphabets, perfect rate adaptation, infinite block lengths, perfect channel state information (CSI), full buffers at each transmitter, a perfect system synchronization, and that all users require the same quality of service (QoS). Our results consider an in-band feeder link between BSs and relay nodes (RNs), which occupies the same spectral resources as the links between RNs and user equipments (UEs). Hence, relay buffers might be empty as a consequence of an unexpected high throughput between a RN and its assigned users.

15.2.3 System and Protocol Description

The previous chapters discussed CoMP algorithms, which involve multiple BSs jointly processing a transmit or receive signal. In this chapter, we apply an extended model, where each communication pair is additionally supported by a RN. We refer to this model as the relay-assisted interference and broadcast channel, which has been investigated among others in [RFL09, SSPS10]. Although a variety of approaches exists for this channel, we choose, based on link level results in [RFL09], one particular approach, which is further explained in the following paragraphs.

This relaying strategy is compared with a CoMP approach based in the downlink on the transmit Wiener filter derived in [JJT+09] and also used in practice in Sections 13.3 and 13.4. All following results are obtained on the basis of at maximum three cooperating BSs and under the assumption of an unlimited backhaul link connecting individual BSs. In the uplink, cooperating BSs apply the QR decomposition $\mathbf{H} = \mathbf{QR}$, which decomposes the channel matrix $\mathbf{H}$ into a unitary matrix $\mathbf{Q}$ and an upper triangular matrix $\mathbf{R}$. After multiplying the received signals with $\mathbf{Q}^H$, the effective channel is given by $\mathbf{R}$, and therefore user i is only interfered by users $j > i$, which can be removed using successive interference cancelation (SIC).

This section evaluates performance results of a relaying protocol, which applies CoMP between BSs and RNs, and applies Han-Kobayashi (HK) coding [HK81] between RNs and UEs. HK coding is illustrated in Fig. 15.6 where each transmitter has two message parts $W_{(1,1)}$ and $W_{(1,\mathcal{D})}$, which are mapped to the channel input signals $X_{(1,1)}$ and $X_{(1,\mathcal{D})}$ with power assignment $P_{(1,1)}$ and $P_{(1,\mathcal{D})}$ (sim-

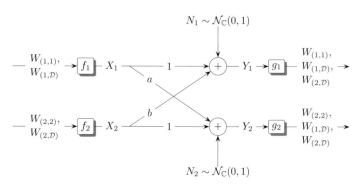

Figure 15.6 Message encoding and decoding of Han-Kobayashi coding. Encoders are denoted by f_1 and f_2, decoders are denoted by g_1 and g_2.

ilarly the second transmitter uses $X_{(2,2)}$ and $X_{(2,\mathcal{D})}$). The first part $W_{(1,1)}$ is called the private part and is only decoded by the first receiver while the second part $W_{(1,\mathcal{D})}$, the common part, is decoded by both receivers as it has the purpose to reduce the interference for the second receiver. Let $P_{(i,i)} + P_{(i,\mathcal{D})} = P_i$, then the channel input at terminal i is given as

$$X_i = \sqrt{P_{(i,i)}} X_{(i,i)} + \sqrt{P_{(i,\mathcal{D})}} X_{(i,\mathcal{D})}.$$

HK coding using all possible power assignments is practically not applicable as it requires a complex and non-convex optimization over all power assignments and a complex joint decoder at the receiver. Etkin et al. simplified the HK coding approach in [ETW06, ETW08] and derived a strategy, which has been proven to be within 1 bits per channel use (bpcu) of channel capacity while being much simpler. This strategy aligns interference power with the noise power at the non-intended receiver. Let the channel between RNs and UEs be as illustrated in Fig. 15.6, i.e.

$$\mathbf{H}_{\mathrm{RN}} = \begin{pmatrix} 1 & b \\ a & 1 \end{pmatrix}, \quad a, b \in \mathbb{C}, \tag{15.12}$$

then the individual power levels are chosen such that

$$a^2 \cdot P_{(1,1)} \leq \sigma_{\mathrm{nn},2}^2 \text{ and } b^2 \cdot P_{(2,2)} \leq \sigma_{\mathrm{nn},1}^2 \tag{15.13}$$

holds, where $\sigma_{\mathrm{nn},i}^2$ denotes the receiver noise covariance at terminal i. Both common messages are jointly decoded, subtracted from the received signal, and then the receiver's own private message is decoded.

In order to reduce the signaling overhead of this approach, we consider in this chapter two simplifications. At first, we apply the approach using the large-scale parameters (pathloss) instead of instantaneous channel states. Secondly, we apply the power assignment

$$P_{(1,1)} = \begin{cases} P_1 & \frac{\sigma_{nn,2}^2}{a^2} \geq \frac{P_1}{2} \\ 0 & \text{otherwise,} \end{cases} \tag{15.14}$$

which implies that we only use either common or private messages [Ros09], which reduces the signaling overhead for each communication pair to 1 bit.

Our numerical analysis additionally compares the previous approaches with a mixed strategy, where a UE is served either by its assigned BS or RN depending on the large-scale signal quality towards both RAPs. In the main lobe direction of the BS, it is preferable to directly serve UEs as the experienced signal strength is likely to exceed the SINR towards the RN. By contrast, UEs close to the cell-edge benefit from the shorter distance towards the RN and are almost exclusively connected to RNs.

The reference case of our numerical analysis is a conventional system without relay nodes and three-fold sectorization, where parts of the available spectrum are reused. More specifically, edge-band resources with a reuse factor of 1/3 are assigned to users with high inter-cell interference, and center-band resources (10% of the overall resources) with reuse 1 are assigned to users with SINR $\geq$ 10 dB.

15.2.4 Trade-Offs in Relay Networks

Among the difficult tasks of comparing relaying-based systems and conventionally deployed systems is the mean of normalization, i.e., which basic property of both systems is kept constant in order to allow for a fair comparison. An intuitive answer is to normalize the required costs such that both systems require the same amount of money to be deployed. This approach is similar to the analysis in [WMS08], where the performance of the system has been kept constant and then the required expenditures of both systems have been compared. Either way, the normalization using the costs of a system allows us to qualify the cost/benefit trade-off, which describes the offered performance gains in a relay-based system depending on its required expenditures.

The second trade-off that we focus on is the energy/benefit trade-off, where we use the overall transmission energy consumption as the means of normalization such that both relay-based system and conventional system consume the same amount of energy. This analysis gives us an insight on how much the energy distribution between base stations and relay nodes affects the performance. Finally, we briefly discuss the computation/transmission energy trade-off, where we assume that both systems use imperfect algorithms, which induce a constant signal-to-noise ratio (SNR) shift. We can then discuss how sensitive the employed algorithms are w.r.t. this SNR shift, and how much additional transmission power must be spent to outweigh the performance loss.

15.2.5 Numerical Evaluation of CoMP and Relaying

In this section, we analyze different trade-offs in the previously introduced relay-based deployment. More specifically, we discuss results for the following four protocols:

1. Multi-cell multiple-input multiple-output (MIMO): Up to three cells cooperatively serve a UE.
2. Conventional: No cooperation.
3. 2 relays, relay-only: All UEs are served using relay nodes.
4. 2 relays, mixed: Depending on the pathloss towards the closest RN and BS, a UE is either cooperatively served by up to three base stations or it is served by a relay node.

15.2.6 Cost/Benefit Trade-Off

The first part of our analysis regards the cost/benefit trade-off of a relay-based deployment, which we measure by normalizing the system-wide deployment costs of both conventional deployments and relay-based deployments. More specifically, let the cost-ratio of a relay node and base station be α_{RN}, then the inter-site distance is increased such that for a particular α_{RN} the system-wide expenditures are the same [RFL10].

Fig. 15.7 shows both the average throughput $\bar{\theta}$ and 5% quantile throughput $\theta_{5\%}$ in uplink and downlink using the aforementioned cost-normalization. In addition, the figure shows the results for unlimited backhaul between multiple sites (dashed lines) as well as for limited cooperation where only base stations at the same site cooperate and relay nodes only use two states for the HK coding. Both uplink and downlink draw different pictures; while in the downlink the CoMP strategy dominates both relay protocols, this reverses in the uplink. In the case of unlimited backhaul, CoMP downlink transmission outperforms relaying for almost all α_{RN}. However, in the case of limited cooperation, the performance of CoMP drops such that for $\alpha_{RN} \leq 0.1$ the worst-user (cell-edge user) performance of relaying is higher than for CoMP transmission. This indicates that in particular the worst users benefit from inter-site cooperation. However, the relaying performance is unaffected by the limitation of the cooperation, which indicates that neither the links between BSs and RNs nor between RNs and UEs are interference-limited. Rather, the higher distance due to the increased costs as well as the half-duplex operation at the relays limit the maximum rates.

In the uplink, the mixed relaying strategy significantly outperforms CoMP in terms of average and worst-user throughput. However, if only relays are used to serve UEs, this performance advantage cannot be kept up but the performance drops due to the worse performance of users in the main lobe direction of the base stations, which should preferably be served by BSs. Hence, this result already

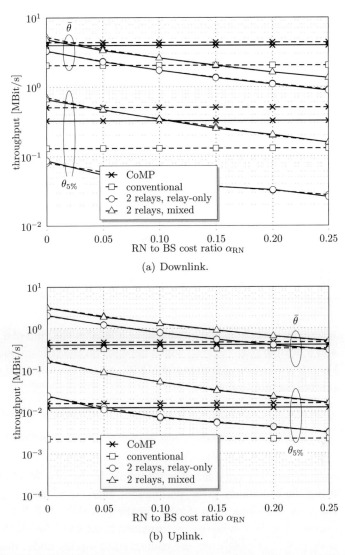

Figure 15.7 Cost/benefit trade-off with normalized deployment costs. Dashed lines indicate full cooperation and solid lines indicate limited cooperation.

suggests that relaying should preferably be employed in the uplink and must be complemented by CoMP for users in the main-lobe direction of the cells.

15.2.7 Energy/Benefit Trade-Off

Using the same methodology, we now analyze the energy/benefit trade-off, where we flexibly assign different transmission power to RNs and BSs and increase the ISD in order to normalize the energy spent by both the CoMP-based and relay-based system. More specifically, we employ the model introduced in [ARFB10],

Outlook

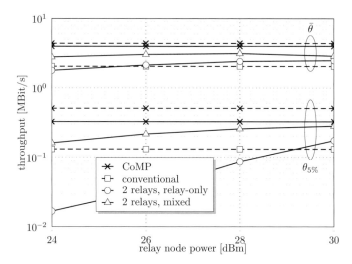

Figure 15.8 Energy/benefit trade-off in the downlink with normalized energy consumption. Dashed lines indicate full cooperation and solid lines indicate limited cooperation.

where the overall energy consumed at a RAP is the sum of a constant part P_{const} and a flexible part $\varepsilon_{\text{Tx}} P_{\text{Tx}}$, which depends on the transmission power P_{Tx}. Using the reference values given in [RFMF10], we use the following values for the overall power consumption at one site

$$\varepsilon_{\text{Tx,site}} = 44 \qquad P_{\text{const,site}} = 59 \text{ dBm}$$
$$\varepsilon_{\text{Tx,RN}} = 4 \qquad P_{\text{const,RN}} = 47 \text{ dBm}.$$

Fig. 15.8 shows again the average and worst-user throughput for the energy/benefit trade-off in the downlink. In the case of an energy-normalization using the previously given parameters, relaying is not anymore able to achieve the same performance as CoMP in the downlink, but it is significantly outperformed. The best relaying performance is achieved for the highest value of relay node power as most users are connected to a relay node. However, the decreased transmission power at the base station implies a significant performance loss for those users in the main lobe direction of the base station. Nonetheless, these results strongly depend on the chosen parameters and relaying will perform much better if the constant energy part at a relay node is reduced. In addition, the uplink performance is not affected, which again emphasizes the ability of CoMP in the downlink to improve data rates and to outperform relaying, while relaying appears to be the preferable choice in the uplink.

15.2.8 Computation/Transmission Power Trade-Off

We finally observe the computation/transmission power trade-off, which qualifies how a decreased computation power (reflected by a worse achievable rate) can

15.2 Relay-Assisted Mobile Communication using CoMP

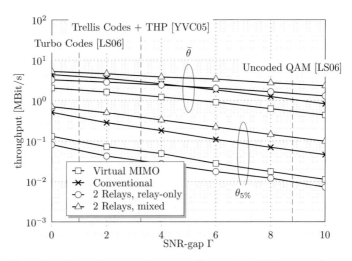

Figure 15.9 Downlink throughput performance for varying SNR-gap values.

be outweighed by an increased transmission power. A similar analysis has been conducted in [RF10], where the computation complexity and transmission power has been related to the system-wide spent energy in a multi-hop relay network. In this analysis, we apply the SNR-gap approximation introduced in [LS06, pp. 66], which multiplies the achieved SNR with a constant algorithm-dependent SNR shift. Hence, the additive white Gaussian noise (AWGN) capacity expression [Sha48, Sha49] is changed to

$$C(\Gamma) = \log_2\left(1 + \frac{\text{SINR}}{\Gamma}\right) \qquad (15.15)$$

using the SNR-gap Γ. This method is particularly useful to approximate the effects of imperfect channel estimation, finite block lengths as well as imperfect modulation and coding schemes (MCSs).

Fig. 15.9 shows the downlink performance depending on Γ and illustrates three exemplary values, i.e., turbo codes, trellis codes and Tomlinson-Harashima Precoding (THP), and uncoded QAM. The average throughput of both relaying strategies decreases with a slightly smaller slope than in the case of CoMP, which reflects the high SINR values achieved by relaying (which are cut at $2^8 - 1$ to ensure that rates do not exceed 8 bpcu). This conclusion is supported by the decline of the worst-user performance, which has the same slope for all protocols. These results demonstrate that there are further energy-saving potentials using power control in relaying. In addition, relaying is less prone to imperfections than CoMP, as it does not use channel state information at the transmitter (CSIT) (in the downlink) and achieves higher SINR values. For instance, based on Fig. 15.9, relaying with uncoded QAM provides higher data rates than conventional transmission without SNR-gap.

15.2.9 Summary

This section discussed how relaying can complement a mobile communication system employing CoMP transmission. In particular, a mixed strategy has been presented, which combines the strengths of both in order to benefit from the interference-cancelation ability of CoMP and the improved power-distribution and path-loss of relaying.

The results show that it appears to be preferable to use CoMP in the downlink, at least for users in the main lobe direction of the base station. However, in the uplink, relaying provides significant performance improvements as the link quality is improved due to the shorter distances between UEs and RAPs. Under a cost-normalization and energy-normalization constraint the conclusions slightly change as then CoMP dominates relaying in the downlink. Nonetheless, relaying still offers more flexibility for the deployment such as in micro-cellular scenarios and indoor scenarios [FHK+10].

15.3 Next Generation Cellular Network Planning and Optimization

Jens Voigt

15.3.1 Introduction

Cellular network planning is the dimensioning of a wireless cellular network in terms of *planning parameters* such as the

- *cell properties* — number of base stations/cells, their position, sectorization,
- *antenna properties* — type, height, tilt, azimuth, and beamwidth of antennas at the cells, and
- *transmitter properties* — transmission power for pilot or reference signals.

Cellular network optimization is the process of (ongoing) improvement of the planning parameters of an already operating cellular network with respect to, e.g., current/projected traffic distributions and densities as well as other measured physical characteristics of the real network, compare, e.g., [LWN02, NDA06].

Within cellular network planning and optimization, mainly *physical characteristics* are used as metrics, as they are directly affected by the planning parameters defined above. The results of a planning/optimization task are evaluated by additional *service characteristics* such as some quality of service (QoS) indicators like mean or cell-edge throughput. The service characteristics directly depend on the planning parameters, but usually have a highly non-linear dependency on them. Also, service characteristics estimation often requires system simulations, which depend on many further modeling assumptions and can require either a high computational effort or further abstracting look-up tables. Due to these drawbacks, service characteristics may not be included in planning and optimization tasks, but only used to compare the results of a planning or optimization task with the status before this task.

In *classical* cellular network planning and optimization, thus without technologies like multiple-input multiple-output (MIMO) and CoMP, the two main physical characteristics are receive power levels and signal-to-interference ratios (SIRs).

The capacity in cellular networks deploying MIMO technology can benefit from multiple antenna element multiplexing or from beamforming capabilities. CoMP transmissions further feature beamforming coordination between sectors or joint transmission from multiple sectors in favorable parts of a cell's coverage area. We propose to include the evaluation of additional physical characteristics as metrics for MIMO and CoMP benefits into *advanced* planning and optimization of cellular radio networks. This section introduces and discusses several approaches.

15.3.2 Classical Cellular Network Planning and Optimization

This section describes a generic approach for cellular network planning and optimization. A typical cellular network planning scenario consists of a set $\mathcal{M}$ of base stations (BSs) $m \in \mathcal{M}$ and a set $\mathcal{K}$ of user equipments (UEs) $k \in \mathcal{K}$. A BS m is placed at a fixed three-dimensional position in (x,y,z). UEs k are modeled as non-overlapping horizontal square planes in (x,y), with each having a lateral size a_{UE} of several meters. Thus, in this section a UE k is identical to a *receiver pixel* k. All receiver pixels are positioned at a common height z over ground. The set of all UEs $\mathcal{K}$ is thus represented by a matrix of receiver pixels with a resolution of a_{UE} in x and y and at a typical value of $z = 1.5$ m.

For every receiver pixel k, the receive power level of pilot or reference signals is used as a metric for coverage, while the signal-to-interference-and-noise ratio (SINR) is used as a metric for quality. These metrics are estimated in an algorithm shown in Fig. 15.10.

First, the pathloss is predicted between all members of $\mathcal{M}$ and all members of $\mathcal{K}$. Then, the receive power per BS-UE pixel ($\mathcal{M} \to \mathcal{K}$) combination $P_{\mathrm{rx},m,k}$ is calculated adding, e.g., directional antenna gains, cable and other losses, and the transmission power to the predicted pathloss value. This step creates a set of receive powers $\mathcal{P}_{\mathrm{rx},m,k} = \{P_{\mathrm{rx},m,k} | m \in \mathcal{M} | k \in \mathcal{K}\}$. For every $k \in \mathcal{K}$, the equation $|\mathcal{P}_{\mathrm{rx},m,k}| = |\mathcal{M}|$ is valid.

The set $\mathcal{K}_{\mathrm{coverage}} \subset \mathcal{K}$ contains all receiver pixels k which have a receive power level $P_{\mathrm{rx},m,k}$ for at least one m that is not below a certain coverage planning threshold $P_{\mathrm{rx, threshold}}$ [dBm], which is usually a minimum receive power level for pilot or reference signals, i.e.

$$\mathcal{K}_{\mathrm{coverage}} = \{k \in \mathcal{K} | \exists m \in \mathcal{M} : P_{\mathrm{rx},m,k} \geq P_{\mathrm{rx, threshold}}\}. \tag{15.16}$$

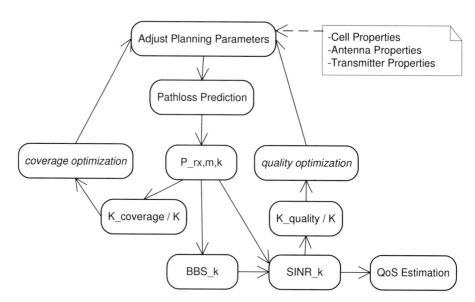

Figure 15.10 Classical network planning and optimization.

A planning scenario is then said to be well planned in terms of coverage in case the ratio $\frac{|\mathcal{K}_{\text{coverage}}|}{|\mathcal{K}|}$ is above a minimum coverage planning target.

Next, the best serving BS BBS_k per user equipment k is determined by selecting the BS m with the highest receive power (e.g., the reference signal received power (RSRP) in LTE) at k. For this selection, the set $\mathcal{P}_{\text{rx},m,k}$ is converted for every $k \in \mathcal{K}_{\text{coverage}}$ into a descending ordered set and further divided into two parts: the top member is the receive power of the best serving base station $P_{\text{rx},\text{BBS}_k}$, while the remaining members go into the set of receive powers $\mathcal{P}_{\text{rx},\mathcal{I}_k} = \{P_{\text{rx},m,k} | m \in \mathcal{I}_k\}$ of the set of interfering cells $\mathcal{I}_k = \{m | m \in \mathcal{M} | m \neq \text{BBS}_k\}$, so $\mathcal{P}_{\text{rx},m,k} = P_{\text{rx},\text{BBS}_k} \cup \mathcal{P}_{\text{rx},\mathcal{I}_k}$. All receiver pixels $k \in \mathcal{K}_{\text{coverage}}$ having the same base station m as BBS_k determine the so-called coverage area of this base station.

Next, the SINR is estimated for every user equipment $k \in \mathcal{K}_{\text{coverage}}$ using

$$\text{SINR}_k = \frac{P_{\text{rx},\text{BBS}_k}}{\sum_{m \in \mathcal{I}_k} P_{\text{rx},m,k} + N_0} \tag{15.17}$$

In (15.17), N_0 represents the thermal noise power at 300 K. The set $\mathcal{K}_{\text{quality}} \subset \mathcal{K}_{\text{coverage}}$ contains all $k \in \mathcal{K}_{\text{coverage}}$ which have a SINR that is not below a certain quality planning threshold $\text{SINR}_{\text{threshold}}$ [dB], which is usually a minimum SINR for pilot or reference signals[3]:

$$\mathcal{K}_{\text{quality}} = \{k \in \mathcal{K}_{\text{coverage}} | \text{SINR}_k \geq \text{SINR}_{\text{threshold}}\}. \quad (15.18)$$

Then, a planning scenario is said to be well-planned in terms of quality in case the ratio $|\mathcal{K}_{\text{quality}}|/|\mathcal{K}|$ is above a minimum quality planning target.

Coverage optimization has the goal to raise $|\mathcal{K}_{\text{coverage}}|/|\mathcal{K}|$ without degrading $|\mathcal{K}_{\text{quality}}|/|\mathcal{K}|$ by adapting planning parameters, compare Fig. 15.10. Similarly, quality optimization has the goal to raise $|\mathcal{K}_{\text{quality}}|/|\mathcal{K}|$ without degrading $|\mathcal{K}_{\text{coverage}}|/|\mathcal{K}|$ by adapting planning parameters, compare Fig. 15.10.

15.3.3 Physical Characterization of Capacity Gains through CoMP

Problem and Motivation

In classical cellular network planning and optimization, the SINR of pilot or reference signals is used as metric for network quality. This metric is based on the Shannon formula for channel capacity which states a dependency of the channel capacity on the SINR. The capacity of a MIMO channel can be estimated by the well-known Foschini formula [FG98] and depends on the SINR *and* some properties of the channel matrix. Consequently, a capacity increase through MIMO can be achieved by two effects:

- A power gain through an increase of the receive signal level and the SINR; this effect is based on an antenna array gain and beamforming.
- Creation of a properly shaped channel matrix due to the usage of space and/or polarization as an additional physical dimension in the signal space (in addition to frequency and time). This effect is called a degree–of–freedom or multiplexing gain, see [TV05, PBT05].

Space and/or polarization as additional physical dimension could also be used to create receive or transmit diversity. Diversity, however, does not increase the (ergodic) capacity of a planning scenario, but decreases the outage probability, see, e.g., [GA02]. Furthermore, diversity gains are usually independent of the environment of a planning scenario. For these reasons, effects based on diversity through CoMP-MIMO technology will not be reflected in advanced cellular network planning and optimization.

However, the capacity increasing effects power gain and/or multiplexing gain should consequently be characterized in advanced cellular network planning and optimization. In this section, we will present several metrics for CoMP cellular network capacity enhancements through:

---

[3] In cellular technologies applying adaptive modulation and coding, the application of multiple $\text{SINR}_{\text{threshold},MCS}$ to reflect the requirements of different modulation and coding schemes (MCSs) might be useful. Then, different sets $\mathcal{K}_{\text{quality},MCS}$ can be evaluated.

1. Power/SINR gain through beamforming, in a non-cooperative setup or in the context of interference coordination or joint signal processing CoMP.
2. Degrees-of-freedom gain through multiplexing: Spatial/polarization multiplexing in a non-cooperative setup or multi-site spatial multiplexing.

Estimation of a Metric for a Degrees-of-Freedom Gain

A CoMP-MIMO channel can be described by a channel matrix $\mathbf{H}_{\mathcal{M}_{C,k}}$. The CoMP cooperating set $\mathcal{M}_{C,k}$ contains all cells $m \in \mathcal{M}$ that cooperate for transmission to k in the case of joint transmission. This set always includes BBS_k. In the case of non-cooperative CoMP, BBS_k is the only member of $\mathcal{M}_{C,k}$. Neglecting multi-user MIMO (MU-MIMO), the CoMP-MIMO channel matrix has the dimension of $\mathbf{H}_{\mathcal{M}_{C,k}}^{[N_{\text{bs}} \cdot |\mathcal{M}_{C,k}| \times N_{\text{ue}}]}$. Here, N_{bs} and N_{ue} represent the number of antenna elements at a cell and a UE, respectively.

The number of degrees-of-freedom for a transmission to k is equal to the number of eigenvalues of the channel matrix $\mathbf{H}_{\mathcal{M}_{C,k}}$, which is equal to its rank: $|\mathcal{L}_k| = \text{rank}(\mathbf{H}_{\mathcal{M}_{C,k}})$. Here, $\mathcal{L}_k$ is the set of all eigenvalues at a receiver pixel k: $\mathcal{L}_k = \{\lambda_{i,k}\}$ and $\lambda_{i,k}$ is the i-th (random) eigenvalue of $\mathbf{H}_{\mathcal{M}_{C,k}} \mathbf{H}_{\mathcal{M}_{C,k}}^H$. The strongest eigenvalue $\lambda_{\max,k}$ of $\mathbf{H}_{\mathcal{M}_{C,k}} \mathbf{H}_{\mathcal{M}_{C,k}}^H$ is $\lambda_{\max,k} = \max(\lambda_{i,k})$.

In the case that $\text{rank}(\mathbf{H}_{\mathcal{M}_{C,k}}) = \min(N_{\text{ue}}, N_{\text{bs}} \cdot |\mathcal{M}_{C,k}|)$, then $\mathbf{H}_{\mathcal{M}_{C,k}}$ is said to be of full rank. Without polarization effects, this is true when all elements of $\mathbf{H}_{\mathcal{M}_{C,k}}$ follow the independently and identically distributed (i.i.d.) complex Gaussian fading model. This model is true for a rich scattering environment with de-correlated elements of the transmitting and receiving multi-element antennas. In real cellular network scenarios, however, the i.i.d. complex Gaussian fading model can in general not be assumed to be valid due to:

- the shape of the environment including reflectors, scatterers, and obstacles for radio wave propagation which determine the number, strength, length, polarization changes, and angular distribution of dominant multi-paths,
- the typical directivity of the transmit antenna elements,
- the small spread of the angle-of-departures for scenarios with transmit antennas well above the average building height of the planning scenario, and
- the correlation of the antenna elements, especially when they have typical small inter-element distances.

Furthermore, a variety of multi-element antenna types, including cross-polarization types, uniform linear or circular arrays, remote radio heads (RRHs), as well as different CoMP schemes with antennas at different positions will be deployed in real cellular networks. Therefore, the rank of $\mathbf{H}_{\mathcal{M}_{C,k}}$ depends on:

1. a *3D environment model* of the cellular network scenario,
2. *antenna array properties*, such as the number and position of antenna elements, the inter-element distance, and polarization properties, as well as
3. *cellular technology properties*, such as the CoMP scheme.

15.3 Next Generation Cellular Network Planning and Optimization

In addition, not all eigenvalues in $\mathcal{L}_k$ can be used for data transmission. For every eigenvalue $\lambda_{i,k}$, a condition ratio $\kappa_{i,k} = \frac{\lambda_{\max,k}}{\lambda_{i,k}}$ can be defined. Then, a threshold ratio κ_{data} exists to define the maximum condition ratio for an eigenvalue to be useful for data transmission: In case of $\kappa_{i,k} > \kappa_{\text{data}}$, then λ_i does not represent a useful transmission through the channel anymore. The set of *useful eigenvalues* $\mathcal{L}_{\text{data},k} \subset \mathcal{L}_k$ can then be defined as

$$\mathcal{L}_{\text{data},k} = \{\lambda_{\text{data},i,k} | \kappa_{i,k} \leq \kappa_{\text{data}}\}. \tag{15.19}$$

The condition ratio threshold κ_{data} should be a planning threshold. Consequently, a metric for a degree-of-freedom / multiplexing gain is the *number of useful eigenvalues* $\Theta_k = |\mathcal{L}_{\text{data},k}|$ at a receiver pixel k with $[1 \leq \Theta_k \leq \text{rank}(\mathbf{H}_{\mathcal{M}_C,k})]$. The set $\mathcal{K}_{\text{multiplexing}} \subset \mathcal{K}_{\text{quality}}$ contains all receiver pixels $k \in \mathcal{K}_{\text{quality}}$[4] which have a Θ_k that is a not below a planning threshold $\Theta_{\text{threshold}}$, which is an expected minimum number of eigenvalues:

$$\mathcal{K}_{\text{multiplexing}} = \{k \in \mathcal{K}_{\text{quality}} | \Theta_k \geq \Theta_{\text{threshold}}\}. \tag{15.20}$$

A planning scenario is then said to be well-planned in terms of multiplexing in case the ratio $|\mathcal{K}_{\text{multiplexing}}|/|\mathcal{K}|$ is above a minimum planning target. Multiplexing optimization has the goal to raise $|\mathcal{K}_{\text{multiplexing}}|/|\mathcal{K}|$ without degrading $|\mathcal{K}_{\text{coverage}}|/|\mathcal{K}|$ and $|\mathcal{K}_{\text{quality}}|/|\mathcal{K}|$ by adapting planning parameters, compare Fig. 15.12.

Algorithm 1: Direct Computation

One solution to estimate the number of useful eigenvalues per receiver pixel k is the direct estimation of the channel matrix $\mathbf{H}_{\mathcal{M}_C,k}$, e.g., through ray tracing in a detailed 3D environment model of the planning scenario, as done in Section 14.2. $\mathbf{H}_{\mathcal{M}_C,k}$ depends on the 3D environment of both, the transmitter and the receiver, and the antenna array types at both ends of the channel. Concrete mathematics for different CoMP schemes and antenna types:

- Non-cooperative CoMP: Cross-polarized antennas (X-pole), uniform linear array (ULA), and RRHs as well as
- Joint signal processing CoMP: Single antenna elements or ULAs

are presented in [VFS09]. The estimation of $\mathbf{H}_{\mathcal{M}_C,k}$ must be followed by an evaluation of the eigenvalues of $\mathbf{H}_{\mathcal{M}_C,k} \mathbf{H}_{\mathcal{M}_C,k}^H$. Then, Θ_k can be found by evaluating (15.19).

---

[4] It can be useful to define a SINR$_{\text{threshold, multiplexing}}$ different from (usually higher than) SINR$_{\text{threshold}}$ and to evaluate a set $\mathcal{K}_{\text{quality, multiplexing}}$ similar to (15.18) to be applied in (15.20), compare also the remarks in the footnote on page 435. Read further in [SSESV06] for discussions on a separate evaluation of the SINR at k and Θ_k at high values of the SINR.

Algorithm 2: Angular Channel Model

The direct computation of Θ_k as described above is computational expensive. In order to be acceptable for planning and optimization purposes, algorithms need to be fast and memory-efficient. Consequently, this paragraph introduces a computationally much less demanding model.

In [TV05, PBT05, Say02, Mol03] another view upon the MIMO channel is introduced: The angular domain. Here, the channel is represented by *orthogonal spatial units*. These spatial units can be determined as follows for the special case of a ULA.

An antenna array is defined as an arrangement of single antenna elements in a plane or optionally in two perpendicular planes in a three-dimensional Cartesian coordinate system. Each antenna array has a physically limited size in each plane, which leads to a limited spatial resolution. For a given carrier frequency, this resolution depends on the number and spacing of the single antenna elements.

In the special case of a ULA with N antenna elements, the size L of the antenna array is $L = N \cdot \delta$, where δ represents the inter-element antenna spacing normalized to the wavelength λ of the carrier frequency. Typical normalized inter-element distances are $\delta = 0.5$ or $\delta = 1$. In [TV05],

$$\mathbf{e}(\phi) = \begin{bmatrix} 1 \\ e^{-j2\pi\delta\cos(\phi)} \\ \vdots \\ e^{-j2\pi\delta(N-1)\cos(\phi)} \end{bmatrix} \tag{15.21}$$

is called the spatial signature at angle ϕ. Two paths through the environment are linearly independent if they arrive at the array (or depart from the array) at two angles ϕ_1 and ϕ_2, for which $\mathbf{e}(\phi_1) \cdot \mathbf{e}(\phi_2)^* = \mathbf{0}$ is valid. For the special case of ULA antennas at both ends of the channel, it can be shown [TV05] that $\Omega = \{\omega = \cos(\phi) | \phi \in \{0, \frac{1}{L}, \ldots, \frac{N-1}{L}\}\}$ is such an orthogonal set of angles ϕ. Each ϕ defines the center of a spatial orthogonal unit. Then, spatial orthogonal units in a plane can be defined by a graph in polar coordinates with

$$\left(\phi, \left| \frac{R(\phi) \cdot \sin(\pi \cdot L \cdot (\cos(\phi) - \omega))}{N \cdot \sin\left(\frac{\pi \cdot L \cdot (\cos(\phi) - \omega)}{N}\right)} \right| \right) \quad \forall \omega \in \Omega, \tag{15.22}$$

wherein $R(\phi)$ represents the electric directional diagram of the antenna in this plane to cope for sectorial and directive antennas. Crossing points between the graphs defined by (15.22) for different ω connected to the center position of the antenna array define the borders between the spatial orthogonal units, see Fig. 15.11.

To determine Θ_k, a matrix can be evaluated for every k, in which the spatial units of the transmit antenna arrays at every $m \in \mathcal{M}_{C,k}$ form the columns, and the spatial units of the receive antenna array at k form the rows.

15.3 Next Generation Cellular Network Planning and Optimization

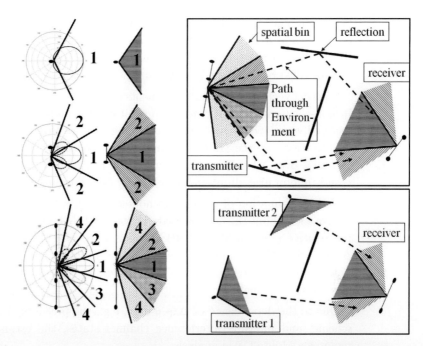

Figure 15.11 Example for determination of spatial orthogonal units in azimuth for ULAs ($\delta = 0.5$; sectorial single antenna elements having a 3 dB bandwidth of $65°$) with one, two, and four elements (left – top, middle, bottom) and occupation estimation for non-cooperative and joint signal processing CoMP (right – top, bottom).

A much simpler ray-tracing with, e.g., a limited number of reflections only and possibly in a limited model of the environment (e.g., a clutter-height model) could then find paths between any of the spatial units representing the transmit and receive antenna arrays, see Fig. 15.11.

An orthogonal unit from which a configurable minimum number (a planning threshold) of paths emanate into or arrive from an environment is referred to as 'occupied'. The number of useful eigenvalues Θ_k corresponds to the minimum number of rows and columns that are occupied [TV05]. Then, (15.20) can be evaluated. See [PT08, PT09] for extensions of this model for cross-polarized antenna types and [HRPV08] for mathematics on non-uniform antenna arrays.

Estimation of a Metric for a Beamforming Power Gain

Beamforming results in an SINR gain through a directive steering of the antennas of BBS_k towards k and/or the interfering signals away from or even nulling k. This resulting SINR gain $\Delta_{SINR,k}$ shall be used as a metric for beamforming. A pre-condition for the application of beamforming is the deployment of uniform linear/circular arrays (ULA/UCA) as antenna types[5]. ULA/UCA antenna types are able to form a beam by the application of steering vectors that cause phase shifts between the single antenna elements. This phase shift results in shaping the electrical beam of the antenna array.

Steering vectors are precoder elements defined by different precoding schemes: In non-cooperative CoMP, maximum ratio transmission (MRT) maximizes the received signal power at k, or zero-forcing can null multi-user interference signals to maximize the SINR at k. In cooperative CoMP schemes, an information exchange over X2 enables to maximize the power gain for the desired signal at k and to minimize all interference at k by beamforming coordination.

For an estimation of $\Delta_{\text{SINR},k}$ in non-cooperative CoMP for MRT precoding, information about the channel state between BBS_k and k (closed-loop) or about the relative position of k to BBS_k (open-loop) is required. The estimation of $\Delta_{\text{SINR},k}$ for MRT in interference coordination and joint transmission CoMP as well as for zero-forcing in all CoMP schemes requires information about the interference channel(s) state in addition.

Due to the dependency of $\Delta_{\text{SINR},k}$ on, e.g., the precoding scheme, the CoMP scheme, and multiple interference channel states, the estimation of $\Delta_{\text{SINR},k}$ requires a Monte-Carlo analysis.

Monte-Carlo Analysis of the SINR Gain
To estimate the power gain through beamforming for a transmission from a cell m to a receiver pixel k by usage of a steering vector $\mathbf{w}_o$, a Gain_{BF} can be defined as

$$\text{Gain}_{\text{BF},m,k,o} = \frac{\|\mathbf{H}_{m,k} \cdot \mathbf{w}_o\|_F^2}{h_{m,k}^2}. \tag{15.23}$$

In (15.23), $\|\mathbf{H}\|_F$ represents the Frobenius norm of $\mathbf{H}$, $h_{m,k}$ is the channel impulse response for a single-input single-output (SISO) channel from cell m to receiver pixel k, and $\mathbf{H}_{m,k}$ is the MIMO channel matrix between them, compare with Section 15.3.3. See, e.g., [VFS09] for mathematical details on how to compute $\mathbf{H}_{m,k}$. $\mathbf{w}_o$ is determined by the precoder element selection strategy of the precoding scheme. This determination can range from a simple selection of pre-defined precoder elements (e.g., unitary precoding) to a direct calculation (e.g., zero-forcing (ZF) or dirty paper coding (DPC)).

In the special case that MRT precoding is done by a limited number of pre-defined precoder elements (e.g. [3GP07b]), the definition of Gain_{BF} in (15.23) is constant for a certain antenna array type (e.g., a 2-port ULA with an inter-element distance of $\delta = 0.5$). Then the Gain_{BF} can be pre-calculated as a matrix

---

[5] Note that this is only true if the term *beamforming* is interpreted as actually steering a beam in a geometrical sense. Often, the term is used for precoding in a more general sense, where any kind of antenna setup may be used, and still an SINR gain can be obtained.

with the resolution of the receiver pixel size a_{UE}. The number of these matrices necessary as antenna array properties for a planning or optimization scenario equals the product of number of antenna array types times precoder elements per array type. Thus, to calculate the Gain$_{\text{BF}}$, again *antenna array properties* and *cellular technology properties* (e.g., a fixed set of precoder elements) are additional parameters.

In order to further estimate $\Delta_{\text{SINR},k}$ resulting from beamforming, the CoMP scheme and precoder element selection strategy need to be included. In all CoMP schemes, we assume a precoding element selection strategy at BBS$_k$ that selects a precoder element $\mathbf{w}_o$ which maximizes the SINR at a receiver pixel k by maximizing the enumerator in (15.23), but does not take care of the implied interference at any other receiver pixel (egoistic strategy [HvR09] — argmax operation). The precoder element selection strategies for the cells in $\mathcal{I}_k$ and $\mathcal{M}_{\text{C},k}$ depend on the CoMP scheme:

1 - Non-cooperative CoMP: Uncoordinated Beamforming

In the case of non-cooperative beamforming, we assume the precoder element selection strategy of the cells in $\mathcal{I}_k$ to use the same egoistic strategy towards their receiver to be served in the same physical resource block (PRB). Thus, from the view of the receiver pixel k served by its BBS$_k$, they use any other (random) precoder element. With $\Psi_{m,k,o} = P_{\text{rx},m,k} \cdot \text{Gain}_{\text{BF},m,k,o}$, the $\Delta_{\text{SINR},k}$ is then equal to:

$$\Delta_{\text{SINR},k} = \frac{\underset{o}{\operatorname{argmax}}\left(\Psi_{\text{BBS}_k,k,o}\right)}{\underset{\substack{m\in\mathcal{I}_k \\ o=\text{any}}}{\sum}\left(\Psi_{m,k,o}\right)+N_0} - \text{SINR}_k. \tag{15.24}$$

2 - Interference Coordination

For interference coordination and joint transmission CoMP, the set $\mathcal{M}$ is for every $k \in \mathcal{K}_{\text{coverage}}$ partitioned into two disjoint subsets, the CoMP coordinating set $\mathcal{M}_{\text{C},k}$ and the interfering set $\mathcal{M}_{\text{I},k} = \{m|m \in \mathcal{M}|m \notin \mathcal{M}_{\text{C},k}\}$. As the CoMP coordination set includes BBS$_k$, the set $\mathcal{M}_{\text{C},k}\backslash\text{BBS}_k$ represents the CoMP coordinating set without BBS$_k$. The precoder element selection strategy for BBS$_k$ was discussed above. $\mathcal{M}_{\text{C},k}\backslash\text{BBS}_k$ and $\mathcal{M}_{\text{I},k}$ use different precoder element selection strategies:

Subset $\mathcal{M}_{\text{I},k}$: The members of $\mathcal{M}_{\text{I},k}$ do not coordinate their scheduling with BBS$_k$, but use any other (random) precoder element like in (15.24).

Subset $\mathcal{M}_{\text{C},k}\backslash\text{BBS}_k$: This subset contains interfering cells that coordinate their precoder element selection with BBS$_k$ to further maximize the SINR at k. SINR maximization through interfering cells means that the members of $\mathcal{M}_{\text{C},k}\backslash\text{BBS}_k$ coordinate their precoder element selection to minimize their interfering receive power at k (altruistic strategy [HvR09] — argmin operation): The antenna beams of inter-cell interfering signals are intentionally steered away from the position of

k, hereby increasing the SINR there. Thus, the precoder element selection in the members of $\mathcal{M}_{C,k}\backslash\text{BBS}_k$ will not use certain precoder elements that would steer their beams towards k: Their set of available precoder elements for their own transmission will be limited. We exemplarily discuss two options for precoder element selection strategies in the cells belonging to $\mathcal{M}_{C,k}\backslash\text{BBS}_k$:

Option 1: All members of $\mathcal{M}_{C,k}\backslash\text{BBS}_k$ do not use the precoder element that would result in the highest interfering power at k, but can choose from any other. The limitation of the set of available precoder elements for the members of $\mathcal{M}_{C,k}\backslash\text{BBS}_k$ is minimal.

$$\Delta_{\text{SINR},k} = \frac{\underset{o}{\text{argmax}}\left(\Psi_{\text{BBS}_k,k,o}\right)}{\sum\limits_{\substack{m\in\mathcal{M}_{C,k},m\neq\text{BBS}_k \\ o\neq o_{\max}}} \Psi_{m,k,o} + \sum\limits_{\substack{m\in\mathcal{M}_{I,k} \\ o=\text{any}}} \Psi_{m,k,o} + N_0} - \text{SINR}_k \qquad (15.25)$$

Option 2: All members of $\mathcal{M}_{C,k}\backslash\text{BBS}_k$ use only those precoder elements that result in the corresponding lowest interfering power at k. The limitation of the set of available precoder elements for the members of $\mathcal{M}_{C,k}\backslash\text{BBS}_k$ is maximal.

$$\Delta_{\text{SINR},k} = \frac{\underset{o}{\text{argmax}}\left(\Psi_{\text{BBS}_k,k,o}\right)}{\sum\limits_{\substack{m\in\mathcal{M}_{C,k} \\ m\neq\text{BBS}_k}} \underset{o}{\text{argmin}}\left(\Psi_{m,k,o}\right) + \sum\limits_{\substack{m\in\mathcal{M}_{I,k} \\ o=\text{any}}} \Psi_{m,k,o} + N_0} - \text{SINR}_k \qquad (15.26)$$

3 - Joint Transmission
In multi-cell joint transmission, the precoder element selection strategy for all cells in $\mathcal{M}_{C,k}$ is the argmax strategy described above. The cells in $\mathcal{M}_{I,k}$ behave like in the interference coordination case. Then, the $\Delta_{\text{SINR},k}$ for joint transmission can be calculated like:

$$\Delta_{\text{SINR},k} = \frac{\sum\limits_{m\in\mathcal{M}_{C,k}} \underset{o}{\text{argmax}}\left(\Psi_{m,k,o}\right)}{\sum\limits_{\substack{m\in\mathcal{M}_{I,k} \\ o=\text{any}}} \Psi_{m,k,o} + N_0} - \text{SINR}_k \qquad (15.27)$$

The randomness in the denominator of (15.24), (15.25), (15.26) and (15.27) should be analyzed in a Monte-Carlo analysis.

With the SINR gains through beamforming estimated by (15.24), (15.25), (15.26) or (15.27), a subset $\mathcal{K}_{\text{quality, BF}} \subset \mathcal{K}$ which contains all $k \in \mathcal{K}_{\text{coverage}}$ that have an SINR above a certain quality planning threshold $\text{SINR}_{\text{threshold}}$ [dB] can be calculated, compare (15.18):

$$\mathcal{K}_{\text{quality, BF}} = \{k \in \mathcal{K}_{\text{coverage}} | \text{SINR}_k + \Delta_{\text{SINR},k} > \text{SINR}_{\text{threshold}}\}. \qquad (15.28)$$

Then, the ratio $\frac{|\mathcal{K}_{\text{quality, BF}}|}{|\mathcal{K}|}$ can be evaluated with $\frac{|\mathcal{K}_{\text{quality}}|}{|\mathcal{K}|}$ to see quality improvements through beamforming.

15.3 Next Generation Cellular Network Planning and Optimization

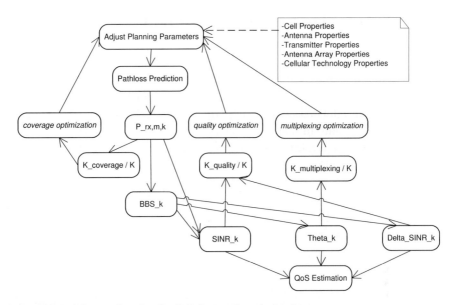

Figure 15.12 Advanced network planning and optimization.

15.3.4 Summary

We propose to include the estimation of capacity enhancements through MIMO and CoMP into advanced planning and optimization of cellular radio networks by the evaluation of additional physical metrics, see Fig. 15.12. As metrics, we suggest an estimation of the *number of useful eigenvalues* Θ as a metric for a *multiplexing gain* and the *SINR gain* Δ_{SINR} as a metric for *beamforming gains*. As a prerequisite for the estimation of these metrics *additional planning parameters* such as *antenna array properties* (e.g., number and position of antenna elements, inter-element distances, cross-polarization properties) and *cellular technology properties* (e.g., CoMP scheme, precoding scheme) as well as *advanced planning thresholds* (e.g., minimum power ratio for an eigenvalue to be useful) are necessary.

This section has outlined computationally expensive, but high quality estimations that require ray-tracing based channel impulse response simulation in a detailed 3D environment model to evaluate the number of useful eigenvalues. Further, system level simulations were discussed that may cover precoder element selection strategies and CoMP coordination specifics to evaluate SINR gains through beamforming. Also, we introduced faster metric estimation algorithms that can go with a rougher 3D environment model or can use pre-calculated beamforming power gain matrices, but are bound to special cases (e.g., a fixed set of linear or circular array antenna types and pre-defined precoder elements).

15.4 Energy-Efficiency Aspects of CoMP

Albrecht Fehske, Patrick Marsch and Gerhard Fettweis

Recently, global warming has shifted the focus of technological advancement to decreasing the carbon footprint of mass market products and services. In this regard, information and communications technologies (ICT) in general and the mobile communications sector in particular represent an increasing share in overall global emissions. While in a business as usual case, the overall ICT footprint will approximately double until 2020, recent studies predict that the footprint of mobile communications might almost triple, ramping up to more than one third of present annual emissions of the whole UK [FMBF10]. The global footprint is predicted to increase on the order of 20-30% within the same period. Today, the largest fraction of more than 45% of the mobile communications industry's carbon footprint is due to the operation of radio access networks, which offer great potential for reduction [FMBF10].

Besides improving the environmental impact of their networks, operators have a strong incentive to make their networks "greener", since this also provides a clear opportunity to cut cost and save money through increased energy efficiency. In western markets, today the energy cost due to network operation amount to about 1% of earnings before interest and tax (EBIT), and a further increase by a factor between two and three can be expected until 2020. In developing countries, the ratio of energy cost and EBIT is much worse, not only due to lower revenue per user, but also due to the use of diesel powered sites, where inefficiency of the generators and high diesel transporting cost often prohibit provision of wireless services. Energy efficient base station equipment will be a key enabler for use of alternative energy sources that allow for much lower operational cost and feasible business models [FMBF10].

As pointed out in Chapter 1, there are two major development paths currently foreseen for mobile communication networks: Use of more base stations and the use of coordination or cooperation between base stations. While both concepts are able to achieve both an increased sum-rate as well as a more uniform distribution of user rates in the cell, they incur different costs in terms of consumed energy. Naturally, densification increases the network's energy need due to increased number of radio access points, and trade-offs between gains in cell throughput and increased energy consumption have been investigated, e.g., in [FRF09, RFMF10]. CoMP schemes in uplink and downlink require additional backhaul connections between cooperating sites as well as additional signal processing at the base stations. This section investigates the trade-offs between attainable gains in cell throughput obtained from CoMP schemes on one hand, and increased power dissipation of base stations due to higher complexity of CoMP processing and additional backhaul requirements on the other. Also, these trade-offs are investigated for different cell sizes. This section is based in parts on [FMF10].

15.4 Energy-Efficiency Aspects of CoMP

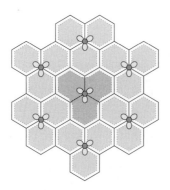 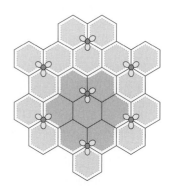

(a) Cooperation size three. (b) Cooperation size seven.

Figure 15.13 Cellular network layout with two exemplary cooperation sizes.

15.4.1 System Model

The network topology is modeled as a setup of M cellular base stations (BSs), co-located in groups of three at a regular grid of $\frac{N_{\rm bs}}{3}$ sites characterized by the inter-site-distance D. We assume the BSs to be equipped with $N_{\rm bs} = 2$ transmit and receive antennas each. We further assume the user terminals to be equipped with a single transmit and receive antenna. As throughout the book, we assume a *cell* to be equivalent to a *sector*, served by a single BS. For given inter-site-distance D, the cell size A calculates as $A = \frac{D^2}{2\sqrt{3}}$.

Depending on the scenario under study, up to $N_{\rm c}$ BSs are allowed to jointly transmit and receive user data. We assume all BSs to be connected by a full mesh of backhaul links allowing them to exchange information, as further detailed in Subsection 15.4.4. CoMP techniques are able to strongly increase sum and cell-edge throughputs in a cellular system [MF07c, MF07b], however, they introduce a certain amount of overhead into the system, which can be categorized as

- additional pilots,
- additional backhauling, and
- additional signal processing.

Modeling gains and energy needs involved in CoMP schemes is the concern of the following paragraphs.

Propagation Model

Deterioration of signal quality is commonly assumed to be due to three different causes: pathloss, slow fading, also referred to as shadowing, and fast fading. In this work, we mainly concentrate on the effects of pathloss and fast fading, and include the effects of slow fading as a margin in the link budget. We employ a signal propagation model as follows

$$P_{\rm rx}(r,\phi) = K\left(r\right)^{-\lambda} P_{\rm tx}, \qquad (15.29)$$

where P_{tx}, P_{rx}, r, ϕ, and λ denote transmit and receive power, propagation distance, relative angle between transmitter and receiver, and pathloss exponent, respectively. We assume the parameter K to be composed of three factors

$$K = K(r, \phi) = U \cdot V \cdot W(\phi). \tag{15.30}$$

Factor U incorporates the impact of user terminal and base station antenna heights, carrier frequency, and propagation environment. Propagation loss due to outdoor-to-indoor propagation is captured in the factor V. The antenna pattern depending on the relative angle ϕ between transmitter and receiver is modeled by the term $W(\phi)$. For simulative investigations, we employ the propagation parameter values specified in [3GP10d]. In order to capture the fast fading component of the channel, we follow a standard procedure and sample a channel coefficient from a complex Gaussian distribution with zero mean and unit variance independently for each transmit and receive antenna pair in the setup.

Transmission Equations

As in previous parts of this book, we consider an orthogonal frequency division multiplex (OFDM) system and observe a single sub-carrier for uplink (UL) and downlink (DL) transmission. All observed effects are than translated to a full system through simple scaling. Note that in the following, we assume the user equipments (UEs) and the BSs to be equipped with a single transmit and receive antenna and two transmit and receive antennas, respectively. A generalization to arbitrary numbers of receive antennas at the BSs is straightforward.

Uplink

Assuming that each BS has assigned exactly one UE to the observed sub-carrier and each UE's signal is received at a cluster of N_c cooperating BSs, we can state the uplink transmission equation for each OFDM symbol as

$$\mathbf{y} = \mathbf{W}^H \left(\mathbf{HP}_x^{\frac{1}{2}} \mathbf{x} + \mathbf{GP}_u^{\frac{1}{2}} \mathbf{u} + \mathbf{n} \right), \tag{15.31}$$

where $\mathbf{x} \in \mathbb{C}^{N_c \times 1}$, $\mathbf{y} \in \mathbb{C}^{N_c \times 1}$, $\mathbf{W} \in \mathbb{C}^{2N_c \times N_c}$, $\mathbf{H} \in \mathbb{C}^{2N_c \times N_c}$, $\mathbf{u} \in \mathbb{C}^{(M-N_c) \times 1}$, and $\mathbf{n} \in \mathbb{C}^{2N_c \times 1}$ model transmitted and received symbol vectors of UEs inside the cooperation cluster, the receive filter matrix, the channel matrix, the symbol vector of UEs outside the cluster, and a noise term with $E\{\mathbf{nn}^H\} = \sigma_n^2$, respectively. The matrices $\mathbf{H} \in \mathbb{C}^{2N_c \times N_c}$ and $\mathbf{G} \in \mathbb{C}^{2N_c \times (M-N_c)}$ contain the channel coefficients between the receiving BS antennas and the terminals that are processed by the cluster BS and those that are not, respectively. We assume $E\{\mathbf{xx}^H\} = \mathbf{I}$ and $E\{\mathbf{uu}^H\} = \mathbf{I}$, i.e., the powers of the signals from cluster and non-cluster users are assumed to be contained in the diagonal matrices $\mathbf{P}_x$ and $\mathbf{P}_u$, respectively.

We assume $\mathbf{W}$ to be a minimum mean square error (MMSE) filter matrix. The average achievable rate in *bits per channel use per cell* can then be bounded as

$$R_{\text{ul}} \leq \frac{1}{N_c} \log_2 \left| \mathbf{I} + \left(\mathbf{\Phi}_{\text{i}} + \sigma_n^2 \mathbf{I} \right)^{-1} \cdot \mathbf{HP}_x \mathbf{H}^H \right|, \tag{15.32}$$

where the term $\boldsymbol{\Phi}_{\bar{\text{i}}} = \mathbf{GP}_u \mathbf{G}^H$ denotes the covariance matrix of the signals coming from interfering UEs outside the cluster.

Downlink

The transmission equation for N_c BSs transmitting jointly to N_c users can be stated as

$$\mathbf{y} = \mathbf{H}^H \mathbf{W} \mathbf{x} + \mathbf{L}^H \mathbf{T} \mathbf{u} + \mathbf{n}, \tag{15.33}$$

where $\mathbf{W} \in \mathbb{C}^{2N_c \times N_c}$ and $\mathbf{T} \in \mathbb{C}^{2(M-N_c) \times (M-N_c)}$ denote the transmit filters of the BSs inside and outside the cluster, respectively. The matrix $\mathbf{L} \in \mathbb{C}^{2(M-N_c) \times N_c}$ contains the channel coefficients from all BSs outside the considered cluster to the cluster UEs. In analogy to the uplink receive filter, we assume an MMSE transmit filter $\mathbf{W}$ to be applied jointly to all cooperating antennas, i.e.

$$\mathbf{W} = \sqrt{\beta} \cdot \mathbf{H} \left(\mathbf{H}^H \mathbf{H} + \frac{\text{tr}\{\boldsymbol{\Phi}_{\bar{\text{i}}} + \sigma_n^2 \mathbf{I}\}}{N_c \cdot P_{\text{dl}}} \mathbf{I} \right)^{-1}. \tag{15.34}$$

The term P_{dl} denotes the maximum transmit power per BS, and the factor $\beta = \frac{N_c P_{\text{dl}}}{\text{tr}\{\mathbf{W}\mathbf{W}^H\}}$ ensures that this limit is kept after application of the filter. The matrix $\boldsymbol{\Phi}_{\bar{\text{i}}} = \mathbf{L}^H \mathbf{T} \mathbf{T}^H \mathbf{L}$ refers to the interference coming from BSs outside the cluster. For simplicity, the transmit filter $\mathbf{T}$ at all other BS is assumed to be $\sqrt{P_{\text{dl}}/(2 \cdot (M - N_c))}$ in all elements, i.e., to map an equal amount of the symbol energy to each of the transmit antennas of each BS outside the considered cluster. The average downlink rate in *bit per channel use per cell* for a cluster of N_c cooperating BSs can now be stated as

$$R_{\text{dl}} \leq \frac{1}{N_c} \sum_{k=1}^{N_c} \log_2 \left(1 + \frac{|\mathbf{h}_k^H \mathbf{w}_j|^2}{\sum_{j \neq k} |\mathbf{h}_k^H \mathbf{w}_j|^2 + \sigma_k^2 + \sigma_n^2} \right), \tag{15.35}$$

where

$$\sigma_k^2 = \frac{P_{\text{dl}}}{2} \cdot \mathbf{l}_k^H \mathbf{1}_{2(M-N_c) \times 2(M-N_c)} \mathbf{l}_k \tag{15.36}$$

denotes the received power from interfering BSs outside the cluster and $\mathbf{1}_{K \times L}$ denotes a $K \times L$ matrix with all elements equal to one.

15.4.2 Effective Transmission Rates

While (15.32) and (15.35) state the achievable rates *per channel use*, a certain amount of rate has to be invested into signaling overhead due to pilots and channel state information (CSI) feedback.

Uplink

Uplink pilot overhead scales linearly with the number of cooperating BSs, as orthogonal pilots are needed for all UEs in the cluster to enable joint detection. If downlink joint transmission (JT) is performed, CSI must be fed back in the uplink. Since CSI needs to be generated per transmit antenna, the CSI feedback rate scales linearly with the total number of downlink transmit antennas within a cluster. Thus, for N_c cooperating BSs, the achievable rate per channel use reduces on average to

$$R_{\text{ul,eff}} = R_{\text{ul}} \cdot (1 - \rho \cdot N_c) - 2N_c \cdot \rho \cdot q. \tag{15.37}$$

Here, the term ρ models the average pilot density in time and frequency, which is governed by the average coherence time and frequency expected by the system. The term q denotes the amount of feedback bits employed to feed back CSI of acceptable quality. It becomes clear from (15.37) that there is an upper limit on the cluster size N_c, since at some point the effective uplink rate will be zero.

Downlink

Similar to the uplink case, the achievable downlink rate per channel use is degraded due to pilots, where we need both orthogonal pilots such that the terminals can estimate and feed back the channel matrix, as well as precoded, stream-specific pilots for equalization and decoding, as pointed out in Section 9.1.2. Given two transmit and one receive antenna at BS and UE side, respectively, we need $3N_c$ orthogonal pilot sequences. In addition, some OFDM symbols are reserved for control information of various kinds, so that the effective downlink rate is given by

$$R_{\text{dl,eff}} = R_{\text{dl}} \cdot \left(1 - \rho \cdot 3N_c - \delta\right), \tag{15.38}$$

where δ models the average control overhead.

15.4.3 Backhauling

As stated above, the usage of multi-cell joint detection and joint transmission requires information exchange between all sites involved, where the rate of exchange depends on the cooperation size. In this work, we are not directly concerned with backhaul data rates, but rather look at the amount of additional energy that is required to provide a certain backhaul rate.

Uplink

In order to jointly detect N_c user signals received at N_c BSs, the sampled receive signals have to be forwarded to one master site, where processing is performed, as considered in Section 6.1. The average required backhaul in *bit per channel use and cell* can be calculated as

$$\beta_{\text{ul}} = \frac{(N_c - \xi) \cdot 2 \cdot z}{N_c}, \tag{15.39}$$

where z denotes the number of quantization bits applied to each complex symbol. We have to consider that cooperating BSs which are co-located do not require backhaul infrastructure. This is taken care of by the term ξ, which denotes the maximum number of cells within the cooperation cluster being located at the same site. Only the remaining cooperating cells need then forward quantized signals to this site. Note that we assume that not only data, but also pilot symbols are quantized and forwarded, such that an additional exchange of channel knowledge among BSs is not required. The detected data are then forwarded into the system by the master site. Since this distribution effects primarily the backbone, i.e, connection of the BSs to the core network, rather than the BS themselves, it is not considered here.

Downlink
In the downlink, the CSI feedback decoded centrally in the uplink has to be distributed among the cooperating BSs. The overall backhaul effort in *bit per channel use and cell* can be stated as

$$\beta_{\mathrm{dl}} = (\chi_{\mathrm{c}} - 1) \cdot 2 N_{\mathrm{c}} \cdot \rho \cdot q, \qquad (15.40)$$

where χ_{c} denotes the number of sites involved in the cluster. In addition, each of the sites involved in a cooperation must be provided with the data bits of all jointly served UEs. However, for similar reasons as the uplink case, the communication with the core network is not considered here.

15.4.4 Energy Consumption of Cellular Base Stations

A simple model of the long term base station energy consumption is given in [ARFB10]. In this work, the impact of energy needed for signal processing and transmit power on the total energy dissipation is assumed to be linear. In order to also capture backhauling energy needs, we propose to extend this model to the general form

$$P_{\mathrm{BS}} = a \cdot P_{\mathrm{tx}} + b \cdot P_{\mathrm{sp}} + c \cdot P_{\mathrm{bh}}, \qquad (15.41)$$

where P_{BS}, P_{tx}, P_{sp} and P_{bh} denote the average consumed energy per base station, the radiated power per base station, the signal processing power per base station, and the power due to backhauling, respectively. The coefficients a, b, and c model effects that scale with the corresponding power type such as amplifier and feeder losses, cooling, or battery backup [ARFB10]. In the following, we briefly review the three power types. Generally, the investigation presented in this study assumes full load conditions, i.e., the dependency of energy consumption on load conditions or potential sleep modes is not considered.

Transmission
Transmit power effects the overall base station power consumption through the efficiency of the power amplifier, the cooling equipment, as well as battery backup

Table 15.1. LTE-based system model parameters.

Parameter	Value
Air Interface	
Carrier frequency	2.4 GHz
Bandwidth	10 MHz
Sub-carrier spacing B_{sc}	15 kHz
Sub-carriers occupied	600
Pilot density per channel use ρ	$\frac{8}{168}$
Control overhead per channel use δ	$\frac{3}{14}$
Quantization bits per complex pilot q	8
Quantization bits per complex symbol z	16
Pathloss Model	
Pathloss exponent λ	3.76
Pathloss $10\log(U)$	-128.1 dB
Indoor penetration loss $10\log(V)$	-20 dB
Antenna gain at main lobe $10\log\left(W(0)\right)$	14 dBi
Power Model	
a	7.35
b	2.9
c	1
p_{sp}	58 W

required for operation. The average transmit power per base station scales with the inter-site-distance D according to the pathloss model (15.29) as

$$10\log(P_{tx}) = 10\log(P_{min}) + 10\log(U) + 10\log(V) + 10\lambda\log\left(\sqrt{C}\frac{D}{2}\right), \tag{15.42}$$

where P_{min} is the receiver sensitivity at the mobile, and the term $10\log U + 10\log(V) + 10\log\frac{D}{2}$ is the pathloss at the cell-edge in dB for a given inter-site-distance D. Note that (15.42) assumes that $P_{tx} \geq P_{min}$ holds for the fraction C of the cell area and that the BSs are centered in their cells.

Signal Processing

Base band digital signal processing is performed in all cellular base stations. The complexity of the operations and the energy consumption depends amongst others on the employed air interface as well as the amount of cooperation between base stations. We assume that about 10% of the overall analog and digital processing power are due to uplink channel estimation and between 1% and 10% are due to uplink and downlink multiple-input multiple-output (MIMO) processing. The former scales linearly with N_c due to the increasing number of estimated links. Assuming an MMSE filter operation, the latter requires N_c^3 operations, however, the computation is performed only once per cooperation cluster, such that average MIMO processing per base station only scales quadratically with N_c. With a base value of p_{sp}, the signal processing power per sector as a function

15.4 Energy-Efficiency Aspects of CoMP

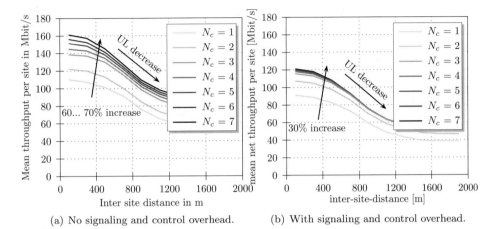

(a) No signaling and control overhead. (b) With signaling and control overhead.

Figure 15.14 Throughput per site for different inter-site-distances [FMF10]. © 2010 IEEE.

of different cooperation sizes scales between

$$P_{\rm sp} = p_{\rm sp} \cdot \left(0.89 + 0.1 N_{\rm c} + 0.01 N_{\rm c}^2\right), \text{ and} \tag{15.43}$$

$$P_{\rm sp} = p_{\rm sp} \cdot \left(0.80 + 0.1 N_{\rm c} + 0.1 N_{\rm c}^2\right). \tag{15.44}$$

Backhauling

Reflecting the state-of-the-art in most cellular networks, we model backhaul as a collection of wireless microwave links of 100 Mbit per second capacity and a power dissipation of 50 W each. Thus, for a given average backhaul requirement per base station $c_{\rm bh}$, the additional backhaul power computes as

$$P_{\rm bh} = \frac{50 {\rm W}}{100 {\rm Mbit/s}} c_{\rm bh}. \tag{15.45}$$

15.4.5 System Evaluation

In this subsection, we evaluate a cellular network model of varying density employing different cooperation sizes. We discuss results obtained by Monte Carlo simulations, which are performed as follows. For each Monte Carlo sample, a single user is randomly placed into each of the 57 observed cells, and the channel matrix stated in (15.31) and (15.33) is obtained via the pathloss model (15.29) and independent rayleigh fading samples with key parameters summarized in Table 15.1.

We assume power control in the uplink, where a target receive power density at the BS side is chosen for each value of inter-site-distance D, so that at most 5% of all UEs are operating above their power limit of 20 dBm. In the downlink,

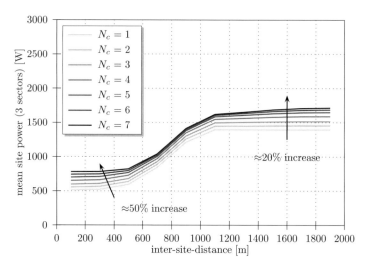

Figure 15.15 Total power dissipation per site as a function of inter-site-distance [FMF10]. © 2010 IEEE.

the total transmit power per BS is a function of D as stated in (15.42), which is then invested equally into all sub-carriers.

We generally apply cooperation in a flexible way, where the cooperation size is chosen for each fading realization such that the sum-rate of uplink and downlink is maximized. Since we assume full buffer users, all BS are under full load.

Net Site Throughput
Fig. 15.14 compares the combined up and downlink throughput before and after subtraction of pilot, control and feedback overhead as in (15.37) and (15.38). In both cases, the curves are decreasing for larger site distances due to the uplink, where UE powers are limited. In Fig. 15.14(b), about 30% gain from a cooperation size of 7 can be observed for site distances below 500 m and above 1000 m. Note that in both cases a significant performance improvement is visible if the cooperation size is increased from 2 to 3, as for $N_c = 2$ there is typically still one dominant interferer outside the cluster [MF07c, MF07b].

Site Powers
Fig. 15.15 and Fig. 15.16 display the corresponding total power dissipation per site and power fractions due to transmission, processing, and backhauling according to the power model given in Section 15.4.4. We observe from Fig. 15.16 two regimes of energy consumption: For small site distances, it is governed by signal processing and backhauling, whereas for large inter-site-distances, the transmission power becomes equally significant due to the over linear growth of the pathloss in (15.42). If large cooperation clusters are used for small site distances, where the transmission power is low, the energy consumption due to CoMP *in addition* to other processing per site can be as high as 50% of total power, even

15.4 Energy-Efficiency Aspects of CoMP

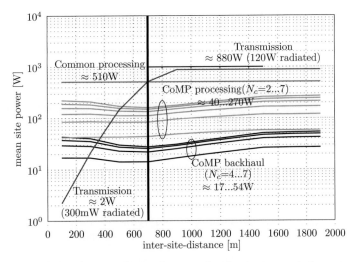

Figure 15.16 Fractions of power dissipation per site due to transmission, processing, and backhauling for different cooperation sizes. MIMO processing is assumed to be 1% of overall processing [FMF10]. © 2010 IEEE.

if MIMO processing accounts for only 1% of the base line processing power. Note in this regard, that the power values in Fig. 15.16 not only incorporate the functional block, but also include the proportionate overheads due to power supply and conversion, battery backup, and cooling.

Bit per Joule Efficiencies

Fig. 15.17 depicts the overall bit per Joule efficiency of the system for two cases, where the fraction of power due to MIMO processing was 1% and 10% of the base line processing power consumption. In the first case, depicted in Fig. 15.17(a), we observe that with increasing site distances the bit per Joule efficiency decreases due to increasing transmit powers and decreasing sum throughputs. For site distances below 500 m, the bit per Joule efficiency increases for cooperation sizes up to four BSs and decreases with higher cooperation sizes compared to no cooperation. For large site distances over 1000 m, the bit per Joule efficiency is always increased through cooperation mostly due to the uplink, where array gain mitigates low received powers from distant UEs. For cooperation sizes larger than three, spectral efficiency gains are rather small (compare Fig. 15.14). In comparison, the additional energy consumption is larger due to the quadratic increase of the MIMO processing fraction on one hand and the additional backhaul power on the other (see Fig. 15.15).

15.4.6 Summary

The bit per Joule efficiency of the system is only moderately affected by the use of CoMP schemes, with potential gains of 10% and 20% for small and large site

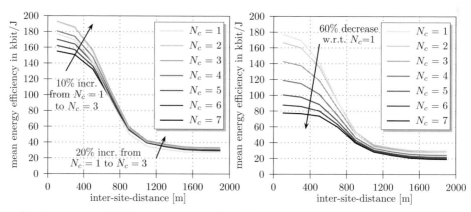

(a) Assumption: MIMO processing is 1% of baseline power.

(b) Assumption: MIMO processing is 10% of baseline power.

Figure 15.17 Bit per Joule efficiency of CoMP for different inter-site-distances [FMF10]. © 2010 IEEE.

distances, respectively, if MIMO processing power is on the order of 1% of the total processing power. From a bit per Joule perspective and for the power model used in this research, cooperation between non co-located base stations does not appear beneficial due to (1) the diminishing additional spectral efficiency gains for cluster sizes above three, (2) the quadratically increasing power consumption of UL and DL MIMO processing and (3) the additionally required backhauling power for non co-located clusters. For this reason, highest energy efficiency gains are observed when all cooperating BS are co-located throughout all site distances.

For the power model used in this research, where base station transmit powers are adjusted to the network density, the contributions of processing and backhauling power dominate power related to transmission for small site distances. This fact should be kept in mind, considering recent research activities focusing strongly on optimization of power amplifiers for relatively high transmit powers of 20 W and above, in particular in the light of decreasing site distances for future generations of mobile technology.

16 Summary and Conclusions

16.1 Summary of this Book

Patrick Marsch

In this book, coordinated multi-point (CoMP) has been investigated from the point of view of a multitude of authors, some of which have been working in this field since a decade. Clearly, the different opinions stated throughout the book emphasize that CoMP is still a controversially discussed topic, and it is still too early to draw final conclusions. Nevertheless, we now want to briefly review the contributions in this book and try to extract a more or less conclusive *big picture*.

16.1.1 Most Promising CoMP Schemes and Potential Gains

First, let us summarize the key CoMP schemes discussed in this book, which were categorized according to the extent of base station (BS) cooperation involved:

Non-cooperative, but interference aware transceiver techniques have been investigated in Section 5.1. Here, LTE Release 8 style precoding is used in the downlink, in conjunction with multi-cell interference estimation at the receiver side, which was also covered in Section 10.2. The results predict large gains from using smart multi-cell pilot and channel estimation schemes, even though no cooperation between BSs is required.

Interference coordination schemes, in comparison, are based on some limited extent of information exchange between BSs. In Section 5.2, multi-cell *interference prediction* through the exchange of scheduling tables and *coordinated scheduling* by a central unit were introduced, yielding average spectral efficiency gains of more than 20%, according to system level simulation results in Section 14.3. Though only a minimal backhaul *capacity* is required, these schemes are very sensitive to backhaul *latency*. Regarding the downlink, *coordinated scheduling and beamforming* (CS/CB) was addressed in Sections 5.3 and 14.4.3, where BSs perform a trade-off between maximizing the signal towards terminals in the own cell and avoiding inter-cell interference. As shown through simulation results in Section 14.4, such schemes provide only limited gains for particular antenna configurations in homogeneous networks, but appear to be an interesting option for heterogeneous deployments.

Finally, **multi-cell joint signal processing** promises the largest CoMP gains, but at the price of a large backhaul capacity in the uplink, and strong synchronization requirements and a complex system design in the downlink.

Regarding the *uplink*, Section 4.3.1 motivates the usage of *centralized* uplink joint detection (see also Section 6.1) from a backhaul efficiency point of view, but Section 11.2 focuses on *decentralized* schemes, as they offer a better compatibility to an LTE Release 8 system. While simulations in Section 14.3 predict average gains of multi-cell joint detection of 40%, large-scale field trials described in Section 13.2 have indicated about 10% or 33% average cooperation gain, depending on the number of antennas per base station. Both Sections 13.1 and 13.2 point out the benefit of successive interference cancelation (SIC) in non-cooperative detection. Section 6.2 proposes iterative uplink cooperation schemes with various advantages, but the major downside of a large latency.

Regarding the *downlink*, multi-cell joint transmission has been investigated in Sections 6.3 and 6.4. While simulations in Section 14.4 foresee an average gain of about 20% for two-cell cooperation, field trial results in Sections 13.3 and 13.4 report larger gains, for example 40-50% in Section 13.4. This can be explained through the fact that so far no background interference has been considered in the trials. In general, downlink joint signal processing requires less backhaul capacity than the uplink, but is more challenging in terms of synchronization, channel state information (CSI) feedback and the potential exchange of CSI among BSs, as discussed in Section 13.3. An interesting field for further research are precoding techniques that are robust towards a different extent of CSI at different BSs, see Section 6.4. In general, due to CSI feedback delay, downlink joint transmission may most likely only be applied to fairly static terminals.

Clearly, neither the field trial results from Chapter 13, nor system level simulation results provided in Chapter 14 allow to fully predict the gain of CoMP schemes in commercial deployments, as pointed out in Section 13.5. However, the results provide an insight into the *order of magnitude* of potential gains.

In general, it appears that *interference coordination* schemes are indeed an attractive option, as they offer substantial performance gains at a limited effort. While *joint signal processing* schemes potentially offer even larger gains, their success will strongly depend on how enablers such as synchronization and backhaul can be provided in an economically efficient way. The question when certain schemes will appear in commercial systems will also depend strongly on how easily these can be embedded into the LTE-A roadmap. For instance, a scheme based on a potential proprietary solution will clearly appear earlier than one that requires changes in standardization. However, as CoMP is the only technology so far that effectively addresses the issue of cell-edge data rates, it seems clear that CoMP will always have its firm place in future mobile communications alongside technologies such as heterogeneous deployments and relaying concepts.

16.1.2 Key Challenges Identified

While field trial results in Chapter 13 have proven that various CoMP concepts discussed in this book do work in practice, their usage in practical and commercial systems may be subject to various challenges, as identified in Chapters 7-12:

Clustering can be seen as a prerequisite of any CoMP concept, i.e. one generally has to decide which set of cells may potentially cooperate, such that various overhead issues (backhaul, signaling etc.) connected to CoMP are kept within reasonable bounds. However, it has been shown in Sections 7.1 and 7.2, respectively, that promising static or dynamic clustering concepts are available that can most likely be used in future communications systems in a very straightforward way. Hence, clustering is here *not* considered to be a major obstacle within the roadmap to CoMP.

Synchronization issues, however, can be seen as a clear limitation of CoMP. It has been shown in Section 8.2 that signal propagation delays between multiple terminals and cooperating base stations may lead to inter-symbol and inter-carrier interference, rendering complex equalization necessary. This problem may especially arise when applying CoMP in rural areas with large base station distances. Further, Section 13.3 has emphasized the importance of having precisely synchronized base station oscillators for downlink joint transmission. This is possible with current off-the-shelf equipment, but may be considered a problematic cost issue in a practical deployment.

Channel knowledge is a key requirement for all CoMP schemes in uplink and downlink. Section 9.1 has shown that legacy channel estimation techniques can also be applied to estimate inter-cell links or interference. However, the fact that weak multi-cell links cannot be accurately estimated, and that pilot overhead increases linearly in the cooperation size, limits CoMP to scenarios of moderate cooperation size. Also the feedback of channel or interference information from terminals to the base station side is doable with legacy techniques, as shown in Section 9.2, but CoMP gains have to be carefully traded against potentially significant feedback overhead. The design of reference signals and channel feedback will most likely require changes in standardization.

Hardware implementation is mainly challenging in the way that future systems have to support a significantly larger number of transmission modes than before, in addition to many potential cooperation modes. Sections 10.1 and 10.2, however, have shown that both transmitter and receiver algorithms can be implemented in a flexible and efficient way. Hardware implementation issues are hence *not* considered to be severe issues connected to CoMP.

Scheduling, signaling and an adaptive usage of CoMP is definitely a large field for future research, where the key question is how CoMP can be applied efficiently among the right terminals and base stations at the right time. Sections 11.1 and 11.2 have shown that promising user grouping and signaling concepts exist that can be used for CoMP, while Section 11.3 has ventured into the novel topic of Ad-Hoc CoMP. Especially signaling for CoMP can be seen as a challenge within LTE-A, and may require changes in standardization. One example is the usage of uplink joint detection in conjunction with hybrid automatic repeat request (HARQ).

Backhaul in fact poses one of the main issues connected to CoMP. As shown in Sections 6.1, 12.1 and 12.2, uplink joint detection may require vast amounts of inter-BS backhaul capacity. This suggests a selective and adaptive usage of different cooperation concepts, see Section 4.3.2. Compared to this, the backhaul requirements of downlink CoMP are rather negligible for reasonably chosen cooperation sizes. In general, the backhaul *capacity* itself might not be the main issue, but rather the fact that this has to be provided at low, possibly guaranteed *latency*. While Section 12.3 has shown that existing backhaul technologies can principally provide the desired capacity and latency, the success of CoMP will depend on the economical attractiveness of such solutions. Furthermore, standardization activities may be required concerning the X2 interface.

Terminal challenges have not been highlighted in this book, as basically all discussed CoMP schemes have shown to require no (in the case of most uplink schemes) or rather moderate changes (in the case of the downlink) to the terminal side. However, the general increase in mobile data rates and required processing power through the overall set of technologies within, e.g., LTE-A will also render a continuous research on improved terminal architectures necessary.

To summarize, one can state that there is no major show-stopper connected to the usage of CoMP in LTE-Advanced, but some standardization changes will be required (e.g. connected to pilot concepts, uplink HARQ, or the X2 interface). Furthermore, some aspects may be critical from a cost point of view (in particular backhaul and synchronization of base stations in frequency).

16.2 Conclusions

Gerhard Fettweis

As Marconi opened the world of opportunities which lie within wireless communications more than a century ago, today's world is embracing this technology with full swing. We have experienced a world where the data rate provided by cellular communications systems has grown at the same rate as Moore's law, i.e. by one order of magnitude every five years. This enormous growth has only been able to happen through many innovations entering systems and their implementation. Obviously, as the demand for more data rate at improved coverage and fairness remains, the physical limits of the capacity of systems need to be explored by carefully engineering the boundary conditions.

Digital cellular communication link capacity is limited through the signal-to-interference-and-noise ratio (SINR) which is being encountered. To maximize the SINR, the first digital cellular systems were designed around minimizing interference through frequency planning. The following (CDMA) systems improved the SINR by maximizing the received signal power through coherent signal combining, referred to as soft(er) hand-off. However, this can also come at the expense of increasing the interference. Therefore, base station placement and antenna downtilt are major means for optimization of the total system capacity.

With the introduction of multiple-input multiple-output (MIMO), the multi-antenna technology can be used not only for parallel communication of streams for rate enhancement, but also for beam steering to ensure that interference is minimized and the signal is steered into the direction of interest. The same can be done at the receiving end, respectively.

The new chapter within wireless communication being opened by coordinated multi-point (CoMP) transmission and reception is that interference is managed as a signal-of-interest and henceforth minimized through multiple means, either by avoidance through smart transmission concepts, or by avoidance or cancelation at the receiving points. CoMP technology therefore can reduce the amount of interference over non-CoMP by orders of magnitude. It opens a new path for multi-cell wireless technology development. While today CoMP technology is driven by cellular communications, it is not necessarily restricted to this area of application. The force of CoMP lies behind its possibility to maximize SINR by minimizing the interference, not by maximizing the signal strength. As the scientific community is only starting to understand the principles behind CoMP, its practical limits cannot be foreseen at this point in time.

16.2.1 About this Book

This book has presented the principle technologies which are behind CoMP, and its practical implementation challenges, as seen today. It is the first comprehensive book on CoMP technology which covers the span from theoretical limits and principles all the way to practical limits and field trial measurements. We have tried to bridge the gap between theory and practice in such a way that each researcher or developer personally siding with either theory or practice shall be enabled to get a bigger picture of the power behind CoMP, and shall be enabled to understand this picture. As CoMP entails an understanding of network setup, channel characteristics, link design, synchronization and channel estimation, signal processing, media access, radio resource management, and more, it is a technology which fundamentally must be taken into account at all layers of future wireless communication systems to be able to unleash its potential for increasing system and link capacity, as well as fairness and coverage.

As researchers, the editors of this book have been active in the area of CoMP during the last decade. We have learned from theoretical analysis as well as from field measurements that CoMP can be a very complex technology which can only unfold its full power when carefully designed into a system. However, if this has been done, the results presented in this book clearly show the potential of this technology.

Today, the capacity of deployed cellular communications systems is interference limited. In theory, CoMP could change the picture by moving from interference limited to noise limited communications. This books shall be a basis for making many further results come to fruition in standards and products.

16.2.2 CoMP's Place in the LTE-Advanced Roadmap and Beyond

The general idea underlying CoMP of interference management, coordination, avoidance, and cancelation can be deployed along manifold different technological paths. A high performance CoMP, e.g. incorporating distributed MIMO, requires all levels of a system to be considered. It therefore will take time to enter into a firm release within the standardization roadmap. First precursors as multi-user MIMO are been seen to enter standardization in Release 10 of 3GPP LTE. Further advancements as coordinated scheduling for interference avoidance and coordination can then be expected to enter thereafter. However, full distributed MIMO with interference cancelation will take some more years on the path towards LTE-Advanced.

Since the current cellular communications roadmap is seen to be dominated by one single standardization body, namely 3GPP, it can be expected that the standardization procedure will become slowed down as all forces try to influence one single entity. CoMP can be seen as the new technology path for moving from interference limited designs towards noise limited designs. It can be expected that today's understanding will be enhanced largely within the near future. Hence, a competing new system proposal will most likely be arising outside 3GPP within the next decade, probably with a new coherent approach of maximizing the gains being provided by CoMP.

Beyond cellular communications, CoMP technology can be expected to become a major part of other future high capacity multi-cellular systems, as e.g. WiFi.

Certainly, many more challenges exist beyond the six categories mentioned in the summary section above, which must be addressed for CoMP to make its major inroad into wireless communications standards. The chapters of this book have given the reader an insight into the new exciting field of CoMP. The field of research has one major differentiating challenge, as mentioned in the above subsections: The use of sophisticated CoMP technology requires the analysis of its impact at a system level as well as at a link level and its signal processing required. Clearly, a deep understanding of the physical layer, all the way to its hardware implementation and measurement experience of test bed setups, as well as radio resource management and network setup, must be considered as a whole. We hope that this book has given the reader an insight and understanding of this challenge and is a basis for further advancement of CoMP technology development.

References

[3GP04] 3GPP2 C.S0022-A v1.0. Position determination service for cdma2000 spread spectrum system, technical specification, Mar. 2004.

[3GP06a] 3GPP R1-062483, Philips. Comparison between MU-MIMO codebook-based channel reporting techniques for LTE downlink, Oct. 2006.

[3GP06b] 3GPP R1-070346, Philips. Comparison of MU-MIMO feedback schemes with multiple UE receive antennas, Jan. 2006.

[3GP06c] 3GPP TR 25.814 V7.0.0. Technical Specification Group Radio Access Network; Physical layer aspects for evolved Universal Terrestrial Radio Access (UTRA), Sept. 2006.

[3GP07a] 3GPP TR 25.996 V7.0.0. Spatial channel model for multiple input multiple output (MIMO) simulations (Release 7), July 2007.

[3GP07b] 3GPP TS 36.211 V8.0.0. E-UTRA - Physical channels and modulation (Release 8), Sept. 2007.

[3GP07c] 3GPP TS 36.423 V8.8.0. E-UTRAN: X2 application protocol (X2AP), 2007.

[3GP08a] 3GPP R1-083759, Alcatel-Lucent. UE PMI feedback signalling for user pairing/coordination, Sept. 2008.

[3GP08b] 3GPP TS 25.943 V8.0.0. Deployment aspects (release 8), Dec. 2008.

[3GP09a] 3GPP R1-091444, Qualcomm Europe. Update on E-IPDL Performance, Mar. 2009.

[3GP09b] 3GPP R1-091782, Alcatel-Lucent. Estimation of extended PMI feedback signaling required for user intra-cell and inter-cell coordination, May 2009.

[3GP09c] 3GPP R1-093016. Consideration on performance of coordinated beamforming with PMI feedback, Apr. 2009.

[3GP09d] 3GPP TS 25.305 V9.0.0. Stage 2 functional specification of User Equipment (UE) positioning in UTRAN (Release 9), Dec. 2009.

[3GP09e] 3GPP TS 36.201 V9.8.0. E-UTRA - LTE physical layer general description, Mar. 2009.

[3GP09f] 3GPP TS 36.213 V8.8.0. Evolved universal terrestrial radio access (EUTRA): Physical layer procedures, Sept. 2009.

[3GP10a] 3GPP. Overview of 3GPP Release 8, Apr. 2010.

[3GP10b] 3GPP R1-100944, Alcatel-Lucent. Performance of coordinated beamforming with multiple PMI feedback, Feb. 2010.

[3GP10c] 3GPP TR 25.214 V9.2.0. Universal mobile telecommunications system (umts); physical layer procedures (FDD), technical specification (release 9), Mar. 2010.

[3GP10d] 3GPP TR 36.814 v9.0.0. Further advancements for E-UTRA - physical layer aspects, Mar. 2010.

[3GP10e] 3GPP TR 36.902 v9.2.0. E-UTRAN - Self-configuring and self-optimizing network (SON) use cases and solutions (Release 9), June 2010.

[3GP10f] 3GPP TS 23.271 V9.3.0. Digital cellular telecommunications system (phase 2+); umts; lte; functional stage 2 description of location services (lcs), technical specification (release 9), Mar. 2010.

[3GP10g] 3GPP TS 29.281 v9.3.0. General Packet Radio System (GPRS) Tunnelling Protocol User Plane (GTPv1-U), June 2010.

[3GP10h] 3GPP TS 36.101 V8.10.0. Evolved universal terrestrial radio access (EUTRA): User equipment (ue) radio transmission and reception, June 2010.

[3GP10i] 3GPP TS 36.300 V9.3.0. E-UTRAN - Overall Description (Release 9), Mar. 2010.

[AA09] I.H. Azzam and R.S. Adve. Linear Precoding for Multiuser MIMO Systems with Multiple Base Stations. In *IEEE Int. Conf. on Comms. (ICC'09)*, June 2009.

[ABG+06] H. Akhavan, V. Badrinath, T. Geitner, H. Lennertz, Y. Sha, T. Utano, and B. West. Next Generation Mobile Networks Beyond HSPA, Dec. 2006. www.ngmn.org.

[ABI10] ABI Research. Wireless subscriber forecasts, June 2010.

[ADF+09] D. Astély, E. Dahlman, A. Furuskär, Y. Jading, M. Lindström, and S. Parkvall. LTE: The evolution of mobile broadband. *IEEE Comms. Magazine*, 47(4):44–51, Apr. 2009.

[ADT07] S. Avestimehr, S. Diggavi, and D. Tse. Wireless network information flow. In *Allerton Conf. on Comm., Control, and Computing (ALLERTON'07)*, Sept. 2007.

[AEH08] E. Aktas, J. Evans, and S. Hanly. Distributed Decoding in a Cellular Multiple-Access Channel. *IEEE Trans. on Wireless Comms.*, 7(1):241–250, Jan. 2008.

[Ahl71] R. Ahlswede. Multi-way communication channels. In *IEEE Int. Symp. on Inf. Theory (ISIT'71)*, Sept. 1971.

[Ahl74] R. Ahlswede. The capacity region of a channel with two senders and two receivers. *The Annals of Probability*, 2(5):805–814, 1974.

[Ahn08] K.S. Ahn. Performance analysis of MIMO-MRC systems with channel estimation error in the presence of cochannel interferences. *IEEE Signal Proc. Letters*, 15:445–448, 2008.

[And05] J.G. Andrews. Interference cancellation for cellular systems: A contemporary overview. *IEEE Trans. on Wireless Comms.*, 12(2):19–29, Apr. 2005.

[APM02] A. Algans, K.I. Pedersen, and P.E. Mogensen. Experimental Analysis of the Joint Statistical Properties of Azimuth Spread, Delay Spread, and Shadow Fading. *IEEE Journal on Selected Areas in Comms.*, 20(3):523–531, Apr. 2002.

[ARFB10] O. Arnold, F. Richter, G. Fettweis, and O. Blume. Power consumption modeling for different base station types in heterogeneous cellular networks. In *Future Network and Mobile Summit*, June 2010.

[BAS+05] K. Brueninghaus, D. Astely, T. Sälzer, S. Visuri, A. Alexiou, S. Karger, and G.A. Seraji. Link performance models for system level simulations of broadband radio access systems. In *IEEE Int. Symp. on Personal, Indoor and Mobile Radio Comms. (PIMRC'05)*, Sept. 2005.

[BC07] S. Bavarian and J.K. Cavers. Reduced Complexity Distributed Base Station Processing in the Uplink of Cellular Networks. In *IEEE Global Telecomms. Conf. (GLOBECOM'07)*, Nov. 2007.

[Ber74] P. Bergmans. A simple converse for broadcast channels with additive white Gaussian noise. *IEEE Trans. on Inf. Theory*, 20(2):279–280, Mar. 1974.

[BH06] F. Boccardi and H. Huang. Zero-forcing precoding for the MIMO broadcast channel under per-antenna power constraints. In *IEEE Workshop on Signal Proc. Advances in Wireless Comms. (SPAWC'06)*, June 2006.

[BH07a] F. Boccardi and H. Huang. A Near-Optimum Technique using Linear Precoding for the MIMO Broadcast Channel. In *IEEE Int. Conf. on Acoustics, Speech and Signal Proc. (ICASSP'07)*, Apr. 2007.

[BH07b] F. Boccardi and H. Huang. Limited downlink network coordination in cellular networks. In *IEEE Int. Symp. on Personal, Indoor and Mobile Radio Comms. (PIMRC'07)*, Sept. 2007.

[BH10] R. Bhagavatula and R.W. Heath Jr. Adaptive bit partitioning for multicell intercell interference cancellation with delayed limited feedback. *IEEE Trans. on Signal Proc.*, July 2010. submitted for publication, available at http://arxiv.org/pdf/1010.3033.

[BHA08] F. Boccardi, H. Huang, and A. Alexiou. Network MIMO with reduced backhaul requirements by MAC coordination. In *Asilomar Conf. on Signals, Systems and Computers (ASILOMAR'08)*, Oct. 2008.

[BHD+05] D.S. Baum, J. Hansen, G. Del Galdo, M. Milojevic, J. Salo, and P. Kyösti. An interim channel model for beyond-3G systems: extending the 3GPP spatial channel model (SCM). In *IEEE Vehicular Technology Conf. (VTC'05 Spring)*, May 2005.

[BMWT00] P.W. Baier, M. Meurer, T. Weber, and H. Troger. Joint transmission (JT), an alternative rationale for the downlink of time division CDMA using multi-element transmit antennas. In *IEEE Int. Symp. on Spread Spectrum Techniques and Appl. (ISSSTA'00)*, Sept. 2000.

[BP07] F. Berggren and B.M. Popovic. A non-hierarchical cell search scheme. In *IEEE Wireless Comms. and Networking Conf. (WCNC'07)*, Mar. 2007.

[BPG+09] G. Boudreau, J. Panicker, N. Guo, R. Chang, N. Wang, and S. Vrzic. Interference coordination and cancellation for 4G networks. *IEEE Comms. Magazine*, 47(4):74–81, Apr. 2009.

[BS02] H. Boche and M. Schubert. A general duality theory for uplink and downlink beamforming. In *IEEE Vehicular Technology Conf. (VTC'02 Fall)*, Sept. 2002.

[BT08] G. Bresler and D. Tse. The two-user Gaussian interference channel: a deterministic view. *European Trans. on Telecomms.*, 19(4):333–354, June 2008.

[BTC06] F. Boccardi, F. Tosato, and G. Caire. Precoding Schemes for the MIMO-GBC. In *Int. Zurich Seminar on Comms. (ZSC'06)*, Feb. 2006.

[BV04] S. Boyd and L. Vandenberghe. *Convex Optimization*. Cambridge University Press, 2004.

[BX05] B. Borloz and B. Xerri. Subspace signal-to-noise ratio maximization: the constrained stochastic matched filter. In *Int. Symp. on Signal Proc. and Its Appl. (ISSPA'05)*, Aug. 2005.

[Cal04] D.B. Calvo. *Fairness Analysis of Wireless Beamforming Schedulers*. PhD thesis, Technical University of Catalonia, 2004.

[Car75] A. Carleial. A case where interference does not reduce capacity. *IEEE Trans. on Inf. Theory*, 21(5):569–570, Sept. 1975.

[CC07] P.W.C. Chan and R.S. Cheng. Capacity maximization for zero-forcing MIMO-OFDMA downlink systems with multiuser diversity. *IEEE Journal on Wireless Comms. Systems*, 6(5):1880–1889, May 2007.

[CDG00] S. Catreux, P.F. Driessen, and L.J Greenstein. Simulation results for an interference-limited multiple input multiple output cellular system. *IEEE Comms. Letters*, 4:334 – 336, Nov. 2000.

[CG79] T. Cover and A. El Gamal. Capacity theorems for the relay channel. *IEEE Trans. on Inf. Theory*, 25(5):572–584, Sept. 1979.

[Cha85] D. Chase. Code combining–a maximum-likelihood decoding approach for combining an arbitrary number of noisy packets. *IEEE Trans. on Comms.*, 33(5):385–393, May 1985.

[CHHT10] C.-B. Chae, I. Hwang, R.W. Heath Jr., and V Tarokh. Interference-aware coordinated beamforming in two-cell systems. 2010. technical report, available at http://dash.harvard.edu/handle/1/3293263.

[Cim85] L.J. Cimini. Analysis and simulation of a digital mobile channel using orthogonal frequency division multiplexing. *IEEE Trans. on Inf. Theory*, 33(7):665–675, July 1985.

[Cis10] Cisco. Cisco Visual Networking Index (VNI) Global Mobile Data Traffic Forecast Update 2009-2014. Feb. 2010.

[CJ08] V.R. Cadambe and S.A. Jafar. Interference alignment and spatial degrees of freedom for the k user interference channel. In *IEEE Int. Conf. on Comms. (ICC'08)*, May 2008.

[CJCU07] P.M. Castro, M. Joham, L. Castedo, and W. Utschick. Robust precoding for multiuser MISO systems with limited-feedback channels. In *Int. ITG Workshop on Smart Antennas (WSA'07)*, Mar. 2007.

[CKH10] C.B. Chae, S. Kim, and R.W. Heath Jr. Network Coordinated Beamforming for Cell-Boundary Users: Linear and Nonlinear Approaches. *IEEE Journal of Selected Topics in Signal Proc.*, 3(6):1094–1105, 2010.

[CLRS01] T.H. Cormen, C.E. Leiserson, R.L. Rivest, and C. Stein. *Introduction to Algorithms*. The MIT Press, 2001.

[CM04] L.-U. Choi and R.D. Murch. A Transmit Preprocessing Technique for Multiuser MIMO Systems Using a Decomposition Approach. *IEEE Trans. on Wireless Comms.*, 3(1):20 – 24, Jan. 2004.

[CMIH08] C.-B. Chae, D. Mazzarese, T. Inoue, and R.W. Heath Jr. Coordinated beamforming for the multiuser MIMO broadcast channel with limited feedforward. *IEEE Trans. on Signal Proc.*, 56(12):6044–6056, Dec. 2008.

[CMJH08] C.-B. Chae, D. Mazzarese, N. Jindal, and R.W. Heath Jr. Coordinated beamforming with limited feedback in the MIMO broadcast channel. *IEEE Journal on Selected Areas in Comms.*, 26(8):1505–1515, Oct. 2008.

[Cos83] M. Costa. Writing on dirty paper. *IEEE Trans. on Inf. Theory*, 29:439–441, May 1983.

[CRA+08] F.D. Calabrese, C. Rosa, M. Anas, P.H. Michaelsen, K.I. Pedersen, and P.E. Mogensen. Adaptive Transmission Bandwidth Based Packet Scheduling for LTE Uplink. In *IEEE Vehicular Technology Conf. (VTC'08 Fall)*, Sept. 2008.

[CS03] G. Caire and S. Shamai (Shitz). On the achievable throughput of a multiantenna Gaussian broadcast channel. *IEEE Trans. on Inf. Theory*, 49(7):802–811, June 2003.

[CT06] T.M. Cover and J.A. Thomas. *Elements of Information Theory*. Wiley, 2006.

[CTB96] G. Caire, G. Taricco, and E. Biglieri. Capacity of bit-interleaved channels. *Electronics Letters*, 32(12):1060–1061, Juni 1996.

[DB07] G. Dietl and G. Bauch. Linear precoding in the downlink of limited feedback multiuser MIMO systems. In *IEEE Global Telecomms. Conf. (GLOBECOM'07)*, Nov. 2007.

[dCS09] A. del Coso and S. Simoens. Distributed compression for MIMO coordinated networks with a backhaul constraint. *IEEE Trans. on Wireless Comms.*, 8(9):4698 – 4709, Sept. 2009.

[DEFVF07] V. Degli-Esposti, F. Fuschini, M. Vitucci, and G. Falciasecca. Measurement and Modelling of Scattering From Buildings. *IEEE Trans. on Antennas and Propagation*, 55(1):143 – 152, Jan. 2007.

[DEGd+04] V. Degli-Esposti, D. Guiducci, A. de'Marsi, P. Azzi, and F. Fuschini. An Advanced Field Prediction Model Including Diffuse Scattering. *IEEE Trans. on Antennas and Propagation*, 52(7):1717 – 1728, July 2004.

[DH08] D.R. Brown III and H.V. Poor. Time-slotted round-trip carrier synchronization for distributed beamforming. *IEEE Trans. on Signal Proc.*, 56(11):5630–5643, Nov. 2008.

[DLU09] G. Dietl, O. Labrèche, and W. Utschick. Channel vector quantization for multiuser mimo systems aiming at maximum sum rate. In *IEEE Global Telecomms. Conf. (GLOBECOM'09)*, Nov. 2009.

[DLZ06] P. Ding, D. J. Love, and M. D. Zoltowski. Multiple antenna broadcast channels with limited feedback. In *Int. Conf. on Acoustics, Speech, and Signal Proc. (ICASSP'06)*, May 2006.

[DMF10] F. Diehm, P. Marsch, and G. Fettweis. The FUTON Prototype: Proof of Concept for Coordinated Multi-Point in Conjunction with a Novel Integrated Wireless/Optical Architecture. In *IEEE Wireless Comms. and Networking Conf. (WCNC'10)*, Apr. 2010.

[DMP04] H. Dai, A.F. Molisch, and H.V. Poor. Downlink capacity of interference-limited MIMO systems with joint detection. *IEEE Trans. on Wireless Comms.*, 3(2):442 – 453, Mar.

2004.

[DP03] H. Dai and H.V. Poor. Asymptotic spectral efficiency of multicell MIMO systems with frequency-flat fading. *IEEE Trans. on Signal Proc.*, 51(11):2976–2988, Nov. 2003.

[DPM05] D.R. Brown III, G. Prince, and J.A. McNeill. A method for carrier frequency and phase synch. of two autonomous cooperative transmitters. In *IEEE Workshop on Signal Proc. Advances in Wireless Comms. (SPAWC'05)*, June 2005.

[DPSB08] E. Dahlman, S. Parkvall, J. Skold, and P. Beming. *3G Evolution: HSPA and LTE for Mobile Broadband*. Academic Press, 2008.

[DS04] Goran Dimic and Nicholas D. Sidiropoulos. Low-complexity downlink beamforming for maximum sum capacity. In *Int. Conf. on Acoustics, Speech, and Signal Proc. (ICASSP'04)*, May 2004.

[DS05] G. Dimić and N. D. Sidiropoulos. On Downlink Beamforming With Greedy User Selection: performance Analysis and a Simple New Algorithm. *IEEE Trans. on Comms.*, 53(10):3857 – 3868, Oct. 2005.

[dS08] A. del Coso and S. Simoens. Distributed compression for the uplink channel of a coordinated cellular networkwith a backhaul constraint. In *IEEE Workshop on Signal Processing Advances in Wireless Comms. (SPAWC'08)*, July 2008.

[DSR98] D. Divsalar, M.K. Simon, and D. Raphaeli. Improved parallel interference cancellation for CDMA. *IEEE Trans. on Comms.*, 46(2):258–268, Feb. 1998.

[DY08] H. Dahrouj and W. Yu. Coordinated beamforming for the multi-cell multi-antenna wireless system. In *Conf. on Inf. Sciences and Systems (CISS'08)*, Mar. 2008.

[EC91] W.H.R. Equitz and T.M. Cover. Successive refinement of information. *IEEE Trans. on Inf. Theory*, 37(2):269 – 275, Mar. 1991.

[EC05] A. Ekbal and J.M. Cioffi. Distributed transmit beamforming in cellular networks - a convex optimization perspective. In *IEEE Int. Conf. on Comms. (ICC'05)*, May 2005.

[ETW06] R. Etkin, D.N.C. Tse, and H. Wang. Gaussian interference channel capacity to within one bit: the symmetric case. In *IEEE Inf. Theory Workshop (ITW'06)*, Oct. 2006.

[ETW08] R.H. Etkin, D.N.C. Tse, and H. Wang. Gaussian interference channel capacity to within one bit. *IEEE Trans. on Inf. Theory*, 54(12):5534–5562, Dec. 2008.

[FDGH07] M. Fuchs, G. Del Galdo, and M. Haardt. Low-Complexity Space-Time-Frequency Scheduling for MIMO Systems With SDMA. *IEEE Trans. on Vehicular Technology*, 56(5):2775–2784, Sept. 2007.

[FDH07] M. Fuchs, G. Del Galdo, and M. Haardt. Low-complexity space-time-frequency scheduling for MIMO systems with SDMA. *IEEE Trans. on Vehicular Technology*, 56(5):2775–2784, Sept. 2007.

[Fed99] Federal Communications Commission (FCC). FCC 99-245: Third Report and Order, 1999. http://www.fcc.gov/911/enhanced/.

[FG98] G. J. Foschini and M. J. Gans. On Limits of Wireless Communications in a Fading Environment When Using Multiple Antennas. *Wireless Personal Comms.*, 6(3):311–335, Mar. 1998.

[FHG09] L. Falconetti, C. Hoymann, and R. Gupta. Distributed uplink macro diversity for cooperating base stations. In *Int. Workshop on LTE Evolution (at ICC'09)*, June 2009.

[FHK+05] G. Foschini, H.C. Huang, K. Karakayali, R.A. Valenzuela, and S. Venkatesan. The value of coherent base station coordination. In *Conf. on Inf. Sciences and Systems (CISS05)*, Mar. 2005.

[FHK+10] G. Fettweis, J. Holfeld, V. Kotzsch, P. Marsch, E. Ohlmer, Z. Rong, and P. Rost. Field trial results for lte-advanced concepts. In *Int. Conf. on Acoustics, Speech, and Signal Proc. (ICASSP'10)*, Mar. 2010.

[FKR+09] G. Fodor, C. Koutsimanis, A. Rácz, N. Reider, A. Simonsson, and W. Müller. Intercell interference coordination in OFDMA networks and in the 3GPP long term evolution system. *Journal of Comms.*, 7:445–453, Aug. 2009.

[FMBF10] A. Fehske, J. Malmodin, G. Biczok, and G. Fettweis. The global carbon footprint of mobile communications - the ecological and economic perspective. *IEEE Comms. Magazine*, Nov. 2010.

[FMd09] FMVC5S series datasheet, February 2009.

[FMDS10] P. Frank, A. Müller, H. Droste, and J. Speidel. Cooperative interference-aware joint scheduling for the 3GPP LTE uplink. In *IEEE Int. Symp. on Personal, Indoor and Mobile Radio Comms. (PIMRC'10)*, Sept. 2010.

[FMF10] A.J. Fehske, P. Marsch, and G. Fettweis. Bit per Joule Efficiency of Cooperating Base Stations in Cellular Networks. In *IEEE Global Telecomms. Conf. (GLOBECOM'10)*, Dec. 2010.

[FRF09] A. J. Fehske, F. Richter, and G. P. Fettweis. Energy efficiency improvements through micro sites in cellular mobile radio networks. In *Workshop on Green Communications (at GLOBECOM'09)*, Dec. 2009.

[FU98] G.D. Forney and G. Ungerboeck. Modulation and coding for linear Gaussian channels. *IEEE Trans. on Inf. Theory*, 44(6):2384–2415, Oct. 1998.

[FVJF10a] R. Fritzsche, J. Voigt, C. Jandura, and G. Fettweis. Comparing Ray Tracing Based MU-CoMP-MIMO Channel Predictions with Channel Sounding Measurements. In *European Conf. on Antennas and Propagation (EuCAP'10)*, Apr. 2010.

[FVJF10b] R. Fritzsche, J. Voigt, C. Jandura, and G. Fettweis. Verifying Ray Tracing Based CoMP–MIMO Predictions with Channel Sounding Measurements. In *Int. ITG/IEEE Workshop on Smart Antennas (WSA'10)*, Feb. 2010.

[G.707] G.707/Y.1322 Standardization Group. Network node interface for the synchronous digital hierarchy (SDH), Jan. 2007.

[GA02] D. Gesbert and J. Akthar. Breaking the barriers of Shannon's capacity: An overview of MIMO wireless systems. *Telenor's Journal Telektronikk*, 1:53–64, Jan. 2002.

[GA10] J. Giese and M. A. Amin. Performance Upper Bounds for Coordinated Beam Selection in LTE-Advanced. In *Int. ITG/IEEE Workshop on Smart Antennas (WSA'10)*, Feb. 2010.

[Gas04] M. Gastpar. The wyner-ziv problem with multiple sources. *IEEE Trans. on Inf. Theory*, 50(11):2762–2768, Nov. 2004.

[GDV02] M. Gastpar, P.L. Dragotti, and M. Vetterli. The distributed Karhunen-Loeve transform. In *IEEE Workshop on Multimedia Signal Processing*, Dec. 2002.

[GG92] A. Gersho and R.M. Gray. *Vector quantization and signal compression*. Kluwer Academic Publishers, 1992.

[GJJV03] A. Goldsmith, S.A. Jafar, N. Jindal, and S. Vishwanath. Capacity limits of MIMO channels. *IEEE Journal on Selected Areas in Comms.*, 21(5):684 – 702, June 2003.

[GK00] P. Gupta and P.R. Kumar. The capacity of wireless networks. *IEEE Trans. on Inf. Theory*, 46(2):388–404, Mar. 2000.

[GKH+07] D. Gesbert, M. Kountouris, R.W. Heath Jr, C.-B. Chae, and T. Salzer. Shifting the MIMO paradigm: From single user to multiuser communications. *IEEE Signal Proc. Magazine*, 24(5):36–46, Sept. 2007.

[GMF09] M. Grieger, P. Marsch, and G. Fettweis. Uplink Base Station Cooperation by Iterative Distributed Interference Subtraction. In *IEEE Int. Symp. on Personal, Indoor and Mobile Radio Comms. (PIMRC'09)*, Sept. 2009.

[GMF10a] M. Grieger, P. Marsch, and G. Fettweis. Ad Hoc Cooperation for the Cellular Uplink with Capacity Constrained Backhaul. In *IEEE Int. Conf. on Comms. (ICC'10)*, May 2010.

[GMF10b] M. Grieger, P. Marsch, and G. Fettweis. Progressive uplink base station cooperation for backhaul constrained cellular systems. In *IEEE Workshop on Signal Proc. Advances in Wireless Comms. (SPAWC'10)*, June 2010.

[GMF10c] M. Grieger, P. Marsch, and G. Fettweis. Uplink ad hoc cooperation by distributed equalization under a constrained backhaul. In *IEEE Int. Conf. on Wireless Inf. Technology and Systems (ICWITS'10)*, Aug. 2010.

[GMFC09] M. Grieger, P. Marsch, G. Fettweis, and J.M. Cioffi. Compressed Interference Forwarding: A New Scheme for Uplink Base Station Cooperation Under a Capacity Constrained Backhaul Infrastructure. In *IEEE Global Telecomms. Conf. (GLOBECOM'09)*, Nov. 2009.

[GMRF10] M. Grieger, P. Marsch, Z. Rong, and G. Fettweis. Field trial results for a coordinated multi-point (CoMP) uplink in cellular systems. In *Int. ITG/IEEE Workshop on Smart Antennas (WSA'10)*, Feb. 2010.

[Gud91] M. Gudmundson. Correlation model for shadow fading in mobile radio systems. *Electronics Letters*, 27(23):2145–2146, Nov. 1991.

[GVL96] G.H. Golub and C.F. Van Loan. *Matrix computations*. Johns Hopkins University Press, 1996.

[GWBL95] B.E. Gschwendtner, G. Wölfle, B. Burk, and F.M. Landstorfer. Ray Tracing vs. Ray Launching in 3D Microcell Modelling. In *European Personal and Mobile Comms. Conf. (EPMCC'95)*, Nov. 1995.

[Ham10] K.A. Hamdi. Precise Interference Analysis of OFDMA Time-Asynchronous Wireless Ad-hoc Networks. *IEEE Trans. on Wireless Comms.*, 9(1):134 – 144, Jan. 2010.

[Hay02] S. Haykin. *Adaptive Filter Theory*. Prentice Hall, 2002.

[HBO08] D. Hammarwall, M. Bengtsson, and B. Ottersten. Acquiring partial CSI for spatially selective transmission by instantaneous channel norm feedback. *IEEE Trans. on Signal Proc.*, 56(3):1188–1204, Mar. 2008.

[HFG+04] T. Haustein, A. Forck, H. Gabler, C.V. Helmolt, V. Jungnickel, and U. Kruger. Implementation of adaptive channel inversion in a real-time mimo system. In *IEEE Int. Symp. on Personal, Indoor and Mobile Radio Comms. (PIMRC'04)*, Sept. 2004.

[HFG09] C. Hoymann, L. Falconetti, and R. Gupta. Distributed uplink signal processing of cooperating base stations based on IQ sample exchange. In *IEEE Int. Conf. on Comms. (ICC'09)*, June 2009.

[HJ99] R.A. Horn and C.R. Johnson. *Matrix analysis*. Cambridge University Press, 1999.

[HJ10] J. Han and D.-K. Jeong. A practical implementation of IEEE 1588-2008 transparent clock for distributed measurement and control systems. *IEEE Trans. on Instrumentation and Measurement*, 59(2):433–439, Feb. 2010.

[HK81] T. Han and K. Kobayashi. A new achievable rate region for the interference channel. *IEEE Trans. on Inf. Theory*, 27(1):49–60, Jan. 1981.

[HKF10] J. Holfeld, V. Kotzsch, and G. Fettweis. Order-recursive precoding for cooperative multi-point transmission. In *Int. ITG/IEEE Workshop on Smart Antennas (WSA'10)*, Feb. 2010.

[HKK+07] L. Hentilä, P. Kyösti, M. Käske, M. Narandzic, and M. Alatossava. MATLAB implementation of the WINNER phase II channel model ver1.1. Online available at https://www.ist-winner.org/phase_2_model.html, Dec. 2007.

[HKR97a] P. Hoeher, S. Kaiser, and P. Robertson. Pilot-symbol-aided channel estimation in time and frequency. *Multi-carrier spread-spectrum*, pages 169–178, 1997.

[HKR+97b] P. Hoeher, S. Kaiser, P. Robertson, et al. Two-dimensional pilot-symbol-aided channel estimation by Wiener filtering. In *Int. Conf. on Acoustics, Speech, and Signal Proc. (ICASSP'97)*, Apr. 1997.

[HM72] H. Harashima and H. Miyakawa. Matched-Transmission Technique for Channels With Intersymbol Interference. *IEEE Trans. on Comms.*, 20(4):774780, Aug. 1972.

[Ho80] Y.C. Ho. Team decision theory and information structures. *Proc. of the IEEE*, 68(6):644–654, June 1980.

[HP05a] J. Hahn and E. Powers. Implementation of the GPS to Galileo time offset (GGTO). In *IEEE Int. Frequency Control Symp. and Precise Time and Time Interval (PTTI) Systems and Appl. Meeting*, Aug. 2005.

[HP05b] R.W. Heath Jr and A. J. Paulraj. Switching between Diversity and Multiplexing in MIMO Systems. *IEEE Trans. on Comms.*, 53(6):962–968, June 2005.

[HRF06] R. Habendorf, I. Riedel, and G. Fettweis. Reduced Complexity Vector Precoding for the Multiuser Downlink. In *IEEE Global Telecomms. Conf. (GLOBECOM'06)*, Nov. 2006.

[HRPV08] D. Huang, V. Raghavan, A.S.Y. Poon, and V.V. Veeravalli. Angular Domain Processing for MIMO Wireless Systems with Non–Uniform Antenna Arrays. In *Asilomar Conf. on Signals, Systems and Computers (ASILOMAR'08)*, Oct. 2008.

[HS09] H. Huang and D. Samardzija. Determining backhaul bandwidth requirements of Network MIMO. In *European Signal Proc. Conf. (EUSIPCO'09)*, Aug. 2009.

[HSJ+05] T. Haustein, S. Schiffermüller, V. Jungnickel, M. Schellmann, T. Michel, and G. Wunder. Interpolation and Noise Reduction in MIMO-OFDM - A Complexity Driven Perspective. In *Int. Symp. on Signal Proc. and Its Appl. (ISSPA'05)*, Aug. 2005.

[HSP01] R.W. Heath Jr, S. Sandhu, and A. Paulraj. Antenna selection for spatial multiplexing systems with linear receivers. *IEEE Comms. Letters*, 5(4):142 – 144, Apr. 2001.

[HT09] H. Holma and A. Toskala. *LTE for UMTS-OFDMA and SC-FDMA Based Radio Access*. Wiley, 2009.

[HTB03] B.M. Hochwald and S. Ten Brink. Achieving near-capacity on a multiple-antenna channel. *IEEE Trans. on Comms.*, 51(3):389–399, Mar. 2003.

[HTJ08] Y. Hadisusanto, L. Thiele, and V. Jungnickel. Distributed Base Station Cooperation via Block-Diagonalization and Dual-Decomposition. In *IEEE Global Telecomms. Conf. (GLOBECOM'08)*, Nov. 2008.

[H.V94] H.V. Poor. *An Introduction to Signal Detection and Estimation*. Springer-Verlag, 1994.

[HV04] H. Huang and S. Venkatesan. Asymptotic downlink capacity of coordinated cellular networks. In *Asilomar Conf. on Signals, Systems and Computers (ASILOMAR'04)*, Nov. 2004.

[HvHJ+02] T. Haustein, C. von Helmolt, E. Jorswieck, V. Jungnickel, and V. Pohl. Performance of MIMO systems with channel inversion. In *IEEE Vehicular Technology Conf. (VTC'02 Spring)*, May 2002.

[HVKS03] H. Huang, S. Venkatesan, A. Kogiantis, and N. Sharma. Increasing the peak data rate of 3G downlink packet data systems using multiple antennas. In *IEEE Vehicular Technology Conf. (VTC'03 Spring)*, April 2003.

[HvR09] P. Hosein and C. van Rensburg. On the Performance of Downlink Beamforming with Synchronized Beam Cycles. In *IEEE Vehicular Technology Conf. (VTC'08 Spring)*, May 2009.

[HYK+10] X. Hou, C. Yang, B. Kiong, et al. Impact of Channel Asymmetry on Performance of Channel Estimation and Precoding for Downlink Base Station Cooperative Transmission. *IEEE Trans. on Comms.*, 2010. submitted, available at http://arxiv.org/abs/1006.3855.

[I+09] R. Irmer et al. Multisite Field Trial for LTE and Advanced Concepts. *IEEE Comms. Magazine*, 47(2):92–98, Feb. 2009.

[IAL+07] R. Irmer, S. Abeta, G. Liu, J. Krmer, T. Slzer, E. Jacks, G. Wannemacher, and A. Buldorini. Next Generation Mobile Networks Radio Access Performance Evaluation Methodology, June 2007. www.ngmn.org.

[IEE08a] IEEE standard for a precision clock synchronization protocol for networked measurement and control systems, 2008.

[IEE08b] IEEE 802.3 Standardization Group. Part 3: Carrier sense multiple access with Collision Detection (CSMA/CD) Access Method and Physical Layer Specifications, Dec. 2008.

[IEE09] IEEE 802.3 Standardization Group. Amendment 1: Physical Layer Specifications and Management Parameters for 10 Gb/s Passive Optical Networks, Sept. 2009.

[IEE10a] IEEE 802.16 Standardization Group. Advanced air interface, standard draft v6.0, May 2010.

[IEE10b] IEEE 802.3 Standardization Group. Amendment 4: Media Access Control Parameters, Physical Layers and Management Parameters for 40 Gb/s and 100 Gb/s Operation,

[IET07] IETF RFC 4960. Stream Control Transmission Protocol, Sept. 2007. June 2010.

[IHJ08] A. Ibing, Y. Hadisusanto, and V. Jungnickel. Scaleable network multicast for cooperative base stations. In *Int. Conf. on Comm. System Software and Middleware (COMSWARE'08)*, Jan. 2008.

[Inf10] Informa Telecoms and Media. Global Mobile Forecasts to 2014, Worldwide Market Analysis, Strategic Outlook and Forecasts to 2014. Jan. 2010.

[IST06] IST-4-027756 WINNER II. D6.13.7 Test scenarios and calibration issue 2, Dec. 2006.

[IST07a] IST-4-027756 WINNER II. D1.1.2: Winner ii channel models, 2007.

[IST07b] IST-4-027756 WINNER II. D2.2.3. Modulation and coding schemes for the WINNER II system, Nov. 2007.

[ITU06a] ITU-T G.7042/Y.1305 Standardization Group. Link capacity adjustment scheme (LCAS) for virtual concatenated signals, Mar. 2006.

[ITU06b] ITU-T G.993.2 Standardization Group. Very high speed digital subscriber line transceivers 2 (VDSL2), Feb. 2006.

[ITU08a] ITU. What really is a Third Generation (3G) Mobile Technology, 2008. http://www.itu.int/ITU-D/imt-2000/DocumentsIMT2000/What_really_3G.pdf.

[ITU08b] ITU-T G.7041/Y.1303 Standardization Group. Generic framing procedure (GFP), Oct. 2008.

[ITU08c] ITU-T G.984.3 Standardization Group. Gigabit-capable Passive Optical Networks (G-PON): Transmission convergence layer specification, Mar. 2008.

[ITU10a] ITU-R. Itu-r wp 5d third workshop on IMT-Advanced, focussed on candidate technologies and evaluation, Oct. 2010. http://groups.itu.int/Default.aspx?tabid=574.

[ITU10b] ITU-T G.8262/Y.1362 Standardization Group. Timing characteristics of a synchronous Ethernet equipment slave clock (EEC), Jul. 2010.

[ITU10c] ITU-T G.987.1 Standardization Group. 10-Gigabit-capable passive optical networks (XG-PON): General requirements, Jan. 2010.

[ITU10d] ITU-T G.987.2 Standardization Group. 10-Gigabit-capable passive optical networks (XG-PON): Physical media dependent (PMD) layer specification, Jan. 2010.

[ITU10e] ITU-T G.987.3 Standardization Group. 10-Gigabit-capable passive optical networks (XG-PON): Transmission convergence layer specification, not yet approved, 2010.

[ITU10f] ITU-T G.993.5 Standardization Group. Self-FEXT cancellation (vectoring) for use with VDSL2 transceivers, Apr. 2010.

[ITU10g] ITU-T G.998.3 Standardization Group. Multi-pair bonding using time-division inverse multiplexing, Jan. 2010.

[Jak74] W.C. Jakes. *Microwave Mobile Communications*. Wiley, 1974.

[JC94] W.C. Jakes and D.C. Cox. *Microwave Mobile Communications*. Wiley, 1994.

[JFH05] V. Jungnickel, A. Forck, and T. Haustein. 1 Gbit/s MIMO-OFDM transmission experiments. In *IEEE Vehicular Technology Conf. (VTC'05 Fall)*, Sept. 2005.

[JFH+06] V. Jungnickel, A. Forck, T. Haustein, C. Juchems, and W. Zirwas. *MIMO System Technology for Wireless Communications*, chapter 11, Gigabit mobile communications using real-time MIMO-OFDM signal processing. Taylor & Francis, 2006.

[JFJ+10] V. Jungnickel, A. Forck, S. Jaeckel, F. Bauermeister, S. Schiffermüller, S. Schubert, S. Wahls, L. Thiele, T. Haustein, W. Kreher, J. Müller, H. Droste, and G. Kadel. Field Trials using Coordinated Multi-Point Transmission in the Downlink. In *Int. Workshop on Wireless Distributed Networks (WDN'10, at PIMRC'10)*, Sept. 2010.

[JH09] H.J. Choi S.-S. Kim J.S. Han, J.-H. Jang. Enhanced spatial covariance matrix estimation for asynchronous inter-cell interference mitigation in MIMO-OFDMA system. In *IEEE Vehicular Technology Conf. (VTC'09 Spring)*, Apr. 2009.

[JHJvH01] V. Jungnickel, T. Haustein, E. Jorswieck, and C. von Helmolt. A MIMO WLAN based on linear channel inversion. *IEE Seminar on MIMO*, Dec. 2001.

[JHJvH02] V. Jungnickel, T. Haustein, E. Jorswieck, and C. von Helmolt. On linear pre-processing in multi-antenna systems. In *IEEE Global Telecomms. Conf. (GLOBECOM'02)*, Nov. 2002.

[Jin06] N. Jindal. A feedback reduction technique for MIMO broadcast channels. In *IEEE Int. Symp. on Inf. Theory (ISIT'06)*, July 2006.

[JJ10] S. Jäckel and V. Jungnickel. Estimating MIMO capacities from broadband measurements in a cellular network. In *European Conf. on Antennas and Propagation (EuCAP'10)*, Apr. 2010.

[JJT+09] V. Jungnickel, S. Jaeckel, L. Thiele, L. Jiang, U. Kruger, A. Brylka, and C. von Helmolt. Capacity measurements in a cooperative MIMO network. *IEEE Trans. on Vehicular Technology*, 58(5):2392–2405, June 2009.

[JKI+04] V. Jungnickel, U. Krüger, G. Istoc, T. Haustein, and C. von Helmolt. A MIMO system with reciprocal transceivers for the time-division duplex mode. In *IEEE Int. Symp. on Antennas and Propagation (APS '04)*, June 2004.

[JL03] J. Jang and K.B. Lee. Transmit power adaptation for multiuser OFDM systems. *IEEE Journal on Selected Areas in Comms.*, 21(2):171–178, Jan. 2003.

[JLD08] E.A. Jorswieck, E.G. Larsson, and D. Danev. Complete characterization of the Pareto boundary for the MISO interference channel. *IEEE Trans. on Signal Proc.*, 56(10):5292–5296, Oct. 2008.

[JMT+09] V. Jungnickel, K. Manolakis, L. Thiele, T. Wirth, and T. Haustein. Handover sequences for interference-aware transmission in multicell MIMO networks. In *Int. ITG/IEEE Workshop on Smart Antennas (WSA'09)*, Feb. 2009.

[JRV+05] N. Jindal, W. Rhee, S. Vishwanath, S.A. Jafar, and A. Goldsmith. Sum power iterative waterfilling for multiple-antenna Gaussian broadcast channels. *IEEE Trans. on Inf. Theory*, 51(4):1570–1580, Apr. 2005.

[JSF+07] V. Jungnickel, M. Schellmann, A. Forck, H. Gäbler, S. Wahls, A. Ibing, K. Manolakis, T. Haustein, W. Zirwas, J. Eichinger, E. Schulz, C. Juchems, F. Luhn, and R. Zavrtak. Demonstration of virtual MIMO in the uplink. In *IET Smart Antennas and Cooperative Comms. Seminar (IET'07)*, Oct. 2007.

[JST+09] V. Jungnickel, M. Schellmann, L. Thiele, T. Wirth, T. Haustein, O. Koch, E. Zirwas, and E. Schulz. Interference aware scheduling in the multiuser MIMO-OFDM downlink. *IEEE Comms. Magazine*, 47:56–66, June 2009.

[JTJ10] S. Jäckel, L. Thiele, and V. Jungnickel. Interference limited MIMO measurements. In *IEEE Vehicular Technology Conf. (VTC'10 Spring)*, May 2010.

[JTS+08a] S. Jing, D.N.C. Tse, J.B. Soriaga, J. Hou, J.E. Smee, and R. Padovani. Multicell downlink capacity with coordinated processing. *EURASIP Journal on Wireless Comms. and Networking*, 2008:1–19, Jan. 2008.

[JTS+08b] V. Jungnickel, L. Thiele, M. Schellmann, T. Wirth, W. Zirwas, T. Haustein, and E. Schulz. Implementation Concepts for Distributed Cooperative Transmission. In *Asilomar Conf. on Signals, Systems and Computers (ASILOMAR'08)*, Oct. 2008.

[JTW+09] V. Jungnickel, L. Thiele, T. Wirth, T. Haustein, S. Schiffermüller, A. Forck, S. Wahls, S. Jaeckel, S. Schubert, C. Juchems, F. Luhn, R. Zavrtak, H. Droste, G. Kadel, W. Kreher, J. Mueller, W. Stoermer, and G. Wannemacher. Coordinated Multipoint Trials in the Downlink. In *Broadband Wireless Access Workshop (BWAWS'09, at GLOBECOM'09)*, Nov. 2009.

[JUN05] M. Joham, W. Utschick, and J.A. Nossek. Linear transmit processing in MIMO communications systems. *IEEE Trans. on Signal Proc.*, 53(8):2700–2712, Aug. 2005.

[JVG04] N. Jindal, S. Vishwanath, and A. Goldsmith. On the duality of Gaussian multiple-access and broadcast channels. *IEEE Trans. on Inf. Theory*, 50(5):768–783, May 2004.

[JWS+08] V. Jungnickel, T. Wirth, M. Schellmann, T. Haustein, and W. Zirwas. Synchronization of cooperative base stations. In *IEEE Int. Symp. on Wireless Comms. Systems (ISWCS'08)*, Oct. 2008.

[K07] D. Kühling. Design and realtime implementation of mac-layer hardware and software components for a 3gpp lte base station. Diplomarbeit, TU Berlin, 2007.

[Kam08] K.-D. Kammeyer. *Nachrichtenübertragung*. B.G. Teubner, 2008.

[Kay93] S.M. Kay. *Fundamentals of statistical signal processing: estimation theory*, volume 1. Prentice Hall, 1993.

[KC07] M. Kobayashi and G. Caire. Joint beamforming and scheduling for a multi-antenna downlink with imperfect transmitter channel knowledge. *IEEE Journal on Selected Areas in Comms.*, 25(7):1468–1477, Sept. 2007.

[KDA+10] K. Kusume, G. Dietl, T. Abe, H. Taoka, and S. Nagata. System level performance of downlink MU-MIMO transmissions for 3GPP LTE-Advanced. In *IEEE Vehicular Technology Conf. (VTC'10 Spring)*, May 2010.

[KF08] S. Khattak and G. Fettweis. Low Backhaul Distributed Detection Strategies for an Interference Limited Uplink Cellular System. In *IEEE Vehicular Technology Conf. (VTC'08 Spring)*, May 2008.

[KF10] V. Kotzsch and G. Fettweis. Interference Analysis in Time and Frequency Asynchronous Network MIMO OFDM Systems. In *IEEE Wireless Comms. and Networking Conf. (WCNC'10)*, Apr. 2010.

[KFV06] M.K. Karakayali, G.J. Foschini, and R.A. Valenzuela. Network coordination for spectrally efficient communications in cellular systems. *IEEE Trans. on Wireless Comms.*, 13(4):56–61, Aug. 2006.

[KFVY06] M.K. Karakayali, G.J. Foschini, R.A. Valenzuela, and R.D. Yates. On the maximum common rate achievable in a coordinated network. In *IEEE Int. Conf. on Comms. (ICC'06)*, June 2006.

[KHF09] V. Kotzsch, J. Holfeld, and G. Fettweis. Joint Detection and CFO Compensation in Asynchronous Multi-User MIMO OFDM Systems. In *IEEE Vehicular Technology Conf. (VTC'09 Spring)*, Apr. 2009.

[KJRF10] V. Kotzsch, C. Jandura, W. Rave, and G. Fettweis. On Timing Constraints and OFDM Parameter Design for Cooperating Base Stations. In *Int. ITG/IEEE Workshop on Smart Antennas (WSA'10)*, Feb. 2010.

[KP74] R. G. Kouyoumjian and P. H. Pathak. A uniform geometrical theory of diffraction for an edge in a perfectly conducting surface. *Proc. of the IEEE*, 62:1448–1461, Nov. 1974.

[KRF07] S. Khattak, W. Rave, and G. Fettweis. On the Impact of User Positions on Multiuser Detection in Distributive Antenna Systems. In *European Wireless Conf. (EW'07)*, Apr. 2007.

[KRF08] S. Khattak, W. Rave, and G. Fettweis. Distributed iterative multiuser detection through base station cooperation. *EURASIP Journal on Wireless Comms. and Networking*, 2008, Jan. 2008.

[KZM07] W. Kiess, S. Zalewski, and M. Mauve. Improving System Clock Precision With NTP Offline Skew Correction. In *Mediterranean Ad Hoc Networking Workshop (MedHoc-Net'07)*, June 2007.

[LAK99] W. Lewandowski, J. Azoubib, and W.J. Klepczynski. GPS: primary tool for time transfer. *Proc. of the IEEE*, 87(1):163–172, Jan. 1999.

[Lan08] M. Landmann. Limitations of Experimental Channel Characterisation. PhD thesis, Technische Universität Ilmenau, 2008. ISBN (E-Book): 978-3-640-20944-6, available to download at http://www.grin.com.

[Lar07] E.G. Larsson. Model-averaged interference rejection combining. *IEEE Trans. on Comms.*, 55(2):271–274, Feb. 2007.

[LBC+04] J. Liu, A. Bourdoux, J. Craninckx, P. Wambacq, B. Come, S. Donnay, and A. Barel. OFDM-MIMO WLAN AP front-end gain and phase mismatch calibration. In *IEEE Radio and Wireless Conf. (RAWCON '04)*, Sept. 2004.

[LBG80] Y. Linde, A. Buzo, and R. Gray. An Algorithm for Vector Quantizer Design. *IEEE Trans. on Comms.*, 28(1):84–95, Jan. 1980.

[LHSH04] D.J. Love, R.W. Heath Jr, W. Santipach, and M.L. Honig. What is the value of limited feedback for MIMO channels? *IEEE Comms. Magazine*, 42(10):54–59, Oct. 2004.

[Lia72] H.H.J. Liao. *Multiple Access Channels*. PhD thesis, Hawaii University, Honolulu, 1972.

[LJSL09] B. Lee, H. Je, O.S. Shin, and K. Lee. A novel uplink MIMO transmission scheme in a multicell environment. *IEEE Trans. on Wireless Comms.*, 8(10):4981–4987, Oct. 2009.

[LKL09] J. Lindblom, E. Karipidis, and E.G. Larsson. Selfishness and altruism on the MISO interference channel: The case of partial transmitter CSI. *IEEE Comms. Letters*, 13(9):667–669, Sept. 2009.

[LKT07] M. Landmann, W. Kotterman, and R.S. Thomä. On the Influence of Incomplete Data Models on Estimated Angular Distributions in Channel Characterisation. In *European Conf. on Antennas and Propagation (EuCAP'07)*, Nov. 2007.

[LP01] L. Litwin and M. Pugel. The principles of OFDM. *RF Design*, 1:30–48, 2001.

[LS06] B. Li and G. Stüber. *Orthogonal Frequency Division Multiplexing for Wireless Communications*. Birkhäuser, 2006.

[LST+07] M. Landmann, K. Sivasondhivat, J.-I. Takada, I. Ida, and R. Thom"a. Polarization Behavior of Discrete Multipath and Diffuse Scattering in Urban Environments at 4.5 GHz. *EURASIP Journal on Wireless Comms. and Networking*, 2007, Jan. 2007.

[LWN02] J. Laiho, A. Wacker, and T. Novosad. *Radio Network Planning and Optimisation for UMTS*. Wiley, 2002.

[LZ06] K.B. Letaief and Y.J. Zhang. Dynamic multiuser resource allocation and adaptation for wireless systems. *IEEE Wireless Comms. Magazine*, 13(4):38–47, Aug. 2006.

[LZLG08] Q. Li, J. Zhu, Q. Li, and C. Georghiades. Efficient spatial covariance estimation for asynchronous co-channel interference suppression in MIMO-OFDM systems. *IEEE Trans. on Wireless Comms.*, 7(12):4849–4853, Dec. 2008.

[MAMK08] M.A. Maddah-Ali, A.S. Motahari, and A.K. Khandani. Communication over MIMO X channels: Interference alignment, decomposition, and performance analysis. *IEEE Trans. on Inf. Theory*, 54(8):3457–3470, Aug. 2008.

[Mar10] P. Marsch. *Coordinated Multi-Point under a Constrained Backhaul and Imperfect Channel Knowledge*. PhD thesis, Technische Universität Dresden, 2010. Ph.D. thesis.

[MBM07] R. Mudumbai, G. Barriac, and U. Madhow. On the feasibility of distributed beamforming in wireless networks. *IEEE Trans. on Wireless Comms.*, 6(5):1754–1763, May 2007.

[MBQ04] M. Meurer, P.W. Baier, and W. Qiu. Receiver orientation versus transmitter orientation in linear MIMO transmission systems. *EURASIP Journal on Applied Signal Proc.*, 2004:1191–1198, Jan. 2004.

[MC06] Y. Mostofi and D.C. Cox. Mathematical Analysis of the Impact of Timing Synchronization Errors on the Performance of an OFDM System. *IEEE Trans. on Comms.*, 54(2):226 – 230, Feb. 2006.

[McC07] W. McCoy. Overview of 3GPP LTE Physical Layer: White Paper by Dr. Wes McCoy, 2007.

[MDMH09] R. Mudumbai, D.R. Brown III, U. Madhow, and H.V. Poor. Distributed transmit beamforming: Challenges and recent progress. *IEEE Comms. Magazine*, 47(2):102–110, Feb. 2009.

[Med] Medav GmbH, Uttenreuth, Germany. RUSK channel sounder family.

[Med00] M. Medard. The effect upon channel capacity in wireless communications of perfect and imperfect knowledge of the channel. *IEEE Trans. on Inf. Theory*, 46(3):933–946, May 2000.

[Mei] Meinberg. Available online at http://www.meinberg.de/german/specs/gpsopt.htm.

[MF07a] P. Marsch and G. Fettweis. A Decentralized Optimization Approach to Backhaul-Constrained Distributed Antenna Systems. In *IST Mobile and Wireless Communications Summit (IST'07)*, July 2007.

[MF07b] P. Marsch and G. Fettweis. A Framework for Optimizing the Uplink Performance of Distributed Antenna Systems under a Constrained Backhaul. In *IEEE Int. Conf. on Comms. (ICC'07)*, June 2007.

[MF07c]   P. Marsch and G. Fettweis. A decentralized optimization approach to backhaul-constrained distributed antenna systems. In *IST Mobile and Wireless Communications Summit (IST'07)*, July 2007.

[MF08a]   P. Marsch and G. Fettweis. On Backhaul-Constrained Multi-Cell Cooperative Detection based on Superposition Coding. In *IEEE Int. Symp. on Personal, Indoor and Mobile Radio Comms. (PIMRC'08)*, Sept. 2008.

[MF08b]   P. Marsch and G. Fettweis. On Base Station Cooperation Schemes for Downlink Network MIMO under a Constrained Backhaul. In *IEEE Global Telecomms. Conf. (GLOBECOM'08)*, Nov. 2008.

[MF08c]   P. Marsch and G. Fettweis. On Base Station Cooperation Schemes for Uplink Network MIMO under a Constrained Backhaul. In *Int. Symp. on Wireless Personal Multimedia Comms. (WPMC'08)*, Sept. 2008.

[MF08d]   P. Marsch and G. Fettweis. On the rate region of a multi-cell MAC under backhaul and latency constraints. In *IEEE Wireless Comms. and Networking Conf. (WCNC'08)*, Mar. 2008.

[MF09a]   P. Marsch and G. Fettweis. On Downlink Network MIMO under a Constrained Backhaul and Imperfect Channel Knowledge. In *IEEE Global Telecomms. Conf. (GLOBECOM'09)*, Nov. 2009.

[MF09b]   P. Marsch and G. Fettweis. On Uplink Network MIMO under a Constrained Backhaul and Imperfect Channel Knowledge. In *IEEE Int. Conf. on Comms. (ICC'09)*, June 2009.

[MF10]   A. Müller and P. Frank. Cooperative interference prediction for enhanced link adaptation in the 3GPP LTE uplink. In *IEEE Vehicular Technology Conf. (VTC'10 Spring)*, May 2010.

[MF11]   P. Marsch and G. Fettweis. Uplink CoMP under a Constrained Backhaul and Imperfect Channel Knowledge. *IEEE Trans. on Wireless Comms.*, 2011.

[MFF10]   P. Marsch, A. Fehske, and G. Fettweis. Increasing Mobile Rates while Minimizing Cost per Bit - Cooperation vs. Denser Deployment. In *Int. Symp. on Wireless Comms. Systems (ISWCS'10)*, Sept. 2010.

[MGF10]   P. Marsch, M. Grieger, and G. Fettweis. Field Trial Results on Different Uplink Coordinated Multi-Point (CoMP) Concepts in Cellular Systems. In *IEEE Global Telecomms. Conf. (GLOBECOM'10)*, Dec. 2010.

[MGF11]   P. Marsch, M. Grieger, and G. Fettweis. Large Scale Field Trial Results on Different Uplink Coordinated Multi-Point (CoMP) Concepts in an Urban Environment. In *IEEE Wireless Comms. and Networking Conf. (WCNC'11)*, Mar. 2011.

[MHMB05] R. Mudumbai, J. Hespanha, U. Madhow, and G. Barriac. Scalable feedback control for distributed beamforming in sensor networks. In *IEEE Int. Symp. on Inf. Theory (ISIT'05)*, Sept. 2005.

[MHMB10] R. Mudumbai, J. Hespanha, U. Madhow, and G. Barriac. Distributed transmit beamforming using feedback control. *IEEE Trans. on Inf. Theory*, 56:411 – 426, Jan. 2010.

[Mil91]   D.L. Mills. Internet time synchronization: the network time protocol. *IEEE Trans. on Comms.*, 39(10):1482–1493, Oct. 1991.

[Mil03]   D.L. Mills. A brief history of NTP time: memoirs of an Internet timekeeper. *ACM SIGCOMM Computer Comm. Review*, 33(2):9–21, Apr. 2003.

[Mil09]   S. Milijevic. An introduction to synchronized ethernet. *EETimes 3/3/2009*, 2009.

[Min99]   N. Minar. A survey of the NTP network, December 1999.

[MJH06]   T. Mayer, H. Jenkac, and J. Hagenauer. Turbo base-station cooperation for intercell interference cancellation. In *IEEE Int. Conf. on Comms. (ICC'06)*, June 2006.

[MK09]   A.S. Motahari and A.K. Khandani. Capacity bounds for the Gaussian interference channel. *IEEE Trans. on Inf. Theory*, 55(2):620–643, Feb. 2009.

[MK10]   T.F. Maciel and A. Klein. On the performance, complexity, and fairness of suboptimal resource allocation for multiuser MIMO-OFDMA systems. *IEEE Trans. on Vehicular Technology*, 59(1):406–419, Jan. 2010.

[MKF06]   P. Marsch, S. Khattak, and G. Fettweis. A framework for determining realistic capacity bounds for distributed antenna systems. In *IEEE Inf. Theory Workshop (ITW'06)*, Oct. 2006.

[MKP07]   M. Morelli, C.C.J. Kuo, and M.O. Pun. Synchronization techniques for orthogonal frequency division multiple access (OFDMA): A tutorial review. *Proc. of the IEEE*, 95(7):1394–1427, July 2007.

[MLG06]   H.G. Myung, J. Lim, and D.J. Goodman. Single carrier FDMA for uplink wireless transmission. *IEEE Vehicular Technology Magazine*, 1(3):30–38, Sept. 2006.

[MMK07]   A. Maltsev, R. Maslennikov, and A. Khoryaev. Comparative analysis of spatial covariance matrix estimation methods in OFDM communication systems. In *IEEE Int. Symp. on Signal Proc. and Inf. Technology (ISSPIT'06)*, Aug. 2007.

[MMT08]   X. Mao, A. Maaref, and K.H. Teo. Adaptive Soft Frequency Reuse for Inter-Cell Interference Coordination in SC-FDMA Based 3GPP LTE Uplinks. In *IEEE Global Telecomms. Conf. (GLOBECOM'08)*, Nov. 2008.

[MN00]   R. Mathar and T. Niessen. Optimum positioning of base stations for cellular radio networks. *Wireless Networks*, 6:421–428, 2000.

[MOJ10] K. Manolakis, C. Oberli, and V. Jungnickel. Synchronization requirements for OFDM-based cellular networks with coordinated base stations: Preliminary results. In *Int. OFDM Workshop (InOWo '10)*, Sept. 2010.

[Mol01] A.F. Molisch. Internal memo, at&t labs - research, 2001. see also A.F. Molisch, M.V. Clark, H. Dai, M.Z. Win, and J.H. Winters, "Method and apparatus for reducing interference in multiple-input-multiple-output (MIMO) systems", US Patent 2003/0214917 A1 (provisional filing 2001, publication 2003).

[Mol03] A.F. Molisch. Effect of far scatterer clusters in MIMO outdoor channel models. In *IEEE Vehicular Technology Conf. (VTC'03 Spring)*, Apr. 2003.

[MRB06] J. Medbo, M. Riback, and J.E. Berg. Validation of 3GPP spatial channel model including WINNER wideband extension using measurements. In *IEEE Vehicular Technology Conf. (VTC'06 Fall)*, Sept. 2006.

[MRF10] P. Marsch, P. Rost, and G. Fettweis. Application Driven Joint Uplink-Downlink Optimization in Wireless Communications. In *Int. ITG/IEEE Workshop on Smart Antennas (WSA'10)*, Feb. 2010.

[MS09] M. Mehlhose and S. Schiffermueller. Efficient fixed-point implementation of linear equalization for cooperative MIMO systems. In *European Signal Proc. Conf. (EUSIPCO'09)*, Aug. 2009.

[MSDU09] C. Mensing, S. Sand, A. Dammann, and W. Utschick. Data-aided location estimation in cellular OFDM communications systems. In *IEEE Global Telecomms. Conf. (GLOBECOM'09)*, Nov. 2009.

[MSEA03] K. K. Mukkavilli, A. Sabharwal, E. Erkip, and B. Aazhang. On beamforming with finite rate feedback in multiple-antenna systems. *IEEE Trans. on Inf. Theory*, 49(10):2562–2579, Oct. 2003.

[MSLT07] E. Matskani, N.D. Sidiropoulos, Z.-Q. Luo, and L. Tassiulas. Joint multiuser downlink beamforming and admission control: a semidefinite relaxation approach. In *Int. Conf. on Acoustics, Speech, and Signal Proc. (ICASSP'07)*, Apr. 2007.

[MTH97] D.L. Mills, A. Thyagarajan, and B.C. Huffman. Internet timekeeping around the globe. In *Precision Time and Time Interval (PTTI) Appl. and Planning Meeting*, Dec. 1997.

[MWMR06] R. Mudumbai, B. Wild, U. Madhow, and K. Ramchandran. Distributed beamforming using 1 bit feedback: from concept to realization. In *Allerton Conf. on Comm., Control, and Computing (ALLERTON'06)*, Sept. 2006.

[MZC06] M. Mohseni, R. Zhang, and J.M. Cioffi. Optimized transmission for fading multiple-access and broadcast channels with multiple antennas. *IEEE Journal on Selected Areas in Comms.*, 24(8):1627 – 1639, Aug. 2006.

[NAV02] M. Nakamura, Y. Awad, and S. Vadgama. Adaptive control of link adaptation for high speed downlink packet access (HSDPA) in W-CDMA. In *Int. Symp. on Wireless Personal Multimedia Comms. (WPMC'02)*, Oct. 2002.

[NDA06] M. J. Nawrocki, M. Dohler, and H. A. Aghvami. *Understanding UMTS Radio Network Modelling, Planning and automated Optimisation: Theory and Practice*. Wiley, 2006.

[NEHA08] B.L. Ng, J.S. Evans, S.V. Hanly, and D. Aktas. Distributed Downlink Beamforming With Cooperative Base Stations. *IEEE Trans. on Inf. Theory*, 54(12):5491–5499, Dec. 2008.

[NK02] V.D. Nguyen and H.-P. Kuchenbecker. Intercarrier and Intersymbol Interference Analysis of OFDM Systems on Time-Invariant Channels. In *IEEE Int. Symp. on Personal, Indoor and Mobile Radio Comms. (PIMRC'02)*, Sept. 2002.

[NKJ+11] M. Narandzic, M. Käske, S. Jäckel, C. Sommerkorn, C. Schneider, and R.S. Thomä. Variation of estimated Large–Scale MIMO Channel Properties between repeated Measurements. In *IEEE Vehicular Technology Conf. (VTC'11 Spring)*, May 2011.

[NKK+10] M. Narandzic, W. Kotterman, M. Käske, C. Schneider, G. Sommerkorn, A. Hong, and R.S. Thomä. On a characterisation of large-scale channel parameters for distributed (multi-link) MIMO ? the impact of power level differences. In *European Conf. on Antennas and Propagation (EuCAP'10)*, Apr. 2010.

[NKM+07] M. Narandzic, P. Kyösti, J. Meinilä, L. Hentilä, M. Alatossava, T. Rautiainen, Y.L.C. de Jong, C. Schneider, and R.S. Thoma. Advances in WINNER wideband MIMO system-level channel modelling. In *European Conf. on Antennas and Propagation (EuCAP'07)*, Nov. 2007.

[NMK+07] M. Noda, M. Muraguchi, Tran Gia Khanh, K. Sakaguchi, and K. Araki. Eigenmode Tomlinson-Harashima Precoding for multi-antenna multi-user MIMO broadcast channel. In *Int. Conf. on Inf., Comms. and Signal Proc. (ICICS'07)*, Dec. 2007.

[NST+07] M. Narandzic, C. Schneider, R.S. Thomä, T. J"ams"a, P. Ky"osti, and X. Zhao. Comparison of SCM, SCME, and WINNER Channel Models. In *IEEE Vehicular Technology Conf. (VTC'07 Spring)*, Apr. 2007.

[OSC+10] V. Oksman, H. Schenk, A. Clausen, J.M. Cioffi, M. Mohseni, G. Ginis, C. Nuzman, J. Maes, M. Peeters, K. Fisher, and P.-E. Eriksson. The ITU-T's new G.vector standard proliferates 100 Mb/s DSL. *IEEE Comms. Magazine*, Oct. 2010.

[PA10] A. Papadogiannis and G. C. Alexandropoulos. The Value of Dynamic Clustering of Base Stations for Future Wireless Networks. In *IEEE Int. Conf. on Fuzzy Systems (FUZZ'10)*, July 2010.

[PBT05] A. S. Y. Poon, R. W. Brodersen, and D. N. C. Tse. Degrees of Freedom in Multiple–Antenna Channels: A Signal Space Approach. *IEEE Trans. on Inf. Theory*, 51(2):523–536, Feb. 2005.

[PD10] R. Preuss and D.R. Brown III. Two-way carrier synchronization for coherent coordinated multi-cell downlink transmission. *IEEE Trans. on Signal Processing*, 2010. submitted.

[PDF+01] S. Parkvall, E. Dahlman, P. Frenger, P. Beming, and M. Persson. The evolution of WCDMA towards higher speed downlink packet data access. In *IEEE Vehicular Technology Conf. (VTC'01 Spring)*, May 2001.

[PDF+08] S. Parkvall, E. Dahlman, A. Furuskar, Y. Jading, M. Olsson, S. Wanstedt, and K. Zangi. LTE-Advanced - Evolving LTE towards IMT-advanced. In *IEEE Vehicular Technology Conf. (VTC'08 Fall)*, Sept. 2008.

[PGH08] A. Papadogiannis, D. Gesbert, and E. Hardouin. A Dynamic Clustering Approach in Wireless Networks with Multi-Cell Cooperative Processing. In *IEEE Int. Conf. on Comms. (ICC'08)*, May 2008.

[PHG09] A. Papadogiannis, E. Hardouin, and D. Gesbert. Decentralising Multicell Cooperative Processing: A Novel Robust Framework. *EURASIP Journal on Wireless Comms. and Networking*, 2009, Feb. 2009.

[PHS05] C.B. Peel, B.M. Hochwald, and A.L. Swindlehurst. A vector-perturbation technique for near-capacity multiantenna multiuser communication-part I: channel inversion and regularization. *IEEE Trans. on Comms.*, 53(1):195–202, Jan. 2005.

[PNG03] A. Paulraj, R. Nabar, and D. Gore. *Introduction to space-time wireless communications*. 2003.

[Pon64] C. Pon. Retrodirective array using the heterodyne technique. *IEEE Trans. on Antennas and Propagation*, 12(2):176–180, Mar. 1964.

[PSS04] G. Primolevo, O. Simeone, and U. Spagnolini. Effects of imperfect channel state information on the capacity of broadcast OSDMA-MIMO systems. In *IEEE Workshop on Signal Proc. Advances in Wireless Comms. (SPAWC'04)*, July 2004.

[PT08] A. S. Y. Poon and D. N. C. Tse. Polarization Degrees-of-Freedom. In *IEEE Int. Symp. on Inf. Theory (ISIT'08)*, July 2008.

[PT09] A. S. Y. Poon and D. N. C. Tse. Degree-of-Freedom Gain from using Polarimetric Antenna Elements. *IEEE Trans. on Inf. Theory*, Apr. 2009. submitted for publication.

[PTW10] N. Prasad, A. Tajer, and X. Wang. Robust beamforming for multi-cell downlink transmission. In *IEEE Int. Symp. on Inf. Theory (ISIT'10)*, Jun. 2010.

[PW09] N. Palleit and T. Weber. Obtaining Transmitter Side Channel State Information in MIMO FDD Systems. In *IEEE Int. Symp. on Personal, Indoor and Mobile Radio Comms. (PIMRC'09)*, Sept. 2009.

[PW10] N. Palleit and T. Weber. Frequency Prediction of the Channel Transfer Function in Multiple Antenna Systems. In *Int. ITG/IEEE Workshop on Smart Antennas (WSA'10)*, Feb. 2010.

[Rad62] R. Radner. Team decision problems. *The Annals of Mathematical Statistics*, pages 857–881, 1962.

[Rak09] IT2200B datasheet, April 2009.

[Rap02] T. S. Rappaport. *Wireless Communications: Principles and Practice*. Prentice Hall, 2002.

[RB74] D.C. Rife and R.R. Boorstyn. Single-tone parameter estimation from discrete-time observations. *IEEE Trans. on Inf. Theory*, 20(5):591–598, Sept. 1974.

[RCP09] S.A. Ramprashad, G. Caire, and H.C. Papadopoulos. Cellular and Network MIMO architectures: MU-MIMO spectral efficiency and costs of channel state information. In *Asilomar Conf. on Signals, Systems and Computers (ASILOMAR'09)*, Nov. 2009.

[RF10] P. Rost and G. Fettweis. On the transmission-computation-energy tradeoff in multi-hop relay networks. In *IEEE Workshop on Green Communications (at GLOBECOM'10)*, Dec. 2010.

[RFL09] P. Rost, G. Fettweis, and J.N. Laneman. Opportunities, constraints, and benefits of relaying in the presence of interference. In *IEEE Int. Conf. on Comms. (ICC'09)*, June 2009.

[RFL10] P. Rost, G. Fettweis, and J.N. Laneman. Energy and cost efficient mobile communication using multi-cell MIMO and relaying. *IEEE Trans. on Wireless Comms.*, March 2010. submitted for publication.

[RFMF10] F. Richter, A. Fehske, P. Marsch, and G. Fettweis. Traffic demand and energy efficiency in heterogeneous cellular mobile radio networks. In *IEEE Vehicular Technology Conf. (VTC'10 Spring)*, May 2010.

[Ric05] A. Richter. *Estimation of Radio Channel Parameters: Models and Algorithms*. PhD thesis, Technische Universität Ilmenau, 2005.

[Ros09] P. Rost. *Opportunities, Benefits, and Constraints of Relaying in Mobile Communication Systems*. PhD thesis, Technische Universität Dresden, Dresden, Germany, 2009.

[RRMF10] I. Riedel, P. Rost, P. Marsch, and G. Fettweis. Creating Desirable Interference by Optimized Sectorization in Cellular Systems. In *IEEE Global Telecomms. Conf. (GLOBECOM'10)*, Dec. 2010.

[SAH+04] S.T.Chung, A.Lozano, H.C. Huang, A.Sutivong, and J.M. Cioffi. Approaching the MIMO Capacity with a Low-Rate Feedback Channel in V-BLAST. *EURASIP Journal on Applied Signal Proc.*, 2004:762–771, Jan. 2004.

[Sat78] H. Sato. An outer bound to the capacity region of broadcast channels. *IEEE Trans. on Inf. Theory*, 24(3):374–377, May 1978.

[Say02] A.M. Sayeed. Deconstructing Multi–Antenna Fading Channels. *IEEE Trans. on Signal Proc.*, 50:2563–2579, Oct. 2002.

[Say08] A. H. Sayed. *Adaptive Filters*. Wiley, 2008.

[SB04] M. Schubert and H. Boche. Solution of the multiuser downlink beamforming problem with individual SINR constraints. *IEEE Trans. on Vehicular Technology*, 53(1):18–28, Jan. 2004.

[SBM+04] G.L. Stuber, J.R. Barry, S.W. McLaughlin, Ye Li, M.A. Ingram, and T.G. Pratt. Broadband MIMO-OFDM wireless communications. *Proc. of the IEEE*, 92(2):271–294, Feb. 2004.

[SBO06] R. Stridh, M. Bengtsson, and B. Ottersten. System evaluation of optimal downlink beamforming with congestion control in wireless communication. *IEEE Trans. on Wireless Comms.*, 5(4):743–751, Apr. 2006.

[Sch09] M. Schellmann. *Multi-user MIMO-OFDM in practice: Enabling spectrally efficient transmission over time-varying channels*. PhD thesis, Technical University of Berlin, 2009.

[SDAD02] A. Stamoulis, S.N. Diggavi, and N. Al-Dhahir. Intercarrier Interference in MIMO OFDM. *IEEE Trans. on Signal Proc.*, 50(10):2451 – 2464, Oct. 2002.

[Sem10] ACS9550 product brief, February 2010.

[SG01] T.A. Schonhoff and A.A. Giordano. *Detection and Estimation Theory*. Addison-Wesley, 2001.

[SH02] Q.H. Spencer and M. Haardt. Capacity and downlink transmission algorithms for a multi-user MIMO channel. In *Asilomar Conf. on Signals, Systems and Computers (ASILOMAR'02)*, Nov. 2002.

[SH09] D. Samardzija and H. Huang. Determining backhaul bandwidth requirements for network MIMO. In *European Signal Proc. Conf. (EUSIPCO'09)*, Aug. 2009.

[Sha48] C.E. Shannon. A mathematical theory of communication. *Bell System Technical Journal*, 27:379–423 and 623–656, 1948.

[Sha49] C.E. Shannon. Communication in the presence of noise. *Institute of Radio Engineers*, 37(1):10–21, 1949.

[SJ06] S. Schiffermüller and V. Jungnickel. Practical Channel Interpolation for OFDMA. In *IEEE Global Telecomms. Conf. (GLOBECOM'06)*, Nov. 2006.

[SJ09] M. Schellmann and V. Jungnickel. Multiple CFOs in OFDM-SDMA uplink: Interference analysis and compensation. *EURASIP Journal on Wireless Comms. and Networking*, 2009, Jan. 2009.

[SKC09] X. Shang, G. Kramer, and B. Chen. A New Outer Bound and the Noisy-Interference Sum-Rate Capacity for Gaussian Interference Channels. *IEEE Trans. on Inf. Theory*, 55:689 – 699, Feb. 2009.

[SM03] M. Speth and H. Meyr. Synchronization Requirements for COFDM Systems with Transmit Diversity. In *IEEE Global Telecomms. Conf. (GLOBECOM'03)*, Dec. 2003.

[Smi86] J. Smith. *Modern Communications Circuits*. McGraw-Hill, 1986.

[SMM+08] S. Sand, C. Mensing, Y. Ma, R. Tafazolli, X. Yin, J. Figueiras, J. Nielsen, and B.H. Fleury. Hybrid data fusion and cooperative schemes for wireless positioning. In *IEEE Vehicular Technology Conf. (VTC'08 Fall)*, Sept. 2008.

[SRIA06] R.A. Shafik, S. Rahman, R. Islam, and N.S. Ashraf. On the error vector magnitude as a performance metric and comparative analysis. In *Int. Conf. on Emerging Technologies (ICET'06)*, Nov. 2006.

[SSBN+06] O. Somekh, O. Simeone, Y. Bar-Ness, A.M. Haimovich, and S. Shamai. Distributed multi-cell zero-forcing beamforming in cellular downlink channels. In *IEEE Global Telecomms. Conf. (GLOBECOM'06)*, Nov. 2006.

[SSESV06] J. Salo, P. Suvikunnas, H. M. El-Sallabi, and P. Vainikainen. Some Insights into MIMO Mutual Information: The High SINR Case. *IEEE Trans. on Wireless Comms.*, 5(11):2997–3001, Nov. 2006.

[SSH04] Q.H. Spencer, A.L. Swindlehurst, and M. Haardt. Zero-forcing methods for downlink spatial multiplexing in multiuser MIMO channels. *IEEE Trans. on Signal Proc.*, 52(2):461–471, Feb. 2004.

[SSPS09a] A. Sanderovich, O. Somekh, H.V. Poor, and S. Shamai. Uplink macro diversity of limited backhaul cellular network. *IEEE Trans. on Inf. Theory*, 55(8):3457–3478, Aug. 2009.

[SSPS09b] O. Simeone, O. Somekh, H.V. Poor, and S. Shamai. Downlink Multicell Processing with Limited-Backhaul Capacity. *EURASIP Journal on Advances in Signal Proc.*, 2009, Feb. 2009.

[SSPS09c] O. Simeone, O. Somekh, H.V. Poor, and S. Shamai. Local base station cooperation via finite-capacity links for the uplink of simple cellular networks. *IEEE Trans. on Inf. Theory*, 55(1):190–204, Jan. 2009.

[SSPS10] O. Somekh, O. Simeone, V. Poor, and S. Shamai. Cellular systems with non-regenerative relaying and cooperative base stations. *IEEE Trans. on Wireless Comms.*, 9(8):2654–2663, Aug. 2010.

[SSS07a] A. Sanderovich, O. Somekh, and S. Shamai. Uplink macro diversity with limited backhaul capacity. In *IEEE Int. Symp. on Inf. Theory (ISIT'07)*, June 2007.

[SSS+07b] S. Shamai, O. Somekh, O. Simeone, A. Sanderovich, B.M. Zaidel, and V. Poor. Cooperative Multi-Cell Networks: Impact of Limited-Capacity Backhaul and Inter-Users Links. In *Joint Workshop on Comms. and Coding (JWCC'07)*, Oct. 2007.

[SSSP08] S. Shamai, O. Simeone, O. Somekh, and H.V. Poor. Joint Multi-Cell Processing for Downlink Channels with Limited-Capacity Backhaul. In *Inf. Theory and Appl. Workshop (ITA'08)*, Jan. 2008.

[STB09] S. Sesia, I. Toufik, and M. Baker. *LTE, The UMTS Long Term Evolution: From Theory to Practice*. Wiley, 2009.

[STJ08] M. Schellmann, L. Thiele, and V. Jungnickel. Low-complexity Doppler compensation in mobile SIMO-OFDM systems. In *Asilomar Conf. on Signals, Systems and Computers (ASILOMAR'08)*, Oct. 2008.

[STJH07] M. Schellmann, L. Thiele, V. Jungnickel, and T. Haustein. A Fair Score-Based Scheduler for Spatial Transmission Mode Selection. In *Asilomar Conf. on Signals, Systems and Computers (ASILOMAR'07)*, Nov. 2007.

[STK05] A.H. Sayed, A. Tarighat, and N. Khajehnouri. Network-based wireless location: challenges faced in developing techniques for accurate wireless location information. *IEEE Signal Proc. Magazine*, 22(4):24–40, July 2005.

[STKL01] F. Shad, T.D. Todd, V. Kezys, and J. Litva. Dynamic Slot Allocation (DSA) in indoor SDMA/TDMA using a smart antenna basestation. *IEEE/ACM Trans. on Networking*, 9(1):69–81, Feb. 2001.

[SW73a] D. Slepian and J. Wolf. Noiseless coding of correlated information sources. *IEEE Trans. on Inf. Theory*, 19(4):471–480, July 1973.

[SW73b] D. Slepian and J.K. Wolf. A coding theorem for multiple access channels with correlated sources. *Bell Systems Technical Journal*, 52(7):1037–1076, 1973.

[SZ01] S. Shamai and B. Zaidel. Enhancing the cellular downlink capacity via co-processing at the transmitter end. In *IEEE Vehicular Technology Conf. (VTC'01 Spring)*, May 2001.

[TA84] J.N. Tsitsiklis and M. Athans. On the complexity of decentralized decision making and detection problems. In *IEEE Conf. on Decision and Control (CDC'84)*, Dec. 1984.

[TBB+10] L. Thiele, F. Boccardi, C. Botella, T. Svensson, and M. Boldi. Scheduling-Assisted Joint Processing for CoMP in the Framework of the WINNER+ Project. In *Future Network and Mobile Summit*, June 2010.

[TBH08] M. Trivellato, F. Boccardi, and H. Huang. Zero-forcing vs. unitary beamforming in multiuser MIMO systems with limited feedback. In *IEEE Int. Symp. on Personal, Indoor and Mobile Radio Comms. (PIMRC'08)*, Sept. 2008.

[TBT08] M. Trivellato, F. Boccardi, and F. Tosato. On transceiver design and channel quantization for downlink multiuser MIMO systems with limited feedback. *IEEE Journal on Selected Areas in Comms.*, 26(8):1494–1504, Oct. 2008.

[TC94] J.A. Tague and C.I. Caldwell. Expectations of useful complex Wishart forms. *Multidimensional Systems and Signal Processing*, 5(3):263–279, 1994.

[TC04] Y.M. Tsang and R.S. Cheng. Rate maximization in space-division multiple access (SDMA)/multi-input-single-output (MISO)/OFDM systems under QoS and power constraints. In *IEEE Int. Conf. on Comms. (ICC'04)*, June 2004.

[Tel99] E. Telatar. Capacity of multi-antenna Gaussian channels. *European Trans. on Telecomms.*, 10(6):585–595, Nov. 1999.

[THR+01] R.S. Thomä, D. Hampicke, A. Richter, G. Sommerkorn, and U. Trautwein. MIMO vector channel sounder measurement for smart antenna system evaluation. *European Trans. on Telecomms.*, 12(5):427–438, Sep. 2001.

[TJMW01] W. Turin, R. Jana, C. Martin, and J. Winters. Modeling wireless channel fading. In *IEEE Vehicular Technology Conf. (VTC'01 Fall)*, Oct. 2001.

[Tom71] M. Tomlinson. New automatic equaliser employing modulo arithmetic. *Electronics Letters*, 7(5/6):138139, Mar. 1971.

[Ton05] D. Tonks. IEEE 1588 in telecommunications applications. In *Third Int. Telecom. Sync. Forum (ITSF'05)*, Oct. 2005.

[TP02] Y.S. Tu and G.J. Pottie. Coherent cooperative transmission from multiple adjacent antennas to a distant stationary antenna through AWGN channels. In *IEEE Vehicular Technology Conf. (VTC'02 Spring)*, May 2002.

[TSS+08] L. Thiele, M. Schellmann, S. Schiffermüller, V. Jungnickel, and W. Zirwas. Multi-Cell Channel Estimation using Virtual Pilots. In *IEEE Vehicular Technology Conf. (VTC'08 Spring)*, May 2008.

[TSSJ08] L. Thiele, M. Schellmann, S. Schiffermüller, and V. Jungnickel. Multi-cell channel estimation using virtual pilots. In *IEEE Vehicular Technology Conf. (VTC'08 Spring)*, May 2008.

[TST05] U. Trautwein, C. Schneider, and R.S. Thomä. Measurement based performance evaluation of advanced MIMO transceiver designs. *EURASIP Journal on Applied Signal*

[TSWJ08] L. Thiele, M. Schellmann, T. Wirth, and V. Jungnickel. On the value of synchronous downlink MIMO-OFDMA systems with linear equalizers. In *IEEE Int. Symp. on Wireless Comms. Systems (ISWCS'08)*, Oct. 2008.

[TSWJ09] L. Thiele, M. Schellmann, T. Wirth, and V. Jungnickel. Interference-aware scheduling in the synchronous cellular multi-antenna downlink. In *IEEE Vehicular Technology Conf. (VTC'09 Spring)*, Apr. 2009.

[TSZJ07] L. Thiele, M. Schellmann, W. Zirwas, and V. Jungnickel. Capacity scaling of multiuser MIMO with limited feedback in a multi-cell environment. In *Asilomar Conf. on Signals, Systems and Computers (ASILOMAR'07)*, Nov. 2007.

[TUBN06] P. Tejera, W. Utschick, G. Bauch, and J.A. Nossek. Subchannel allocation in multiuser Multiple-Input-MultipleOutput systems. *IEEE Trans. on Inf. Theory*, 52(10):4721–4733, Oct. 2006.

[TV05] D. Tse and P. Viswanath. *Fundamentals of wireless communication*. Cambridge University Press, 2005.

[TW05] E. Tuomaala and H. Wang. Effective SINR approach of link to system mapping in OFDM/multi-carrier mobile network. In *IEEE Int. Conf. on Mobile Technology, Appl. and Systems (MTAS'05)*, Nov. 2005.

[TWB+09] L. Thiele, T. Wirth, K. Börner, M. Olbrich, V. Jungnickel, J. Rumold, and S. Fritze. Modeling of 3D Field Patterns of Downtilted Antennas and Their Impact on Cellular Systems. In *Int. ITG/IEEE Workshop on Smart Antennas (WSA'09)*, Feb. 2009.

[TWH+09] L. Thiele, T. Wirth, T. Haustein, V. Jungnickel, E. Schulz, and W. Zirwas. A unified feedback scheme for distributed interference management in cellular systems: Benefits and challenges for real-time implementation. In *European Signal Proc. Conf. (EUSIPCO'09)*, Aug. 2009.

[TWS+09] L. Thiele, T. Wirth, M. Schellmann, Y. Hadisusanto, and V. Jungnickel. MU-MIMO with Localized Downlink Base Station Cooperation and Downtilted Antennas. In *IEEE Int. Workshop on LTE Evolution (at ICC'09)*, June 2009.

[TZ03] D.N.C. Tse and L. Zheng. Diversity and multiplexing: A fundamental tradeoff in multiple-antenna channels. *IEEE Trans. on Inf. Theory*, 49(50):1073–1096, May 2003.

[TZGO07] Y. Tsai, G. Zhang, D. Grieco, and F. Ozluturk. Cell search in 3GPP long term evolution systems. *IEEE Vehicular Technology Magazine*, 2(2):23–29, June 2007.

[Ven07] S. Venkatesan. Coordinating base stations for greater uplink spectral efficiency in a cellular network. In *IEEE Int. Symp. on Personal, Indoor and Mobile Radio Comms. (PIMRC'07)*, Sept. 2007.

[Ver98] S. Verdu. *Multiuser detection*. Cambridge University Press, 1998.

[VFS09] J. Voigt, R. Fritzsche, and J. Schueler. Optimal Antenna Type Selection in a real SU–MIMO Network Planning Scenario. In *IEEE Vehicular Technology Conf. (VTC'09 Fall)*, Sep. 2009.

[VGL03] S.A. Vorobyov, A.B. Gershman, and Z.Q. Luo. Robust adaptive beamforming using worst-case performance optimization: A solution to the signal mismatch problem. *IEEE Trans. on Signal Proc.*, 51(2):313–324, Feb. 2003.

[VHLV09] S. Venkatesan, H. Huang, A. Lozano, and R. Valenzuela. A WiMAX-based implementation of network MIMO for indoor wireless systems. *EURASIP Journal on Advances in Signal Proc.*, 2009, Feb. 2009.

[VHW+08] V. Venkatkumar, T. Haustein, H. Wu, E. Schulz, T. Wirth, A. Forck, S. Wahls, and V. Jungnickel. Field trial results on multi-user MIMO downlink OFDMA in typical outdoor scenario using proportional fair scheduling. In *Int. ITG Workshop on Smart Antennas (WSA'08)*, Feb. 2008.

[Vod10] Vodafone. Vodafone Group Financial Results, May 2010. www.vodafone.com.

[VT03] P. Viswanath and D.N.C. Tse. Sum capacity of the vector Gaussian broadcast channel and uplink-downlink duality. *IEEE Trans. on Inf. Theory*, 49(8):1912–1921, Aug. 2003.

[VTL02] P. Viswanath, D.N.C. Tse, and R. Laroia. Opportunistic beamforming using dumb antennas. *IEEE Trans. on Inf. Theory*, 48(6):1277–1294, June 2002.

[VVH03] H. Viswanathan, S. Venkatesan, and H. Huang. Downlink capacity evaluation of cellular networks with known-interference cancellation. *IEEE Journal on Selected Areas in Comms.*, 21(5):802–811, June 2003.

[WBBJ10] M. Wiese, H. Boche, I. Bjelakovic, and V. Jungnickel. Downlink with partially cooperating base stations. In *IEEE Workshop on Signal Proc. Advances in Wireless Comms. (SPAWC'10)*, June 2010.

[WBOW00] T. Weber, P.W. Baier, J. Oster, and M. Weckerle. Performance enhancement of time division CDMA (TD-CDMA) by multistep joint detection. In *Int. Conf. on Telecomms. (ICT00)*, May 2000.

[WE71] S.B. Weinstein and P.M. Ebert. Data transmission by frequency-division multiplexing using the discrete fourier transform. *IEEE Trans. on Comms.*, 19(5):628–634, Oct. 1971.

[WES06] A. Wiesel, Y.C. Eldar, and S. Shamai (Shitz). Linear precoding via conic optimization for fixed MIMO receivers. *IEEE Trans. on Signal Proc.*, 54(1):161–176, Jan. 2006.

[WFGV98] P.W. Wolniansky, G.J. Foschini, G.D. Golden, and R.A. Valenzuela. V-BLAST: An architecture for realizing very high data rates over the rich-scattering wireless channel.

[WG99] In *URSI Int. Symp. on Signals, Systems, and Electronics (ISSSE'98)*, Sept. 1998.
D. Wulich and L. Goldfeld. Reduction of peak factor in orthogonal multicarrier modulation by amplitude limiting and coding. *IEEE Trans. on Comms.*, 47(1):18–21, Jan. 1999.

[WG00] Z. Wang and G.B. Giannakis. Wireless Multicarrier Communications. *IEEE Signal Proc. Magazine*, 17(3), May 2000.

[WH02] J.P. Woodard and L. Hanzo. Comparative study of turbo decoding techniques: An overview. *IEEE Trans. on Vehicular Technology*, 49(6):2208–2233, Nov. 2002.

[WHE08] Deliverable D3.1 WHERE. Intermediate: Location based optimisation for phy algorithms/protocols, Dec. 2008.

[Win84] Jack Winters. Optimum Combining in Digital Mobile Radio with Cochannel Interference. *IEEE Journal on Selected Areas in Comms.*, 2(4):528–539, July 1984.

[WIN03] WINNER. Ist-2003-507581 winner d5.4 final report on link level and system level channel models v1.4. https://www.ist-winner.org, 2003.

[WIN05] I. WINNER. IST-2003-507581 Assessment of advanced beamforming and MIMO technologies, 2005.

[WIN09] WINNER. IST-Winner. http://www.ist-winner.org, Jan. 2009.

[WJA08] T. Wirth, V. Jungnickel, and A. Forck et al. Realtime multi-user multi-antenna downlink measurements. In *IEEE Wireless Comms. and Networking Conf. (WCNC'08)*, Mar. 2008.

[WJF+08] T. Wirth, V. Jungnickel, A. Forck, S. Wahls, V. Venkatkumar, T. Haustein, and H. Wu. Polarization dependent MIMO gains on multiuser downlink OFDMA with a 3GPP LTE air interface in typical urban outdoor scenarios. In *Int. ITG Workshop on Smart Antennas (WSA'08)*, Feb. 2008.

[WMS08] M. Werner, P. Moberg, and P. Skillermark. Cost assessment of radio access network deployments with relay nodes. In *IST Mobile and Wireless Communications Summit (IST'08)*, June 2008.

[WMSL02] T. Weber, I. Maniatis, A. Sklavos, and Y. Liu. Joint transmission and detection integrated network (JOINT), a generic proposal for beyond 3G systems. In *Int. Conf. on Telecomms. (ICT'02)*, June 2002.

[WMZ05a] T. Weber, M. Meurer, and W. Zirwas. Improved Channel Estimation Exploiting Long Term Channel Properties. In *Int. Conf. Conference on Telecomms. (ICT'05)*, May 2005.

[WMZ05b] T. Weber, M. Meurer, and W. Zirwas. Subspace Based Channel Estimation. In *COST 273 TD(05)004*, Jan. 2005.

[WP99] X. Wang and H.V. Poor. Iterative (turbo) soft interference cancellation and decoding for coded CDMA. *IEEE Trans. on Comms.*, 47(7):1046 – 1061, July 1999.

[WSJ+10] G. Wunder, J. Schreck, P. Jung, H. Huang, and R. Valenzuela. Rate Approximation: A new Paradigm for Multiuser MIMO Downlink Communications. In *IEEE Int. Conf. on Comms. (ICC'10)*, May 2010.

[WSS06] H. Weingarten, Y. Steinberg, and S. Shamai. The Capacity Region of the Gaussian Multiple-Input Multiple-Output Broadcast Channel. *IEEE Trans. on Inf. Theory*, 52(9):3936–3964, Sept. 2006.

[WT08] I-H. Wang and D.N.C. Tse. Gaussian interference channels with multiple receive antennas: Capacity and generalized degrees of freedom. In *Allerton Conf. on Comm., Control, and Computing (ALLERTON'08)*, Sept. 2008.

[WT09a] I.H. Wang and D.N.C. Tse. Interference mitigation through limited receiver cooperation. *IEEE Trans. on Inf. Theory*, 2009. submitted, available at http://arxiv.org/abs/0911.2053.

[WT09b] I.H. Wang and D.N.C. Tse. Interference mitigation through limited receiver cooperation: Symmetric case. In *IEEE Inf. Theory Workshop (ITW'09)*, Oct. 2009.

[WT10] I.H. Wang and D.N.C. Tse. Interference mitigation through limited transmitter cooperation. *IEEE Trans. on Inf. Theory*, 2010. submitted for publication, available at http://arxiv.org/abs/1004.5421.

[WWWS09] X. Wei, T. Weber, A. Wolfgang, and N. Seifi. Joint Transmission with Significant CSI in the Downlink of Distributed Antenna Systems. In *IEEE Int. Conf. on Comms. (ICC'09)*, June 2009.

[WXBD09] M.M. Wang, L. Xiao, T. Brown, and M. Dong. Optimal Symbol Timing for OFDM Wireless Communications. *IEEE Trans. on Wireless Comms.*, 8(10):5328 – 5337, Oct. 2009.

[WZ76] A. Wyner and J. Ziv. The rate-distortion function for source coding with side information at the decoder. *IEEE Trans. on Inf. Theory*, 22(1):1–10, Jan. 1976.

[XLC04] Z. Xiong, A.D. Liveris, and S. Cheng. Distributed source coding for sensor networks. *IEEE Signal Proc. Magazine*, 21(5):80–94, Sept. 2004.

[XSX07] F. Xiangning, C. Si, and Z. Xiaodong. An Inter-Cell Interference Coordination Technique Based on Users' Ratio and Multi-Level Frequency Allocations. In *Int. Conf. on Wireless Comms., Networking and Mobile Computing (WiCom'07)*, Sept. 2007.

[YG05] T. Yoo and A. Goldsmith. Optimality of zero-forcing beamforming with multiuser diversity. In *IEEE Int. Conf. on Comms. (ICC'05)*, May 2005.

[YG06a] T. Yoo and A. Goldsmith. Capacity and power allocation for fading MIMO channels with channel estimation error. *IEEE Trans. on Inf. Theory*, 52(5):2203–2214, May 2006.

[YG06b] T. Yoo and A. Goldsmith. On the optimality of multiantenna broadcast scheduling using zero-forcing beamforming. *IEEE Journal on Selected Areas of Comms.*, 24(3):528 – 541, Mar. 2006.

[YL07] W. Yu and T. Lan. Transmitter optimization for the multi-antenna downlink with per-antenna power constraints. *IEEE Trans. on Signal Proc.*, 55:2646–2660, June 2007.

[YRBC01] W. Yu, W. Rhee, S. Boyd, and J. Cioffi. Iterative water-filling for Gaussian vector multiple access channels. In *IEEE Int. Symp. on Inf. Theory (ISIT'01)*, June 2001.

[YVC05] W. Yu, D.P. Varodayan, and J.M. Cioffi. Trellis and convolutional precoding for transmitter-based interference presubtraction. *IEEE Trans. on Inf. Theory*, 53(7):1220–1230, July 2005.

[ZA10] J. Zhang and J.G. Andrews. Adaptive spatial intercell interference cancellation in multicell wireless networks. *IEEE Journal on Selected Areas in Comms., Special issue on Coop. Comms. in MIMO Cellular Networks*, 28:1455 – 1468, Dec. 2010.

[Zar08] B. Zarikoff. *Frequency Synchronization Techniques for Coordinated Multibase MIMO Wireless Communication Systems.* PhD thesis, Simon Fraser University, 2008.

[ZD04] H. Zhang and H. Dai. Cochannel interference mitigation and cooperative processing in downlink multicell multiuser MIMO networks. *EURASIP Journal on Wireless Comms. and Networking*, 2004(2):222–235, Dec. 2004.

[ZDZ04] H. Zhang, H. Dai, and Q. Zhou. Base station cooperation for multiuser MIMO: Joint transmission and BS selection. In *Conf. on Inf. Sciences and Systems (CISS04)*, Mar. 2004.

[ZG10a] R. Zakhour and D. Gesbert. On the value of data sharing in constrained-backhaul network MIMO. In *Int. Zurich Seminar on Comms. (ZSC'10)*, Mar. 2010.

[ZG10b] R. Zakhour and D. Gesbert. Team decision for the cooperative MIMO channel with imperfect CSIT sharing. In *Inf. Theory and Appl. Workshop (ITA'10)*, Jan. 2010.

[ZHKA09] J. Zhang, R.W. Heath Jr, M. Kountouris, and J.G. Andrews. Mode switching for the multi-antenna broadcast channel based on delay and channel quantization. *EURASIP Journal on Advances in Signal Proc.*, 2009:1–15, Feb. 2009.

[ZKJ+07] W. Zirwas, J. H. Kim, V. Jungnickel, M. Schubert, T. Weber, A. Ahrens, and M. Haardt. *Distributed Antenna Systems*, chapter 10, Distributed Organization of Cooperative Antenna Systems, pages 279–311. Auerbach, 2007.

[ZL04] Y.J. Zhang and K.B. Letaief. Multiuser adaptive subcarrier-and-bit allocation with adaptive cell selection for OFDM systems. *IEEE Trans. on Wireless Comms.*, 3(5):1566–1575, Sept. 2004.

[ZMM+08] H. Zhang, N.B. Mehta, A.F. Molisch, J. Zhang, and H. Dai. Asynchronous Interference Mitigation in Cooperative Base Station Systems. *IEEE Trans. on Wireless Comms.*, 7(1):155 – 165, Jan. 2008.

[ZMS+09a] W. Zirwas, W. Mennerich, M. Schubert, L. Thiele, V. Jungnickel, and E. Schulz. chapter Cooperative Transmission Schemes. CRC Press, Taylor and Francis Group, 2009.

[ZMS+09b] W. Zirwas, W. Mennerich, M. Schubert, L. Thiele, V. Jungnickel, and E. Schulz. *Long Term Evolution: 3GPP LTE radio and cellular technology*, chapter 7, Cooperative Transmission Schemes. Auerbach, 2009.

[ZSK+06] W. Zirwas, E. Schulz, J. H. Kim, V. Jungnickel, and M. Schubert. Distributed Organization of Cooperative Antenna Systems. In *European Wireless Conf. (EW'06)*, Apr. 2006.

[ZT03] L. Zheng and D. Tse. Diversity and multiplexing: A fundamental tradeoff between in multiple antenna channels. *IEEE Trans. on Inf. Theory*, 49(5):1073–1096, May 2003.

[ZY08] L. Zhou and W. Yu. Gaussian Z-interference channel with a relay link: Achievable rate region and asymptotic sum capacity. In *Int. Symp. on Inf. Theory and Its Appl. (ISITA'08)*, Dec. 2008.

Index

antenna
  downtilt, 118, 152, 205, 332, 370
  pattern, 112, 156, 219, 252, 370, 380, 401, 446

backhaul, 25, 244
  capacity, 32, 55, 87, 102, 109, 130, 251, 262, 268, 277, 291, 293, 332, 375, 387, 391, 396, 406, 444
  latency, 55, 63, 69, 108, 116, 163, 275, 300, 303, 307, 333, 341, 396, 404, 458
  signaling, 64
  technologies, 301
  topology, 300
broadcast channel, 74, 108, 131

clustering, 71, 94, 115, 208, 294, 333, 385, 457
  dynamic, 118, 148, 259, 391, 397
  static, 141
computational complexity, 57, 83, 97, 129, 265, 317
cost efficiency, 148, 162, 191, 242, 268, 303, 340, 377, 423, 444, 457
CS/CB, 396
CSI
  exchange, 57, 122, 245, 292, 298, 331, 340, 449
  feedback, 67, 69, 110, 116, 122, 166, 203, 208, 251, 332, 353, 355, 369, 397, 447, 449
  feedback latency, 166
  hierarchical, 124, 355
  quantization, 71, 79

deployment
  heterogeneous, 401
  homogeneous, 401
detection
  ML, 83
  PIC, 84, 98
  SIC, 84, 314, 324, 425
  sphere detector, 84

effective channel, 44, 110, 183, 202, 220, 229, 251, 343, 354, 398
energy efficiency, 423
equalization, 14, 19
  IRC, 30, 44, 52, 228, 264, 323, 324, 356, 389
  MMSE, 45, 74, 84, 98, 211, 230, 315, 324, 344
  MRC, 30, 45, 52, 74, 229, 264, 389
  ZF, 98, 191
exchange
  CSI, 57, 122, 245, 292, 298, 331, 340, 449
  data, 331
  IQ samples, 83, 297, 332, 392, 448
  precoded signals, 130
  scheduling information, 56, 63
  soft coded bits, 90, 298, 392
  user data, 121, 130, 299, 333

fairness, 48, 57
feedback
  CQI, 43, 116
  CSI, 67, 69, 110, 116, 122, 166, 203, 208, 251, 332, 353, 355, 369, 397, 447, 449
  decision, 239
  PMI, 43, 51, 396
  WCI, 396, 398

geometry factor, 50, 141, 153, 351, 358, 375, 384

Hadamard matrices, 46, 322
HARQ, 52, 66, 116, 257, 259, 268, 275, 299, 315, 338, 368, 370, 389

interference channel, 78, 131, 278
interpolation gain, 198, 236

latency
  backhaul, 63, 69, 108, 116, 163, 275, 300, 303, 307, 333, 341, 396, 404, 458
  CSI exchange, 333
  CSI feedback, 166

decoding, 34, 84, 239, 266, 274, 315
signaling, 116, 333

macro diversity, 93
multiple access channel, 81, 132

pilots, *see* reference signals
power control, 30, 51, 62, 92, 178, 207, 246, 258, 391, 426
precoding, 14, 19, 43
  DFT-based, 44
  DPC, 111, 135
  EOC, 114
  ICIN, 71
  linear, 123
  MET, 114
  THP, 109
  Wiener filter, 221
  ZF, 113, 221

reference signals, 26, 46, 64, 197, 233, 258, 292, 321, 389, 415, 432, 448
  cell- and antenna-specific, 116, 204, 331, 336, 347, 354, 448
  LTE Release 8 CRS, 200, 236, 419
  positioning, 418
  stream-specific, 119, 204, 331, 343, 354, 448

S1 interface, 159, 340
scheduling, 56, 246, 256, 292, 294, 388, 398, 403
  proportional fair, 58, 256
  score-based, 46
self organizing network (SON), 157
signaling
  delay, 297
  for clustering, 157
  latency, 333
source coding, 32, 87
spatial channel model, 48, 73, 112, 152
standardization impact, 92, 157, 205, 209, 260, 317, 458
superposition coding, 20, 30, 32, 92, 131, 284
synchronization, 45, 161, 221, 321, 331, 334, 356, 377
  channels, 413
  in frequency, 162, 181, 407, 419, 458
  in time, 162, 173, 412
  local, 94
  network, 334, 415

X2 interface, 63, 116, 151, 301, 307, 333, 406, 440

Zadoff-Chu sequences, 205, 322